MW00592215

S. Pardue

FEEDSTUFF EVALUATION

These titles are now out of print but are available in microfiche editions

Feedstuff Evaluation

JULIAN WISEMAN and D.J.A. COLE
University of Nottingham School of Agriculture

BUTTERWORTHS
London Boston Singapore Sydney Toronto Wellington

 PART OF REED INTERNATIONAL P.L.C.

All rights reserved. No part of this publication may be reproduced in any material form (including photocopying or storing it in any medium by electronic means and whether or not transiently or incidentally to some other use of this publication) without the written permission of the copyright owner except in accordance with the provisions of the Copyright, Designs and Patents Act 1988 or under the terms of a licence issued by the Copyright Licensing Agency Ltd, 33–34 Alfred Place, London, England WC1E 7DP. Applications for the copyright owner's written permission to reproduce any part of this publication should be addressed to the Publishers.

Warning: The doing of an unauthorised act in relation to a copyright work may result in both a civil claim for damages and criminal prosecution.

This book is sold subject to the Standard Conditions of Sale of Net Books and may not be re-sold in the UK below the net price given by the Publishers in their current price list.

First published 1990

© Contributors, 1990

British Library Cataloguing in Publication Data

Feedstuff evaluation.
    1. Livestock. Feedingstuffs. Composition
    I. Wiseman, Julian   II. Cole, D. J. A. (Desmond James Augustus), *1935–*
    636.0855
    ISBN 0-408-04971-5

Library of Congress Cataloging-in-Publication Data

Feedstuff evaluation/[edited by] Julian Wiseman and D.J.A. Cole
    p. cm.
    Includes index.
    ISBN 0-408-04971-5 :
    1. Feeds–Evaluation. I. Wiseman, J. (Julian) II. Cole, D. J. A.
SF97.F38   1990
636.08′55–dc20   90-1920
                                             CIP

Phototypeset by Scribe Design, Gillingham, Kent
Printed in Great Britain at The University Press, Cambridge

PREFACE

This volume represents the proceedings of the 50th University of Nottingham Easter School in Agricultural Sciences held at Sutton Bonington in July 1989. The accurate evaluation of feedstuffs for livestock is of fundamental importance to the overall efficiency of animal production. Initially, systems of expressing the nutritive value of feeds were considered, as such an approach is essential if comparative estimates are to have any meaning. Modifications to feeding value as influenced by animal factors including intake and palatability, were discussed as, ultimately, the nutritive value of ingested food may be viewed in terms of animal responses. Specific dietary ingredients, being plant polysaccharides, fats, minerals and vitamins, were considered subsequently. Prediction of the nutritive value of compound feeds and individual feeds through classical wet chemistry and the more recent NIR is assuming considerable importance in the rapid evaluation of diets. Associated with these developments is an appreciation of the relevance of both inter- and intra-laboratory variation in determinations. Finally, the need to collate information into an interactive data-base is being actively pursued. It is evident that safety of animal feeds is becoming an increasingly topical issue and the last session considered the relevance of naturally-occurring toxic factors, residues, mycotoxins and, finally, animal pathogens.

It is hoped that the contents of the proceedings will have a wide appeal to all those involved in every aspect of nutrient supply to animals.

ACKNOWLEDGEMENTS

The contributions of those who presented papers at the conference, together with their efforts in preparing written versions for inclusion in the proceedings, is gratefully acknowledged. Individual sessions were chaired by: Dr D.J.A. Cole, Dr B.G. Vernon, Dr G. Emmans, Dr G. Norton, Dr K.N. Boorman, D.I. Givens, J. Lowe, Professor P.J. Buttery.

The following provided financial support as an invaluable contribution towards the expenses of speakers:

David Patton Limited
Nutec Limited
Vitafoods Limited
Carrs Farm Foods Limited
Beechams Animal Health
International Association of Fish Meal Manufacturers
Perstorp Analytical Limited
J. Bibby Agriculture Limited
Peter Hand (GB) Limited
Eurolysine
Butterworth Scientific Limited
Scientific and Medical Products Limited
AGM Systems Limited
Roche Products Limited
Favor Parker Limited
Pauls Agriculture Limited
Preston Farmers Limited
Rumenco
BP Nutrition (UK) Limited
Pentland Scotch Whisky Research Limited
Forum Feeds
W.J. Oldacre Limited
BOCM–Silcock Limited

Mrs Jose Newcombe with administration, Dr P.C. Garnsworthy who designed the registration software, Chris Mills with audio-visual equipment, Chris Wareham, Jayne Powles together with Sylvia Bateman and Faculty catering and administrative services all contributed to the smooth running of the conference.

Related Titles

BIOTECHNOLOGY IN GROWTH POPULATION Edited by R.B. Heap, C.G. Prosser, G.E. Lamming
THE CALF 5th Edition Volume 1 Management of Health J.H.B. Roy
LEANESS IN DOMESTIC BIRDS Edited by B. Leclercq and C.C. Whitehead
NEW TECHNIQUES IN CATTLE PRODUCTION Edited by C.J.C. Phillips
NUTRIENT REQUIREMENTS OF POULTRY AND NUTRITIONAL RESEARCH 19th Poultry and Nutritional
   Research Edited by C. Fisher
OUTLINE OF CLINICAL DIAGNOSIS IN CATTLE A.H. Andrews
OUTLINE OF CLINICAL DIAGNOSIS IN THE GOAT J. Matthews
OUTLINE OF CLINICAL DIAGNOSIS IN THE HORSE P.J.N. Pinsent
OUTLINE OF CLINICAL DIAGNOSIS IN SHEEP J.C. Hindson–Agnes, C. Winter
PIG PRODUCTION IN AUSTRALIA Edited by J.A.A. Gardener, A.C. Dunkin, L.C. Lloyd
RECENT ADVANCES IN ANIMAL NUTRITION – 1990 24th Feed Manufacturers Conference Edited by W.
   Haresign, D.J.A. Cole
RECENT ADVANCES IN TURKEY SCIENCE 21st Poultry Science Symposium Edited by C. Nixey and T.C. Grey
STRUCTURE AND FUNCTION OF DOMESTIC ANIMALS W. Bruce Currie

CONTENTS

1

COMPARISON OF ENERGY EVALUATION SYSTEMS OF FEEDS FOR RUMINANTS

Y. VAN DER HONING and A. STEG
Research Institute of Livestock Feeding and Nutrition (IVVO), Postbox 160, 8200 AD Lelystad, The Netherlands

Abbreviations used in the text

CH_4	= methane energy
DE	= digestible energy
dE	= energy digestibility
dO	= digestibility of organic matter
DXL	= digestible lipids
DXP	= digestible crude protein
DXF	= digestible crude fibre
DXX	= digestible N-free extract
FCM	= fat-corrected milk (4%)
FL	= level of feeding
FRG	= Federal Republic of Germany
FU	= fodder unit
GB	= Great Britain
GDR	= German Democratic Republic
IE	= gross energy
k_f	= efficiency of utilization of ME for fattening
k_g	= efficiency of utilization of ME for growth
k_m	= efficiency of utilization of ME for maintenance
k_{mg}	= overall efficiency of utilization of ME
k_l	= for lactation
ME	= gross energy minus losses of faeces, methane and urine
NE_g	= net energy for growth
NE_l	= net energy for lactation
NE_m	= net energy for maintenance
rse	= residual standard error
SE_K	= starch equivalent
TDN	= total digestible nutrients (digestible organic matter plus 1.25 digestible ether extract)
UFL	= Unité fourragère lait
XP	= crude protein

1

Introduction

To describe or predict the performance of farm animals effective feed evaluation systems are required, which generate information necessary to formulate diets of optimum quality. Haecker (1914) described the necessary knowledge as follows:

'In order to determine the actual net nutrients required to produce a given animal product, the composition of the product should be known as well as the composition and the available nutrients in food which is to be fed for its production, so that the nutrients in the ration might be provided in the proportions needed by the animal'.

Feed evaluation systems should be simple. This requirement is in great conflict with accuracy of prediction of responses over a wide range of variation of rations and a correct modelling of the underlying physiological processes in the farm animals. Most systems applied on a large scale in practice are a reasonable compromise between simplicity and accuracy of prediction.

Animal production is very much dependent on the quantity of energy consumed. Systems have been developed for animal nutrition in practice and have been in use since the beginning of this century (Breirem, 1969, pp. 656–677).

Current energy evaluation systems are simplified models to describe the nutrient requirement of animals for a target production on one hand and to indicate the potential of the feeds to those requirements on the other.

Recently, alternative approaches (for example those based upon mechanistic modelling) to overcome the weakness of our current systems has been given increasing attention in research studies (Webster, Dewhurst and Waters, 1988; Baldwin and Miller, 1988). Because the practical application of these alternatives is not likely in the forseeable future a detailed comparison between the current energy evaluation systems is still useful.

It should be emphasized that feed evaluation systems have a much wider significance than the formulation of adequate rations to achieve the desired animal performance. They contribute to the farmer and feed industry and also to the management of least-cost strategy of feeding of farm animals and the purchase-policy of feedstuffs for least-cost formulation of concentrate mixtures. Moreover, they play a role in finding the best systems of grassland management and fodder conservation. In addition wider issues of agricultural policy on, for example, utilizing national feed resources in an efficient way, reducing adverse side-effects to the environment and planning future alternatives in animal production as a result of changing public opinion and development of consumer markets is partly dependent on a correct feed evaluation.

Characteristic features of current energy evaluation systems for ruminants will be discussed briefly and some information on interrelationships between systems given. Different ways of comparison will be discussed.

The demands to have one common system of feed evaluation in several countries will increase substantially with developments in Europe as planned for 1992 and onwards. Some aspects of the future trends will also be given attention in this paper.

Some historical aspects of energy evaluation of feeds and feeding standards

In the history of feed evaluation, since Albert Thaer (1752–1828) introduced the concept of hay equivalents as measures of relative value based on determining the materials in feed extractable with water (and other solvents), the Weende analysis of feedstuffs, developed by Henneberg and Stohmann (1864) in the nineteenth century, has been important in the description of feedstuffs. Within the last 40 years new methods of analysis have improved the description of fibrous components, carbohydrates, proteins and lipids. However, in the previous century scientists had already realized that information from feeding trials and chemical analysis of feeds was not sufficient to understand energy metabolism and that energy losses should be measured more accurately.

MEASUREMENT OF ENERGY CONVERSION

According to Maynard *et al.* (1979) the first real balance experiment with a dairy cow was conducted by Boussingault in 1839, without however measuring gaseous losses. Knowledge of energy metabolism has been improved by various techniques for example calorimetry. During the late part of the nineteenth and the early part of the twentieth century extensive energy studies were carried out by Rubner, Kuhn, Kellner, Armsby and co-workers in respiration chambers according to the Pettenkofer principles. Møllgaard, Fingerling, Wood, Benedict, Kleiber, Breirem, Crasemann, Nehring, and many others extended these studies.

DEVELOPMENT OF FEEDING STANDARDS

The first standards were based on digestible nutrients, derived from feeding trials described by Wolff in 1864 (Maynard *et al.*, 1979). Atwater brought the Wolff standards to the attention of the American workers, which resulted in the publication by Armsby in 1880 of his book, '*Manual of Cattle Feeding*'. In 1898 tables showing the average composition of American feeds, digestion coefficients for protein, crude fibre, ether extract and nitrogen-free extract and the Wolff–Lehman standards were published by Henry in his book, '*Feeds and Feeding*'. The intakes of digestible nutrients were added, together with digestible ether extract multiplied by 2.25, as a sum of nutrients (TDN).

DEVELOPMENT OF NET ENERGY SYSTEMS

Kellner's work in Germany (Kellner, 1905) based on net energy for fattening resulted in the use of net energy systems in Europe, such as the starch equivalent and the Scandinavian fodder unit, which was modified to be used for dairy cattle by Møllgaard (1929) after evaluation of a great number of feeding trials with lactating cattle.

Since the 1960s the factorial approach as proposed by Blaxter (1962a) of splitting the total requirement into various parts (e.g. for maintenance and physical activity, for milk production, for body gain, for wool growth, etc.) has been adopted by several scientists and used to develop new and revised systems. An EAAP Working Group on Feed Evaluation under the leadership of Van Es attempted to formulate a new European standard system for energy requirements of ruminants in the mid-seventies, but did not succeed. However, Van Es was able to secure a good deal of agreement on the central relationships now in use in the majority of the new and revised systems. In this chapter the comparison of feed evaluation systems will be focussed on these modern systems.

Essential features of current energy feed evaluation systems

TYPE OF INFORMATION REQUIRED

The value of feedstuffs for an animal cannot be assessed from its gross energy value as such. The utilizable portion consists of the absorbable components as only these can be metabolized in the animals' tissues and organs. However, its net effect depends on the efficiency of utilization of these absorbed components in the intermediary metabolism. Accordingly there are two factors arising: (1) the potential of feedstuffs; and (2) the requirements of animals and utilization of feed. Various factors affect one or both aspects of feed conversion.

Knowledge of the potential of feedstuffs and the restrictions to utilizing that potential is important to allow the prediction of the contribution of a given quantity of a feedstuff in a ration. The nutritive value of feeds is measured for example by their voluntary intake, digestibility, chemical composition and presence of anti-nutritional factors. Such data can be assembled to tables of feed composition and nutritive value expressed per kg of feed, as fed to the animal or per kg of dry matter.

Secondly, information is needed on the requirement for energy and nutrients for the various classes of ruminant livestock and for various levels of animal production (meat, milk, wool, reproduction). This requirement should include data about voluntary feed intake and indicate effects of short- or long-term deficits or surpluses of nutrients (Bickel, 1988).

TYPE OF ENERGY LOSSES AND ITS MEASUREMENT

The utilization of feeds from an energetic point of view is accompanied by four kinds of losses: in faeces, in urine, gaseous losses (mainly as methane) and heat (Figure 1.1). The magnitude of all four kinds of losses depends, at least partly, on the type of feed. In general the largest variation is found in faecal and in heat losses.

A large part of the heat losses is dependent to an extent on the feed but is due mainly to the inefficient utilization of absorbed nutrients. Moreover the energy required for maintenance is measured totally as heat. As indicated in Figure 1.1 heat losses also vary in relation to type and level of production and therefore it is difficult to assume that these heat losses are a constant proportion of feedstuffs.

Although the variation in losses in urine (3–7% of gross energy, GE) and as methane (5–10%) are small compared with that in faecal losses (15–50%) it has become common practice to rank feedstuffs at least in terms of their content of metabolizable energy (ME = gross energy minus losses in faeces, urine, CH_4),

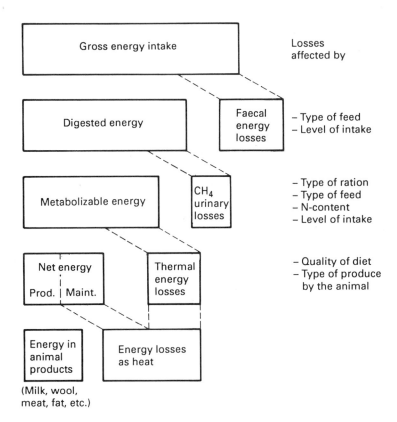

Figure 1.1 Diagram of energy model and factors in energy losses

measured under standard conditions (generally at a maintenance level of feeding). Accordingly ME is a currently accepted measurement of dietary energy evaluation, representing reasonable approximation of the total amount of energy available for metabolism.

Measurement of faecal losses and energy in urine can be undertaken comparatively easily and accurately by collecting of daily excreta. For measurement of methane a respiration chamber is required and measurement of heat losses also need a direct or indirect calorimeter. However, methane losses can be predicted reasonably satisfactory by using the equation presented by Blaxter and Clapperton (1965). This was recently confirmed by Edwards (1988) for grass silages.

The net energy in animal products, such as milk and wool, can be determined accurately by measuring its combustion value. From the difference between energy consumed and the net production total losses can be derived. However, in live animals the energy deposition has to be derived from the energy balance as the difference between input and output of energy. Owing to accumulation of errors, energy balance has a large standard deviation.

STANDARDIZATION OF DIGESTIBILITY MEASUREMENTS

Knowledge of the apparent digestibility of the organic components in feeds is of major importance and the first essential parameter in current energy evaluation systems. Most digestibility coefficients have been derived from trials with sheep, fed at around the maintenance requirement or predictions are made aiming at that level. These data can be converted to cattle because of the great similarities in digestive capacity between cattle and sheep as recently confirmed by Meissner and Roux (1989). However, for a good comparison of values of digestible energy or digestible organic matter, sufficient standardization in the conduct of digestion trials is necessary. Digestibility coefficients should be measured in ruminants under 'normal' conditions, so that rumen fermentation, ruminating and other digestive processes are not disturbed. A minimum amount of 'structure' in the form of long hay is given when deriving the digestibility of feedstuffs, which cannot be fed as a single feedstuff.

An increased feed intake generally depresses digestibility of organic matter with a mean value of approximately 3 units for each incremental increase in intake over maintenance intake (Van der Honing, 1975). This value is in line with the 4% reduction in digestibility as contained in the Nutrient Requirements of Dairy Cattle (NRC, 1978, 1988).

SYSTEMS BASED ON DE AND ME

Few systems are based solely on digestible nutrients or digestible energy. The widely used TDN-system (Total Digestible Nutrients = digestible organic matter plus 1.25 digestible ether extract) is an example of a system based on digestible nutrients. In the USA DE (digestible energy) is also in use and the relationship between both is: 1 kg TDN = 4.40 g Mcal DE (NRC, 1978, 1988).

The current systems in Sweden and Great Britain (GB) are based on metabolizable energy (ME). In this way variation in urinary and methane energy losses are taken into account which provides a more precise basis compared with DE.

In other systems ME is usually used as an intermediary step in the calculation of the net energy value. This intermediary step is a logical approach since the partitioning of metabolizable energy is dependent more on the type of animal and the production level than on the individual feedstuffs.

ME as a percentage of DE is assumed to increase the greater the digestibility. Faecal losses increase at a higher level of feeding, but are partly compensated for by lower losses in urine and combustible gases.

The ME/DE ratio, according to the literature (Van Es and Van der Honing, 1977), increases at a higher level of feeding from 0.81–0.82 at maintenance up to 0.87 at 3–4 times maintenance. This is due mainly to reduction in relative methane and urinary losses. Part of the reduced losses in methane and urine may be attributed to a higher proportion of concentrates in the ration and/or the ground and pelleted form of part of the ration (Van der Honing, 1975). In the French system ME/DE is negatively corrected for crude fibre and crude protein content of the feed (Andrieu and Demarquilly, 1987) and similar effects were calculated from our own data.

Table 1.1 shows the equations to calculate ME from digestible nutrients. The coefficients vary substantially between systems. Differences in coefficients reflect the inaccuracy of prediction from digestible nutrients. The reasons may be found in analytical methods (including errors, changes with time, alternative extraction or pretreatment with for example lipids), variation in composition of the feedstuffs tested and the procedure for digestion trials.

Table 1.1   EQUATIONS USED FOR THE CALCULATION OF ME* AS AN INTERMEDIARY STEP OR AS A FINAL EQUATION IN THE SYSTEM CONCERNED. VALUES OF COEFFICIENT RELATIVE TO (d) WITHIN PARANTHESES

Country	System code	a	b	c	d	e
Germany (GDR)	EFr	17.7 (120)	37.9 (256)	13.4 (91)	14.8 (100)	
Sweden (S)	ME (S)	18.0/18.8 (116–121)	20.9/36.81 (135–237)	12.1 (78)	15.5 (100)	
United Kingdom (GB)	ME(GB)	15.2 (95)	34.2 (214)	12.8 (80)	15.9 (100)	
Germany (FRG)	NE (D)	15.2 (95)	34.2 (214)	12.8 (80)	15.9 (100)	†
USA	NE (US)	18.6 (100)	41.9 (225)	18.6 (100)	18.6 (100)	−1883
Netherlands (NL)	VEM	15.9 (109)	37.7 (257)	13.8 (94)	14.6 (100)	†

*See appendix 1 of the report of Van der Honing and Steg (1984).
†Correction for mono- and disaccharides if >8%.
ME (in kJ/kg DM) = a.DXP + b.DXL + c.DXF + d.DXX + e
DXP, DXL, DXF, DXX in g/kg dry matter.

The equations of GB and Federal Republic of Germany (FRG) were taken from the information of Schiemann *et al.* (1971a), but in that publication several equations are published. GB and FRG have chosen a different equation than that used by the authors in the German Democratic Republic (GDR). In addition the Dutch equation was to a great extent based on the data of Kellner-Fingerling and of Schiemann *et al.* (1971a). However, data obtained from cattle by Van Es have also been included (Van Es and Van der Honing, 1977). The lack of uniformity in coefficients applied is confusing. Schiemann *et al.* (1971a) argued that a general equation, applicable to rations, should be used and took that one obtained with sheep. However, they neglected a term related to the metabolic body weight of the animal which, according to Van Es and Van der Honing (1977), is not acceptable because the omission of this negative term resulted in an overestimation of ME. GB and FRG used the (less accurate) equation derived from cattle trials with concentrate ingredients. Both Schiemann and Van Es concluded from their studies that separate equations for forages and for concentrates were not needed. The Dutch equation was a weighted average of the coefficients of four equations, after the coefficient of digestible lipid (DXL) was fixed at 9.00. As ME for cattle is predicted mainly from sheep data the coefficient for digestible crude protein (DXP) is lowered by 7% in view of the lower digestibility of crude protein (XP) in cattle as demonstrated by Schiemann *et al.* (1971b).

Table 1.1 also shows the relative values of the coefficients of DXP, DXL and digestible crude fibre (DXF) compared with digestible N-free extract (DXX, set to 100%). The values for DXP of GDR and Sweden are much higher than that of GB and FRG. The Dutch value is intermediate but still higher than the USA with respect to TDN.

Prediction of DE or ME from variables other than digestible nutrients has been analysed from our data from sheep digestibility trials. A wide range of feedstuffs illustrated that DE can be calculated from GE and digestibility of organic matter (dO) with a residual standard error (rse) of 300 kJ/kg organic matter, which is slightly better than from digestible nutrients. ME was predicted less accurately from GE and dO (rse = 500 kJ) than DE, but this was comparable with the prediction of ME from digestible nutrients.

SYSTEMS BASED ON NET ENERGY

Net energy systems following the factorial approach use partial efficiencies of utilization of ME dependent on the type of production; for example k_m for maintenance, k_1 for lactation, k_f for fattening, k_g for growth. The magnitude of these coefficients increases with a higher digestibility or metabolizability of the ration as shown by Blaxter (1962a), which is also presented in Figure 1.2. k_m and k_1 are almost parallel, so that a constant ratio of 1.2 between k_m and k_1 can be used as suggested by Van Es. However, this is more difficult for k_m and k_f or k_g as discussed by MacHardy (1966).

Heat losses can vary quite substantially: dependent partly on the quality of the ration consumed, and decreasing with higher metabolizability, but to a greater extent dependent on the type and the level of production.

Partitioning of energy cannot be predicted easily from the current systems, because no prediction of nutrient-availability is achieved. Evidence obtained in recent years shows that ration manipulation, such as grinding a large proportion of the forage, increasing the concentrates-to-forage ratio or enlarging the proportion of easily fermentable components in the ration, may improve body energy retention more than milk-yield (and milk-fat in particular). Partitioning of nutrients can also be changed by a number of growth and production promoting substances, for example bovine somatotropin and anabolic agents. The feed conversion ratio is improved by these components due to an increased level of production or a lower feed consumption.

However, it is difficult to understand how without any additional treatment of feedstuffs in such a situation the potential nutritive value would have changed. One reason for better feed conversion could be a change in the energy concentration of live weight gain, which is not accounted for. Another reason might be that the animals are able to use the feed more fully to realize their potential. Feed evaluation should provide solid evidence upon the potential nutritive value of feeds. The ration fed and the production level of the animal will determine whether the animal is able to make full use of it.

ENERGY EVALUATION SYSTEMS OF FEEDS IN PRACTICE

With the current systems reasonably uncomplicated tables and equations are available for farmers, feed compounders, agricultural students, extension services

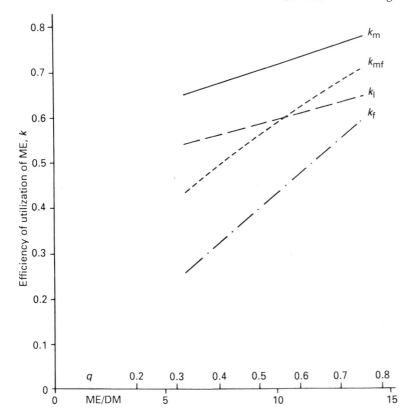

Figure 1.2 Utilization of metabolizable energy for maintenance (k_m), lactation (k_l), fattening (k_f) and maintenance + fattening (k_{mf}) as a function of ME concentration in DM (ME/DM) and gross energy (q), ME/GE

and agricultural research. Additivity of the value of individual feedstuffs is assumed as a rule, which seems to hold in most cases, although with some exceptions due to associative effects (Sauvant *et al.*, 1987; Vermorel, Coulon and Journet, 1987; Bickel, 1988).

The requirements of the animal have to be defined clearly for each type of activity or production. To solve the animal-feed interactions, which cannot simply be attributed to one of the two aspects mentioned above, corrections may be applied to the energy standards or recommended allowances. Some correction factors will for different systems be described in more detail.

In the past, feed evaluation systems have been derived from feeding trials with the target animal at the desired level of production and rations commonly in use in practice. Baldwin and Miller (1989) characterized this approach as empirical models, in which input–output trials were used to parametize equations. Systems derived from energy balance studies, using animals confined in respiration chambers have introduced causal or deterministic elements in feeding systems and might be more precise in more extreme circumstances due to the great variety of feeds tested. However, validation of such systems under conditions in practice is a necessary step in getting these systems introduced and in convincing potential users of the applicability of the system in practice. Instead of data from new feeding trials, Van der Honing, Steg and Van Es (1977) and De Brabander *et al.* (1982) tested the system proposed in The Netherlands on data from earlier feeding trials in Denmark, The Netherlands and Belgium. Additional validation after new comparative feeding trials are conducted seems appropriate.

SOME CORRECTIONS AND THEIR INCLUSION IN THE SYSTEMS

The effect of type of production on the efficiency of utilization of ME resulted in different systems for lactating cows and for growing animals (beef production) in most countries. Within these systems, for dairy cows in particular, the level of feeding and the effect of the quality of the diet are factors to be considered.

In systems for growing ruminants both factors also play a role. The variation in feeding level is less apparent, but the effect of the level of production on the overall efficiency (k_{mg}) is more pronounced and should be taken into account.

Effect of plane of nutrition

Most current systems use information obtained at the maintenance level of feeding to calculate energy values of feeds. At a higher level of feeding the net result of higher faecal and lower methane and urine energy losses should be corrected. Although the correction is variable and probably non-linear at high levels of feeding a mean value for a linear correction has been chosen in the Dutch, Swiss and GB dairy systems (ARC, 1980): 1.8% less ME per multiple of maintenance and this correction has been applied to the requirement of dairy cows, so that the nutritive value of the table can be used.

In the Dutch system, requirements are corrected beyond a standard level of production of 15 kg FCM. The FRG-system has increased the requirement per kg of FCM by 0.07 MJ, which is approximately 2% per multiple of maintenance.

In the GDR-system the nutritive value of a ration is corrected depending upon the energy digestibility (dE) of the ration. In the French system, updated by Vermorel *et al.* (1987), a requirement correction is calculated for rations, in which *Unite fourragère lait* (UFL) supply is calculated from table values. This 'interaction value' includes a correction for the level of feeding as well as for the proportion of concentrates. In the NRC-system net energy for maintenance, growth and lactation (NE_m, NE_g and NE_l) values reflect data at production levels of intake. For the calculation of NE_l from TDN, above maintenance on average a 4% reduction in digestibility per multiple of maintenance level of feeding is assumed. Feeding values are tabulated at 3 times maintenance (NRC, 1988).

For growing cattle in the Dutch/Swiss system no plane of nutrition correction has been applied because the correction would be very small, fluctuating between 0.996 and 0.986 for a level of feeding (FL) between 1.2 and 1.8, respectively. In the French system a correction of 5% is only applied to the energy value of forage maize (Vermorel *et al.*, 1987).

In the starch equivalent (SE_K) and fodder unit (FU) systems the effect of plane of nutrition cannot be identified as such, but because requirements generally have been derived from results of feeding trials, it should have been accounted for.

Effect of ration quality

The parameter to describe feed quality can be dE, metabolizability (q) or ME-concentration (ME/DM). Because of the small variation in q or ME/DM of dairy rations in the GB-system (MAFF, 1975) constant values were taken for $k_m = 0.72$ and $k_1 = 0.62$. The efficiency for growth, k_g, varies with ME/DM.

In the Dutch, Swiss, French, FRG and GB (ARC, 1980) dairy systems a correction for q has been included to evaluate the whole range of poor- and high-quality feedstuffs reasonably correctly: $k_1 = 0.6 + 0.024(q-57)$ (Van Es, 1977) or $k_1 = 0.35q-0.42$ (ARC, 1980) respectively.

In the SE_K- and FU-systems corrections by value number or crude fibre deductions had a similar purpose. The TDN system and others, where such corrections are neglected, overestimate the value of poor feeds and underestimate that of energy-rich feeds.

Effect of animal production level on energy value for growth and fattening

McHardy (1966) tried to solve the problem of varying proportions of ME utilized for maintenance or growth, resulting in varying net energy values of a feed. He introduced the concept of the animal production level (APL), which is the ratio of the total net energy ($NE_m + NE_g$) to the net energy for maintenance (NE_m). Net energy requirements of animals are supposed to be independent of q or ME/DM, whereas ME-requirements depend upon q or ME/DM, making ration formulation rather complex and time consuming. Using the concept of APL, a function can be derived to calculate the total efficiency of ME-utilization for maintenance and growth/fattening (k_{mg}) as follows:

$$k_{mg} = \frac{k_m \times k_g \times APL}{k_g + k_m (APL - 1)}$$

Whilst this general theory was concerned with rations, McHardy (1966) suggested that the ME-value of individual feedstuffs could be transformed to feed NE_{mg} values and used in a purely additive manner to formulate a ration of desired NE_{mg} content, without incurring any significant error. This system was adopted by GB for growing ruminants. After a French proposal, it was decided in the Dutch/Swiss and French system, after consideration of the variation in k_{mg} as affected by APL and q, to choose a single level of APL, 1.5, at which to calculate NE_{mg} values. Vermorel *et al.* (1987) have shown that recent French data confirm that $APL = 1.5$ is a good estimate for energy intake by beef cattle. However, the calculation of energy requirements was adapted substantially (Geay *et al.*, 1987). In the systems mentioned deviations in live weight gain (LWG) from $APL = 1.5$ were corrected in the net energy requirements accordingly. To use NE_1 values at low levels of APL and moderate LWG for rearing dairy cattle requires additional corrections. These are small where the values for k_{mg} approximate to those for k_1 (Van der Honing and Alderman, 1988).

Other corrections

According to ARC (1980) the value of the maintenance requirement for bulls is 15% higher than for castrates and heifers. Energy deposition per kg LWG is assumed to be 15% less for bulls and large breeds compared with medium ones and 15% higher for heifers and for small breeds in order to correct for differences in protein/fat ratio in LWG.

In the revision of the French system (Vermorel, Coulon and Journet, 1987) the maintenance requirements for dairy cows in loose houses or at pasture have been increased by 10% and 20%, respectively.

Safety margins in conversion requirements into standards are used in some systems to prevent negative effects of underestimating the feed intake and/or nutritive value of rations (for example GDR-systems; ARC, 1980; French system).

Comparisons of systems

Differences between systems in the predicted level of animal production from a ration may be the result of a variety of reasons, such as differences in composition of products with identical names; differences in analysed composition of feedstuffs and assumed or measured digestibility; variations in calculation of ME and its predicted utilization; differences in tabulated requirements for maintenance and/or production; variation in corrections or safety margins in standards.

DIFFERENCES BETWEEN SYSTEMS AND TABLE VALUES OF BARLEY

A stepwise comparison to study the effects of several sources of variation between systems is not easy and it is more important to know the net effects of differences between systems on the precision of prediction of animal performance. However, to compare between systems and to develop conversion coefficients or equations to transfer values from one system to another it was thought better to exclude errors due to differences in composition and digestibility of feedstuffs (Van der Honing and Steg, 1984). Therefore the same input data of digestible nutrients in feedstuffs was used to calculate the feeding values according to different systems. To ensure a good insight into the differences between the systems a wide range of feedstuffs is required to ensure that deviations in particular cases were also shown. In Table 1.2 a small number of feedstuffs (after Van der Honing and Steg, 1984) demonstrates this.

To improve the comparison between systems the energy values of feedstuffs have been calculated relative to the energy value of a standard ration (500 g/kg roughages, 500 g/kg concentrates) in that system. The standard ration (SR) was composed of 200 g/kg good quality grass hay, 200 g/kg wilted grass silage, 100 g/kg high dry matter maize silage and concentrate ingredients (75 g/kg maize grain, 50 g/kg barley, 50 g/kg maize gluten feed, 100 g/kg beetpulp, 50 g/kg manioc, 40 g/kg cane molasses, 50 g/kg coconut expeller, 50 g/kg extracted soybean meal, 25 g/kg rapeseed meal) and 10 g/kg of a mineral + vitamin mixture.

Averages of the relative values in that study varied slightly but the standard deviations, even without animal fat, were appeciably different: high for SE_K, low for net energy lactation in the USA (NEL(US)), TDN and Energiewert Fett Rinder (EFr) and intermediate for Dutch dairy feed unit (VEM), UFL, metabolizable energy in Great Britain and Sweden, respectively (ME(GB) and ME(S)). A higher standard deviation is caused in general by lower values for the poorly digestible roughages and higher values for the well digestible concentrates.

Table 1.2   ENERGY VALUES OF DIFFERENT ENERGY EVALUATION SYSTEMS*
RELATIVE TO STANDARD RATION (COMPOSITION IN TEXT)

Feedstuff	SE_K	TDN	EFr	ME (S)	ME (GB)	NEL (US)	VEM	UFL
Fresh grass: early cut	108.6	104.3	103.7	105.0	102.6	104.5	107.4	105.7
Wilted grass silage	87.6	93.1	94.5	88.2	91.0	92.5	88.5	90.2
Grass hay: good qual.	77.8	88.8	88.5	86.4	86.3	87.9	84.8	87.0
Maize silage (dough)	97.8	95.6	95.9	91.5	95.2	95.2	95.8	91.5
Barley straw	31.5	60.9	61.2	54.9	56.8	58.0	53.0	51.9
Fodder beet	98.4	113.2	110.1	114.2	116.7	114.0	113.5	99.8†
Barley (grain)	126.9	113.5	112.5	117.5	117.1	114.5	117.9	118.7
Beet pulp	95.5	107.1	103.4	105.7	107.3	107.5	107.7	107.0
Cane molasses	79.8	94.0	92.0	95.5	98.1	93.6	91.8	95.4
Brewers grains	89.3	93.2	102.8	95.1	93.1	92.7	94.1	85.8
Manioc	124.2	107.9	105.5	109.4	112.3	108.4	111.3	108.6
Rapeseed meal solv. extr.	100.0	93.9	89.0	104.3	94.7	93.4	94.7	94.9
Soybean meal solv. extr.	124.8	114.3	104.4	128.1	114.6	115.2	120.1	121.6
Fat (veg. origin)	355.1	285.0	463.5	305.5	284.2	297.9	369.0	303.8

*SE_K = starch equivalent according to Kellner; TDN = total digestible nutrients; EFr = energy value
cattle in GDR-system; ME (S) = metabolizable energy in Swedish system; ME (GB) = ME in system of
Great Britain in Bull. 33; NEL (US) = system NE$_1$ in USA; VEM = Dutch feed unit system; UFL =
French feed unit system.
†115.8 according to UFL updated by Vermorel, Coulon and Journet, 1987.

In the SE_K, which is a net energy fattening system, and the fodder unit system, values of forages for milk production are underestimated. In these systems molasses, beet pulp and fodder beet also have lower values due to a deduction for sugar. The same applied to fodder beets in the French system, but in the second issue of UFL no deduction is made at limited inclusion of fodder beet in dairy rations (Vermorel, Coulon and Journet, 1987). In the ME(S)-system the values for protein-rich feedstuffs differ from those in other systems due to a high coefficient for protein in the prediction equation.

For roughages TDN-, EFr- and NEL(US)-values are in general higher than VEM-values. It should be appreciated that NEL(US) is calculated from TDN: NEL(US) = -0.12 + 0.0245 TDN. Concentrate ingredients generally have a higher value in VEM than in TDN or NEL(US). Differences between VEM and ME(GB) are small. In this comparison the TDN-system is in surprisingly good agreement with the modern systems.

Specific products such as maize silage and brewers'grains have a considerably lower value in UFL than in VEM, due mainly to differences in the calculation of energy value.

It has to be emphasized that variations in composition and digestibility of feedstuffs may add considerable inaccuracies to the comparison as made by Van der Honing and Steg (1984). Nevertheless between countries and areas, where feedstuffs are grown, account should be taken of these differences, for forages in particular. The latter can be harvested in various stages of maturity and lignification is moderately dependent on climatic conditions. For concentrates this variation will be less and therefore a comparison of table values of a ceral, such as barley grain, might be informative. Table 1.3 shows results of ME and/or NE values for dairy cattle in tables of the various countries. For ME the highest value is 7% more than the lowest value, but for NE a larger difference of 20% is found.

Table 1.3   ME AND NE₁ VALUES OF BARLEY GRAIN FOR DAIRY CATTLE (MJ/kg DM) FROM NATIONAL FEED TABLES

	ME	*NE*	*Table issue*
Denmark	12.7 (100)	9.1 (117)	8th., 1983
Germany (GDR)	13.1 (103)	7.3 (95)	5th., 1986
France	13.3 (105)	8.4 (108)	1987
Germany (FRG)	13.1 (103)	8.3 (107)	1982
Netherlands	12.6 (= 100)	7.8 (= 100)	33rd, 1988
Sweden	13.5 (107)	–	1976
United Kingdom	12.8 (101)	–	1986
USA, Israel	13.6 (107)	8.0 (103)	1976

Differences in k_1 between systems contribute to the larger difference in NE than in ME values.

CONVERSION OF ENERGY VALUES FROM ONE SYSTEM TO ANOTHER

As a result of an increasing interest in the transformation of given energy values Van der Honing and Steg (1984) have also concentrated their study on the accuracy of conversion of energy values between current energy systems for dairy cattle. Multiple regression analysis, using the chemical composition according to the Weende analysis of feedstuffs as additional variables, was used to improve the accuracy of conversion. Table 1.4 presents the accuracy of prediction of ME(GB) from other systems. Provided a good reproducibility of chemical analysis and digestibility measurements, these data show that ME(GB) can be predicted accurately (CV < 2%) from TDN, EFr, VEM, ME(S) and NEL(US), when additional information on Weende characteristics is available. Conversion from the starch equivalent and related systems, however, was less precise.

Table 1.4   ACCURACY OF PREDICTION OF ME(GB)-VALUES (= y) FROM ENERGY VALUES IN OTHER SYSTEM (= X_1) AND CHEMICAL COMPOSITION (XP, XL, XF) OF 31 FEEDSTUFFS (VAN DER HONING AND STEG, 1984)

X_1	*Sign. coeff. for*	*Residual coefficient of variation (%)*	
		Only X_1	*X_1 + sign. variables*
SEk	XF	7.1	4.5
TDN	XP, XF	1.8	0.6
EFr	XP, XL, XF	5.4	0.9
VEM	XP, XF	2.6	1.5
UFL	XP, XF	4.1	3.3
ME(S)	XP	4.0	1.8
NEL(US)	XP, XF	1.8	0.7

COMPARISON OF QUANTITIES OF FEED NEEDED TO ACHIEVE A GIVEN
ANIMAL PERFORMANCE

Because errors in the energy values of feeds can be corrected by the formulation of
energy standards, a comparison of energy systems should include a comparison of
predicted animal performance and not be restricted solely to energy values of
feedstuffs. Van der Honing and Steg (1984) have calculated the milk yield of a
550 kg cow to be expected from various quantities of a standard ration (SR, for
ingredients see before) and the milk yield of a marginal increase in dry matter
intake. A summary is presented in Table 1.5. Variations in kg SR required for
maintenance (4.2–5.3 kg DM) caused part of the differences observed in Table 1.5,
which are about 15% between extremes for 20 kg SR.

Table 1.5   POTENTIAL MILK YIELD (kg FCM), INCLUDING MAINTENANCE BY A 550 kg
COW OF 20 kg DRY MATTER OF A STANDARD RATION (SR, SEE TEXT) AND MILK YIELD
OF MARGINAL INTAKE

System	FCM/20 kg DM of SR		FCM/kg DM of SR	
	kg	*Relative to ME(GB)*	*kg*	*Relative to ME(GB)*
SE$_K$	35.9	112	2.34	109
TDN	33.7	105	2.30	107
EFr	32.9	102	2.16*	100
VEM	31.1	97	2.16*	100
UFL	31.0	97	2.25	105
ME(GB)	32.1	100	2.15	100
ME(S)	34.1	106	2.28*	106
NEL(US)	34.1	106	2.32	108

*Without additional corrections

By excluding SE$_K$, this difference within the modern systems is reduced to 10%.
The magnitude of these differences is still striking because the calculations are
based upon a rather balanced ration of roughages and concentrates at a specific
level of intake. It demonstrates that the uncertainties about the energy
requirements of lactating cows seems to be greater than differences in energy value
of feedstuffs between systems. In an alternative way this was also shown by Van der
Honing and Alderman (1988). They calculated within systems the theoretical
amount of barley, which would meet the formulated requirements of a dairy cow
producing 20 kg FCM or a growing bull, gaining 1 kg daily. The results are
presented in Table 1.6. In the presented systems the quantity of barley needed for
20 kg of milk varies between 11.7 and 13.2 kg DM, on average 12.4 kg if those
values based on fattening values 'Denmark' are omitted. Thus a difference of about
10% between modern systems is noticed here as well. The calculation for a growing
bull shows a larger variation (5.0–6.9 kg DM). This is understandable because
retained energy per kg LWG varies widely across Europe due to differences in
breeds and feeding systems. These figures also demonstrate the inadequacy of
expressing energy requirement per kg LWG without accounting for factors that
affect the energy content of LWG, such as age and sex of the animal. Geay *et al.*

Table 1.6   QUANTITY OF BARLEY (kg DM) NEEDED TO MEET ENERGY REQUIREMENTS (AFTER VAN DER HONING AND ALDERMAN, 1988)

	600 kg cow, 20 kg FCM/day	*400 kg bull, LWG = 1000 g/day*
Denmark	10.41 (= 100)	5.0 (= 100)
France	11.71 (112)	5.3 (106)*
Germany GDR	11.87 (114)	6.7 (131)†
Germany FRG	12.65 (121)	5.4 (108)
Netherlands	12.55 (120)	6.0 (120)
Sweden	11.99 (115)	6.5 (130)
United Kingdom	13.20 (127)	6.4 (128)
USA, Israel	12.83 (123)	–
USA (NRC, 1988)	12.63 (121)	6.5 (130)

*4.6 according to Vermorel, Coulon and Journet (1987)
†Excluding 10% extra for loose housing conditions

(1987) have introduced in the adapted French energy system corrections for the content of alimentary tract, body fat content as dependent on the empty-body weight (EBW) and type of animal, fat content of LWG as related to EBW-gain and protein deposition from fat-corrected EBW-gain. Although such corrections are helpful for a better prediction of the performance of beef cattle it still remains a major difficulty and weakness in all systems to relate energy gain from characteristics of a live beef animal.

Future developments in feed evaluation

Tendencies within the EC will stimulate international trade in feedstuffs. In addition to this improved quality control and guarantees on the energy declaration to be given by feed compounders will be required and as a consequence, methods for accurate prediction of nutritive value are needed.

These developments may stimulate more uniformity in feed evaluation systems. In addition organizations such as the OECD and FAO will welcome a common unit for the energy value of feeds. Aside from energy, the protein resources are also a vital element of EG, FAO and OECD.

From the point of view of the users it is clearly necessary to define a unit, which can be easily understood by non-specialists. Because metabolizable energy will be used for different types of animal production (for example milk yield, wool production, meat production, maintenance), with different efficiencies of utilization, one net energy unit is not feasible. Therefore Van der Honing and Alderman (1988) proposed as a common energy feed unit the quantity of metabolizable energy in 1 kg of barley. The average ME of barley in the tables of those countries or systems, where measurements *in vivo* have been performed, is close to 13 MJ ME per kg DM of barley (see Table 1.3). There is some variation between countries due to small differences in the composition and digestibility of nutrients in barley.

The striking differences in the coefficients of the equations to calculate ME from digestible nutrients are hard to justify. It should be possible to arrive at a common equation to be applied in modern systems which hopefully can be organized by EC or the EAAP.

The increasing interest in accurate description of the nutritive potential of feedstuffs, the improvemnent of methods for chemical, physical (for example NIRS, NMR, pyrolytic MS) and *in-vitro* digestion analyses and information technology will certainly stimulate reliable (accurate and rapid) methods to predict digestibility, energy and protein values of feeds. However, such methods require a great number and a wide range of variation of reference samples of feedstuffs, which have been measured in digestibility trials with sheep or cattle. Individual institutes have too limited sample stock with *in vivo* data for adequate calibration of modern prediction methods and exchange of such samples of feedstuffs is necessary to improve prediction with in vitro methods and NIRS. Steg *et al.* (1988) took the initiative to arrange such an exchange to obtain a large pool of reference samples. Energy evaluation in practice may also profit from systematic collection, storage and distribution of information on variation in composition and digestibility of feedstuffs and special properties known for application in feeding of farm animals. The International Network of Feed Information Centres serves as a coordinating organization for improving the exchange of information.

The weak basis for evaluation systems for growing and fattening ruminants requires substantial improvement. Data on body composition and the energy (fat and protein) content of live weight-gain particularly are lacking and the available information shows a large variability. New methods to quantify the body composition of live animals and measure the composition of live weight-gain during growth are urgently needed to improve energy evaluation for growing ruminants.

Current energy systems are criticized for being insufficiently predictive of animal performance under variable conditions and, in particular, with rations that do not guarantee optimal rumen fermentation. Nutrient supply and factors involved need to be quantified much better in the near future. Progress by several research groups developing alternative models and collecting relevant data should result in new systems, which can improve the management of animal production in practice in the future.

References

Andrieu, J. and Demarquilly, C. (1987) Valeur nutritive des fourrages: tables et prevision. *Bulletin Technique C.R.Z.V., INRA*, **70**, 61–73

ARC (1980) *The Nutrient Requirements of Ruminant Livestock*, Commonwealth Agricultural Bureaux, Farnham Royal, Slough

Baldwin, R.J. and Miller, P.S. (1989) *Energy Metabolism of Farm Animals*. (eds Y. Van der Honing and W.H. Close) EAAP-Publication, nr. 43. pp. 239–241

Beever, D.E. (1988) In *Ruminant Feed Evaluation and Utilization*. (eds B.A. Stark, J.M. Wilkinson and D.I. Givens) Chalcombe Publications, Marlow, pp. 1–12

Bickel, H. (1988). III. Feed evaluation and nutritional requirements. III. 1. Introduction. *Livestock Production Science*, **19**, 211–216

Blaxter, K.L. (1962a) *The Energy Metabolism of Ruminants*, Hutchinson London

Blaxter, K.L. (1962b) Progress in assessing the energy value of feedingstuffs for ruminants. *Journal of Research of the Agricultural Society*, **123**, 7

Blaxter, K.L. and Clapperton, J.L. (1965) Prediction of the amount of methane produced by ruminants. *British Journal of Nutrition*, **19**, 511–522

Breirem, K. (1969) In *Handbuch der Tierenährung, I. Algemeine Grundlagen* (eds W. Lenkeit, K. Breirem and E. Crasemann) Hamburg, Verlag Paul Parey, pp. 656–677

De Brabander, D.L., Ghekiere, P.M., Aerts, J.V., Buysse, F.X. and Moermans, R.J. (1982) Tests of 6 energy evaluation systems for dairy cows. *Livestock Production Science*, **9**, 457–469

Dewhurst, R.J. and Webster, A.G.F. (1989) *Energy Metabolism of Farm Animals* (eds Y. van der Honing and W.H. Close), EAAP-Publication nr. 43, pp.223–225

Edwards, R.A. (1988) In *Ruminant Feed Evaluation and Utilization*. (eds B.A. Stark, J.M. Wilkinson and D.I. Givens), 1988, Chalcombe Publications, Marlow, pp. 25–40

Es, A.J.H. Van and Honing, Y. Van der (1977) Het nieuwe energetische voederwaarderingssysteem voor herkauwers: wijze van afleiding en uiteindelijk voorstel. *Report IVVO nr* **92**, 1–48

Es, A.J.H. Van (1978) Feed evaluation for ruminants. I. The systems in use from May 1977 onwards in the Netherlands. *Livestock Production Science*, **5**, 331–345

Es, A.J.H. Van and Meer, J.M. Van der (1980) *Methods of Analysis for Predicting the Energy and Protein Value of Feeds for Farm Animals*. Proceedings Workshop, Lelystad.

Geay, Y., Micol, D., Robelin, J., Berge, Ph. and Malterre, C. (1987) Recommendations alimentaires pour les bovins en croissance et a l'engrais. *Bulletin Technique C.R.Z.V. Theix, INRA*, **70**, 173–184

Haecker, T.L. (1914) *Investigations in Milk-Production*. Minnesota Agriculture Experimental Station. Bulletin nr. 140.

Henneberg, W. and Stohmann, F. (1864) *Beiträge zur Begründung einer rationellen Fütterung der Wiederkäuer II. Braunschweig*, **29**, 48

Honing, Y. Van der (1975) Intake and utilization of energy of rations with pelleted forages by dairy cows. *Agricultural Research Reports*, **836**, 1–156

Honing, Y. Van der, Steg, A. and Es, A.J.H. Van (1977) Feed evaluation for dairy cows: tests on the system proposed in the Netherlands. *Livestock Production Science*, **4**, 57–67

Honing, Y. van der and Steg, A. (1984) Relationships between energy values of feedstuffs predicted with thirteen feed evaluations systems for dairy cows. *Report IVVO nr*, **160**, 1–51

Honing, Y. van der and Alderman, G. (1988) III. Feed evaluation and nutritional requirements. III. 2. Ruminants. *Livestock Production Science*, **19**, 217–278

Kellner, O. (1905) *Die Ernähung der landwirtschaftlichen Nutztiere*, 1st edn. Berlin, Verlag Paul Parey

Kirchgessner, M., Schwab, W. and Müller, H.L. (1989) *Energy Metabolism of Farm Animals* (eds Y. Van der Honing and W.H. Close), EAAP-Publication nr. 43, pp. 143–146

MAFF (1975). *Energy Allowances and Feeding Systems for Ruminants*. Technical Bulletin No. 33. Ministry of Agriculture, Fisheries and Food (1975) London, HMSO

Maynard, L.A., Loosli, J.K., Hintz, H.F. and Warner, R.G. (1979) *Animal Nutrition*, (ed. R.C. Zappa), 7th edn, McGraw-Hill Book Company

McHardy, F.V. (1966) Simplified ration formulation. In *Ninth International Congress of Animal Production*, Edinburgh (Abstract), p. 25

Meer, J.M. van der (1984) CEC-Workshop on methodology of feedingstuffs for ruminants. European '*in-vitro*' ringtest 1983. *Report IVVO nr.* 155.
Meissner, H.H. and Roux, C.Z. (1989) *Energy Metabolism of Farm Animals*. (eds Y. Van der Honing and W.H. Close), EAAP-Publications nr. 43, pp. 359–361
Møllgaard, H. (1929) *Fütterungslehre des Milchviehs* Hannover , M. & H. Schaper
NRC (1978) *Nutrient Requirements of Dairy Cattle*, 5th revised edition. Publication NRC, 1978. Washington DC, National Academy Press
NRC (1988) *Nutrient Requirements of Dairy Cattle*, 6th revised edition. Publication NRC, 1988. Washington DC, National Academy Press
Sauvant, D., Aufrère, J., Michalet-Doreau, B., Giger, S. and Chapoutot, P. (1987) Valeur nutritive des aliments concentrés simples: Tables et prévision. *Bulletin Technique C.R.Z.V. Theix, INRA*, **70**, 75–89
Schiemann, R., Nehring, K., Hoffmann, L., Jentsch, W. and Chudy, A. (1971a) *Energetische Futterbewertung und Energienormen*, Berlin, VEB/DLV
Schiemann, R., Jentsch, W. and Wittenberg, H. (1971b) Zur Abhängigkeit der Verdaulichkeit der Energies und der Nährstoffe von der Höhe der Futteraufnahme und der Rationszusammensetzung bei Milchkühen. *Archiv für Tierernährung*, **21**, 223–240
Steg, A., Meer, J.M. Van der, Smits, B. and Hindle, V.A. (1988) *Predicting the Digestibility of Feedstuffs: Recent Developments*. Annual Report IVVO 1987, Lelystad, pp. 21–29
Steingass, H., Graff, C. and Menke, K.H. (1986) Einfluss der Fettbestimmungsmethode auf die scheinbare Verdaulichkeit und den energetischen Futterwert von Futtermitteln für Wiederkäuer und Schweine. *Landwirtschaftliche Forschungen*, **39**, 206–214
Tamminga, S. and Vuuren, A.M. Van (1988) Formation and utilization of end products of lignocellulose degradation in ruminants. *Animal Feed Science and Technology*, **21**, 141–159
Vermorel, M., Coulon, J.B. and Journet, M. (1987) Révision du système des unites fourragères (UF). *Bulletin Technique C.R.Z.V. Theix, INRA*, **70**, 9–18
Webster, A.J.F., Dewhurst, R.J. and Waters, C.J. (1988) In *Recent Advances in Animal Nutrition* (eds W. Haresign and D.J.A. Cole), Butterworths, London.

2

THE EVALUATION OF FEEDS THROUGH CALORIMETRY STUDIES

W.H. CLOSE
AFRC Institute for Grassland and Environmental Research, Church Lane, Shinfield, Reading RG2 9AQ

Introduction

Systems of feed evaluation for farm animals are based upon knowledge of two separate sets of information: the requirement of the animal for the nutrient in question, on the one hand, and the ability of the feed, or combination of dietary ingredients, to meet this requirement, on the other. It is generally assumed that energy yielding components are the most limiting dietary ingredients and that the extent to which animals convert feed into usable products is primarily dependent upon the efficiency of dietary energy utilization. It is not surprising, therefore, that considerable attention has been directed towards the development of systems for expressing both the energy requirements of animals and the energy value of feeds.

A key component in the development of these systems is knowledge of the heat produced by the animal as a consequence of living, since this represents a major proportion of the dietary energy intake of animals. The measurement of heat production or energy exchange is fundamental to the determination of the energy requirements of animals and to the assessment of the diet to meet those requirements. This paper is therefore concerned with the measurement of heat production, principally by calorimetry, those factors that influence it and the way in which the observations may be used to facilitate feed evaluation. Although the principles involved are common to all species, specific attention is directed to non-ruminant animals, since aspects of the determination of the energy needs and evaluation of feeding systems for ruminants have been reported in the previous chapter (see Chapter 1).

Energy evaluation: Why calorimetry?

The gross energy (GE) describes the energy contained in the feed when it is completely oxidized and indicates the energy value of the feed, as fed (Figure 2.1). However, there are considerable losses associated with the digestion and metabolism of the nutrients in the feed before they become available for accretion into body tissue. A first step is the determination of the digestible energy (DE) content since this takes account of the energy contained in the undigested fractions

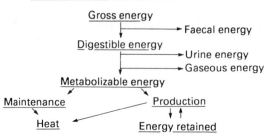

Figure 2.1 Schematic representation of the partition of energy in animals

of the feed lost in the faeces. The use of DE as a measure of the energy value of the feed is used in many feed evaluation systems (ARC, 1981; De Boer and Bickel, 1988) but it takes no account of the contribution of the various dietary nutrients that supply energy and hence of the different efficiencies with which energy may be utilized. This has led to the estimation of the metabolizable energy (ME) content which takes account of the energy losses in both urine and methane. This classification is more representative of the chemical composition of the feed and provides a better basis on which to evaluate systems of energy. It is also a useful basis on which to estimate the energy requirements of the animal since it can be conveniently partitioned into a component for maintenance (ME_m) and a component for production (ME_p), that is: $ME = ME_m + ME_p$. Similarly it may be partitioned into energy lost as heat production (HP) and energy retained (ER) within the body: $ME = HP + ER$.

This heat produced is a function not only of the animal *per se* but also of the feed it consumes and of the rate at which tissues or products are formed within the body. Of the total heat loss of the animal only that associated with the feed, that is the heat increment of feeding, is truly wasteful and the deduction of this component from the ME intake gives the net energy (NE) value of the feed. The net energy represents the true energy available to the animal for productive purposes and takes account of the losses in the metabolism of absorbed nutrients which varies not only with the source of energy, but also with the form stored, that is as protein, fat or carbohydrate. Knowledge of those animal, dietary and climatic factors which influence heat production, and hence net energy, have therefore direct implications for both the evaluation and utilization of dietary energy in animals.

Measurement of heat exchange

PRINCIPLES, SYSTEMS AND APPLICATION

Determination of the heat of combustion of the food ingested and the excreta produced are relatively simple and difficulties associated with providing estimates of energy retention in nutrient balance trials are predominantly associated with the measurement of energy expenditure.

Both indirect and direct types of calorimeter have been developed for measuring energy expenditure but most of the progress has been made with the former.

Indirect methods are based on the relationship between the amount of heat produced from the oxidation of food, or body constituents, and the amount of oxygen consumed, carbon dioxide produced and the nitrogen excreted in the urine. The general equation is (Brouwer, 1965):

$$H = 16.18\ O_2 + 5.02\ CO_2 - 2.17\ CH_4 - 5.99\ N \tag{2.1}$$

where H is the heat production (kJ/day), O_2 is the rate of oxygen consumption (1/day), CO_2 is the rate of carbon dioxide production (1/day), CH_4 is the rate of methane production (1/day) and N is the nitrogen excreted in the urine (g/day). The equation is based on the combustion of starch, protein and fat, with methane and urea being produced as end-products. Respiration chambers for measuring energy expenditure may be of the open-circuit, closed-circuit or total collection type depending upon the system of ventilation and method of gaseous exchange employed.

Direct calorimeters measure the heat emitted by an animal and are usually designed around two basic types; the adiabatic or isothermal system where the heat produced by the animal is removed and measured, or the conduction or gradient-layer type in which the heat produced is allowed to flow across the walls of the calorimeter and the rate of flow measured. The main advantage of 'the gradient-layer calorimeter is that it measures instantaneous changes in heat loss whereas others react more slowly. Direct calorimeters are more expensive to construct and since the measurement of respiratory exchange is technically simpler a large number of indirect systems have been developed to measure the energy expenditure of both laboratory and farm animals and man. A most comprehensive account of the principles, development, design and application of the various systems of calorimetry, and other methods of measuring energy expenditure, has been provided by Blaxter (1962) and McLean and Tobin (1987).

Other methods to determine heat production and energy retention are based on the nutrient balance procedure in which the content of carbon, nitrogen and energy in food and excreta is measured, or from the comparative slaughter technique where differences in the energy, nitrogen and fat content of representative animals at the begnning and end of the experimental period are measured and the rates of change calculated. Balance and calorimetry procedures do not involve the destruction of the animal and can be repeated regularly with the same animals and may include several nutritional treatments. They may be used to measure several small changes in energy expenditure unlike the comparative slaughter procedure which only provides a measure of the mean rate of response of the animal throughout the experiment.

The accuracy and errors associated with the different measurements of energy expenditure have been discussed by Blaxter (1967, 1971). It was observed that there was little difference in the rate of heat production of the same sheep measured by direct or indirect calorimetry and concluded that provided great care was taken in balance trials, there was no reason to suppose that energy retention measured by the various methods was inaccurate or biased in any significant way. However the degree of agreement in the estimates of heat production measured by calorimetry and the comparative slaughter procedure is less good, with differences ranging between -2 to $+10\%$ (Schiemann, Chudy and Herceg, 1969; Pullar, Brockway and McDonald, 1967; McCracken and Rao, 1989). These differences suggest that the measurements of energy expenditure, even in the best designed trials, may be confounded by differences in activity, stress and nervousness between animals confined in calorimeters or housed in conventional surroundings and exposed to normal activity patterns (Blaxter, 1967; Webster, 1989).

This raises the question of the reliability and biological relevance of results obtained in calorimeter trials. Blaxter (1967) has commented that 'a respiration chamber is no doubt a fearful place for an animal to enter on the first occasion' and 'there may well be an adaptation of the animal to the conditions of the experiment'. It was demonstrated subsequently that the fasting metabolism of untrained adult sheep declined by 17% in a period of four weeks following introduction into the chamber, whereas fully-trained animals exhibited no such effects over the same period. The untrained animal lay more at the end of the period but the difference between lying and standing was insufficient to explain the decrease in energy expenditure. Blaxter (1974) interpreted the results as an increased stress associated with confinement but one which the animal gradually accommodates and concluded that the results from experiments which do not provide sufficient adaptation, and of measurements of energy expenditure of unrestrained animals living in field conditions, should be carefully interpreted. Once an animal has however adapted to its new living conditions, then the estimates of heat production

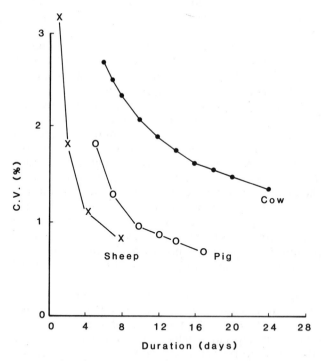

Figure 2.2   The influence of the duration of the trial on the coefficient of variation (C.V.%) associated with the measurement of energy balance in farm animals. Values taken from the results of (1) Blaxter (1967) on the measurement of heat losses in sheep (x), (2) Schneider and Flatt (1975) on the measurement of the digestibility of energy in lactating cows (●) and (3) Verstegen, Mount and Close (1973) on the measurement of the metabolizability of energy in growing pigs (○)

of both pigs (Gray and McCracken, 1980) and penned lambs (Webster, Smith and Brockway, 1972) measured on the first day of confinement within the calorimeter were within the normal range of variation and provided valid estimates of energy expenditure.

Even though an animal has become trained to the conditions and protocol of the experiment, the analytical and sequential errors which can occur in the determination of energy expenditure may be considerable. From a series of experiments Blaxter (1967) showed that the error associated with the estimates of energy retention in mature sheep was about 1% of the intake, but that this was only attained where long collection periods were used. The errors associated with the determination of heat production and methane were 1.0 and 2.4%, respectively. Since the faecal error is one of the largest components, collection periods of 7–10 days are normally used for poultry and pigs, compared with 10–14 days for ruminants, with heat production usually measured for 2–4 days duration within each period. However, as indicated in Figure 2.2, the longer the period of collection the lower the accumulative error and the greater the accuracy of the measurements.

Metabolic heat production

The heat that an animal produces is predominantly a function of its body size, that is the maintenance heat production, and the quantity and quality of the ration provided, the so-called heat increment of feeding. This division allows calculation of both the energy requirements of the animal and the energetic efficiency associated with tissue deposition as a result of feeding. The maintenance heat production represents an animal in a state of equilibrium neither gaining nor losing energy so that the intake of dietary energy exactly balances the animal's heat output. The heat arising from the accretion of tissue or products within the body represents the work done in their deposition and varies with the level of feeding. The partition of heat production into meaningful physiological or metabolic components is both complex and controversial, and includes components associated with basal or fasting metabolism, heat of digestion, absorption and accretion, heat of fermentation, heat resulting from the formation of products within the body, extra-thermoregulatory heat, heat arising from activity and heat of waste formation and excretion (Baldwin and Bywater, 1984). All of these are influenced by a number of factors so that at any given level of feeding, and as heat production increases, there is a reduction in the rate of energy retention and hence in energetic efficiency, that is in the net energy value of the feed.

The capacity of a feed to sustain maintenance and to promote energy gain is illustrated in Figure 2.3 in relation to ME intake. As ME increases there is a concomitant increase in heat production (HP), but the increase in ME exceeds the increase in HP, so that the animal has the capacity to retain energy (ER). The efficiency with which energy is retained, $\Delta ER/\Delta ME$, is greater below the maintenance energy requirement (k_m) than above (k_g). Partition of total energy retention into protein and fat components allows simultaneous determination of the energetic efficiency both of protein (k_p) and of fat (k_f) deposition. Thus the total heat production associated with the metabolism of dietary energy is the sum of the energy costs of maintenance and the increment in heat associated with the deposition of tissue calculated as $(1-k_g)$ or partitioned into that associated with

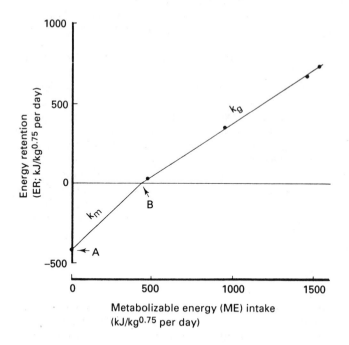

Figure 2.3 The relationship between energy retention (ER) and metabolizable energy (ME) intake in the growing pig (data of Close, Mount and Brown, 1978). A, represents the basal or fasting heat production, B represents the maintenance heat production. Below B, the efficiency of energy utilization ($\triangle$ER/$\triangle$ME) represents that for maintenance (k_m) whereas above B it represents that for production (k_g) (from Close and Verstegen, 1981, courtesy of Livestock Production Science)

protein deposition $(1-k_p)$ and fat deposition $(1-k_f)$. This is known as the 'factorial approach' to the determination of energy and feed requirements and ensures that both animal energy requirements and feed energy values are expressed in compatible terms.

BODY STATE AND HEAT PRODUCTION

The maintenance energy requirement represents a considerable proportion of an animal's energy intake, varying from 0.35 for growing pigs to 0.57 for laying hens (Van Es, 1972). It is primarily dependent upon the weight of the animal, but varies with age, breed, sex, physiological state and health. For comparative purposes it is normally expressed to some power of body weight, usually 0.75 (Kleiber, 1961). For example, estimates of the maintenance heat production in pigs range between 350 and 700 kJ/kg$^{0.75}$ per day (ARC, 1981) but may vary dependent upon the method of determination, the previous nutritional state of the animal and its body composition (Koong *et al.*, 1982; Armstrong and Blaxter, 1984). Webster (1981) estimated that some 45% of total heat production was associated with the gastro-intestinal organs indicating that the energy expenditure by the more metabolically active tissue is much higher than the energy expenditure associated with the carcase *per se* (Baldwin *et al.*, 1980). A considerable proportion of the maintenance heat may be associated with protein synthesis since this is higher in gastrointestinal tissue than in muscle and Reeds, Wahle and Haggarty (1982) calculated that it alone contributed up to 20% of the total heat production in young pigs. Webster (1981) attributed the variations in maintenance energy requirement to change in body composition and expressed the values for animals of varying physiological states relative to their body protein content. They ranged between 1290 and 2305 kJ/kg body protein$^{0.75}$ per day but the variation between animals of the same species, but of different physiological state, disappeared, suggesting that the maintenance energy requirement is better expressed as a function of lean mass than of body weight since fat metabolism contributes little to heat production compared with protein. Differences in metabolic rate of the same mass of tissue may therefore be attributed to differences in the fractional rate of synthesis and turnover of the major body constituents.

Basal energy expenditure has been further categorized by Baldwin *et al.* (1980) into service functions necessary for the whole organ and costs necessary for the existence of individual cells and tissues. The service functions included the work of the kidneys and heart, nervous and endocrine control and respiration and accounted for 36–50% of basal energy expenditure. Tissue and cellular functions, associated with ion transport and protein and lipid resynthesis (turnover), represented 40–50% of the basal energy expenditure. The 30–40% of basal energy expenditure attributed to ion transport alone is somewhat higher than the value of 20–25% reported by Gregg and Milligan (1987). Nevertheless these results suggest that the variations in maintenance energy requirement may be attributed to changes in the basal energy expenditure of the various tissues within the body and demonstrates that the underlying sources of variation and the metabolic components of basal heat production must be quantified if differences resulting from the age of the animal, its plane of nutrition, physiological state and body composition are to be interpreted.

DIET AND HEAT PRODUCTION

Both the level and the composition of the feed influence the rate of heat production of animals. Increasing the level of feeding increases heat production (Figure 2.3) but as the increment in heat output is less than the increment in energy intake, energy is retained within the body. The increase in heat production associated with feeding is classified as the heat increment of feeding and arises from the consumption and digestion of the food and the absorption and accretion of nutrients into body tissue. When an animal begins to eat, its rate of heat production increases to 40–80% above its basal level but decreases soon after the meal is consumed, even though it may take from 6 to 12 hours for it to decline to its pre-prandial level. This elevation in heat may thus be attributed to the energy cost of eating *per se* and varies with the type of diet consumed, from 1.5 kJ/kg per hour for pelletted and concentrated diets to 2.7 kJ/kg per hour for dried forages (Webster, 1978; Curtis, 1983). In ruminants there is also a cost arising from grazing (2.1 kJ/kg per hour) and ruminating (1.05 kJ/kg per hour). Osuji, Gordon and

Webster (1975) suggested that 50% of the heat associated with feeding was attributable to the processes of digestion and absorption *per se*, that is in the digestive tract, and this represented between 20 and 30% of the total heat production. The fermentation heat loss contributes 6–8% of the total heat loss in ruminants but so far there are no published values for the pig (Longland, Close and Low, 1989).

The efficiency with which dietary energy is absorbed and utilised by the animal varies not only with the type of product formed but also with the relative contribution of the various energy substrates to the total intake. The energy requirement for protein deposition is lower than that for fat and it is to be expected that young growing animals depositing mainly lean tissue will have a lower net efficiency of energy utilization, and a relatively higher heat production, than a mature animal depositing predominantly fat. In pigs, for example, the preferred value for the energetic efficiency of protein and fat deposition was 0.54 and 0.74, respectively (ARC, 1981). Millward, Garlick and Reeds (1976) calculated that the ME from dietary fat was deposited as lipid with an efficiency of 0.99, from dietary carbohydrates 0.85 and from dietary protein 0.69, in agreement with the theoretical calculations of Blaxter (1962) and Armstrong (1969). The higher efficiency for fat suggests that a large proportion of the dietary fatty acids can be incorporated directly into body fat, avoiding the energy losses associated with the degradation and subsequent resynthesis of specific fatty acids. Those high efficiencies are in contrast with those associated with the digestion of fibrous materials.

It has long been recognized that animals respond differently to different types of feed, so that on a similar ME basis concentrate diets provide a greater rate of energy retention above maintenance and a lower rate of heat production than roughages (Benedict and Ritzman, 1927). This additional heat results from differences in the digestibility of the materials, in the energy substrate provided, and in the relative efficiencies with which the energy sources are utilized within the body. Cereal-based diets provide substantial quantities of starch, sugars, proteins and fatty acids, which are enzymatically digested in the small intestine and utilized more efficiently than the products of fermentation. The major energy substrate associated with roughage diets are the volatile fatty acids (VFAs), acetate, propionate and butyrate, and they result predominantly from fermentation in the rumen, or the hindgut of non-ruminant animals, although the capacity for the latter is limited. However, an inevitable energy loss associated with the digestion of fibrous materials is that occurring as methane and Hungate (1966) calculated that as much as 0.18 of the energy value of hexoses could be lost in the rumen, although values for pigs are considerably below this. Recently Müller and Kirchgessner (1982, 1987) demonstrated that the amount of methane energy produced from fermentable carbohydrate in pigs was between 10% and 11% of the additional energy consumed but this loss represented no more than 3% of the total energy intake.

The type of volatile fatty acid formed can influence the heat increment of feeding, with the utilization of acetate being less and giving rise to larger quantities of heat than either butyrate or propionate. Blaxter and colleagues (quoted by Blaxter, 1967) measured the efficiency of utilization of various molar proportions of VFAs at levels above and below the animal's maintenance energy requirement. They found that acetate was utilized less efficiently than either propionate or butyrate and that the values were lower above compared with below maintenance; acetate 0.33–0.59, propionate 0.56–0.87, butyrate 0.62–0.76. These values were considerably less than those of 0.55–0.99 determined when glucose was infused directly into the rumen. When mixtures of VFAs were used the net efficiency of energy utilization varied depending upon the molar proportions, so that the higher the molar proportion of acetate the lower the net efficiency and the higher the heat increment. Ørskov and Allen (1966) and McCrea and Lobley (1982) have suggested that the differences in the net efficiency of utilization of the VFAs may be less than anticipated, and suggested that a possible explanation of the varying net efficiency of acetate utilization originally determined may have resulted from a varying supply of glucose as a pre-cursor of NADPH for fatty acid synthesis. In this respect it is interesting to note that recent estimates of the efficiency of energy utilization of VFAs infused into the caecum of pigs gave estimates of 0.65 for acetate, 0.71 for propionate and 0.67 for butyrate (Gädeken, Breves and Oslage, 1989). Estimates of the net efficiency of energy utilization of the various metabolic substrates are compared in Table 2.1 for several species.

Table 2.1 SOME ESTIMATES OF THE EFFICIENCY WITH WHICH ENERGY FROM VARIOUS SOURCES IS USED FOR PRODUCTION IN SEVERAL SPECIES

	Ruminant	Pig	Poultry
Glucose	0.5–0.7	0.74	
Starch	0.64	0.76	0.78
Fat	0.58	0.86	0.78
Casein	0.5–0.65	0.65	
Acetate	0.2–0.6	0.6–0.7	
Propionate	0.5–0.6	0.6–0.7	
Butyrate	0.6–0.7	0.6–0.7	

Values taken from Nehring et al. (1965), Ørskov and Allen (1966), Burlacu et al. (1969), Blaxter (1971) and Gädeken, Breves and Oslage (1989)

CLIMATE AND HEAT PRODUCTION

Most measurements of energy balance are determined within the zone of thermal neutrality where the environment imposes least thermoregulatory demand upon the animal. However, the environment determines the extent and efficiency with which dietary energy is utilized for maintenance and thermoregulation, on the one hand, and energy retention or growth, on the other. Thus the environment to which the animals are exposed can influence the net energy available for both maintenance and production. This topic will only briefly be considered since it has been extensively reviewed by Blaxter (1962), Close (1981), Curtis (1983), Mount (1968), Webster (1974, 1983).

At temperatures below and above the zone of thermal neutrality, heat production increases with direct effect upon both the rate and efficiency of energy utilization. The temperature at the lower end of the zone is called the critical temperature and this represents the lowest environmental temperature at which optimum utilization of dietary energy for body gain occurs. However, it is not just temperature per se which affects the heat exchange of animals; other components of the environment such as wind speed, radiation, relative humidity, the absence or presence of litter mates, floor type and bedding can have a major effect on the partition of energy intake. The thermal environment therefore comprises a complex of factors which interact to determine the environmental heat demand and there have been several attempts to formulate a scale of effective environmental temperatures which take account of all these factors and which can be referred to as a single parameter reflecting this heat demand. For example, Mount (1975) has calculated that the effective environmental temperature of a group of pigs decreased from 20°C when they were housed within thermal neutrality to 13°C when air speed exceeded 50 cm/s and 6.5°C when lying on a cold, wet, concrete floor in a building where the temperature of the walls was 3°C cooler than that of the air. Similarly Webster (1974) demonstrated that the effective critical temperature of a well-fed beef cow in different cold environments ranged between −21°C and +2°C and in sheep with different coat thickness Alexander (1974) calculated values between −40°C and +13°C.

At temperatures below the lower or effective critical temperature, an animal must raise its rate of heat production in order to maintain homeothermy and it achieves this by increasing its feed intake, by altering its posture and activity, by shivering, by vasoconstriction and by improving its body insulation. However the higher the level of feeding the greater the heat increment of feeding which can be used to substitute for the extra thermoregulatory heat demand of the colder environment, with the effect of reducing critical temperature. In practice the question arises as to whether the additional heat associated with feeding substitutes entirely for the extra thermoregulatory heat. Graham *et al.* (1957) exposed shorn sheep to different feeding levels and observed that the lower critical temperature fell from 32°C to 22°C with increase in feeding level, but, that at temperatures below the lower critical level, heat production was independent of feed intake, that is the heat increment associated with feeding exactly substituted for the increased thermal demand of the environment. Below the critical temperature, the heat produced by the animal was independent of feeding level with the result that the increase in energy retention was similar to the increase in ME intake, that is a net efficiency of 1.0. This suggested that heat production below thermal neutrality was a function of the body core temperature of the animal and thermal insulation. In the pig, on the other hand, the extra heat associated with additional feeding did not exactly substitute for cold thermogenesis, so that the net efficiency was only 0.8, that is each additional MJ of ME resulted in an 0.8 MJ increase in energy retention (Close, 1978; Close and Mount, 1978). This is, however, higher than the normally accepted value of 0.7 determined within thermoneutrality (ARC, 1981). However it is important to note that although the net efficiency of dietary energy utilized may be higher, the maintenance energy requirement is also increased so that the net energy requirement of the animals in a cold environment will be higher than that within thermal neutrality.

Within the zone of thermal neutrality heat production is minimal and independent of environmental temperature. However as environmental temperature is increased above the upper limit of this zone, body temperature begins to rise with a concomitant increase in heat production. At this point the environment limits the ability of the animal to balance heat output against internal and external heat loss. If this persists the animal may die although it can control its rate of heat output to some degree by reducing feed intake, by sweating or panting, by behavioural means such as wallowing and by vasodilation. Several practical measures, such as the provision of wallows, shades and sprinklers, have been used to reduce the heat load on the animal under hot conditions and thus improve its productivity.

Various attempts have also been made to improve feed intake and the efficiency of dietary energy utilization in animals under hot conditions by varying the nutrient composition of the feed or by appropriate 'packaging'. Nutrients differ in their heat increment of feeding and thus diets of low heat increment, especially those based on fat, have been used in attempts to enhance feed intake and to improve energetic efficiency and growth rate above that normally observed in hot conditions (Stahly and Cromwell, 1979; McGlone, Stansbury and Tribble, 1988). Fibrous diets, on the other hand, have high heat increments of feeds, and this makes them more suitable for feeding under cold conditions since the additional heat may be used to compensate for the extra thermal heat demanded of the animal (Noblet, Le Dividich and Bikawa, 1985). Processing of diets may also be effective and pelleting has been shown to improve the digestibility of pig diets by as much as 10% (Vanschoubroek, Couckle and Van Spaendonck, 1967). Similarly MacLeod, Jewitt and Anderson (1987) demonstrated that the heat increment of feeding was 15% lower in cockerels fed by intubation compared with self-feeding and this led to a 3% increase in the efficiency of energy utilization for maintenance.

ACTIVITY AND HEAT PRODUCTION

Additional factors that can contribute to differences in heat production are those associated with activity, movement and locomotion of animals. In a free-ranging animal the energy costs of standing, changing position, walking, grazing and state of alertness are additional to those activities spent in confinement in a calorimeter and may contribute as much as 10% of fasting metabolism (ARC, 1980). They therefore need to be considered if energy needs of farm animals are to be accurately applied in practice.

Wenk and van Es (1976) demonstrated that the degree of physical activity in chickens was related to their feed intake; under *ad-libitum* feeding conditions, activity accounted for 7–10% of the ME intake compared with 15% when restricted feeding was practised. When the animal's maintenance energy requirement was corrected for physical activity it was reduced by 37%, from 505 to 319 kJ ME/kg bodyweight$^{0.75}$ per day. Verstegen *et al.* (1982) also demonstrated a 22% reduction in the maintenance energy requirements of pigs when corrected for zero activity, but found that the net efficiency of energy utilization was increased by 15%. This suggests that the heat increment of feeding may vary with the level of activity of the animal, and since this is dietary-dependent, with feeding level.

These values suggest that normal activity patterns account for some 15–20% of the maintenance energy requirements of animals (Verstegen *et al.*, 1987). However, this is likely to be higher if animals are housed and fed in conditions which induce abnormal behavioural patterns. Cronin *et al.* (1986) measured the energy expenditure of tethered and group-housed sows engaged in varying degrees of stereotypic behaviour. Stereotypes and excessive drinking accounted for between 24% and 86% of the total activity of the animals, and resulted in an additional energy expenditure of between 4% and 23% of total ME intake. Unnecessary or abnormal behaviour patterns can therefore significantly influence the use that an animal makes of its feed and these results highlight the errors in nutrient evaluation that may arise in animals that are stressed and not properly adapted to the conditions and protocol of the experiment.

Nutrient provision and animal requirements

A major objective in the feeding of farm animals is to devise systems which will allow the formulation of rations to supply the nutrients needed to meet animal requirements. This requires knowledge on two separate sets of information. The first relates to the prediction of the ability of the feed or combination of ingredients to give the desired level of animal performance. The second relates to the nutritional needs of the animal and the extent to which the requirements change at the different levels of performance during both growth and reproduction. Both of these are of fundamental importance to the scientific feeding of farm animals and it is obviously outside the scope of this review to consider all those factors that may influence their application in practice. However a brief attempt will be made to consider aspects of feed evaluation and how the prediction of the nutritive value of feeds from simple and reliable chemical measurements may be applied in practice to devise feeding regimes to meet the animal's nutrient requirements. Consideration will only be given to pigs since both poultry and ruminants have been considered elsewhere in this publication (Chapters 1 and 3).

Energy evaluation systems for pigs are expressed on a DE, ME or NE basis and direct *in vivo* determination of various ingredients and compound feeds have been made by digestibility, nutrient balance and calorimetry studies. These procedures are lengthy, laborious and expensive and require specialized animal facilities. As a result there has been considerable interest in the prediction of the energy values of feeds according to their chemical composition (Farrell, 1979; Morgan and Whittemore, 1982; Morgan *et al.*, 1987; Wiseman and Cole, 1983). There is nothing new in this concept and the basic approach is to determine the energy value of the feed and relate this to variations in its chemically-defined constituents, mostly by regression equations. Although simple, this procedure is not foolproof and there is concern whether the samples of feeds studied adequately reflect the spectrum of feeds used in practice and whether observations on one class of animal may be unreservedly applied to others.

The evaluation of feeds on the basis of their DE content is convenient, but it does not provide an indication of the real energy value of the absorbed nutrients. It assumes that the energy value of the individual components of the diet are additive and that the energy yield from different types of ingredients can be measured as a single value. No distinction is made between digestion and absorption of nutrients pre- and post-ileum, that is whether enzymatically digested in the small intestine or fermented in the large intestine (Müller and Kirchgessner, 1986), or between possible variations in the energy value associated with different energy substrates (Just, Fernandez and Jørgensen, 1983). As a result the prediction of DE overestimates protein-rich and fibrous feeds but under-predicts the value for fat. Similar deficiencies also apply to the evaluation of feeds on the basis of ME, but consideration of energy losses in both urine and methane allow a more appropriate prediction of dietary energy yield, especially of protein-rich feeds. In both the prediction of DE and ME the largest source of variation is that associated with the fibrous or non-starch polysaccharide component of the diet and the inclusion of values for crude fibre (CF) and neutral-detergent fibre (NDF) in the regression

Table 2.2 EXAMPLES OF PREDICTION EQUATIONS FOR DETERMINING THE ENERGY VALUE OF PIG FEEDS (kJ/kg) FROM DIGESTIBLE NUTRIENT CONTENTS (g/kg)

System	Regression coefficients				Source
	dCP	dEE	dCF	dNFE	
DE	24.2	39.4	18.4	17.0	Schiemann *et al*. 1971
ME	21.0	37.4	14.4	17.1	Scheimann *et al*. 1971
NE	10.7	35.8	12.4	12.4	Schiemann *et al*. 1971
DE	23.9	36.3	21.1	16.7	Just, 1982
ME	21.5	37.7	19.7	17.3	Just, 1982
NE	NE = 0.75 ME − 1.88(kJ)				

Table 2.3 MATCHING NUTRIENT SUPPLY TO ANIMAL REQUIREMENTS. CALCULATION OF THE ENERGY CONTENT OF THE FEED BASED ON THE PREDICTIVE EQUATION PRESENTED IN TABLE 2.2 AND RESPECTIVE FEEDING LEVELS FOR PIGS AT KNOWN RATES OF PERFORMANCE

(1) Dietary energy content (MJ/kg DM)			(2) Feeding level (kg/day) based on		Source
DE	ME	NE	ME	NE	
15.4	14.8	10.3	2.03	2.07	Schiemann *et al*. 1971
15.1	15.0	9.5	2.00	2.24	Just, 1982

1. DE, ME and NE are calculated from the equations presented in Table 2.2 for a conventional cereal-based diet containing 200 g CP, 40 g EE, 40 g CF and 690 g NFE/kg D.M., the respective digestibility coefficients are 0.78, 0.60, 0.10 and 0.90.
2. The NE and ME requirements have been calculated for a 60 kg pig gaining 1 kg/day, containing 160 g protein and 250 g fat/day.

ME requirement based on a maintenance energy requirement of $0.44 \, MJ/kg^{0.75}$ per day, and an energy content of protein and fat of 23.8 and 39.7 MJ/kg, respectively. The values for k_p and k_f were 0.54 and 0.74, respectively (ARC, 1981).

The ME requirement is therefore:

$(0.44 \times 21.6) + (3.81/0.54) + (9.93/0.74) = 30.0 \, MJ/day$

NE requirement based on a maintenance energy requirement of $0.35 \, MJ/kg^{0.75}$ per day and an energy content of protein and fat of 23.8 and 39.7 MJ/kg, respectively.

The NE requirement is therefore:

$(0.35 \times 21.6) + (3.81) + (9.93) = 21.3 \, MJ/day$

equations gave consistently better results. Kirchgessner and Roth (1983) improved the prediction of ME even further by including a correction for fermentable substances and sugars and this has been officially adopted in some countries (DLG, 1984). The evaluation of feeds and the application of feeding standards for pigs on both the DE and ME systems are practised in many countries, and the relative merits of the different systems, have been discussed by Henry, Vogt and Zoiopoulos (1988).

The prediction or determination of DE and ME values makes no allowance for differences in the use of energy depending upon the type of substrate supplied and in the efficiency with which these are utilized by different classes of animals. These limitations are overcome in the evaluation of feeds based on NE since both the heat increment of feeding and the change in maintenance energy requirements are calculated. Thus all factors that influence energy metabolism as an animal grows, reproduces and lactates are considered. The two most widely applied NE systems for pigs are those of Schiemann *et al.* (1971) and Just (1982). The former relates the net energy value of the feed to the contents of digestible dietary nutrients whereas the latter calculates it from the ME value of the feed (Table 2.2). A comparison of the application of the two systems in practice is presented in Table 2.3 for a 60 kg pig gaining 1 kg body weight per day. It demonstrates that the feeding requirements based on ME are similar but those based on NE are different and indicative of the varying experimental conditions under which the measurements were made (Just, 1982). However, they show how studies on feed evaluation and animal requirements may be integrated to devise feeding strategies to ensure optimal animal performance.

Conclusions

During the past 50 years the measurement of heat production by calorimetry, and other means, has facilitated our understanding of animal metabolism so that the determination of requirements and the evaluation of feeds have been based on sound scientific principles. Despite these advances there are still many problems to be solved about the energy metabolism of animals and the efficiency with which nutrients are utilized for the many metabolic processes within the body. Thus classical calorimetry will continue to play a major role in describing nutrient responses and in evaluating feeds but in the future a more important role will be in integrating whole-body metabolism to events at the organ and cellular level, including physiological and neuro-endocrine regulation. This will help to explain the variation in metabolism associated with changes in physiological state, plane of nutrition and body composition and predict more accurately the requirements of new generations of animals produced as a result of genetic manipulation and biotechnological practices.

Methods for the measurement of energy expenditure have in the past relied predominantly on calorimetry to give accurate and consistent results. However these procedures require the animals to be confined in controlled conditions within chambers which precludes the measurements of effects associated with animals freely-living within their natural environment. For this reason new techniques for the measurement of energy expenditure by isotope dilution techniques and nuclear magnetic resonance spectroscopy are being developed (James, Haggarty and McGaw, 1988). Provided they are accurate with a precision of 2–3%, they will find application in practice. At present however such consistency cannot be attained so that the classical estimation of energy expenditure by whole calorimetry is still the preferred method.

References

Agricultural Research Council (1980) *The Nutrient Requirements of Ruminant Livestock.* Commonwealth Agricultural Bureaux, Slough.

Agricultural Research Council (1981) *The Nutrient Requirements of Pigs.* Commonwealth Bureaux, Slough.

Alexander, G. (1974) Heat loss from sheep. In *Heat Loss from Animals and Man* (eds J.L. Monteith and L.E. Mount), Butterworths, London, pp. 173–203

Armstrong, D.G. (1969) Cell bioenergetics and energy metabolism. In *Handbuch der Tierernährung – I* (eds W. Lenkeit, K. Breirem and E. Casemann), Paul Parey, Berlin and Hamburg, pp. 385–414

Armstrong, D.G. and Blaxter, K.L. (1984) Maintenance requirements: implications for its use in feed evaluation systems. In *Herbivore Nutrition in the Subtropics and Tropics* (eds F.M.C. Gilchrist and R.I. Mackie), Science Press, Craighall, S.A, pp. 631–647

Baldwin, R.L. and Bywater, A.C. (1984) Nutritional energetics of animals. *Annual Review of Nutrition*, **4**, 101–114

Baldwin, R.L., Smith, N.E., Taylor, J. and Sharp, M. (1980) Manipulating metabolic parameters to improve growth rate and milk secretion. *Journal of Animal Science*, **51**, 1416–1428

Benedict, F.G. and Ritzman, E.G. (1927) The metabolic stimulus of feed in the case of steers. *Proceedings of the National Academy of Science, USA*, **13**, 125–140

Blaxter, K.L. (1962) *The Energy Metabolism of Ruminants.* Hutchinson, London

Blaxter, K.L. (1967) Techniques in energy metabolism studies and their limitations. *Proceedings of the Nutrition Society*, **26**, 86–96

Blaxter, K.L. (1969) The efficiency of energy transactions in ruminants. In *Energy Metabolism of Farm Animals* (eds K.L. Blaxter, G. Thorbek and J. Kielanowski), EAAP Publications No. 12, Oriel Press, Newcastle-upon-Tyne, pp. 21–28

Blaxter, K.L. (1971) Methods of measuring energy expenditure of animals and interpretations of results obtained. *Federation Proceedings*, **30**, 1436–1443

Blaxter, K.L. (1974) Adjustments of the metabolism of sheep to confinement. In *Energy Metabolism of Farm Animals* (eds K.H. Menke, H.J. Lantzch and J.R. Reichl), EAAP Publication No. 14. Dokumentationstelle, Universität Hohenheim, pp. 115–118

Brouwer, E. (1965) Report of sub-committee on constants and factors. In *Energy Metabolism* (ed. K.L. Blaxter), EAAP Publication No. 11. Academic Press, London and New York, pp. 441–443

Burlacu, Gh., Grossu, D., Marinescu, G., Baltac, M. and Grunca, D. (1969) Efficiency of utilization of the energy of starch in birds. In *Energy Metabolism of Farm Animals* (eds K.L. Blaxter, G. Thorbek and J. Kielanowski), EAAP Publication No. 12, Oriel Press, Newcastle-upon-Tyne, pp. 369–375

Close, W.H. (1978) The effects of plane of nutrition and environmental temperature on the energy metabolism of the growing pig. 3. The efficiency of energy utilization. *British Journal of Nutrition*, **40**, 433–438

Close, W.H. (1981) The climatic requirements of the pig. In *Environmental Aspects of Housing for Farm Animals* (ed. J.A. Clark), Butterworths, London, pp. 149–166

Close, W.H. and Mount, L.E. (1978) The effects of plane of nutrition and environmental temperature on the energy metabolism of the growing pig. I. Heat loss and critical temperature. *British Journal of Nutrition*, **40**, 413–421

Close, W.H., Mount, L.E. and Brown, D. (1978) The effects of plane of nutrition and environmental temperature on the energy metabolism of the growing pig. 2. Growth rate, including protein and fat deposition. *British Journal of Nutrition*, **40**, 423–431

Close, W.H. and Verstegen, M.W.A. (1981) Factors influencing thermal losses in non-ruminants: a review. *Livestock Production Science*, **8**, 449–463

Cronin, G.M., Van Tartwijk, J.M.F.M., Van der Hel, W. and Verstegen, M.W.A. (1986) The influence of degree of adaptation to tether-housing by sows in relation to behaviour and energy metabolism. *Animal Production*, **42**, 257–268

Curtis, S.E. (1983) *Environmental Management in Animal Agriculture*. Iowa University Press, Iowa

Deutsche Landwirtschafts-Gessellschaft (1984) *DLG-Futterwerttabellen für Schweine*. DLG-Verlag, Frankfurt-am-Main

De Boer, F. and Bickel, H. (1988) Livestock feed resources and feed evaluation in Europe. *Livestock Production Science*, **19**, 1–408

Farrell, D.J. (1979) Energy systems for pigs and poultry: a review. *Journal of the Australian Institute of Agricultural Science*, **45**, 21–34

Gädeken, D., Breves, G. and Oslage, H.J. (1989) Efficiency of energy utilisation of intracaecally infused volatile fatty acids. In *Energy Metabolism of Farm Animals* (eds Y. Van der Honing and W.H. Close), EAAP Publication No. 43. Pudoc, Wageningen, pp. 115–118

Graham, N.McC., Wainman, F.W., Blaxter, K.L. and Armstrong, D.G. (1959) Environmental temperature, energy metabolism and heat regulation in sheep. 1. Energy metabolism in closely clipped sheep. *Journal of Agricultural Science, Cambridge*, **52**, 13–24

Gray, R. and McCracken, K.J. (1980) Effect of confinement in a respiration chamber and changes in temperature and plane of nutrition on heat production of 25 kg pigs. *Journal of Agricultural Science, Cambridge*, **95**, 123–133

Gregg, V.A. and Milligan, L.P. (1987) Thyroid induction of thermogenesis in cultured rat hepatocytes and sheep liver. In *Energy Metabolism of Farm Animals* (eds P.W. Moe, H.F. Tyrrell and P.J. Reynolds), EAAP Publication No. 32. Rowman and Littlefield, USA, pp. 10–13

Henry, Y., Vogt, H. and Zoiopoulos, P.E. (1988) Pigs and poultry. *Livestock Production Science*, **19**, 299–354

Hungate, R.E. (1966) *The Rumen and its Microbes*. Academic Press, New York.

James, W.P.T., Haggarty, P. and McGaw, B.A. (1988) Recent progress in studies on energy expenditure: are the new methods providing answers to the old questions. *Proceedings of the Nutrition Society*, **47**, 195–208

Just, A. (1982) The net energy value of balanced diets for growing pigs. *Livestock Production Science*, **8**, 541–555

Just, A., Fernandez, J.A. and Jørgensen, H. (1983) The net energy value of diets for growth in pigs in relation to the fermentative processes in the digestive tract and the site of absorption of the nutrients. *Livestock Production Science*, **10**, 171–186

Kirchgessner, M. and Roth, F.X. (1983) Schätzgleichungen zur Ermittlung der Energetischen Futterwertes von Mischfuttermitteln für Schweine. *Zeitschrift für Tierphysiologie, Tierernährung und Futtermittelkunde*, **50**, 270–275

Kleiber, M. (1961) *The Fire of Life*, Wiley, New York

Koong, L.J., Nienaber, J.A., Pekas, J.C. and Yen, J.T. (1982) Effects of plane of nutrition on organ size and fasting heat production in pigs. *Journal of Nutrition*, **112**, 1638–1642

Longland, A.C., Low, A.G. and Close, W.H. (1989) Contribution of carbohydrate fermentation to energy balance in pigs. In *Digestive Physiology in the*

Pig. Proceedings of the 4th International Symposium (eds L. Buraczewska, S. Buraczewski, B. Pastuszewska and T. Zebrowska), Institute of Animal Physiology and Nutrition, Jabbonna, Poland, pp. 108–119

McCracken, K.J. and Rao, S.D. (1989) Protein: energy interaction in boars of high lean deposition potential. In *Energy Metabolism of Farm Animals* (eds Y. Van der Honing and W.H. Close), EAAP Publication No. 43. Pudoc, Wageningen, pp. 13–16

McCrea, J.C. and Lobley, G.E. (1982) Some factors which influence thermal energy losses during the metabolism of ruminants. *Livestock Production Science*, **9**, 447–456

McGlone, J.J., Stansbury, W.F. and Tribble, L.F. (1988) Management of lactating sows during heat stress: effects of water drip, snout coolers, floor type and a high energy-density diet. *Journal of Animal Science*, **66**, 883–891

McLean, J.A. and Tobin, G. (1987) *Animal and Human Calorimetry*. Cambridge University Press, Cambridge.

MacLeod, M.G., Jewitt, T.R. and Anderson, Julie E.M. (1987) Energy utilization and physical activity in intubated and self-fed male domestic fowl. *Proceedings of the Nutrition Society*, **46**, 148A

Millward, D.J., Garlick, P.J. and Reeds, P.J. (1976) The energy cost of growth. *Proceedings of the Nutrition Society*, **35**, 339–350

Morgan, C.A. and Whittemore, C.T. (1982) Energy evaluation of feeds and compounded diets for pigs. A review. *Animal Feed Science and Technology*, **7**, 387–410

Morgan, C.A., Whittemore, C.T., Phillips, P. and Crooks, P. (1987) The prediction of energy value of compounded pig foods from chemical analysis. *Animal Feed Science and Technology*, **17**, 81–107

Mount, L.E. (1968) *The Climatic Physiology of the Pig*. Edward Arnold, London

Mount, L.E. (1975) The assessment of the thermal environment in relation to pig production. *Livestock Production Science*, **2**, 381–392

Müller, H.L. and Kirchgessner, M. (1982) Effect of straw and cellulose on heat production and energy utilization in pigs. In *Energy Metabolism of Farm Animals* (eds A. Ekern and F. Sundstøl), EAAP Publication No. 29. The Agricultural University of Norway, Aas-NLH, pp. 229–232

Müller, H.L and Kirchgessner, M. (1986) Some aspects of energy utilization in pigs. *Pig News and Information*, **7**, 419–424

Müller, H.L and Kirchgessner, M. (1987) Energy utilization of pectin and alfalfa meal in pigs. In *Energy Metabolism of Farm Animals* (eds P.W. Moe, H.F. Tyrrell and P.J. Reynolds), EAAP Publication No. 32, Rowman and Littlefield, USA, pp. 268–271

Nehring, K., Schiemann, R., Hoffmann, L., Klippel, W. and Jentsch, W. (1965) Utilization of the energy of cellulose and sucrose by cattle, sheep and pigs. In *Energy Metabolism* (ed K.L Blaxter), EAAP Publication No. 11, Academic Press, London, pp. 249–268

Noblet, J., Le Dividich, J. and Bikawa, T. (1985) Interaction between energy level in the diet and environmental temperature on the utilization of energy in growing pigs. *Journal of Animal Science*, **61**, 452–459

Ørskov, E.R. and Allen, D.M. (1966) Utilization of salts of volatile fatty acids by grazing sheep. 1. Acetate, propionate and butyrate as sources of energy for young grazing lambs. *British Journal of Nutrition*, **20**, 295–305

Osuji, P.O., Gordon, J.G. and Webster, A.J.F. (1975) Energy exchanges associated with eating and rumination in sheep given grazing diets of different physical form. *British Journal of Nutrition*, **34**, 59–71

Pullar, J.D., Brockway, J.M. and McDonald, J.D. (1967) A comparison of direct and indirect calorimetry. In *Energy Metabolism of Farm Animals* (eds K.L. Blaxter, G. Thorbek and J. Kielanowski), EAAP Publication No. 12, Oriel Press, Newcastle-upon-Tyne, pp. 415–421

Reeds, P.J., Wahle, K.W.J. and Haggarty, P. (1982) Energy costs of protein and fatty acid synthesis. *Proceedings of the Nutrition Society*, **41**, 155–159

Schiemann, R., Chudy, A. and Herceg, O. (1969) Der Energieaufwand für die Bildung von Körperprotein beim Wachstum nach Modellversuchen an Ratten. 1. Mitterlung. Versuche mit Voller als Nahrüngsprotein. *Archiv für Tierernahrung*, **19**, 395–407

Schiemann, R., Nehrung, K., Hoffman, L., Jentsch, W. and Chudy, A. (1971) *Energetische Futterbewertung und Energienormen*, VEB. Deutscher Landwirtschaftsverlag, Berlin.

Schneider, B.H. and Flatt, W.P. (1975) *The Evaluation of Feeds Through Digestibility Experiments*. University of Georgia Press, Athens, Georgia

Stahly, T.S. and Cromwell, G.L. (1979) Effect of environmental temperature and dietary fat supplementation on the performance and carcass characteristics of growing swine. *Journal of Animal Science*, **49**, 1478–1488

Van Es, A.J.H. (1972) Maintenance. In *Handbuch der Tierernährung–II* (eds W. Lenkeit and K. Breirem), Paul Parey, Hamburg and Berlin, pp. 1–54

Vanschoubroek, F., Coucke, L. and van Spaendonch, R. (1967) The quantitive effects of pelleting feed on the peformance of piglets and fattening pigs. *Nutrition Abstracts and Reviews*, **41**, 1–9

Verstegen, M.W.A., Mount, L.E. and Close W.H. (1973) The optimum duration of metabolic balance experiments with groups of pigs. *Proceedings of the Nutrition Society*, **32**, 72A.

Verstegen, M.W.A., van der Hel, W., Brandsma, H.A. and Kanis, E. (1982) Heat production of groups of growing pigs as affected by weight and feeding level. In *Energy Metabolism of Farm Animals* (eds A. Ekern and F. Sunstøl), EAAP Publication No. 29, The Agricultural University of Norway, Aas-NLH, pp. 218–221

Verstegen, M.W.A., Henken, A.M. and Van der Hel, W. (1987) Influence of some environmental, animal and feeding factors on energy metabolism in growing pigs. In *Energy Metabolism of Farm Animals* (eds M.W.A. Verstegen and A.M. Henken), Martinus Nijhoff Publishers, Dordrecht, pp. 70–86

Webster, A.J.F. (1974) Heat loss from cattle with particular emphasis on the effects of cold. In *Heat Loss from Animals and Man* (eds J.L. Monteith and L.E. Mount), Butterworths, London, pp. 205–231

Webster, A.J.F. (1978) Prediction of the energy requirements for growth in cattle. *World Review of Nutrition and Dietetics*, **30**, 189–226

Webster, A.J.F. (1981) The energetic efficiency of metabolism. *Proceedings of the Nutrition Society*, **40**, 121–128

Webster, A.J.F. (1983) Nutrition and the thermal environment. In *Nutritional Physiology of Farm Animals* (eds J.A.F. Rook and P.C. Thomas), Longman, London and New York, pp. 639–669

Webster, A.J.F. (1989) Energy utilization during growth and reproduction. In *Energy Metabolism of Farm Animals* (eds Y. Van der Honing and W.H. CLose), EAAP Publication No. 43, Pudoc, Wageningen, pp. 85–88

Webster, A.J.F., Smith, J.S. and Brockway, J.M. (1972) Effects of isolation, confinement and competition for feed on the energy exchanges of growing lambs. *Animal Production*, **15**, 189–201

Wenk, C. and Van Es, A.J.H. (1976) Energy metabolism of growing chickens as related to their physical activity. In *Energy Metabolism of Farm Animals* (ed M. Vermorel), EAAP Publication No. 19, G. de Bussac, Clermont-Ferrand, pp. 189–192

Wiseman, J. and Cole, D.J.A. (1983) Predicting the energy value of pig feeds. In *Recent Advances in Animal Nutrition–1983*, (ed William Haresign), Butterworths, London, pp. 59–70

3

APPARENT AND TRUE METABOLIZABLE ENERGY OF POULTRY DIETS

J.M. MCNAB
Institute of Grassland and Animal Production, Poultry Department, Roslin, Midlothian EH25 9PS

Introduction

Energy, representing the link between the biological and physical, has fascinated scientists for many years. Historically, the elucidation of the role of energy in the physiology of respiration provided the basis for the science of nutrition. In addition, energy intake is implicated in the physiology of appetite and satiety and in the control of food consumption. For economic reasons it is important to be able to describe the energy content of feedingstuffs, and over the past 25 years metabolizable energy (ME) has become accepted almost universally as the preferred measure of the energy content of poultry diets. The establishment of the relationship between the ME content of a diet and its intake and relating the concentrations of other nutrients to the dietary ME value has improved substantially the precision with which poultry are fed.

Although ME is generally considered to be a property of a diet, it is really a characteristic of an animal to which the diet is given. The measurement of ME relates to the complete diet and values for dietary components must, in most cases, be obtained from a comparison of data derived from feeding two or more appropriate diets. In the substitution method proposed by Hill and Anderson (1958) the dietary ingredient to be assayed is substituted for an ingredient of known ME value (usually glucose) in a basal diet to yield a test diet. By comparing the energy balances (gross energy eaten − gross energy excreted) of both basal and test diets and the known ME value of glucose, the ME value of the ingredient can easily be calculated. In the replacement method (Sibbald and Slinger, 1963a) the ingredient for which the ME is required is exchanged for a known amount of basal diet. By determining the ME values of both basal and test diets it is possible to calculate the ME of the ingredient. Both these assays were expected to result in identical values for the test ingredient, the assumption being that ME values of ingredients are additive. In truth little progress can be made if this assumption is not upheld, but the acceptance that both approaches should generate indentical values may not be correct.

Although it has invariably been referred to as metabolizable energy these two bioassays yield values which perhaps more correctly should be referred to as *apparent* metabolizable energy (AME). AME is the gross energy of the food minus the energy lost as faeces, urine and combustible gases when that food is eaten

41

(Harris, 1966); because gaseous losses from poultry are very small they are almost always ignored and this convention is followed here. The energy metabolized is only apparent because, of the energy excreted when the food is consumed, only part has been derived directly from the food; thus only part consists of undigested and unmetabolized dietary residues. Part of the excreta is of endogenous origin, having been derived from the bird, and has come to be known as the endogenous energy loss (EEL). Part of the EEL is faecal in origin and is generally reported as consisting of sloughed-off gut lining, bile excretions and unabsorbed enzymes: part is urinary and consists primarily of the excretory products of nitrogen metabolism.

Values based on this AME system have been widely used by practising nutritionists and form the basis for many data banks. It was only occasionally that concern was expressed on the reliability of the assay. For example, the ME (or AME) values derived for rapeseed meal (Rao and Clandinin, 1970), dehydrated alfalfa meal (Vohra and Kratzer, 1970), rye (Macaulife and McGinnis, 1971) and guar meal (Verma, 1977) all tended to decline as their dietary inclusion concentration increased. Tenable hypotheses to account for these observations were seldom put forward. Other unresolved problems associated with ME (AME) determinations were differences sometimes detected between species (Slinger *et al.*, 1964; Hill *et al.*, 1968; Fisher and Shannon, 1973; Leeson *et al.*, 1974; Sugden, 1974), strains (Sibbald and Slinger, 1963b; Slinger *et al.*, 1964; Bayley *et al.*, 1968; Foster, 1968; Proudman *et al.*, 1970; Marsh and Biely, 1971) and experimental animals of different ages (Renner and Hill, 1960; Lockhart *et al.*, 1963; Zelenka, 1968; Lodhi *et al.*, 1969, 1970; Rao and Clandinin, 1970). Evaluation of rapeseed meal has appeared to present unique problems in as much as the composition of the basal diet (Rao and Clandinin, 1970) and the acclimatization of the birds to the diet (Lodhi *et al.*, 1969) both affected the value ascribed to its (A)ME.

Therefore, although it is generally undisputed that the (A)ME system has played an invaluable role in the development of poultry nutrition, there remained some variation in the values derived, used and recommended in different laboratories. A possible explanation for at least part of this uncertainty lay in an experiment carried out by Guillaume and Summers (1970). They showed that (A)ME values derived for diets fed to adult cockerels were profoundly affected by the amount of food eaten during the assay. The lower the food consumption, the lower the (A)ME value of the diet. This effect was attributed to the contribution made to the excreted energy by the EEL. It was not until 5 years later that Sibbald (1975) drew attention to the significance of this observation and argued that most of the anomalies which had been associated with assays designed to derive (A)ME values were a direct consequence of the effects of EEL. He devised a novel bioassay in which the EEL was determined directly, and argued that, in future, the energy status of both diets and feedingstuffs should be expressed in terms of their true metabolizable energy (TME) contents (Sibbald, 1976).

Since then a rather ineffectual debate has been conducted on the advantages of one system of evaluation over the other (see for example, Sibbald and Wolynetz, 1987; Hartel, 1987; Farrell *et al.*, 1988). It might have been hoped that the introduction of an effective rapid assay based on sound scientific principles would have resulted in the fulfilment of the laudable object of technique standardization. There appears to have been two main reasons why this has not occurred. Firstly, rapid assays almost invariably require starved birds to be used and this has proved controversial; secondly, it is clear that when published methods were adopted some problems were experienced in some laboratories and this has resulted in the introduction of a number of major and minor variations. Consequently there are now probably more methods being applied to the derivation of ME values than at any time in the past, and the prospects for establishing a single standard procedure is probably as remote a goal as it has ever been.

It would be wrong to present the debate as simply one between AME and TME. For some years it has been common practice to 'correct' AME values determined in balance experiments for changes in the nitrogen status of the bird during the period of the trial. The rationale for this adjustment is that the complete catabolism of protein stored in the body results in the need to dispose of the nitrogen it contained. In poultry, nitrogen is excreted mainly as uric acid which contains energy and, for each gram of nitrogen excreted as uric acid, 34.4 kJ of energy are lost from the body and appear in the urine (excreta). However, a bird which is storing protein (e.g. broiler or layer) is spared the energy cost of excreting nitrogen and less uric acid–and, hence, energy–appears in the droppings. Thus the same diet (or feedingstuff), when given to different birds, may have a different ME value because of differences in the amount of ingested protein the bird has retained. To make ME values independent of the conditions under which they were derived, it has become very common to correct them to what they might have been under standard conditions. The most frequently, although not exclusively, used standard is one where the birds are in nitrogen equilibrium (where nitrogen retention has been zero). The principle of nitrogen correction has often been criticized because a diet or feedingstuff is penalised (as an energy source) when it is promoting the retention of protein, often the objective of animal production. However, because the function of an ME system is to evaluate the energy status of feedingstuffs rather than their ability to promote protein synthesis, correction to nitrogen equilibrium can be justified. In any case, failure to make a correction is tantamount to correcting to the nitrogen retention prevailing in the assay. In the same way as AME values are corrected to nitrogen equilibrium (AME_N) so too should TME values (TME_N).

Theoretical Aspects

The central assumption made in all assays for ME is that the energy voided as excreta is linearly related to the energy input. In the TME system the intercept value of this line is positive and corresponds to the EEL. Statistically it is the energy excreted at zero energy input. The complement of the slope of this line $(1-b)$, yields the TME value of the feedingstuff as TME = $(1-b)$ × gross energy.

AME values can be described in similar ways from this model by joining a given energy balance with the origin. It can be readily seen that the slopes of such lines yield the AME values and that these will vary at different levels of input, a constant maximum value being approached asymptotically. For almost all feedingstuffs the linearity of this basic relationship seems intuitively reasonable and it has been demonstrated experimentally for a wide range of foods. Non-linearity has never been demonstrated directly but it is implied in a report that claimed TME values depend on input (Kussaibati *et al.*, 1982). Non-linear relationships may also be found with fats and materials which result in digestive disorders.

From this model it is possible to derive a relationship between AME and TME algebraically (Jonsson and McNab, 1983), although it is evident from their definitions. Thus:

$$AME = TME - \frac{EEL}{Food\ Intake}$$

This equation shows that for a given TME value, AME depends on EEL per unit of food intake. Variations in this ratio clearly explain the effects of food intake on AME values and may also offer an explanation for the effects of strain, species and bird age on AME values, although direct and unequivocal proof of this has still to be demonstrated.

The most important conclusion to be drawn from this relationship, however, is that provided information is available on energy balance, food intake and EEL then results can be expressed either in terms of AME or TME. Provided reasonable assumptions can be made on EEL it does offer a means whereby results from determinations of AME can be combined with those on TME.

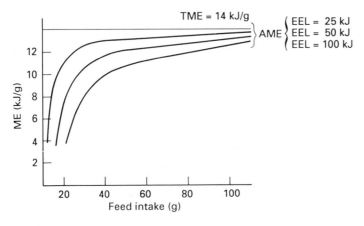

Figure 3.1   Relationship between apparent and true metabolizable energy values

This theoretical model can also be used to demonstrate the effect on determined ME values of variations or uncertainty in the different components of the relationship. For example, Figure 3.1 shows AME values for different food intakes when the TME = 14 kJ/g and EEL ranges from 25 to 100 kJ/day. It can be seen that, between 80 and 120 g/day, a range of food intakes likely to be found with adult male birds, AME will vary between 13.69 and 13.79 kJ/day at the lower end of the EEL scale but between 13.38 and 13.58 kJ/g if EEL is 50 kJ/day. This underlines that variations in EEL are important in the derivation of AME and not only TME. This has not always been clearly stated when the two systems are being compared.

Methods for Determining ME

By separating the question of which ME system to use from that of experimental technique, different assays should be judged on how well they provide the three essential pieces of information: energy balance, food intake and EEL. Other factors which may influence the choice will be speed, cost and, perhaps, convenience. Three general types of experiment have been identified as follows (Fisher and McNab, 1987):

1. Traditional assays which involve preliminary feeding periods to establish a 'state of equilibrium'. Differences in the contents of the digestive tract between the beginning and end of the assay period ('end-effects') are controlled by trying to ensure that they are the same. In most cases complete diets must be fed and substitution methods (described earlier) must be used for ingredients.
2. Rapid assays, using starvation before and after allowing the birds free access to the diet to control the end-effects. Again complete diets and substitution methods for ingredients must be used in most cases.
3. Rapid assays, as above, but using tube-feeding to place the test material directly into the crop of birds. These methods almost invariably avoid the need for substitution, most ingredients being fed as received.

Whilst many variations are found within these three general groups the classification provides a convenient framework within which the many procedural details can be discussed.

Energy Balance and Food Intake

Food presentation and the accurate measurement of energy intake are arguably the most challenging aspects of ME determinations. When birds are given access to food *ad libitum*, a procedure which still seems to be most widely accepted, great care is required to avoid food loss, to prevent separation of the dietary components, to take changes in moisture content into account and to take representative samples. These are difficult to control in a consistent way but specially designed systems have been described and used with apparent success (Terpstra and Janssen, 1975).

Such free-feeding methods are used in type 1 assays which form the greater part of the literature on ME determination in poultry. Farrell (1978) proposed that the advantages of a rapid assay, type 2, could be obtained by training birds to consume sufficiently large intakes in 1 h after a 23 h starvation period. In this assay equal quantities of basal diet and test ingredient were combined and pelleting was recommended to maintain intakes across a range of ingredients. Several labortories (Muztar and Slinger, 1980b; Jonssen and McNab, 1983; Parsons *et al.*, 1984; Kussaibati and Leclercq, 1985) have reported difficulties in maintaining satisfactory intakes but notwithstanding this, the assay has had its adherents. Apart from speed (cost) and amount of material required the technique offers few if any adavantages, over type 1 assays in terms of the accurate measurement of food intake or energy balance (McNab and Fisher, 1981).

It is beyond debate that the presentation of the food by tube in type 3 assays permits the most accurate means of measuring energy intake, food spillage and changes in dry matter content both being avoided. However, because the dose size is reduced, problems may occur in terms of achieving representative samples. The only real disadvantages of the technique are the obvious limits on dose size and, perhaps, attitudes to a procedure frequently referred to as 'force-feeding'. Experience in this laboratory is that, with practice, the procedure is very rapid (15–30 s/bird to feed 50 g of most feedingstuffs) and that there is little evidence of more stress beyond that involved in handling. Skill is required, however, and experience must be attained, although this is easily achieved by most operators. The use of slurry feeding as a way of reducing stress has been suggested (Wehner and Harrold, 1982) but it is the experience of the author that slurry feeding invariably takes considerably longer. Finely divided, hygroscopic or very bulky ingredients may present problems but, with experience, these can all generally be overcome. In this laboratory glucose monohydrate is fed routinely; this can present problems and granulation is carried out to reduce difficulties.

Excreta collection is another simple task which can be difficult to do well in routine experiments. When trays placed under the cages are used to collect the droppings, by far the most common procedure, the problems include adherence of the excreta to feathers, contamination with scurf, fermentation losses and perhaps rare but can occur and contamination with regurgitated material can also take place, can be surprisingly difficult to detect and almost impossible to take into account. Sibbald (1986) lists sensible precautions to be taken; frequent collection (12 hourly) as in Dale *et al.*, (1985) are the sorts of devices which might be judged beneficial but are labour-intensive and reduce the benefits of a low cost assay. Alternatives to collection trays have been discussed earlier (Fisher and McNab, 1987) and it has been concluded that trays cannot be avoided in routine experiments.

Minimizing end effects

In assessing the reliability of data from type 3 assays it is important to remember that, because inputs are small, any imprecision or uncertainty will have a potentially greater effect on the value derived. In the ideal ME system only excreta derived from the intake recorded should be debited against that energy intake. In type 1 assays where it is customary to carry out the balance over several days and where food intakes are often several hundred grams, discrepancies or difference in gut-fill at the beginning and end of the experiment were considered to cancel each other out. Although this was accepted it was not entirely satisfactory because changes in intake of rapidly growing birds or in response to unpalatable ingredients may be more likely to cause systematic bias rather than random error. However, with the much smaller intakes used in assays of types 2 and 3 great care must be taken to ensure that the digestive tract is empty of residues at both the beginning and end of the assay. Factors which are likely to influence the amount of material remaining in the gastro-intestinal tract are the nature of previous diet, the period for which it is removed, the nature of the test feedingstuff and the amount given, the length of the collection period, water intake and random variation in caecal evacuation. Sibbald (1976) originally proposed 24 h starvation and 24 h collection periods (24 h + 24 h assay) but now (Sibbald, 1986) proposes 24 h + 48 h for routine use. In this laboratory 48 h + 48 h is routinely used (McNab and Fisher, 1984; McNab and Blair, 1988). The longer period is clearly more stressful and factors such as bird size and glucose feeding come into consideration.

Sibbald (1982) had shown that 12 h starvation before feeding was insufficient to clear the digestive tract of residues but that extension beyond 24 h had only a small effect on derived TME values. Direct investigation however, shows a measurable difference between residues remaining after 24 h and 48 h starvation (Table 3.1). These observations and the logic of equalizing the pre- and post-feeding starvation periods encourages the use of the 48 h + 48 h assay and adjustment of other factors to deal with the increased stress. In general, it has been found that clearance rates are variable between feedingstuffs and amounts fed (Sibbald, 1982) and it seems appropriate that a constant maintenance diet of well-digested components should be used, although this is not a very critical issue (Shires *et al.*, 1979). To some extent a correction is made for carry-over of energy from the previous diet in calculating TME in the assay of Sibbald because it seems reasonable to expect that a comparable error will occur in both the fed and negative control birds.

The time required to ensure complete clearance of a feedingstuff, especially when single ingredients are fed in type 3 assays, is a complex and largely unresolved issue. The original proposal of 24 h (Sibbald, 1976; Farrell, 1978) is now known to be too short and all data collected under these conditions are unreliable and should be ignored. Farrell (1981) now recommends 32 h and Sibbald (1986) 48 h for routine use. Some TME values comparing data derived from 48 h and 72 h collections when the amount fed was 50 g are shown in Table 3.2. These data suggest that for some ingredients (e.g. blood meal) 48 h is insufficient to allow all undigested residues to be voided. The results of Sibbald and Morse (1983), and the work of Sibbald elsewhere, suggest that the use of lower intakes alleviates the problem but at the cost of both reduced accuracy and increased influence of endogenous effects. The experience of this laboratory is that incomplete clearance is a problem with high protein, and especially finely divided animal products: materials of low density which result in the crop being packed very full can also result in problems, where wetting of the feedstuff in the crop may be a factor. Palatability may also be involved because, for example, when blood meal-fed birds are given water, distaste seems to be experienced and regurgitation can occur. The

Table 3.1   RESIDUES REMAINING IN THE GASTROINTESTINAL TRACT OF STARVED COCKERELS

Starvation period (h)	*Residues* (g)	*Total energy* (kJ/g)
24	1.59 ± 0.56 (0.83–2.58)	20.25
48	0.17 ± 0.08 (0.05–0.30)	2.06

Table 3.2   COMPARISON OF TME_N VALUES DERIVED AFTER 48 AND 72 h EXCRETA COLLECTION

Ingredient	*Samples*	*A* TME_N (48 h)	*B* TME_N (72 h)	*B/A* *Rel*
Full-fat soyameal	4	14.44	14.42	1.00
Wheat meal	12	12.77	12.75	1.00
Fish meal	12	13.29	13.06	0.98
Blood meal	5	13.37	12.09	0.90
Meat-and-bone-meal	7	8.59	10.46	0.97
Wheat feed	12	8.59	8.56	1.00
Carrot	1	9.97	9.80	0.98
Cabbage	1	9.81	9.44	0.96
Pea hulls	1	1.79	1.63	0.91

sudden introduction of some ingredients may induce gut stasis; attempts to evaluate coffee residues by tube-feeding had to be abandoned because the food did not pass from the crop. At the present time it is only possible to advise caution, particularly with unfamiliar ingredients, and to look out for food residues being excreted after the end of the balance period in doubtful cases. Routinely extending the collection period to 72 h would provide an empirical solution but at a cost to the stress on the birds. Longer balance periods also result in higher endogenous: exogenous energy ratios and the relative importance of the error introduced by correcting for EEL assumes greater significance.

The assessment of these factors in type 2 assays (Farrell, 1978) is difficult. With high intakes, such as 70–80 g, 32 h collection periods may be too short for certain feedingstuffs; for example, see Sibbald and Morse (1983) for alfalfa and oats. On the other hand the use of complete diets and allowing a feeding period of 1 h may reduce the problem in comparison to tube-feeding single ingredients. It seems reasonable to suppose that because the crop does not become so tightly packed with dry food under these conditions, water intake and the passage rate of the food residues might be more normal.

The importance of water intake in these assays is yet another area where firm conclusions cannot be reached and which may be a significant source of variation. In the laboratory of the author, it has consistently been observed that, despite the ready availability of water, tube-fed birds were rarely seen to be drinking (McNab and Blair, 1988). Yeomans (1987) has recently shown that 90% of water consumption by domestic fowl is associated with voluntary food intake and it may be that lack of access to food reduces the stimulus to drink. Whether low and variable water intakes explain erratic food passage rates and consequently residue clearance is still speculation but Table 3.3 summarizes the findings which led to the routine administration of water (50 ml/bird) during the balance period. This practice also provides an opportunity to palpate the crop and to mix any food residues remaining with water. It has only rarely (e.g. blood meal) led to losses of food from the crop but it does seem reasonable to argue that it will change the relationship between the amounts fed and clearance rates (Sibbald and Morse, 1983), although this has not been examined.

Table 3.3   ENERGY VOIDED COCKERELS STARVED OF FOOD OR FED 25 g SOYABEAN MEAL

Water administration		*Soyabean meal fed* (g)	*Energy excreted* (kJ)	*TME* (kJ/g)
Before feeding +	*After feeding* +	0	89.38 ± 20.38	–
–	–	25	257.30 ± 41.99	11.00
–	+	25	246.88 ± 33.49	11.42
+	+	25	236.39 ± 5.28	11.83

A direct study into the role of water: food ratios on diet digestibility was reported by Van Kampan (1983) and resulted in ambiguous results. A positive relationship was found among birds with free access to water AME (y, %G.E) and the water: food ratio (x)

$$y = 66.38 + 2.97x \quad (P < 0.01, r = 0.49)$$

This is an effect of considerable magnitude but it could not be demonstrated experimentally when water was administered by tube immediately after feeding (a mixture of free- and tube-feeding, feeding time 15 min) in a rapid assay. However, excreta were collected for only 24 h and this may have concealed any treatment effects. More work requires to be carried out on the effect of water consumption on ME values with a range of ingredients and not just with practical diets.

Endogenous Energy Loss (EEL)

Knowledge of the EEL is a prerequisite for the determination of TME and its measurement in AME assays is strongly recommended. It has to be admitted that there are both difficulties and uncertainties in the determination of this component of the excreta. Any errors which cannot be taken into account will result in errors in the values ascribed to both AME and TME. Three methods have been used to derive EEL: starving birds, giving birds a completely metabolizable energy source (e.g. glucose) or by extrapolating to zero intake a line relating energy excretion to energy intake.

Starvation has been the most widely used means for deriving EEL and is the method currently recommended by Sibbald (1986). However, in a starved state individual birds void quite variable amounts of energy. Values ranging from 33 to 82 kJ/24 h (Farrell, 1978) and from 25 to 69 kJ/24 h (Sibbald and Price, 1978) have been reported for the second 24 h period of 48 h of starvation (24 h + 24 h assay). In the laboratory of the author a somewhat wider range has been found, presumably a consequence of the greater stress associated with the 48 h + 48 h assay. Individual values ranged from 47 to 238 kJ/48 h (24 to 119 kJ/24 h) and the average coefficient of variation within an experiment (6 replicates) was 36.8%.

Neither bird weight nor body weight changes appear to explain a significant proportion of the variation (Muztar and Slinger, 1980a; Sibbald and Price, 1978), although a body weight effect was reported by Shires *et al.* (1979). Dale and Fuller (1981) have associated differences in EEL with variations in environmental temperature. In winter, when the mean temperature was 5°C, EEL was 133.9 kJ/48 h whereas in summer (30°C), EEL was 75.3 kJ/48 h. It had been shown earlier (Farrell and Swain, 1977) that temperature and the acclimation of birds to it affected the EEL from starved adult cockerels. A curious interaction between temperature, EEL and TME values has been reported by Yamazaki and Zhang (1982). Although starved adult cockerels excreted 126.5 kJ/48 h at cool temperatures (5–15°C) and 64.6 kJ/48 h at hotter temperatures (25–35°C), the environment did not affect the energy excreted by birds fed 25 g of a proprietary diet. Consequently the TME calculated for the diet at the lower temperature (15.72 kJ/g) differed markedly from that derived for the hotter environment (12.84 kJ/g). It has not proved possible to confirm any effect caused by temperature with birds fed glucose solutions during the pre-feeding period and tube-fed 50 g glucose. Data from the laboratory of the author indicate that at 5°C, EEL was 82.98 kJ/48 h whereas at 35°C it was 84.84 kJ/48 h and TME values of the feedingstuffs tested were unaffected by the temperature.

In the laboratory of the author, with a 48 h + 48 h assay and using birds every 4 weeks, a considerable reduction in EEL and conspicuously less between-bird variation when excreta are collected from birds which have been fed glucose (50 g) rather than starved has been found. The results of Sibbald (1975) with glucose are at variance with this finding and with that of Dale and Fuller (1981), who have observed that birds given 0, 12.5 and 25 g respectively of a glucose: maize (50:50) mixture voided 57.7, 52.1 and 48.9 kJ/24 h. Reasons for these differences are unclear. Age and strain of the bird but not its sex have been shown to affect EEL from starved birds (Miski and Quazi, 1981). These effects were tentatively attributed to differences in body composition and basal metabolic rate.

Recently data from a series of experiments have indicated that EEL may be an artefact of starvation (Hartel, 1986) and that when birds are fed continuously (as they are under practical conditions) TME = AME. Even when food intakes were reduced to 20 g/day, the derivation of a line relating energy excretion to energy intake and extrapolation to zero intake gave intercept values which did not differ from zero. This finding, of course, supports that of Hill and Anderson (1958) who found no effect of food intake on AME values when food intake was reduced to 0.30 of *ad libitum* and of Potter *et al.* (1960) who claimed that, under similarly severe restrictions, AME tended to increase slightly; it was speculated that this improvement could be attributed to increased diet digestibility often associated with lower intakes. These results are in direct contrast to those of Guillaume and Summers (1970) and to earlier findings from the laboratory of the author (Jonsson and McNab, 1983). Whether the apparent contradiction can be ascribed to starvation followed by tube-feeding must remain a matter for conjecture at this stage. Recent work (Farrell *et al.*, 1988) does not resolve this issue but does suggest that EEL is positive.

What is clear is that both the definition and measurement of EEL from poultry require further investigation. At the present time it can either be argued that the uncertainty is limited and, that for all practical purposes, reasonable estimates of EEL can be obtained from one of the procedures described earlier; or it can be decided that the problem is insoluble and that corrections to account for EEL should be ignored. Broadly speaking these are the respective views of those who either argue for the adoption of a type 3 assay or who maintain that classical assays of type 1 should be retained.

Recent collaborative studies with a type 1 assay in some European laboratories have shown its reliability as a means of deriving the AME_N values of diets (Bourdillon *et al.*, 1990). With 3 diets, a basal mix and this diet replaced with either 300 g/kg of wheat or soya bean meals, the 5 participating groups generated AME_N values for all diets with a high degree of precision (Table 3.4). However, when these data were used to derive AME_N values for the wheat and soya bean meals the results are conspicuously less definitive (Table 3.5). This illustrates the difficulty, if not impossibility, of deriving meaningful ME values for raw materials, as opposed to diets, using type 1 assays, even when great care is taken by experienced people.

The simplicity and speed of type 3 assays mean that many more can be carried out with an increase in the amount of information available. For the evaluation of ingredients, which can be assayed directly rather than by dietary substitution, they must be the methods of choice for the future. To fulfil this objective, no effort should be spared to resolve the uncertainty of the size of the EEL. At the moment there seems reasonable grounds for suspecting that many values quoted for EEL are overestimates and that, under the practical conditions of *ad libitum* feeding, EEL_N lies somewhere between 0 and 20 kJ/bird/24 h (with the 48 + 48 h regime in the laboratory of the author), giving 25 g glucose twice during the first 48 h period and feeding 50 g glucose at the start of the balance, an EEL_N value of 20 kJ/24 h between 35 and 40 kJ/48 h is fairly regularly found. If the upper value is assumed to

Table 3.4 COMPOSITION AND AME$_N$ VALUES OF THREE DIETS DETERMINED IN FIVE LABORATORIES

Composition (g/kg)	Diet No.		
	1	*2*	*3*
Basal mix	1000	700	700
Wheat meal	0	300	0
Soyabean meal	0	0	300
Laboratory		AME$_N$ (kJ/g)	
1	13.90	13.91	12.95
2	14.21	13.97	13.13
3	13.94	14.25	12.96
4	13.66	13.84	12.27
5	14.03	14.20	13.10
Mean ± SD	13.95 ± 0.20	14.03 ± 0.18	12.88 ± 0.35

Table 3.5 AME$_N$ VALUES OF WHEAT AND SOYABEAN MEALS DERIVED FROM FEEDING DIETS CONTAINING 300 g/kg OF EACH BY FIVE LABORATORIES

Laboratory	Wheat	Soyabean
1	11.97	9.43
2	11.47	9.24
3	12.20	8.82
4	12.23	8.00
5	12.60	9.85
Mean ± SD	12.09 ± 0.42	9.07 ± 0.70

be correct, it can readily be seen that, with birds eating 100 g/day, the difference between TME$_N$ and AME$_N$ (EEL$_N$:F.I.) is only 0.2 kJ/g. In other words an ingredient with a TME$_N$ value of 15.0 kJ/g would have an AME$_N$ value of 14.8 kJ/g, only 1.3% lower. The effect is, of course, greatest with raw materials of low ME content, although it still can be considered small and much smaller than the 10% difference suggested earlier (Sibbald, 1977). If the conditions adopted for food presentation are confirmed to affect EEL$_N$, as suggested by Hartel (1986) then the differences between AME$_N$ and TME$_N$ may be even less and it is doubtful whether biological assays capable of detecting such small effects could be designed. With food intakes greater than 100 g/day the differences will be even smaller and this may explain why efforts to detect them have largely been unsuccessful.

References

Bayley, H.S., Summers, J.D. and Slinger, S.J. (1968) Effect of heat-treatment on the metabolizable energy of wheat germ meal and other wheat milling by-products. *Cereal Chemistry*, **45**, 557–563

Bourdillon, A., Carre, B., Conan, L., Duperray, J., Huyghebaert, G., Leclercq, B., Lessire, M., McNab, J. and Wiseman, J. (1990) A European reference method for the *in vivo* determination of metabolisable energy with adult cockerels: reproducibility, effect of food intake and comparison with domestic methods. *British Poultry Science* (in press).

Dale, N.M. and Fuller, H.L. (1981) The use of true metabolizable energy (TME) in formulating poultry rations. In *Proceedings of the Georgia Nutrition Conference*, pp. 50–57

Dale, N.M., Fuller, H.L., Pesti, G.M. and Phillips, R.D. (1985) Freeze drying versus oven drying of excreta in true metabolizable, nitrogen-corrected true metabolizable energy and true amino acid availability bioassays. *Poultry Science*, **64**, 362–365

Farrell, D.J. (1978) Rapid determination of metabolizable energy of foods using cockerels. *British Poultry Science*, **19**, 303–308

Farrell, D.J. (1981) An assessment of quick bioassays for determining the true metabolizable energy and apparent metabolizable energy of poultry feedstuffs. *World's Poultry Science Journal*, **37**, 72–83

Farrel, D.J., du Preez, K. and Hayes, J.P. (1988) Measurement of metabolizable energy of poultry feedstuffs in adult cockerels using four different methods. *Proceedings of the Georgia Nutrition Conference*, pp. 94–104

Farrell, D.J. and Swain, S. (1977) Effect of temperature treatments on the heat production of starving chickens. *British Poultry Science*, **18**, 725–734

Fisher, C. and McNab, J.M. (1987) Techniques for determining the metabolizable energy of poultry feeds. In *Recent Advances in Animal Nutrition – 1987* (eds W. Haresign and D.J.A. Cole), Butterworths, London, pp. 3–18

Fisher, C. and Shannon, D.W.F. (1973) Metabolizable energy determinations using chicks and turkeys. *British Poultry Science*, **14**, 609–613

Foster, W.H. (1968) Variation between and within birds in the estimation of the metabolizable energy content of diets for laying hens. *Journal of Agricultural Science, Cambridge*, **71**, 153–159

Guillaume, J. and Summers, J.D. (1970) Maintenance energy requirement of the rooster and influence of plane of nutrition on metabolizable energy. *Canadian Journal of Animal Science*, **50**, 363–369

Harris, L.E. (1966) *Biological Energy Interrealtionships and Glossary of Energy Terms*, N.A.S.-N.R.C. Publication 1411, Washington

Hartel, H. (1986) Influence of food input and procedure of determination on metabolisable energy and digestibility of a diet measured with young and adult birds. *British Poultry Science*, **27**, 11–39

Hartel, H. (1987) Reply to Sibbald and Wolynetz. *British Poultry Science*, **28**, 784–788

Hill, D.C., Evans, E.V. and Lumsden, H.G. (1968) Metabolizable energy of aspen flower buds for captive ruffed grouse. *Journal of Wildlife Management*, **34**, 854–858

Hill, F.W. and Anderson, D.L. (1958) Comparison of metabolizable and productive energy determinations with growing chicks. *Journal of Nutrition*, **64**, 587–603

Jonsson, G. and McNab, J.M. (1983) A comparison of methods for estimating the metabolisable energy of a sample of grass meal. *British Poultry Science*, **24**, 349–359

Kussaibati, R., Guillaume, J. and Leclercq, B. (1982) The effects of age, dietary fat and bile salts, and feeding rate on apparent and true metabolisable energy values in chickens. *British Poultry Science*, **23**, 393–403

Kussaibati, R. and Leclerq, B. (1985) A simplified rapid method for the determination of apparent and true metabolizable energy values of poultry feed. *Archiv fur Geflugelkunde*, **49**, 54–62

Leeson, S., Boorman, K.N., Lewis, D. and Shrimpton, D.L. (1974) Metabolizable energy studies with turkeys: metabolisable energy of dietary ingredients. *British Poultry Science*, **15**, 183–189

Lockhart, W.C., Bryant, R.L. and Bolin, D.W. (1963) Factors effecting the use of classical metabolizable caloric values. *Poultry Science*, **42**, 1285

Lodhi, G.N., Renner, R. and Clandinin, D.R. (1969) Studies on the metabolizable energy of rapeseed meal for growing chickens and laying hens. *Poultry Science*, **48**, 964–970

Lodhi, G.N., Renner, R. and Clandinin, D.R. (1970) Factors affecting the metabolizable energy value of rapeseed meal 2. Nitrogen absorbability. *Poultry Science*, **49**, 991–999

MacAulife, T. and McGinnis, J. (1971) Effect of antibiotic supplements to diets containing rye on chick growth. *Poultry Science*, **50**, 1130–1134

McNab, J.M. and Blair, J.C. (1988) Modified assay for true and apparent metabolisable energy based on tube feeding. *British Poultry Science*, **29**, 697–707

McNab, J.M. and Fisher, C. (1982) The choice between apparent and true metabolisable energy systems – recent evidence. *Proceedings of the 3rd European Symposium on Poultry Nutrition*, pp. 45–55

McNab, J.M. and Fisher, C. (1984) An assay for true and apparent metabolisable energy. *Proceedings of the XVII World's Poultry Science Congress*, pp. 374–376

March, B.E. and Biely, J. (1971) Factors affecting the response of chicks to diets of different protein value: breed and age. *Poultry Science*, **50**, 1036–1040

Miski, A.M.A. and Quazi, S. (1981) Influence of age and sex of growing broiler chicks and body weight of roosters on their endogenous and metabolic energy losses. *Poultry Science*, **60**, 781–785

Muztar, A.J. and Slinger, S.J. (1980a) Effect of body weight and duration of fast on metabolic and endogenous excretion by the mature rooster. *Nutrition Reports International*, **22**, 147–156

Muztar, A.J. and Slinger, S.J. (1980b) An evaluation of the rapid apparent metabolisable energy assay in relation to feed intake using mature cockerels. *Nutrition Reports International*, **22**, 745–750

Parsons, C.M., Potter, L.M. and Bliss, B.A. (1984) A modified voluntary feed intake bioassay for determination of metabolizable energy with Leghorn roosters. *Poultry Science*, **63**, 1610–1616

Potter, L.M., Matterson, L.D., Arnold, A.W., Pudelkiewicz, W.J. and Singsen, E.P. (1960) Studies in evaluating energy content of feeds for the chick. 1. The evaluation of the metabolizable energy and productive energy of alpha cellulose. *Poultry Science* **39**, 1166–1178

Proudman, J.A., Mellen, W.J. and Anderson, D.L. (1970) Utilization of feed in fast- and slow-growing lines of chicken. *Poultry Science*, **49**, 961–972

Rao, P.V. and Clandinin, D.R. (1970) Effect of method of determination on the metabolizable energy value of rapeseed meal. *Poultry Science*, **49**, 1069–1074

Renner, R. and Hill, F.W. (1960) The utilization of corn oil, lard and tallow by chickens of various ages. *Poultry Science*, **39**, 849–854

Shires, A., Robblee, A.R., Hardin, R.T. and Clandinin, D.R. (1979) Effect of previous diet, body weight, and duration of starvation of the assay bird on the true metabolisable energy value of corn. *Poultry Science*, **58**, 602–608

Sibbald, I.R. (1975) The effect of level of feed intake on metabolizable energy values measured with adult roosters. *Poultry Science*, **54**, 1990–1997

Sibbald, I.R. (1976) A bioassay for true metabolizable energy in feedingstuffs. *Poultry Science*, **55**, 303–308

Sibbald, I.R. (1977) The true metabolizable energy system. Part 2. *Feedstuffs*, **49**, 23–24

Sibbald, I.R. (1982) Measurement of bioavailable energy in poultry feedingstuffs. *Canadian Journal of Animal Science*, **62**, 983–1048

Sibbald, I.R. (1986) The T.M.E. system of feed evaluation: methodology feed composition data and bibliography. *Research Branch Contribution 86–4E* Animal Research Centre, Agriculture Canada

Sibbald, I.R. and Morse, P.M. (1983) The effects of feed input and excreta collection time on estimates of metabolic plus endogenous energy losses in the bioassay for true metabolizable energy. *Poultry Science*, **62**, 68–76

Sibbald, I.R. and Price, K. (1978) The metabolic and endogenous energy losses of adult roosters. *Poultry Science*, **57**, 556–557

Sibbald, I.R. and Slinger, S.J. (1963a) A biological assay for metabolizable energy in poultry feed ingredients together with findings which demonstrate some of the problems associated with the evaluation of fats. *Poultry Science*, **42**, 313–325

Sibbald, I.R. and Slinger, S.J. (1963b) The effects of breed, sex, an arsenical and nutrient density on the utilization of dietary energy. *Poultry Science*, **42**, 1325–1332

Sibbald, I.R. and Wolynetz, M.S. (1987) True and apparent metabolisable energy. *British Poultry Science*, **28**, 782–784

Slinger, S.J., Sibbald, I.R. and Pepper, W.F. (1964) The relative abilities of two breeds of chickens and two varieties of turkeys to metabolize dietary energy nitrogen. *Poultry Science*, **43**, 329–333

Sugden, L.G. (1974) Energy metabolized by bantam chickens and blue-winged teal. *Poultry Science*, **53**, 2227–2228

Terpstra, K. and Janssen, W.M.M.A. (1975) Methods for the determination of metabolisable energy and digestibility coefficients of poultry feeds. *Spelderholt Report* 101.75, Spelderholt Institute for Poultry Research, Beekbergen, The Netherlands.

Van Kampen, M. (1983) Effect of water:food intake ratios in laying hens on food metabolisable energy. *British Poultry Science*, **24**, 169–172

Verma, S.V.S. (1977) *The Nutritive Value of Guar Meal for Poultry*. PhD Thesis, University of Edinburgh.

Vohra, P. and Kratzer, F.H. (1970) Metabolisable energy of alfalfa meals. *Proceedings of the XIV World's Poultry Science Congress*, **2**, 513–516

Wehner, G.R, and Harrold, R.L. (1982) Crop volume of chickens as affected by body size, sex and breed. *Poultry Science*, **61**, 598–600

Yamazaki, M. and Zhang, Z. (1982) A note on the effect of temperature on true and apparent metabolisable energy values of a layer diet. *British Poultry Science*, **23**, 447–450

Yeomans, M. (1987) *Control of Drinking in Domestic Fowls*. PhD Thesis, University of Edinburgh

Zelenka, J. (1968) Influence of the age of chicken on the metabolisable energy values of poultry diets. *British Poultry Science*, **9**, 135–142

4

PROTEIN DEGRADATION OF RUMINANT DIETS

W.M. VAN STRAALEN and S. TAMMINGA
Institute of Livestock Feeding and Nutrition Research, Lelystad, Netherlands; and Department of Animal Nutrition, Agricultural University, Wageningen, Netherlands

Introduction

Feed protein ingested by ruminants is subject to extensive microbial degradation in the rumen. Resulting end products like ammonia and/or amino acids are subsequently incorporated to a varying extent in microbial biomass. Together with feed protein escaping degradation in the rumen, this forms the protein supply of the ruminant. Attempts have been made in recent years to bring existing information together in new protein evaluation systems for ruminants (INRA, 1978; ARC, 1980, 1984; Madsen, 1985; NRC, 1985; Verite *et al.*, 1987). The basic concept of these systems is very similar. Protein supply is estimated as the amount of protein absorbed from the small intestine, and which is the sum of feed protein escaping degradation in the rumen and microbial protein formed and subsequently released to the lower tract, both corrected with an appropriate factor for intestinal digestion. This chapter reviews recently obtained information on quantitative aspects of both ruminal and post-ruminal feed protein digestion in ruminants.

Characterization of feed protein

Proteins used in animal nutrition can be classified according to their solubility in water (albumins), a salt solution (globulins), alcohol (prolamines) or dilute alkali (glutelins) or on the basis of their function in the plant (enzymes, structural protein, storage protein). Nitrogen in fresh forages is 70–90% true protein and 10–30% as non-protein nitrogen (Tamminga, 1986). In forages three main groups of proteins occur: fraction 1 leaf protein (75% of total leaf protein), fraction 2 leaf protein (25% of leaf protein) and chloroplast membrane proteins (Mangan, 1982). Fraction 1 consists mainly of chloroplast enzymes, has a high solubility and is rapidly degraded in the rumen. Fraction 2 is a mixture of different proteins from chloroplasts and cytoplasm, with an unknown degradation rate. Chloroplast membrane proteins (mainly chlorophyll) are insoluble and degraded slowly. A small proportion of the proteins is situated in mitochondria and the nucleus for which the degradation rate is unknown. Finally some protein is linked to structural carbohydrates in the cell walls for which degradation is slow (Mangan, 1982).

Crude protein in seeds (grains, oil seeds, pulses) is mainly true protein and can be present in the husk, the pericarp or the seed itself. In the husk and pericarp, structural protein dominates, whereas in the seed the vast majority of protein (80–90%) is storage protein in the aleuron layer and in the endosperm (Ensminger and Olentine, 1978). In addition enzyme protein is found in the germ.

With respect to solubility characteristics large differences exist between different seeds (Boulter and Derbyshire, 1976). In grain seeds 10–20% of the protein is in albumins and globulins. The remaining 80–90% is equally distributed between prolamins and glutelins. Rice and oats are exceptions in that 70–80% of the protein is present in glutelins and only 5–20% in prolamins. In legume seeds 85–100% of the protein is in albumins and globulins, none in prolamins and 0–15% in glutelins.

Animal protein is usually present in enzymes, membranes, transport proteins (e.g. albumins in blood) or muscle (myoglobin). Depending on their origin, proteins in feeds of animal origin vary widely in their degradation properties.

Ruminal degradation of feed protein

MECHANISM OF PROTEIN DEGRADATION IN THE RUMEN

Anaerobic protein degradation in the rumen contains two steps, being hydrolysis of the peptide bond by proteases and peptidases (1) and decarboxylation and/or deamination of amino acids (2). The first step results in peptides and amino acids, and end products of the second step are volatile and branched chain fatty acids (VFA's and BCFA's), CO_2, and NH_3. In the rumen, deamination is the most important pathway of amino acid degradation (Baldwin and Allison, 1983). For a long time proteolysis was assumed to be the rate-limiting step in the degradation of protein (Tamminga, 1979) but recently Chen, Russell and Sniffen (1987) provided evidence that peptide uptake was the rate-limiting step and that peptides rather than amino acids are the main end products of proteolysis.

Many strains of micro-organisms are involved in protein degradation and until recently it was thought that no major strain could survive on protein as the only source of energy and N (Baldwin and Allison, 1983). Nocek and Russell (1988) however, reported the isolation of a *Peptostreptococcus* with a high proteolytic activity that accounted for 10% of total colony counts and was able to grow rapidly on amino acids and peptides as the only source. Protozoa play a far less important role in the digestion of feed protein than bacteria (Baldwin and Allison, 1983; Nocek and Russell, 1988).

MEASURING DEGRADATION IN THE RUMEN

Various methods exist to estimate protein ($N \times 6.25$) degradation in the rumen, both *in vivo* and *in vitro*. A limitation of the *in vivo* method is its indirect methodology. Undergraded feed N in duodenal digesta is estimated as the difference between total N flow and microbial N flow, sometimes corrected for endogenous contaminations. Techniques to measure duodenal digesta flow as well as microbial N have a large error, which in the calculations is transferred to undergraded feed N. *In vivo* measurements require animals equipped with duodenal cannulae, either re-entrant or T-piece. In re-entrant cannulated animals duodenal N flow can be measured and sampled with reasonable accuracy, provided the measuring period is long enough. However, this period is often restricted to 24 h or less and the flow is subsequently corrected for an incomplete recovery of an indigestible marker. With T-piece cannulae the use of markers is the only way to estimate duodenal flow. An additional source of error may then result from samples which are not representative. If this results from a shift in the ratio between solids and liquids this can be corrected by reconstituting the samples with the double marker technique (Faichney, 1975). If the solids in the sample do not represent the solids in duodenal digesta flow this method fails.

Measuring microbial protein also requires markers. Amino acids, assumed to be present in microbes only (diaminopimelic acid, D-alanin, amino-phosphonic acid), or nucleic acids are used. Alternatives are radio isotopes (^{15}N, ^{32}P, ^{35}S). Estimating microbial protein without using markers is possible by comparing the amino acid profile in duodenal content with that in microbial protein and feed protein. All methods have limitiations and their results often do not agree with each other (Siddons, Beever and Nolan, 1982; Theurer, 1982; Demeyer and Tamminga, 1987).

A method to estimate protein degradation in a more direct way is the nylon bag incubation technique (Mehrez and Ørskov, 1977). In this method feed samples included in nylon bags with a pore size of between 30 and 50μm are incubated in the rumen for various lengths of time yielding a degradation curve from which the rate of degradation can be estimated (Ørskov and McDonald, 1979). Combining the rate of degradation with an appropriate rate of passage yields estimates of the 'effective' protein degradation. This method also has a number of weaknesses (Lindberg, 1985; Nocek, 1988). The first limitation is that the contents of the bags are not subjected to particle size reduction through chewing and rumination and become contaminated with microbial protein (Kennedy, Hazlewood and Milligan, 1984; Varvikko and Lindberg, 1985). This results in an overestimate of the rumen escape value which for feedstuffs low in protein may be quite significant (Nocek and Grant, 1987). A second weakness is the assumption that protein washed out of the bags is degraded instantaneously and completely. Recently it was demonstrated (Chen, Sniffen and Russel, 1987) that peptide N leaving the rumen could account for differences in effective degradation of 3–5% units. A further complication is that the shape of the curve does not always follow the pattern of first-order kinetics (Kristensen, Møller and Hvelplund, 1983; Nocek and English, 1986). Finally no agreement has been reached yet on the most appropriate rate of passage out of the rumen. Rate of passage of solids as well as liquids is influenced by level of feed intake, but the passage rate of the latter is usually much higher. Rumen clearance of solids also depends on the size of the particles and their specific weight (Hooper and Welch, 1985). The latter not only depends on the specific weight of the feed itself, but also on microbial activity on their surface (Sutherland, 1986). Degree of digestion also seems to have an influence. Recently it was demonstrated in dairy cows that undigestible material (IADF) was passing out of the rumen at a much faster rate than digestible material (Tamminga *et al.*, 1989).

In the ARC approach (ARC, 1984) passage is restricted to the small particles and rates of 0.02 to 0.08 fractions per hour were suggested for different diets and levels of intake. In the Nordic and French protein evaluation system for ruminants (Madsen, 1985; Verité *et al.*, 1987) a rate constant of 0.08 and 0.6 respectively is suggested for all diets and all levels of intake. In the Cornell Net Carbohydrate/ Protein system for evaluating cattle diets, concentrate ingredients are classified on the basis of their weight and the assumed rumen passage rate at maintenance level of feed intake varies between 1 and 3.5% per hour. In a review Owens and Goetsch (1986) derived regression equations for the passage of liquid, roughage particles and concentrate particles. The equations proposed by them are summarized in Table 4.1.

A number of *in vitro* methods to estimate protein degradation in the rumen have also been developed. Initially N-solubility in rumen fluid was proposed (Wohlt,

Table 4.1 REGRESSION EQUATIONS TO ESTIMATE RATES OF PASSAGE OUT OF THE RUMEN (FROM OWENS AND GOETSCH, 1986)

Y	A	CI	RI	$(RI)^2$
Fluid	4.12	0.77	2.32	
Roughage	0.94	1.34	1.24	
Concentrates	1.30	0.61	4.88	1.25

CI = Concentrate DM intake (% of body weight)
RI = Roughage DM intake (% of body weight)

Table 4.2 PROTEIN DEGRADABILITY CHARACTERISTICS OF CONCENTRATE FEEDSTUFFS DETERMINED AT IVVO

	cp	W	U	k_d	B
Beans	26.5	27.5	0.8	8.24	31
Horse beans	26.3	62.8	0.3	10.82	14
Lupin	34.2	25.5	0.2	12.87	24
Peas	25.2	55.6	0.0	8.95	18
Hominy feed	18.3	27.9	1.7	7.84	32
Rice bran	14.4	32.6	6.6	9.40	31
Wheat middlings	18.4	12.8	7.2	13.44	32
Beet pulp	10.3	23.8	6.4	5.15	44
Brewers grain	24.9	4.6	30.1	5.09	65
Citrus pulp	7.0	40.7	3.3	5.62	32
Corn gluten feed	21.6	44.9	5.5	5.15	32
Palm kernels	10.4	8.0	6.5	2.39	68
Soy beans (raw)	40.6	33.9	0.0	10.24	24
Soy beans (toasted)	39.0	7.4	0.0	6.76	44
Babassu meal	20.2	3.0	9.5	3.35	66
Coconut meal	21.5	13.9	3.2	3.03	58
Cottonseed meal	48.6	13.4	1.8	7.58	39
Groundnut meal	57.0	22.3	1.3	9.39	31
Linseed meal	33.4	17.2	4.1	5.04	48
Nigerseed meal	36.1	9.4	5.4	10.56	36
Palmkernel meal	15.1	8.8	6.7	3.47	60
Rapeseed meal	36.9	21.2	5.9	13.84	29
Ricebran meal	14.2	4.3	17.6	6.18	56
Soybean meal	49.5	6.2	0.1	8.25	40
Sunflowerseed meal	37.2	14.7	3.3	14.68	27
Feathermeal	88.9	13.3	9.5	0.95	76
Meatmeal	60.4	36.2	11.8	1.66	53
Alfalfa meal	16.0	26.0	19.9	5.44	48

cp = crude protein fraction in dm (%).
W = washable crude protein fraction (%).
U = undigestible crude protein fraction (%).

k_d = degradation rate (%/hour).
B = effective bypass protein fraction (%) assuming Kp = 0.06.

Sniffen and Hoover, 1973). The results varied however with the solvent used (Crooker *et al.*, 1978). The method does not give information on rumen degradation characteristics of the non-soluble part of the protein, which may be quite variable. A further development was incubation *in vitro* with rumen fluid using ammonia or amino acid release as an indicator for microbial degradation (Broderick, 1982) or with purified proteolytic enzymes (Pichard and Van Soest, 1977).

Although nylon bag incubations are too complicated and laborious to be used as screening method for large numbers of feedstuffs, because of its directness it is felt that this technique is at present the most reliable method, provided a standardized procedure is used. Presenting results should also be standardized. With nylon bag incubation studies, protein in ruminant feeds can be separated into a washable fraction (W), which can be washed out of the bags without rumen incubation, an undegradable fraction (U), determined with a long-term (10 days) rumen incubation and a degradation rate (k_d) for the unsoluble degradable fraction (D=100−W−U), estimated from the degradation curve and, if relevant, a lag period. It is recommended that data be presented in this way (Table 4.2). The advantage is that effective degradation or bypass (B) can easily be recalculated in case rumen passage rates need adjustment. Figures in Table 4.2 are based on assumed rate of passage of 0.06/hr.

EFFECTIVE DEGRADATION OF PROTEIN IN FEEDSTUFFS

Data on effective protein degradation of concentrate feedstuffs, determined by nylon bag incubations in different laboratories with at least six different feedstuffs each, were collected (W.A.G. Veen, unpublished observations; J. Oskamp, unpublished observations; Ørskov, 1982; Cronjé, 1983; Shibui *et al.*, 1983; De Boever *et al.*, 1984; Madsen and Hvelplund, 1985; Barrio, Goetsch and Owens, 1986; Verité *et al.*, 1987; Erasmus, Prinsloo and Meissner, 1988; Tamminga and Ketelaar, 1988; Suomel *et al.*, 1990). Between laboratory differences were substantial, but the sequence of degradation of the feedstuffs was usually very similar. Regression equations were calculated between the results in each data set and those obtained at IVVO with the same ingredients. In this regression analysis the products of animal origin were excluded because they usually did not fit the regression equation. R squared for the regression equations ranged between 0.63 and 0.96. From these equations corrected degradation values were calculated. This reduced the average coefficient of variation from 32% to 16% (Table 4.3).

Protein escape values for roughages, based on nylon bag incubations are limited (Filmer, 1982; Cronjé, 1983; Madsen and Hvelplund, 1985; Verité *et al.*, 1987). The composition of forages varies much more than that of concentrate feedstuffs and depends on species, maturity, fertilization level, season, soil type and weather conditions. Degradation of forage protein, using nylon bag incubations was studied extensively in our institute. A total of 28 samples of fresh grass (Van Vuuren and Tamminga, unpublished observations), 36 samples of grass silage and 10 samples of grass hay (Tamminga, Ketelaar and Van Vuuren, 1989; Bosch and De Visser, unpublished observations) were studied. Important sources of variation were N content and date of harvest for all forage samples. Silages varied also in dry matter content.

Material was chopped at 1 cm before incubation. Grass samples were incubated in grass fed cows, whereas silage samples and hay samples were incubated in cows

Table 4.3 MEAN BYPASS PROTEIN OF CONCENTRATE FEEDSTUFFS OBTAINED FROM DIFFERENT LABORATORIES, BEFORE AND AFTER CORRECTION FOR LABORATORY INFLUENCE

Feedstuff	cp	n	Before		After	
			B	CV	B	CV
Barley	11	8	28	32	34	12
Corn	10	8	58	24	57	21
Oats	12	5	19	42	24	36
Milo	10	3	58	11	57	10
Rye	11	3	19	26	22	34
Wheat	14	7	23	26	29	25
Beans	26	4	32	42	33	6
Horse beans	29	4	18	40	19	27
Lupin	33	5	14	52	22	25
Peas	24	6	20	28	24	24
Hominy feed	14	2	48	47	38	24
Cassava meal	3	2	32	20	40	26
Rice bran	15	4	31	48	34	6
Rye middlings	16	1	42	–	32	–
Wheat middlings	19	5	24	25	31	15
Wheat bran	18	3	25	40	34	5
Beetpulp (10–15% sugar)	11	7	49	34	49	12
Beetpulp (20–25% sugar)	14	4	43	26	39	25
Brewers grains	27	7	67	19	61	14
Citrus pulp	7	3	37	84	37	29
Corn gluten meal	65	9	75	11	69	7
Corn gluten feed	22	7	24	48	32	13
Cottonseed	18	2	22	55	27	12
Linseed	23	3	28	42	27	43
Rapeseed	21	1	13	–	22	–
Palm kernels	10	1	68	–	68	–
Soyabeans (raw)	39	6	19	50	25	16
Soyabeans (toasted)	40	3	41	55	41	11
Sunflowerseed	19	2	8	10	20	11
Babassu meal	20	2	76	18	62	10
Coconut meal	22	6	60	23	57	7
Cottonseed meal	43	8	44	25	43	11
Groundnut meal	52	7	21	50	26	13
Linseed meal	36	9	40	33	42	16
Nigerseed meal	36	1	36	–	36	–
Palmkernel meal	16	5	61	35	58	11
Rapeseed meal	39	7	29	24	34	17
Rice bran meal	14	2	53	9	54	4
Sesame seed meal	46	2	28	66	33	6
Soyabean meal	50	11	36	32	39	8
Sunflowerseed meal	37	10	23	40	28	19
Soya hulls	16	2	53	26	43	1
Blood meal	89	2	79	7	75	7
Feather meal	90	3	81	6	66	16
Fish meal	73	11	57	18	56	13
Meat meal	60	6	51	14	49	15
Meat and bone meal	52	3	44	24	45	23
Alfalfa meal	17	5	44	33	45	7
Grass meal	13	2	35	47	43	6

cp = crude protein content in the dry matter (%). B = effective protein bypass fraction (%).
n = number of laboratories. CV = coefficient of variation.

Table 4.4 RELATIONSHIPS BETWEEN DEGRADABILITY CHARACTERISTICS, CHEMICAL COMPOSITION AND SEASON OF HARVESTING OF ROUGHAGE FEEDSTUFFS

Roughage	Y	a	b_1 = dm	b_2 = cp	b_3 = day	R^2
Fresh grass	U	10.2		−0.037	0.022	0.62
	k_d	8.9		0.027	0.034	0.44
	B	38.6		−0.080	0.070	0.73
Grass silage	W	81.5	−0.040		−0.093	0.63
	U	19.0	0.006	−0.066	−0.025	0.58
	k_d	2.5	−0.008	0.035		0.43
	B	19.8	0.031	−0.077	0.071	0.81
Grass hay	W	12.5		0.098		0.49
	U	24.2		−0.073		0.81
	k_d	5.9		0.007	−0.033	0.55
	B	50.3		−0.110	0.131	0.80

b_1 = coefficient for dry matter fraction in product (%)
b_2 = coefficient for crude protein fraction in dm (%)
b_3 = coefficient for days elapsed since 1st of May
W = washable crude protein fraction (%)
U = undigestible crude protein fraction (%)
k_d = degradation rate (%/hour)
B = effective protein bypass fraction (%)

fed diets consisting of hay and concentrates. Measurements were on washable fraction (W), undegradable fraction (U) and rate of degradation (k_d). For all forages a ruminal passage rate of 4.5% per hour was assumed.

The results (Table 4.4) showed significant influences on W, U and kd in all forages for N content, NDF content and data of harvest and for silage an additional influence for dry matter content. For fresh grass, grass silage and grass hay an average protein bypass value (B) was found of 29, 22 and 42% at an average N content of 35, 36 and 32 g/kg dm respectively. Again significant influences of N content and day of harvest were observed for all forages with, additionally, an influence of dry matter content for silages. Estimating protein degradation in corn silage was unsuccessful, because of severe contamination of the bag contents with bacteria during the first 24 hours of the incubation. From the disappearance of N between 0 and 48 hours it was estimated that the proportion escaping degradation in the rumen was close to 25%, of which almost half was undegradable (Tamminga and Ketelaar, 1988).

Manipulation of protein degradation in the rumen

As was stated before, protein in ruminant feeds can be characterized as containing an undegradable fraction (U), a (slowly) degradable insoluble fraction (D) and a (rapidly) degradable soluble fraction (W). Degradability of feed protein is thus determined by the fraction U, degradation by the ratio of the rate of degradation (k_d) and the rate of passage out of the rumen (k_p).

Manipulation of degradability is restricted to reducing the size of U, manipulation of degradation can be achieved both by changing the size of U and by changing the ratio k_p/k_d. Means of manipulating degradability of individual feedstuffs include changing growing conditions (forage) or processing (physical, chemical). Manipulation of degradation is possible through the same methods but, additionally, feeding management factors are important. For mixed diets the same manipulation principles are applicable but, in addition, protein degradation of the diet can be manipulated by selection of its ingredients for a high or a low degradability.

MANIPULATION OF PROTEIN DEGRADABILITY

Research at IVVO (Van Vuuren and Tamminga, unpublished observations) illustrated that degradability of protein in forages can be influenced by manipulating the growing conditions, of which level of N fertilization, maturity and season are important factors.

A high N fertilization of grasses, consisting mainly of perennial ryegrass, resulted in a decreased size of U, hence degradability increases. Increasing stage of maturity had the opposite effect as did progression of the season. Similar observations were made for silages and grass hays (Tamminga, Ketelaar and Van Vuuren, 1990). Sun drying, leading to wilted silage or hay did increase the size of U as well and therefore resulted in a decreased degradability. Nocek and Grant (1987) observed a similar tendency for orchard grass and timothy, but alfalfa and clover showed an opposite trend.

During the ensiling process, artificial drying of roughages and processing of seeds, protein can be damaged by heat, leading to less soluble protein and probably to an increased size of fraction U. Acid detergent insoluble N (ADIN) has been proposed as a measure of such damage (Thomas *et al.*, 1982).

MANIPULATION OF PROTEIN DEGRADATION

N fertilization increases crude protein content but also the size of the fraction W and the rate at which fraction D is degraded in the rumen. This leads to a reduced protein escape from the rumen. In grass with a relatively high protein content less protein is bound to the structural carbohydrates (Møller, 1985; Tamminga, Ketelaar and Van Vuuren, 1990). In nylon bag incubation studies Nocek and Grant (1987) observed large differences in rate of protein degradation between grasses (orchard grass, timothy) and legumes (alfalfa, clover); N in legumes disappeared at a much faster rate. Tamminga, Ketelaar and Van Vuuren (1989) found that when the season progressed the degradation of protein in fresh grass and grass silage decreased. However, Beever *et al.* (1986) in experiments with sheep found no difference in protein degradation *in vivo* between ryegrass and clover and season of harvesting.

Grass silage can be made from freshly cut grass or after wilting. During the wilting period plant proteases are active, particularly under moist conditions and a high temperature (McDonald, 1982). Changes in the nitrogen fraction as result of the ensiling process itself depend on the type of fermentation but can be extensive, resulting in a high N-solubility. In a stable silage lactic acid bacteria which have very little proteolytic and deamination activity predominate. In an unstable silage *Clostridia* may proliferate and their activity results in severe protein breakdown (McDonald, 1982). Wilting grass, before ensiling, increases the osmotic pressure and therefore reduces the fermentation of protein in the silage (Tamminga *et al.*, 1990).

Protein breakdown in silage can also be reduced by the application of formic acid, formaldehyde or a combination of both. Addition of formic acid quickly reduces the pH and establishes a stable silage, but has no influence on the protein degradation in the rumen (Chamberlain, Thomas and Wait, 1982; McDonald, 1982). Application of formaldehyde in grass silage has a sterilizing effect on *Clostridia* and decreases protein degradation in the silage as well as in the rumen through the formation of bonds between the formaldehyde and proteins (McDonald, 1982). With fresh forage (lucerne and grass), formaldehyde treatment also resulted in a decreased rumen degradation of the protein (Beever *et al.*, 1987).

Due to proteolysis during wilting, protein in grass hay has a higher soluble fraction than protein in fresh grass. This is more than compensated for by a lower degradation rate as well as a higher undegradable fraction, thus a smaller effective degradation (Tamminga and Ketelaar, 1988) because of the absence of proteolysis during storage. Grass hay has a smaller soluble fraction and degradation rate than grass silage, but a higher undegradable fraction, which also results in a higher protein escape (Janicki and Stallings, 1988; Tamminga, Ketelaar and Van Vuuren, 1990).

Protein degradation of concentrate feedstuffs can be influenced by processing, which normally takes place with cereals and oilseeds and treatment with chemical agents. Oilseed extraction reduced protein degradation in the rumen; expelling was more effective than solvent extraction (Goetsch and Owens, 1985; Stern, Santos and Satter, 1985; Broderick, 1986; Pena, Tagari and Satter, 1986). According to Satter (1986), heat treatment increases both the undegradable and undigestable protein fraction and the maximum intestinal supply of absorbable protein depends on the time and temperature of the treatment. Formaldehyde treatment of soybean meal and rapeseed meal resulted in an increased ruminal escape (Rooke, Brookes and Armstrong, 1983; Crooker *et al.*, 1986). Alcohol treatment changes the structure of proteins, leading to a more hydrophobic character. This also causes a decreased rumen degradation (Lynch *et al.*, 1987; Van der Aar *et al.*, 1984).

Manipulation of protein degradation can also be achieved by feeding management, for example changing the ratio, sequence, level and frequency of feeding concentrate and roughage (Tamminga, 1979).

Feeding roughage rich diets generally results in a faster ruminal protein degradation than diets rich in concentrates which is probably due to a higher ruminal pH that stimulates microbial activity. In concentrate based diets more protein is in small particles, which can escape rumen fermentation when fed in large amounts at a time (Ganev, Ørskov and Smart, 1979; Owens and Bergen, 1983; Zinn and Owens, 1983). However, Tamminga (1982) found no difference in protein escape between diets differing in their roughage to concentrates ratio. This would suggest that diets which are broken down slowly stay longer in the rumen. Within roughage based diets, protein degradation was demonstrated to be higher in animals fed fresh grass than in diets based on hay (Tamminga, unpublished observations). This can be due to the proteolytic activity of plant proteases of fresh grass in addition to the bacterial activity or due to a shift in the microbial population in favour of proteolytic bacteria.

The effect of both frequency of feeding and feed intake level on protein degradation seems restricted to concentrate based diets and becomes evident mainly at a high level of feed intake. More frequent feeding stabilizes rumen fermentation and increasing the feed intake level can increase ruminal passage rate of particles and fluid and alter microbial proteolytic activity (Robinson and Tammínga, 1984).

Feeding concentrate based diets more frequently at a high level of feed intake resulted in an increased protein degradation while, with diets consisting mainly of roughages, no effect was recorded (Tammínga, 1981; Robinson and Sniffen, 1985; McAllan, Lewis and Griffith, 1987). Firkins *et al.* (1986) and Rahnema *et al.* (1987) found no effect of feeding level on protein degradation in experiments with steers. With dairy cows fed concentrate based diets, Tammínga, Van der Koelen and Van Vuuren (1979) observed a decreased protein degradation with increased feed intake, while Robinson, Sniffen and Van Soest (1985) could not detect any effect with roughage based diets.

The effects of protein content of the diet on the effective protein degradation are conflicting. Barney *et al.* (1981) and Kirkpatrick and Kennelly (1987) showed an increased degradation with increasing protein content in the diet, while Forster *et al.* (1983) and Murphy and Kennelly (1987) found no effect.

Intestinal digestion of protein

MECHANISM OF INTESTINAL PROTEIN DIGESTION

Reduction of the degradation of feed protein in the rumen will only be beneficial to the animal when protein escaping degradation in the rumen is absorbed from the small intestine (SI). Apparent absorption gives an estimate of the amount of protein which becomes available for the animal's organs and tissues, true absorption gives an indication of the true protein value of the feed. The difference is endogenous protein, which not only means a direct protein drain from the animal, but also an indirect loss as its replacement needs considerably more protein than is excreted as such. Recently it was stated (NRC, 1985) that faecal excretion of endogenous protein was closely related to the amount of dry matter excreted in the faeces rather than the dry matter ingested and that replacement of the excreted protein would require 1.5 times that amount to be absorbed.

Although the large intestine is believed not to absorb amino acids, protein escaping small intestine digestion can be of value to the host animal through microbial fermentation to NH_3, which can be taken up by the large intestine, and subsequently used by rumen microorganisms.

MEASURING INTESTINAL DIGESTION

Various ways are possible to estimate intestinal digestion, of which the majority are only applicable to the mixture of undegraded feed protein, microbial protein and endogenous protein. Duodenally cannulated animals, preferably with cannulae in the beginning and at the end of the SI, provide information on the apparent absorption of intestinal protein. Alternatives are to infuse protein sources in the abomasum or at the beginning of the SI and to measure the increased faecal protein output (Schwarting and Kaufmann, 1978) or increased ileal protein flow (Hvelplund, 1985). The advantage of such methods is that information on the absorption of protein in individual feedstuffs can be obtained. A limitation is that infusion is not restricted to protein, but that the feedstuffs also contain dry matter other than protein, which causes the release of extra endogenous protein. This is likely to result in an underestimation of the true absorption, particularly if the increased ileal flow is measured. An alternative method is by regression (Van Bruchem *et al.*, 1985).

A recently developed method is the mobile nylon bag technique (Sauer, Jorgensen and Berzins, 1983). With this method small quantities of feedstuffs are included in small nylon bags, incubated in the rumen and, subsequently, introduced into the beginning of the small intestine and, after passage through the intestinal tract, recovered from the faeces or ileum. This technique has a high potential value, because the capacity to evaluate individual feedstuffs is quite high. In research with dairy cows at IVVO, four bags could be introduced in the beginning of the small intestine every 20 minutes over a period of more than 24 hours, resulting in a daily capacity of up to 300 bags.

INTESTINAL PROTEIN DIGESTION OF FEEDSTUFFS

The apparent absorption of total amino acid nitrogen (AAN) from the small intestine ranges from 65 to 75% and is higher than the non-ammonia nitrogen (NAN) absorption (55–70%). It is generally thought that essential amino acids (EAA) are absorbed to a greater extent than non-essential amino acids (NEAA) but literature data are conflicting (Santos, Stern and Satter, 1983; Hvelplund, 1984; Moller, 1985; Stern, Santos and Satter, 1985). From the regression technique (Van Bruchem *et al.*, 1985) it appears that true digestion of protein in concentrate ingredients is usually high. This observation was confirmed by the results of infusion studies (Schwarting and Kaufmann, 1978; Hvelplund, 1985) as well as mobile nylon bag studies (Robinson and Tamminga, 1984; Hvelplund, 1985; De Boer, Murphy and Kennelly, 1987).

Results of intestinal digestion studies on feedstuffs with the mobile nylon bag technique were collected from different laboratories (Hvelplund, 1985; De Boer, Murphy and Kennelly, 1987; Verite *et al.*, 1987; Tamminga and Ketelaar, 1988) and are presented in Table 4.5. In general, roughages show a much lower intestinal protein digestion than concentrate ingredients. This difference probably results from the fact that leaf protein in roughages is already largely degraded in the rumen and the protein escaping is associated with cell walls which cannot be digested in the small intestine and only to a small extent in the large intestine. In concentrate ingredients protein escaping degradation in the rumen is mainly storage protein, not protected by cell walls.

Data obtained with the infusion technique (Schwarting and Kaufmann, 1978; Kaufmann, 1979; Hvelplund, 1985) were consistently lower than the digestibilities obtained by the mobile nylon bag technique, probably due to increased endogenous secretions with the infusion technique. The mobile nylon bag technique can give overestimations when bags are recovered from the faeces because of disappearance of N in the large intestine (Hvelplund, 1985; Voigt *et al.*, 1985). *In vivo* disappearance of NAN and AAN from the large intestine ranges from between 11 and 34 and from −3 to 37% respectively (Santos, Stern and Satter, 1983; Moller, 1985). However, for the majority of feedstuffs the fraction of protein that has not disappeared in the small intestine is rather small and the overestimation limited.

Table 4.5   MEAN PROTEIN DIGESTIBILITY IN THE SMALL INTESTINE DETERMINED BY THE MOBILE NYLON BAG AND INFUSION TECHNIQUE OF FEEDSTUFFS FROM DIFFERENT LABORATORIES

Feedstuff	Mobile nylon bag			Infusion		
	Mean	CV	n	Mean	CV	n
Barley	90	–	1			
Corn	97	–	1			
Wheat	94	–	1			
Horse beans	91	4.0	2			
Lupin	98	–	1			
Wheat middlings	79	8.5	2			
Dried beet pulp	89	–	1			
Brewers grains	93	1.6	2			
Corn gluten feed	89	4.0	2			
Corn gluten meal	99	1.8	2			
Palm kernel	56	–	1			
Rapeseed	50	–	1			
Soyabean (raw)	88	11.7	2			
Soyabean (toasted)	84	–	1			
Coconut meal	92	3.1	3	85	–	1
Corn germ meal	91	–	1			
Cottonseed meal	87	–	1	82	19.8	2
Groundnut meal	97	0.3	2	90	–	1
Linseed meal	88	2.6	2	83	5.1	2
Palmkernel meal	82	7.6	2	54	–	1
Rapeseed meal	79	7.3	4	78	17.0	3
Soyabean meal 45% cp	99	0.3	2			
Soyabean meal 55% cp	98	2.1	3	84	3.6	3
Soyabean/rapeseed meal	97	–	1			
Sunflowerseed meal	91	4.7	3	71	8.0	2
Soya hulls	68	–	1			
Fish meal	92	2.2	3	87	2.7	3
Meatmeal	72	–	1			
Meat and bone meal	65	–	1			
Fresh grass	63	7.2	6			
Alfalfa	73	6.4	3			
Alfalfa hay	64	34.3	2			
Alfalfa silage	85	–	1			
Corn silage	74	–	1			

CV = coefficient of variation.
n   = number of laboratories.

MANIPULATION OF INTESTINAL DIGESTIBILITY

The disappearance of protein from the intestine can be described by digestibility (1 minus the undigestible fraction) and digestion (the fraction which has actually disappeared). Because of the high potential of the intestine to digest and absorb nitrogen, digestion approaches digestibility and therefore the distinction seems rather theoretical. At a high level of feed intake the difference becomes apparent but remains small. At a high level of feed intake Hvelplund (1984) found a decreased apparent N digestion in the small intestine and Tamminga and Ketelaar

(1988) reported a reduced intestinal digestion measured with the mobile nylon bag technique, due to a much shorter transit time.

Frequency of feeding and level of feed intake can alter the protein escape from the rumen. The effect of rumen retention time on intestinal digestion of escaped feed protein is variable and depends on the feedstuff (Hvelplund, 1985; Rooke, 1985; Voigt *et al.*, 1985; De Boer, Murphy and Kennelly, 1987). Results from mobile nylon bag experiments from IVVO indicate that incubation in the rumen for 6 or 18 hours had no significant effect on intestinal digestion. Treatment of feedstuffs with heat or chemical agents can reduce protein degradation in the rumen and also intestinal digestion when feedstuffs are overprotected. Hvelplund (1984, 1985) showed decreased digestion of formaldehyde treated soybean meal and heat damaged fishmeal and Rooke, Brookes and Armstrong (1983) found the same effect with formaldehyde treated soybean meal and rapeseed meal. Heat treatment of whole soybeans on the other hand resulted in an increased small intestine digestion of total N and AAN, which could be due to denaturation of the trypsin inhibitor (Stern, Santos and Satter, 1985). Treatment of roughages with formic acid or formaldehyde had no effect on the apparent total tract digestion of the protein (Chamberlain, Thomas and Wait, 1982).

Conclusion

Measuring feed protein escape from the rumen and subsequent digestion in the intestine should be standardized. When using nylon bag incubations, protein should preferably be described by a washable fraction (W), undegradable fraction (U) and a degradation rate (k_d) of the unsoluble degradable fraction (D=100−W−U). The degradability of protein in feedstuffs is determined by the fraction U; the actual degradation depends on the degradability and the conditions under which degradation takes place, which determine the ratio degradation rate/passage rate (k_d/k_p). When measured under standardized conditions protein degradation in both concentrate ingredients and roughages is variable. For measurement of the intestinal digestion of escape protein the mobile nylon bag technique is at present the most powerful method. Based on limited data, bypass protein from concentrate ingredients has a high digestibility; that of roughages is lower and more variable.

References

Agricultural Research Council (1980) *The Nutrient Requirements of Ruminant Livestock*, Commonwealth Agricultural Bureaux, Slough

Agricultural Research Council (1984) Report of the protein group of the ARC working party on the nutrient requirements of ruminants. In *The Nutrient Requirements of Ruminant Livestock*, Suppl. 1, Commonwealth Agricultural Bureaux, London

Baldwin, R.L. and Allison, M.J. (1983) Rumen metabolism. *Journal of Animal Science*, **57**, Suppl. 2, 461–477

Barney, D.J., Grieve, D.G., MacLeod, G.K. and Young, L.G. (1981) Response of cows to a reduction in dietary crude protein from 17% to 13% during early lactation. *Journal of Dairy Science*, **64**, 25–33

Barrio, J.R., Goetsch, A.L. and Owens, F.N. (1986) Effect of dietary concentrate on *in situ* dry matter and nitrogen disappearance of a variety of feedstuffs. *Journal of Dairy Science*, **69**, 420–430

Beever, D.E., Losada, H.R., Cammell, S.B., Evans, R.T. and Haines, M.J. (1986) Effect of forage species and season on nutrient digestion and supply in grazing cattle. *British Journal of Nutrition*, **56**, 209–225

Beever, D.E., Losada, H.R., Gale, D.L., Spooner, M.C. and Dhanoa, M.S. (1987) The use of monensin or formaldehyde to control the digestion of the nitrogenous constituents of perennial ryegrass (Lolium perenne cv. Melle) and white clover (Trifolium repens cv. Blanca) in the rumen of cattle. *British Journal of Nutrition*, **57**, 57–67

Boulter, D. and Derbyshire, E. (1976) The general properties, classification and distribution of plant protein. In *Plant Proteins*, (ed G. Norton) Butterworths, London, pp. 3–24

Broderick, G.A. (1982) Estimation of protein degradation using *in situ* and *in vitro* methods. In *Protein Requirements for Cattle: Symposium*, (ed F.N. Owens), Stillwater, Oklahoma State University, Stillwater, pp. 72–80

Broderick, G.A. (1986) Relative value of solvent and expeller soybeanmeal for lactating dairy cows. *Journal of Dairy Science*, **69**, 2948–2958

Chamberlain, D.G., Thomas, P.C. and Wait, M.K. (1982) The rate of addition for formic acid to grass at silage and the subsequent digestion of silage in the rumen and intestine of sheep. *Grass and Forage Science*, **37**, 159–164

Chen, G., Russell, J.B. and Sniffen, C.J. (1987) A procedure for measuring peptides in rumen fluid and evidence that peptide uptake can be a rate-limiting step in ruminal protein degradation. *Journal of Dairy Science*, **70**, 2111–1219

Chen, G., Sniffen, C.J. and Russell, J.B. (1987) Concentration and estimated flow of peptides from the rumen of dairy cattle. Effects of protein quantity, protein solubility and feeding frequency. *Journal of Dairy Science*, **70**, 983–992

Cronje, P.B. (1983) Protein degradability of several South African feedstuffs by the artificial fibre bag technique. *South African Journal of Animal Science*, **13**, 225–228

Crooker, B.A., Clark, J.H., Shanks, R.D. and Hatfield, E.E. (1986) Effects of ruminal exposure on the amino acid profile of non treated and formaldehyde treated soybean meal. *Journal of Dairy Science*, **69**, 2628–2657

Crooker, B.A., Sniffen, C.J., Hoover, W.H. and Johnson, L.L. (1978) Solvents for soluble nitrogen measurements. *Journal of Dairy Science*, **61**, 437–447

De Boer, G., Murphy, J.J. and Kennelly, J.J. (1987) Mobile nylon bag for estimating intestinal availability of rumen undegradable protein. *Journal of Dairy Science*, **70**, 977–982

De Boever, J.L., Aerts, J.V., Cottyn, B.G., Vanacker, J.M. and Buysse, F.X. (1984) The in sacco protein degradability vs. protein solubility of concentrate ingredients. *Zeitschrift für Tierphysiologie, Tierernährung und Futtermittelkunde*, **52**, 227–234

Demeyer, D.I. and Tamminga, S. (1987) Microbial protein yield and its prediction. In *Feed Evaluation and Protein Requirement Systems for Ruminants* (eds G. Alderman and R. Jarrige), ECE, Brussels, pp. 129–144

Ensminger, M.E. and Olentine, G.G. (eds) (1978) *Feeds and Nutrition – Complete*, The Ensminger Publishing Company, Clovis, pp. 334

Erasmus, L.J., Prinsloo, J. and Meissner, H.H. (1988) The establishment of a protein degradability data base for dairy cattle using the nylon bag technique. 1.

Protein sources. *South African Journal of Animal Science*, **18**, 23–29

Faichney, G.J. (1975) The use of markers to partition digestion within the gastro-intestinal tract of ruminants. In *Digestion and Metabolism in the Ruminant* (eds I.W. McDonald and A.C.I. Warner), Armidale, University of New England Publishing Unit, Armidale, pp. 277–291

Filmer, D.G. (1982) Assessment of protein degradability of forage. In *Forage Protein in Ruminant Animal Production*, Occasional Publication No. 6. (eds D.J. Thomson, D.E. Beever, and R.G. Gunn), British Society of Animal Production, Thames Ditton, pp. 129–140

Firkins, J.L., Berger, L.L., Merchen, N.R., Fahey, G.C. and Nelson, D.R. (1986) Effects of feed intake and protein on ruminal characteristics and site of digestion in steers. *Journal of Dairy Science*, **69**, 2111–2123

Forster, R.J., Grieve, D.G., Buchanan–Smith, J.G. and MacLeod, G.K. (1983) Effect of dietary protein degradability on cows in early lactation. *Journal of Dairy Science*, **66**, 1653–1662

Ganev, G., Ørskov, E.R. and Smart, R. (1979) The effect of roughage or concentrate feeding and rumen retention time on total degradation of protein in the rumen. *Journal of Agricultural Science, Cambridge*, **93**, 651–656

Goetsch, A.L. and Owens, F.N. (1985) The effects of commercial processing method of cottonseed meal on site and extent of digestion in cattle. *Journal of Animal Science*, **60**, 803–813

Hooper, A.P. and Welch, J.G. (1985) Effects of particle size and forage composition on functional specific gravity. *Journal of Dairy Science*, **68**, 1181–1188

Hvelplund, T. (1984) Intestinal digestion of protein in dairy cows. *Canadian Journal of Animal Science Supplement*, **64**, 193–194

Hvelplund, T. (1985) Digestibility of rumen microbial protein and undegraded dietary protein estimated in the small intestine of sheep and by in sacco procedure. *Acta Agriculturae Scandinavica Supplement*, **25**, 132–144

INRA (1978) *Alimentation des Ruminants*, Versailles, INRA Publications

Janicki, F.J. and Stallings, C.C. (1988) Degradation of crude protein in forages determined by *in vitro* and *in situ* procedures. *Journal of Dairy Science*, **71**, 2220–2448

Kaufmann, W. (1979) Zur Eiweissverdauung bei Wiederkauern im Hinblick auf die faktorielle Berechnung des Eiweissbedarfes. *Zeitschrift für Tierphysiologie, Tierernährung und Futtermittelkunde*, **42**, 326–332

Kennedy, P.M., Hazlewood, G.P. and Milligan, L.P. (1984) A comparison of methods for the estimation of the proportion of microbial nitrogen in duodenal digesta, and of correction for microbial contamination in nylon bags incubated in the rumen of sheep. *British Journal of Nutrition*, **52**, 403–417

Kristensen, E.S., Moller, P.D. and Hvelplund, T. (1983) Estimation of the effective protein degradability in the rumen of cows using the nylon bag technique combined with the outflow rate. *Acta Agriculturae Scandinavica*, **32**, 123–127

Kirkpatrick, B.K. and Kennelly, J.J. (1987) In situ degradability of protein and dry matter from single protein sources and from a total diet. *Journal of Animal Science*, **65**, 567–576

Lindberg, J.E. (1985) Estimation of rumen degradability of feed proteins with the in sacco technique and various *in vitro* methods. *Acta Agriculturae Scandinavica Supplement*, **25**, 64–97

Lynch, G.L., Berger, L.L., Merchen, N.R., Fahey, G.C. and Baker, E.C. (1987) Effects of ethanol and heat treatments of soybean meal and infusion of sodium chloride into the rumen on ruminal degradation and escape of soluble and total soybean meal protein in steers. *Journal of Animal Science*, **65**, 1617–1625

Madsen, J. (1985) The basis for the proposed Nordic protein evaluation system for ruminants. The AAT-PBV system. *Acta Agriculturae Scandinavica Supplement*, **25**, 9–20

Madsen, J. (1986) Influence of feeding level on digestion and protein passage to the duodenum in cows fed high concentrate diets. *Acta Agriculturae Scandinavica*, **36**, 275–285

Madsen, J. and Hvelplund, T. (1985) Protein degradation in the rumen. A comparison between *in vivo*, nylon bag, *in vitro* and buffer measurements. *Acta Agriculturae Scandinavica*, **25**, 103–124

Mangan, J.L. (1982) The nitrogenous constituents of fresh forages. In *Forage Protein in Ruminant Animal Production*, Occasional Publication No. 6, (eds D.J. Thomson, D.E. Beever and R.G. Gunn), British Society of Animal Production, Thames Ditton, pp. 25–40

McAllan, A.B., Lewis, P.E. and Griffith, E.S. (1987) The effects of frequency of feeding on some quantitative aspects of digestion in the rumen of growing steers. *Archives for Animal Nutrition*, **37**, 791–803

McDonald, P. (1982) The effect of conservation processes on the nitrogenous components of forages. In *Forage Protein in Ruminant Animal Production*, Occasional Publication No. 6, (eds D.J. Thomson, D.E. Beever and R.G.Gunn), British Society of Animal Production, pp. 41–49

Mehrez, A.Z. and Ørskov, E.R. (1977) A study of the artificial fibre bag technique for determining the digestibility of feeds in the rumen. *Journal of Agricultural Science, Cambridge*, **88**, 645–650

Moller, P.D. (1985) Results of grass silage based rations on the nitrogen absorption in the gastro-intestinal tract of dairy cows applied to the nordic protein evaluation system. *Acta Agriculturae Scandinavica Supplement*, **25**, 49–63

Murphy, J.J. and Kennelly, J.J. (1987) Effect of protein concentration and protein source on the degradability of dry matter and protein *in situ*. *Journal of Dairy Science*, **70**, 1841–1849

National Research Council (1985) *Ruminant Nitrogen Usage*, Washington, National Academy of Sciences

Nocek, J.E. (1988) *In situ* and other methods to estimate ruminal protein and energy digestibility: A review. *Journal of Dairy Science*, **69**, 77–87

Nocek, J.E. and English, J.E. (1986) *In situ* degradation kinetics: Evaluation of rate determination procedure. *Journal of Dairy Science*, **69**, 77–87

Nocek, J.E. and Grant, A.L. (1987) Characterisation of *in situ* nitrogen and fibre digestion and bacterial nitrogen contamination of hay crop forages preserved at different dry matter percentages. *Journal of Animal Science*, **64**, 552–564

Nocek, J.E. and Russell, J.B. (1988) Protein and energy as an integrated system. Relationship of ruminal protein and carbohydrate available to microbial synthesis and milk production. *Journal of Dairy Science*, **71**, 2070–2107

Oldham, J.D. and Tamminga, S. (1980) Amino acid utilisation by dairy cows. I. Methods of varying amino acid supply. *Livestock Production Science*, **7**, 437–452

Ørskov, E.R. (1982) *Protein Nutrition in Ruminants*, Academic Press, London

Ørskov. E.R. and McDonald, I. (1979) The stimation of protein degradability in the rumen from incubation measurements weighted according to rate of passage. *Journal of Agricultural Science, Cambridge*, **92**, 499–503

Owens, F.N. and Bergen, W.G. (1983) Nitrogen metabolism of ruminant animals: historical perspective, current understanding and future implications. *Journal of Animal Science*, **57**, Suppl. 2, 498–518

Owens, F.N. and Goetsch, A.L. (1986) Digesta passage and microbial protein synthesis. In *Control of Digestion and Metabolism in Ruminants* (eds L.P. Milligan, W.L. Grovum and A. Dobson), Prentice Hall, New York, pp. 196–223

Pichard, G. and Van Soest, P.J. (1977) Protein solubility of ruminant feeds. In *Proceedings of the Cornell Nutrition Conference*, Cornell University, Ithaca, p. 91

Rahnema, S.H., Theurer, B., Garcia, J.A., Young, M.C. and Swingle, R.S. (1987) Site of protein digestion in steers fed sorghum grain diets. 1. Effect of level of feed intake. *Journal of Animal Science*, **64**, 1533–1540

Robinson, P.H. and Sniffen, C.J. (1985) Forestomach and whole tract digestibility for lactating dairy cows as influenced by feeding frequency. *Journal of Dairy Science*, **68**, 857–867

Robinson, P.H., Sniffen, C.J. and Van Soest, P.J. (1985) Influence of level of feed intake on digestion and bacterial yield in the forestomachs of dairy cattle. *Canadian Journal of Animal Science*, **65**, 437–444

Robinson, P.H. and Tamminga, S. (1984) Gegenwärtiger Kentnisstand über die Proteinverdauung und –Absorption bei Wiederkäuern. *Übersicht Tierernärung*, **12**, 119–164

Rooke, J.A. (1985) The nutritive value of feed protein and feed protein resistant to degradation by rumen microorganisms. *Journal of Food Science and Agriculture*, **36**, 629–637

Rooke, J.A., Brookes, I.M. and Armstrong, D.G. (1983) The digestion of untreated and formaldehyde-treatred soya-bean and rapeseed meals by cattle fed a basal silage diet. *Journal of Agricultural Science, Cambridge*, **100**, 329–342

Santos, K.A., Stern, M.D. and Satter, L.D. (1983) Protein degradation in the rumen and amino acid absorption in the small intestine of lactating dairy cattle fed various protein sources. *Journal of Dairy Science*, **58**, 244–255

Satter, L.D. (1986) Protein supply from undegraded dietary protein. *Journal of Dairy Science*, **69**, 2734–2749

Sauer, W.C., Jorgensen, H. and Berzins, R. (1983) A modified nylon bag technique for determining apparent digestibilities of protein in feedstuffs for pigs. *Canadian Journal of Animal Science*, **63**, 233–237

Schwarting, G. and Kaufmann, W. (1978) Die Verdaulichkeit des Proteins beim Wiederkauer. *Zeitschrift für Tierphysiologie Tierernährung und Futtermittelkunde*, **40**, 6–18

Shibui, H., Kawai, T., Katoh, N. and Abe, M. (1983) Degradation properties of feed protein in the rumen of cows fed a high concentrate ration. *Japanese Journal of Zootechnical Science*, **54**, 511–517

Siddons, R.C., Beever, D.E. and Nolan, J.V. (1982) A comparison of methods for the estimation of microbial nitrogen in duodenal digesta of sheep. *British Journal of Nutrition*, **48**, 377–389

Stern, M.D., Santos, K.A. and Satter, L.D. (1985) Protein degradation in rumen and amino acid absorption in small intestine of lactating dairy cows fed heat-treated whole soybeans. *Journal of Dairy Science*, **68**, 45–56

Susmel, P., Stefanen, B., Mills, C.R. and Colitsi, M. (1990) The evaluation of PDI concentrations in some ruminant feedstuffs: a comparison of *in situ* and *in vitro* protein degradability. *Annales de Zootechnie*, **38**, 269–283

Sutherland, T.M. (1986) Particle separation in the forestomachs of sheep. In

Aspects of Digestive Physiology in Ruminants (eds A. Dobson and M.J. Dobson), Comstock Publishing Associates, Ithaca, Chapter 3

Tamminga, S. (1979) Protein degradation in the forestomachs of ruminants. *Journal of Animal Science*, **49**, 1615–1630

Tamminga, S. (1982) Effect of the roughage/concentrate ratio on nitrogen entering the small intestine of dairy cows. *Netherlands Journal of Agricultural Science*, **29**, 273–283

Tamminga, S. (1986) Utilization of naturally occurring NPN-compounds by ruminants. *Archives of Animal Nutrition*, **36**, 169–176

Tamminga, S., Van der Koelen, C.J. and Van Vuuren, A.M. (1979) Effect of level of feed intake on nitrogen entering the small intestine of dairy cows. *Livestock Production Science*, **6**, 255–262

Tamminga, S. and Ketelaar, R. (1988) *Eiwitbestendigheid van voedermiddelen voor herkauwers* (The Resistance against ruminal degradation of feeds for ruminants) Report IVVO no. 192

Tamminga, S., Ketelaar, R. and Van Vuuren, A.M. (1990) Degradation of N in conserved forages in the rumen of dairy cows. *Grass and Forage Science* (in press)

Tamminga, S., Robinson, P.H., Vogt, M. and Boer, H. (1990) Rumen ingesta kinetics of cell wall components in dairy cows. *Animal Feed Science and Technology*, **25**, 89–98

Thomas, J.W., Yu, Y., Middleton, T. and Stallings, C. (1982) In *Protein Requirements for Cattle Symposium* (ed. F.N. Owens), Oklahoma State University, Stillwater, pp. 81–98

Theurer, G.B. (1982) Microbial protein estimation using DAP, AEP and other amino acids as markers. In *Protein Requirements for Cattle: Symposium* (ed F.N. Owens), Stillwater, Oklahoma State University, Stillwater, pp. 10–22

Van Bruchem, J., Bongers, L.J.G.M., Van Walsum, J.D., Onck, W. and Van Adrichem, P.W.M. (1985) Digestion of proteins of varying degradability in sheep. 3. Apparent and true digestibility in the small intestine and ileal endogenous flow of N and amino acids. *Netherlands Journal of Agricultural Science*, **33**, 285–295

Van Der Aar, P.J., Berger, L.L., Fahey, G.C. and Merchen, N.R. (1984) Effect of alcohol treatments of soybean meal on ruminal escape of soybean meal protein. *Journal of Animal Science*, **59**, 483–489

Varvikko, T. and Lindberg, J.E. (1985) Estimation of microbial nitrogen in nylon-bag residues by feed 15N dilution. *British Journal of Nutrition*, **54**, 473–481

Verite, R., Michalet–Doreau, B., Chapoutot, P., Peyraud, J.L. and Poncet, C. (1987) Revision du systeme des Proteines Digestible dans l'Intestine (PDI). *Bulletin Technique CRZV*, **70**, 19–34

Voigt, J., Piatkowski, B., Engelmann, H. and Rudolph, E. (1985) Measurement of the postruminal digestibility of crude protein by the bag technique in cows. *Archiv für Tierernährung Berlin*, **35**, 555–562

Wohlt, J.E., Sniffen, C.J. and Hoover, W.H. (1973) Measurement of protein solubility in common feedstuffs. *Journal of Dairy Science*, **56**, 1052–1057

Zinn, R.A. and Owens, F.N. (1983) Site of protein digestion in steers: predictability. *Journal of Animal Science*, **56**, 707–716

5

ANIMAL PERFORMANCE AS THE CRITERION FOR FEED EVALUATION

J.D. OLDHAM and G.C. EMMANS
Edinburgh School of Agriculture West Mains Road, Edinburgh EH9 3JG

Introduction

Nutritional evaluation of feedstuffs is undertaken for different purposes. The main ones are:

1. To measure the extent to which one feed can replace another to support an animal function (i.e. a relative ranking of feedstuffs).
2. To relate feed attributes to animal functions (i.e. to give absolute values to feeds scaled according to an identified function, e.g. provision of a first-limiting amino acid to support protein gain).
3. To allow the prediction and/or control of animal performance through nutrition.

Central to each of these is the near truism that nutritional values for feedstuffs are only useful in the context of a description of animal function and performance. Animal performance is therefore the essential criterion by which the relative and absolute nutritional values of feedstuffs are to be judged. The units of measurement used for feed values and animal functions must be consistent. The relevant animal functions are: maintenance, activity, growth of body protein and lipid and the secretion of the milk constituents proteins, lipid and lactose.

Systems for assesssing the nutrient and energy needs of animals (e.g. ARC, 1980, 1981) provide the framework through which feed values can be used or assessed and for which the feed values are required. Success in matching animal performance to current estimates of the value of feeds is therefore a reflection both of the accuracy and precision of the estimate of feed value and of the nutritional system within which that value is to be used. Part of this paper must therefore be concerned with an appraisal of the adequacy of current nutritional systems for the achievement of particular animal performances. This applies to energy evaluation.

Success in matching feed values to animal requirements is, however, only a first step in achieving the control of animal performance through nutrition. The subsequent step of becoming able to predict animal performance from a knowledge of feed attributes depends on a satisfactory solution of another set of problems which are to do with the partition of available nutrients and energy amongst various animal functions. This problem has been discussed previously (Oldham and Emmans, 1988, 1989) during which the importance of quantitative descriptions of animal function, and potential, for response prediction to become achieveable, was emphasized. The subject of partition is also relevant within the context of this chapter.

A considerable portion of that considered subsequently relates to ruminants in particular, although reference is made to other species as appropriate. The issues of nutritional accounting in ruminants, which follow largely from the involvement of rumen fermentation in the digestive process, pose particular problems for nutritional assessment and therefore deserve special attention. In general, though, it is not considered necessary to place ruminant nutrition in a separate category as regards general theory. Once the problem of relating food characteristics to nutrient and energy supply is solved for a ruminant animal, the rules for predicting just how these nutrients and energy are used by the animal should be the same in both ruminant and non-ruminant species.

WHAT IS ANIMAL PERFORMANCE?

In this context, the most practically relevant aspects of animal performance are those which are germane to the production of saleable animal products. These are always presented as rates. Thus rates of milk volume production, of the secretion of the major milk constituents (fat, protein, lactose) of change in weights of fat and lean tissue in the body, and of the growth of fibre (wool, hair) are all main aspects of animal performance. The additional processes of maintenance and of the growth of reproductive tissues and the fetus are clearly also relevant, although these are not directly saleable commodities. Food consumption is, arguably *the* key aspect of animal performance, and the evaluation of feedstuffs in terms of their 'intake characteristics' cannot be ignored. However, food consumption has been considered as being beyond the scope of this present commentary, having been emphasized elsewhere (Oldham and Emmans, 1988, 1989) and in other chapters of this book.

BODY PROTEIN AND FAT CHANGES IN GROWTH AND LACTATION

A particular issue in nutritional accounting is to do with our ability to measure relative changes in protein and fat in the body during growth and in lactation. Weight change as a measure of growth has long been recognized as an inexact and frequently misleading measure. Change in the mass of digesta held in the gastrointestinal (GI) tract ('Gut fill') can mislead as to changes in tissue weight. Furthermore, within the tissues variation in the relative proportions of protein (and associated water) and fat which are gained (or lost) can mislead as to the energy content of the weight change. Considerable efforts are now being put into the development of accurate and precise techniques for the measurement of body composition (Lister, 1984; Kallweit *et al.*, 1989) in order to improve estimates of the composition of body weight change in the live animal, but this remains as a major problem area in the definition of performance and hence in the evaluation of feedstuffs when assessed in relation to performance.

In growing animals empty weight change is due to changes in water, ash, protein and lipid weights. The relative amounts of water, lipid and lipid-free dry matter in the empty body vary with age in a predictable manner (Figure 5.1).

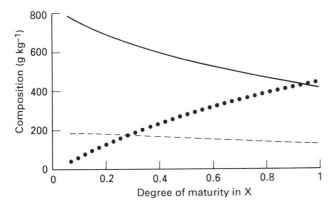

Figure 5.1 A general picture of changes in the relative amounts of water (——), lipid (⋯⋯) and lipid-free dry matter (– – –) in the empty body of growing animals

Energy retention during growth is almost entirely accounted for by protein and lipid changes as glycogen normally accounts for only 1% or less of energy retention.

In growing animals protein and lipid accretion normally occur concurrently – the relationship between the two generally varying with plane of feeding and degree of maturity (Whittemore, 1985). However, protein accretion can occur in association with net catabolism of body lipid and with the animal in near zero or even negative energy balance. Changes in protein:energy proportions in the diet can be used to change, quite sensitively, the protein:lipid proportions in weight gain by pigs (Table 5.1). Such variation is not confined to non-ruminant species. For example, the data in Figure 5.2 show that the fat:protein ratio can vary with nutrition in growing sheep.

Table 5.1 FOOD CONSUMPTION, WEIGHT CHANGE AND BODY LIPID CONTENT OF YOUNG PIGS OFFERED DIETS OF LOW OR HIGH PROTEIN CONTENT FROM 9–16 kg WEIGHT (DATA OF KYRIAZAKIS *et al.*, 1988)

Diet CP content (g/kg)	*134*	*278*
Weight change (g/d)	386	591
Food intake (g/d)	749	666
Body lipid (kg) at 16 kg liveweight	2.97	1.92

Conventional wisdom, as expressed in ARC (1980), recognizes that in growing ruminants the protein and fat contents of weight change, and hence its energy content vary both with the rate of weight change and the weight of the animal (a reflection of its degree of maturity) at the time. No recognition is given to the possibility of variation in the protein:fat proportion in weight change at a given rate of weight change, at a weight. Different forms of relationship to predict the energy content of weight change are used for cattle and sheep:

Cattle $E\Delta_W = (4.1 + 3.32 \times 10^{-2}\,W - 9 \times 10^{-6}W^2)\,(1 - 0.1474\,W)$ MJ/kg
(castrates)

Sheep $E\Delta_W = 44 + 0.32W$ MJ/kg
(castrates)

where $E\Delta_W$ = energy content of weight change (MJ/kg)
        W   = weight (kg).

Separate forms of relationship for cattle and sheep are not supportable, given a general view of growth in ruminants as a class. It is also not supportable for the assumption to be made that weight alone determines the energy content of weight change in sheep (Figure 5.2), nor that energy content of weight change necessarily increases with growth rate in cattle. Gill *et al.* (1987), with growing cattle, have shown that supplementation of silage diets with fishmeal will result in changes both in the rate and nature of growth in cattle such that enhanced growth rate can be achieved with a reduction in the energy content of that growth when measured as MJ per kg weight change.

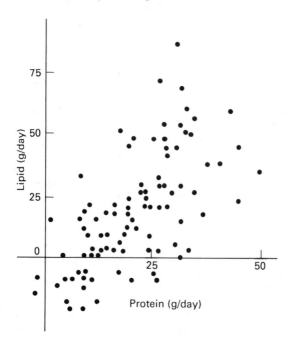

Figure 5.2   Gain (g/d) of protein and fat in the empty bodies of Blackface wether lambs offered diets based on barley, sugar-beet feed and rice hull/dried grass mixtures in different proportions (Data of Emmans, Dingwall and Oldham, unpublished observations.)

Similar problems arise in the assessment of the nature and energy value of weight change during lactation in cattle. Alderman *et al.* (1982) have estimated the energy content of weight change during lactation to be 39.5 ± 11 MJ/kg – a mean value close to that to be expected if only fat were being mobilized and replaced; but with such a high degree of variability that the average value has restricted use. From carbon and nitrogen balance studies with milking cows, Van der Honing (1975) (Figure 5.3) provides an illustration of the extent to which these enormous variations in apparent energy content of liveweight change might be accountable on

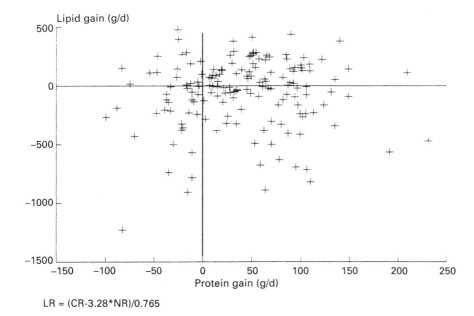

LR = (CR-3.28*NR)/0.765

Figure 5.3 Rates of gain or loss of body protein and body fat of lactating cows. The data (from Van der Honing, 1975, and unpublished observations) are from carbon and nitrogen balances conducted as part of calorimetric assessments of the energy values of a wide range of feeds

the basis of varying changes (both positive and/or negative) in the relative amounts of lipid and protein being deposited or lost from the body. As all combinations of positive and negative gains of protein and fat are possible (with implications for water associated with protein) accounting for the energy conent of weight changes in lactation with one single value (ARC, 1980) is clearly unrealistic.

As discussed below, variations in the proportion of protein and fat associated with liveweight change are important because the energy costs associated with the energy gains as these two constituents are not the same. Hence, evaluation of feeds in terms which do not require these different attributes of performance and their associated costs to be identified explicitly are bound to be flawed, certainly in concept and likely also in precision and accuracy.

Energy value of feeds – the Edinburgh energy system

Metabolizable energy is the most common unit for ascribing energy values to feedstuffs. The manner in which ME is related to animal performance generally does not recognize changes in protein and lipid as separate 'performance products'. Rather, they are combined into an estimate of energy retention. This simplification is not acceptable and it does not allow for identified variations in the relative rates of deposition of these two products. Emmans (1990) has addressed this issue and has proposed a scheme for accounting for the conversion of metabolizable energy into major animal products which accounts satisfactorily for heat production on the

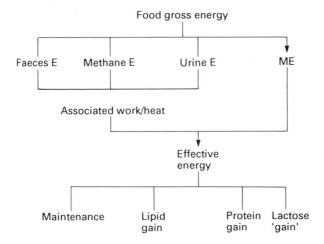

Figure 5.4   A general scheme for accounting for the use of the metabolizable energy of feedstuffs in support of functions of maintenance, the loss of products by excretion and the deposition of major animal products (protein, lipid, lactose)

basis of unit processes and which leads to alternative and more effective means for assessing the nutritional value of feeds, based on descriptors of animal performance.

The system is a development from the well-known description of metabolizable energy content of feedstuffs. The scheme (Figure 5.4) allows for the use of metabolizable energy for purposes of maintenance, of work associated with protein and lipid retention (and for the energy content of the deposited protein and lipid) and for the work associated with processes of digestion and execretion of waste products. Under these three headings the basic elements of the scheme are as follows:

1. *Maintenance* – Emmans and Fisher (1986) have discussed approaches to the assessment of maintenance. A main part of the energy costs associated with tissue 'maintenance' results from the continual process of synthesis, degradation and replacement of those parts of body tissues which 'turn over'. This is particularly the case with body protein for which the process of 'turnover' is substantial, though variable (Reeds, 1989). Fat tissue appears not to 'turn over' at all in animals fed regularly (Metz *et al.*, 1974) although there does appear to be an extent of fatty acid turnover which is obligatory and which might be presumed to represent a degree of turnover of body fat (Konig *et al.*, 1979; Wilson, 1983). The energy cost of maintaining body protein would, however, be expected (on stoichiometric grounds) to exceed that of fat even if their rates of turnover were similar. On these grounds it is biologically unreasonable to expect maintenance to be precisely related to scaled body weight when the composition of that weight of body may vary in its protein and fat contents. Hence, based on developments of scaling rules initiated by Taylor (1981) Emmans and Fisher (1986) have suggested that maintenance should be scaled to a function of current protein mass and potential mature protein mass according to:

Maintenance energy = $1.63 \times P_m^{-0.27} \times P$ MJ/d
where P_m = mature protein mass kg, P = current protein mass kg.

This method of scaling matches closely the more conventional scaling to body weight$^{0.75}$ (or weight$^{0.73}$) in many instances, but can more satisfactorily account for variation in maintenance energy where fatness changes, for example, in dairy cows in early lactation.

2. *Energy costs associated with protein and lipid gain* – There is widespread agreement that the energy cost (as metabolic work done) associated with protein energy accretion is higher than that associated with lipid energy accretion (Kielanowski, 1965; Pullar and Webster, 1977; Reeds *et al.*, 1982).

   The work associated with protein accretion may vary according to the extent to which total protein synthesis exceeds net protein accretion (Reeds *et al.*, 1981). Working values for heat associated with protein and lipid synthesis are 36.5 kJ per g protein deposited and 16.4 kJ per g fat deposited except where feed lipid is used for lipid retention where the value of 4.4 kJ/g is suggested.

3. *Work as heat production associated with the processes of digestion and excretion of products* – In converting gross energy to metabolizable energy, the energy values of faeces voided, methane produced as a result of fermentation, and urine are subtracted from the gross energy of a particular food. The process of fermentation which leads to methane production is heat generating and this heat of fermentation is expressed here as a multiple of methane production. Variation in the stoichiometry of fermentation may lead to variations in heat produced per unit of methane. For example, from consideration of possible rumen stoichiometries, Wolin and Miller (1989) have suggested that methane production per mole of hexose fermented might vary at least over the range of 0.48–0.64 moles/mole and with associated variation in the relative proportions of VFA yielded as the other end-products of the rumen fermentation. The possibility of variation in the heat produced per unit of methane production is recognized, but a value of 0.616 MJ heat/MJ methane produced appears to be a useful working value.

   The products of protein catabolism which appear in urine have both a gross energy conent (a predictable function of nitrogen excretion) and an associated energy cost of synthesizing and excreting those products via the kidneys. The major energy charge here is that due to the heat produced during synthesis of urea, from ammonia (Martin and Blaxter, 1965); an overall estimate of 29.2 kJ/g nitrogen excreted appears to be a fair working value. This would vary with the extent to which urea, once synthesized, recycles to the gastrointestinal tract such that its nitrogen returns to the liver as ammonia for resynthesis into urea. Urea synthesis:excretion ratios can vary up to 3:1 (Oldham and Lindsay, 1983) so that in extreme circumstances the cost of urinary excretion may be substantially higher than that indicated here, although this would generally have only a small effect on the overall estimate of heat production (perhaps ranging from 3% to 8% of total heat production, depending on the extent of urea recycling).

   Estimates of the work, or heat produced, in association with faeces energy excretion is more contentious. Webster (1984) explored the origin of various possible heats of digestion, including energy costs of eating, rumination and of digestion in ruminants. The approach taken here (Emmans, 1990) has been to relate residual heat production (after other processes have been accounted for) to faecal organic matter in cattle and chickens. Both estimates suggest a value of 3.80 kJ/g faecal organic matter excretion as a general estimate of the energy costs associated with this process (see below). This term lead to the ME from feeds of lower digestibility having a lower value to the animal, other things being equal.

The three elements allow the metabolizable energy of the diet, first corrected to zero N retention, to be partitioned:

N-corrected ME $= \text{ME}_n = \text{ME} - (5.63 \times 6.25)$ NR   MJ/d

$\text{ME}_n = 29.2 \, (\text{DN} - \text{NR}) + 0.616 \text{ methane} + 3.8 \text{ FOM}$
$\qquad + 1.63 \, \text{P/P}_m{}^{0.27}$
$\qquad + (36.5 - (0.16 \times 29.2)) \text{ protein retention} + 16.4 \text{ lipid retention}$
$\qquad + (23.8 - 5.63) \text{ protein retention} + 39.6 \text{ lipid retention}$

where DN $=$ digested N, kg/d and NR $=$ positive nitrogen retention, kg/d. (5.1)

The values of the heat increment coefficients – the five values 29.2, 0.616, 3.8, 36.5 and 16.4 in equation 5.1 – were derived from the 15 calorimetric experiments on cattle given in Table 5.2.

Table 5.2   DATA FROM CALORIMETRIC EXPERIMENTS WITH CATTLE USED TO DERIVE COEFFICIENTS IN QUATION 5.1. VALUES ARE FOR INCREMENTS IN HI, UNI, MTHEI, FOMI, PRI, LRI BETWEEN A PAIR OF DIETS

Experiment	HI (kJ)	UNI (g)	MTHEI (kJ)	FOMI (g)	PRI (g)	LRI (g)	Source
1	2056	−13.1	3738	−101	22.8	−17.8	Kellner and Kohler (1900)
2	4525	5.3	2219	328	27.3	44.5	Forbes *et al.* (1928, 1930)
3	6528	36.7	1182	1261	5.5	24.1	Forbes *et al.* (1928, 1930)
4	7212	−22.2	1750	235	39.5	269.8	Kellner and Kohler (1900)
5	7374	9.1	1191	870	25.9	130.4	Fingerling (1944)
6	7407	169.0	−400	52	38.8	65.1	Kellner and Kohler (1900)
7	7480	28.8	2260	725	4.4	144.4	Armsby and Fries (1918)
8	9079	45.9	3479	1531	16.9	−45.2	Forbes *et al.* (1931)
9	9146	−13.2	2346	1076	49.5	126.6	Fingerling *et al.* (1936)
10	9847	−4.1	1541	425	17.4	411.9	Fingerling (1933)
11	10119	176.5	480	1	79.7	106.6	Kellner and Kohler (1900)
12	12779	18.6	3254	545	131.0	208.6	Blaxter *et al.* (1966)
13	15844	12.3	4120	586	64.4	503.2	Nehring *et al.* (1961)
14	23190	54.3	3640	1151	110.9	673.6	Armsby and Fries (1917)
15	27543	32.4	10490	1881	115.4	535.3	Forbes *et al.* (1928, 1930)

HI = heat (kJ); UNI = urinary nitrogen (g); MTHEI = methane energy (kJ); FOMI = faecal organic matter (g); PRI = positive protein retention (g); LRI = positive lipid retention (g)

Using values for heat associated with nitrogen excretion (29.2), methane production (0.616), protein retention (36.5) and lipid retention (16.4) the heat increment unaccounted for can be derived for the 15 experiments in Table 5.2. Figure 5.6 shows the relationship between these residual heat increments and the increments in faecal organic matter, FOM. The regression coefficient is 3.80 (s.e. 0.05) MJ/d. The term for faecal organic matter is equivalent to saying that there is a work of consuming feed organic matter, net of other effects, which is directly proportional to $(1 - D)$ were D is the digestibility of the organic matter.

As the value of the coefficient for FOM was originally estimated from chicken data as 3.8 MJ/kg the derivation of the same number from cattle data suggests some underlying biological rationale for the inclusion of this term.

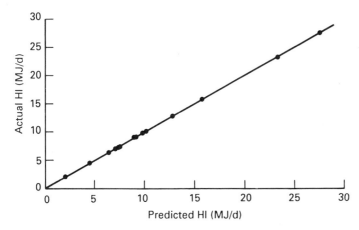

Figure 5.5 Relationship between heat increment predicted from equation 5.1 (see text) and measured heat increment in experiments described in Table 5.2

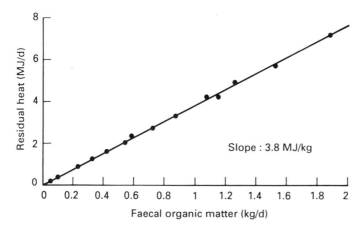

R = H − 29.2 * UN − 0.616 * MTHE − 36.5 * PR − 16.4 * LR

Figure 5.6 Relationship between residual heat increment from data in Table 5.2 and faecal organic matter (FOM). Residual heat increment is actual heat increment minus predicted heat increment (using equation 5.1) minus heat increment associated with FOM in equation 5.1

Effective energy values of feedstuffs

That part of the metabolizable energy content of a feedstuff which is available for the support of maintenance and tissue gain and/or milk secretion is termed 'effective energy' (EE). The effective energy value of the feedstuff can be calculated from other characteristics of the food as follows:

1.   By definition:
EE content = ME content$_n$ −0.616 Methane −3.8 FOM −4.67 DCPC
MJ/kg
where 4.67 = 0.16 × 29.2 and 0.16 is the N content of protein. Methane and
FOM are yields of methane (MJ) and FOM (kg) per kg of food and DCPC is
Digestible Crude Protein Concentration kg/kg; ME content$_n$ = ME concentra-
tion (MJ/kg) adjusted to zero N retention.
2.   As a useful approximation across feeds for ruminants the EE content can be
calculated as:
EE = a.ME$_n$ −b −4.67 DCPC MJ/kg
From publications of the Rowett Feed Evaluation Unit, a and b can be
estimated to be 1.17 and 3.93 respectively.

This relationship has been found to apply usefully across a wide range of feedstuffs
which vary in fat and protein content as well as in digestibility.

The value of feeds to support growth

In ruminants, the efficiency with which metabolizable energy above maintenance is
used for energy retention during growth (k_f) has been estimated to be a function of
the metabolizability (q = ME/GE) of the diet (ARC, 1980). The relationship
between k_f and q is deemed to vary with the category of diet on offer, there being
different relationship defined for forages, aftermaths, mixed diets and pelletted
feedstuffs. The partial efficiency, k_f is not suggested in ARC (1980) to vary with the
nature of growth (as relative rates of protein and fat gain) – except that, by
implication, diets of different q might be supposed to support different relative
rates of protein and fat gain at a particular plane of feeding. This is despite the
widespread understanding that the work efficiency of protein deposition is
substantially less than that for fat deposition.
    The various relationships in ARC (1980) between k_f and q for different classes of
feedstuff reflect the value of these different feed classes for the support of animal
performance. There is evidence that these efficiencies (k_f) are sensitive to protein
supply (MacRae *et al.*, 1985; Gill *et al.*, 1987; Ortigues *et al.*, 1989). Testing these
relationships against performance data is often compromised by a lack of
appropriate data on true energy retention during growth – and especially the
composition of growth (as found by the Working Party which recently evaluated
the ARC, 1980 ME system MAFF, 1988)).
    The value of straw-rich diets for the support of growth (as weight gain) in
Friesian heifers has been found to vary substantially with the inclusion of fishmeal
supplements (Smith *et al.*, 1980) at a given level of digestible organic matter
(DOM) intake. In this instance the composition of weight change would be
expected to have changed, but was not measured. Protein accretion (and associated
water) increased with fishmeal supplementation of silage diets for cattle, with no
variation in fat accretion (Gill *et al.*, 1987) in one experiment where growth studies
were argumented with carcass analysis. In an initial evaluation of barley and
sugar-beet feed for growth in lambs, variations in fat accretion but no variation in
protein accretion as a result of varying the form of dietary carbohydrate in
complete pelletted diets of constant metabolizability have been found (Emmans *et
al.*, 1989).

Each of these examples points to insufficiencies in current description of feeds when tested against animal performance.

Some recognition that the rate of protein deposition is important in determining the partial energetic efficiency of growth has recently been given by Blaxter (1988). In reference to Blaxter and Boynes (1978) interpretation of k_f a new term has been coined, k_{f+p} (the partial energtic efficiency of energy retention for 'growth' (as fat + protein gain) above maintenance) such that:

$k_{f+p} = 0.951 + 0.00037 \ (P/q) - 0.336/q$
where P = protein content of the OM of the diet (g/kg)

This is a step towards a possible revision of the current ARC (1980) ME system which may be helpful to move closer to a proper inclusion of protein accretion as an aspect of animal performance which impinges on the approach to feedstuff evaluation. The Edinburgh energy system described previously does, however, seem to accommodate many of the currently perceived shortfalls in prediction. In order to test it properly, further independent lines of experimentation would be needed which document fully and accurately the necessary component elements.

Associative effects

All of the foregoing assumes that feed values for performance are additive. There are, however, some associative effects between feed values which influence animal performance in ways which are not predictable on the basis of strict additivity. Two particular instances which are worth discussion, are:

1. In ruminants, the effects of specific nutrients in the diet on fibre digestibility and associated food intake.
2. The influence of nutrient balance on nutrient partition.

These will be discussed in turn.

DIETARY NUTRIENT BALANCE, FIBRE DIGESTIBILITY AND FOOD INTAKE IN RUMINANTS

Perhaps the most striking influence of specific dietary nutrients on fibre digestibility in ruminant rations is that which can result from variation in the nitrogen, and especially the rumen degradable nitrogen, content of the diet. Campling *et al.* (1966) provided an elegant example of the manner in which variation in dietary urea supply can affect the digestibility of straw by cattle, with consequent effects on voluntary intake. Hence the presence of one dietary constituent (urea in this instance) has an influence on the nutritional value (digestibility) of another food ingredient (straw in this instance) and, as an associated phenomenon, the amount of that other food constituent which the animal is able to eat. Explanation of the processes by which these phenomena come about lies in the manner in which the rumen microbial population responsible for carbohydrate fermentation varies in its nature and activity according to the form of food eaten and the balance of nutrients available from that food in the rumen. Of particular importance is the form of fermentable carbohydrate (Cheeson and Forsberg, 1989) the amount and form of dietary crude protein (Oldham, 1984) and the provision of particular minerals (Durand and Kawashima, 1980).

The consequence of these interactive processes on the digestibility and intake of feeds can be substantial and important in determining performance of the animal. For example, Oldham (1984) from a simple survey of published data for dairy cows, estimated the average response in food dry matter digestibility and food dry matter intake in dairy cows for each increment of dietary protein concentration over a range of dietary protein concentrations from about 100 to 160 g CP/kg DM. Translated into an estimated response in ME intake, the combined response was estimated as:

$$\Delta ME / \Delta\ CP\% = 14.8\ (0.007\ DMI + 0.19\ DMd + 0.00013)\ MJ/d/unit$$

where DMI = dry matter intake of the lower protein food, DMd is DM digestibility of that food, ΔME = increment in ME intake (MJ/d), $\Delta CP\%$ = increment in concentration (as % units) of crude protein in overall ration dry matter.

This is equivalent to an enhancement of 4–5 MJ ME/increment of 10 g crude protein/kg DM for rations based on grass silage, and 7–8 MJ ME/increment of 10 g CP/kg DM for rations based on maize silage. Changes of this magnitude in ME intake have practical significance for animal performance. At the moment they can only be evaluated through measurements of animal performance, although in the longer term, one might expect such phenomena to become predictable through the medium of adequate models of digestive function which will develop from the current generation of such models, e.g. France *et al.* (1982).

Promotion of the digestibility and ingestion of fibrous foods by ruminants is not just related to the provision of different amounts of dietary crude protein. Small additions of fermentable carbohydrate have been found to enhance intake (and sometimes also digestibility) of forages by dairy cows (Thomas, 1987) and lambs (Eayres *et al.*, 1989) although larger additions of such feeds frequently result in reduced intakes of forages (Thomas, 1987). This effect is usually referred to as 'substitution' or 'replacement' of forage by concentrate. While there are some instances where the substitution effect might be explained on simple additive grounds of supplemental 'fibre' replacing forage 'fibre' on a 1:1 basis (e.g. as appeared to happen for acid detergent fibre in work reported by Aitcheson *et al.*, 1986) this is certainly not a general explanation. Substitution phenomena must therefore be included in the class of interactive effects which have marked relevance for animal performance and for which the criterion of evaluation is, certainly, solely through the measurement of animal performance.

PARTITION

The balance of nutrients made available from food is a major factor which determines the partition of those nutrients between alternative synthetic pathways. Other major partitioning 'factors' are the animal's genotype and its current state (Oldham and Emmans, 1988). The value of a particular food will therefore depend on the profile of nutrients made available from it alone, as a consequence of digestion, together with the profile of nutrients made available from other foods. This concept is now conventionally accepted in relation to the balancing of amino acid profiles, where provision of supplementary amino acids can enhance the value of imbalanced protein sources enormously in terms of their effects on animal performance, especially growth (ARC, 1981). It seems reasonable to extend this principle to any nutrient which relies on the provision of another nutrient to determine its use for particular pathways of metabolism. In the case of amino acid metabolism the main partition is between catabolism (oxidation and/or gluconeogenesis) or synthesis of protein – although partial catabolic products of amino acid degradation might be deposited as fat rather than undergoing complete oxidation (Reeds, 1989; Wahle *et al.*, 1982). Such partitions imply variation in the efficiency of use of the nutrients in question.

An alternative form of partition which need not imply much, if any, variation in energetic efficiency of nutrient retention is that between deposition of nutrients in body tissue or in the constituents of milk. Broster *et al.* (1985) provided a particularly well documented example of the manner in which variation in forage:concentrate ratio in the diet of dairy cows can result in widely different patterns of performance (as measured by milk constituent output and body tissue gain) when digestible energy and crude protein intake were essentially equal between treatments. Variation in the partition of nutrient precursors of fat between milk or body fat synthesis appears to be at the root of this particular form of nutrient partition. The origin of differences in the amounts of milk fat and body fat which are produced seems to depend on the balance of fermentation end-products, especially the volatile fatty acids, acetic (A), propionic (p), and butyric (B) acids (Sutton, 1984). This issue has received a lot of research attention, mainly in relation to the control of 'low milk fat syndrome'. The relative amounts of A, P and B acids yielded on the fermentation of a particular foodstuff are not yet predictable with confidence. Accordingly, it cannot be stated, at least currently, that these (relative) amounts change with feeding circumstances. But it does seem reasonable to suppose that augmentation of the A, P, B 'profile' from one food with that of another could alter the manner in which the overall balance of these products is used either for milk or tissue constituent synthesis. This particular issue of nutrient balance in relation to partition is therefore a good example of, likely, non-additivity in feedstuff values (and one which it is important to solve). Fat partition might also be subject to the balance of dietary metabolizable protein:metabolizable energy in relation to the profile of the nutrients made available from metabolized body tissue (Orskov *et al.*, 1981; Whitelaw *et al.*, 1986). That partition might also be subject to the form and extent of supplementation with dietary lipid (Bines *et al.*, 1978; Storry, 1980).

Not only is the partition of fat precursors between alternative routes of synthesis an important determinant of the nature of animal performance achieved, so too can be the partition of supplementary protein between alternative routes. This might sometimes be a function of the effect of protein supply on amino acid balance (as referred to above). In the case of the dairy cow (Whitelaw *et al.*, 1986) it might also be a reflection of the extent to which potentially labile reserves of protein are replete or depleted (Oldham and Emmans, 1988).

These few particular references to performance studies in which manipulation of dietary composition has resulted in changes in the partition of nutrient use between different aspects of animal performance underline a remaining general problem in relating feed values to animal performance. For lactation in cows, energy accounting by comparing overall net energy retention (milk+tissue) with ME intake using the rules of ARC, 1980 is generally good (MAFF, 1988). But it remains impossible to predict the partition of food ME between alternative routes (milk, body tissue) for net energy retention. 'Hindsight' is, currently, the only way to make such 'predictions'. Hence, the criterion by which the value of such feedstuff mixtures can be assessed remains the measure of animal performance.

Alterations in nutrient balance which influence partition are also likely to influence food consumption (Forbes, 1986). The consequences for animal performances can be substantial. In a series of experiments in which highly digestible diets based on mixtures of casein, sucrose, starch and corn oil, were offered *ad libitum* to lactating rats (Oldham and Friggens, 1989; and Friggens, unpublished observations) the consequences for performance were very wide ranging (Figure 5.7). At similar rates of energy consumption, maternal weight changes and lactational performance (as indexed by pup growth in standard litters) could be quite variable. The effect of substituting carbohydrate for lipid varied according to the level of protein present in the diet. Maternal tissue change and lactational performance were each markedly influenced by food intake as one would expect.

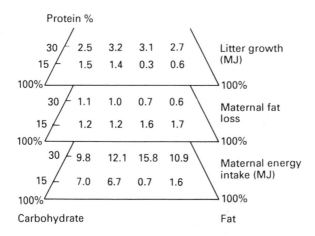

Figure 5.7   Maternal energy intake (MJ), maternal fat loss (MJ) and litter growth (MJ) over the first 10 days post-partum in lactating rats and their litters (of 10 pups) when mothers had *ad libitum* access to diets containing various proportions of protein:carbohydrate:lipid. The data are presented on triangular co-ordinates with the carbohydrate-fat axis at the base and protein content varying on the vertical axis (Data of Friggens, unpublished observations.)

The value of the feed used in these studies was not predictable from chemical descriptors of the foods alone. This, like other examples presented, shows clearly that a combination of chemical evaluation and measurement of animal performance is needed to get a full appraisal of the value of particualr feedstuffs to sustain particular animal performances.

This is not an ideal situation as the purpose of food evaluation is not to assist in the accounting of past events; rather it is to enable the prediction and promotion of future ones. To this end, the improvements which are needed to existing nutritional frameworks are ones which will enable improved prediction of nutrient and energy partition. While nutritional characteristics of feedstuffs constitute a main part of this, the nature of the animal (its genotype and its current state or condition) must also be recognized.

The interpretation of feed value as assessed by animal performance can therefore only be complete if one also has a correct quantitive understanding of the nature of the animal. In evaluating foods, we are therefore inescapably driven to combine knowledge of animal biology and of food chemistry. Each of these separately is interesting but not necessarily useful for the prediction of animal performance. Taken together they can lead to improved nutritional understanding.

References

Agricultural Research Council (1980) *The Nutrient Requirements of Ruminant Livestock*. Commonwealth Agricultural Bureaux, Farnham Royal

Agricultural Research Council (1981) *The Nutrient Requirements of Pigs*. Commonwealth Agricultural Bureaux, Farnham Royal

Aitchison, E.M., Gill, M., Dhanoa, M.S. and Osbourn, D.F. (1986) The effect of digestibility and forage species on the removal of digesta from the rumen and the voluntary intake of hay by sheep. *British Journal of Nutrition*, **56**, 463–476

Alderman, G., Broster, W.H, Strickland, M.J. and Johnson, C.L. (1982) The estimation of the energy value of liveweight change in the dairy cow. *Livestock Production Science*, **9**, 665–673

Armsby, H.P. and Fries, J.A. (1917) Influence of the degree of fatness of cattle upon their utilisation of feed. *Journal of Agricultural Research*, **XI**, 451–472

Armsby, H.P. and Fries, J.A. (1918) Net energy values of alfalfa hay and of starch. *Journal of Agricultural Research*, **XV**, 269–286

Bines, J.A., Brumby, P.E., Storry, J.E., Fulford, Rosemary, J. and Braithwaite, G.D. (1978) The effect of protected lipids on nutrient intakes, blood and rumen metabolites and milk secretion in dairy cows during early lactation. *Journal of Agricultural Science, Cambridge*, **91**, 135–150

Blaxter, K.L. (1988) *Energy Metabolism in Animals and Man*. Cambridge Univeristy Press

Blaxter, K.L., Clapperton, J.L. and Wainman, F.W. (1966) Utilisation of the energy and protein of the same diet by cattle of different ages. *Journal of Agricultural Science, Cambridge*, **67**, 76–75

Broster, W.H., Sutton, J.D., Bines, J.A., *et al.* (1985) The influence of plane of nutrition and diet composition on the performance of dairy cows. *Journal of Agricultural Science, Cambridge*, **104**, 535–537

Broster, W.H., Sutton, J.D., Bines, J.A. (1985) The influence of plane of nutrition and diet composition on the performance of dairy cows. *Journal of Agricultural Science, Cambridge*, **104**, 535–537

Campling, R.C., Freer, M. and Balch, C.C. (1966) Factors affecting the voluntary intake of food by cows. 3. The effect of urea on the voluntary intake of oat straw. *British Journal of Nutrition*, **16**, 115

Chesson, A. and Forsberg, C.W. (1989) Polysaccharide degradation by rumen microorganisms. In *The Rumen Microbial Ecosystem* (ed P.N. Hobson), Elsevier, London, pp. 231–234

Durand, M. and Kawashima, R. (1980) Influence of minerals on rumen microbial digestion. In *Digestive Physiology and Metabolism in Ruminants* (eds. Y. Ruckebusch and P. Thivend), MTP Press, Lancaster, pp. 375–408

Eayres, H., Anderson, D.H. and Oldham, J.D. (1989) The effect of the nature, level and frequency of feeding of supplement on the voluntary forage intake of wether sheep. *Animal Production*, **48**, 635

Emmans, G.C. (1990) An effective energy scale for rationing farm animals. (In preparation)

Emmans, G.C. and Fisher, C. (1986) Problems in nutritional theory. In *Nutrient Requirements of Poultry and Nutritional Research* (eds C. Fisher and K.N. Boorman), London, Butterworths, pp. 9–39

Emmans, G.C., Cropper, M.R., Dingwall, W.S., Brown, H. and Oldham, J.D. (1989) Efficiencies of use of the metabolisable energy from foods based on barley or sugar-beet feed in immature sheep. *Animal Production*, **48**, 634

Fingerling, G. (1933) Der Nahrwert von Kartoffel flochen und Kartoffel schnitzeln. *Laudw. Versuch sstat.*, **114**, 1–112

Fingerling, G. (1944) Der Nahrwert der Kakaoschalen. *Zeitschrift fur Tierernährung und Futtermittelkunde*, **8**, 25–59

Fingerling, G., Eisenkolbe, P., Just, Hientzsch, B. and Kretzsschmann, F. (1936) Aufschliessen des Strohes ohne Chemikolien. *Landwirtschaftlichen Versuchsanstalt Leipzig–Möckern*, **125**, 235–300

Forbes, J.M. (1986) *The Voluntary Food Intake of Farm Animals*. Butterworths, London

Forbes, E.B., Braman, W.W. and Kriss, M. (1928) The energy metabolism of cattle in relation to the plane of nutrition. *Journal of Agricultural Research*, **37**, 253–300

Forbes, E.B., Braman, W.W. and Kriss, M. (1930) Further studies of the energy metabolism of cattle in relation to the plane of nutrition. *Journal of Agricultural Research*, **40**, 37–78

Forbes, E.B., Braman, W.W., Kriss, M. and Swift, R.W. (1931) The metabolisable energy and net energy values of corn meal when fed exclusively and in combination with alfalfa hay. *Journal of Agricultural Research*, **II**, 1015–1026

France, J., Thornley, J.H.M. and Beever, D.E. (1982) A mathematical model of the rumen. *Journal of Agricultural Science, Cambridge*, **99**, 343–353

Gill, M., Beever, D.E., Buttery, P.J., England, P., Gibb, M.J. and Baker, R.D. (1987) *Journal of Agricultural Science, Cambridge*, **108**, 9–16

Kallweit, E., Henning, M. and Groenveld, E. (1989) *Application of n.m.r. Techniques on the Body Composition of Live Animals*, Elsevier, London, ISBN 1–851–66–404–1.

Kellner, O. and Kohler, A. (1900) Untersuchungen uber den stoff-und energgieumsatz des erwachsensen rindes bei ethaaltungs-und productions futter. *Laudw. Vers. Stat. Bd.*, **53**, 1–474

Kielanowski, J. (1965) Estimates of the energy costs of protein deposition in growing animals. In *Energy Metabolism* (ed K.L. Blaxter), Academic Press, London, pp. 13–20

Konig, B.A., Parker, D.S. and Oldham, J.D. (1979) Acetate and palmitate kinetics in lactating dairy cows. *Annales de Recherches Vétérinaires*, **10**, 368–370

Kyriazakis, I., Emmans, G.C. and Whittemore, C.T. (1989) The effect of body composition on diet selection of growing pigs. *Animal Production*, **48**, 627

Lister, D. (1984) *In vivo Measurement of Body Composition in Meat Animals*, Elsevier, London

MacRae, J.C., Smith, J.S., Dewey, P.J.S., Brewer, A.C., Brown, D.S. and Wallar, A. (1985) The efficiency of utilisation of metabolisable energy and apparent absorption of amino acids in sheep given spring- and autumn-harvested dried grass. *British Journal of Nutrition*, **54**, 197–210

MAFF (1988) Report of the Inter-Departmental Working Party ADAS/SAC/ UKASTA on Nutritive Requirements of Ruminant Animals Energy

Martin, A.K. and Blaxter, K.L. (1965) The energy cost of urea synthesis in sheep. In *Energy Metabolism* (ed K.L. Blaxter), Academic Press, London, pp. 83–90

Metz, S.H.M. and Dekker, R.A. (1981) The contribution of fat mobilisation to the regulation of fat deposition in growing Large White and Pietrain pigs. *Animal Production*, **33**, 149–157

Nehring, K., Schiemann, R. and Hoffmann, L. (1961) Die verwertung der futter-energie in abhangigkeit von ernahrung. 3. Mitteilung. Versuch mit Ochsen und Hammeln. *Archiv für Tierernährung*, **II**, 157–204

Oldham, J.D. (1984) Protein-energy interrelationships in dairy cows. *Journal of Dairy Science*, **67**, 1090–1114

Oldham, J.D. and Emmans, G.C. (1988) In *Nutrition and Lactation in the Dairy cow* (ed P.C. Garnsworthy), Butterworths, London, pp. 76–96

Oldham, J.D. and Emmans, G.C. (1989) Prediction of responses to required nutrients in dairy cows. *Journal of Dairy Science*, **72**, 3212–3229

Oldham, J.D. and Friggens, N.C. (1989) Sources of variability in lactational performance. *Proceedings of the Nutrition Society*, **48**, 33–43

Oldham, J.D. and Lindsay, D.B. (1983) Interrelationshps between protein-yielding and energy-yielding nutrients. In *Fourth International Symposium on Protein Metabolism and Nutrition* (les Colloques de l'INRA, No. 16), pp. 183–209

Orskov, E.R., Reid, G.W. and McDonald, I. (1981) The effects of protein degradability and food intake on milk yield and composition in cows in early lactation. *British Journal of Nutrition*, **45**, 547–555

Ortigues, I., Smith, T., Oldham, J.D. and Gill, M. (1989) The effects of fishmeal on growth and calorimetric efficiency in heifers offered straw-based diets. In *Energy Metabolism of Farm Animals* (eds Y. Van der Honing and W.H. Close), Pudoc, Wageningen, EAAP Publication No. 43.

Pullar, J.D. and Webster, A.J.F. (1977) The energy cost of fat and protein deposition in the rat. *British Journal of Nutrition*, **37**, 355–363

Reeds, P.J. (1989) In *Animal Growth Regulation* (eds Campion, D.R., Houseman, G.J. and Martin, R.J.), Plenham Press, New York

Reeds, P.J., Fuller, M.F., Cadenhead, A., Lobley, G.W. and McDonald, J.D. (1981) Effect of changes in the intakes of protein and non-protein energy on whole body protein turnover in growing pigs. *British Journal of Nutrition*, **45**, 539–546

Reeds, P.J., Wahle, K.W.J. and Haggarty, P. (1982) Energy costs of protein and fatty acid synthesis. *Proceedings of the Nutrition Society*, **41**, 155–159

Smith, T., Broster, W.H. and Siviter, V.W. (1980) An assessment of barley straw and oathulls as energy sources for growing cattle. *Journal of Agricultural Science, Cambridge*, **95**, 677–686

Storry, J.E. (1980) Influence of nutritional factors on the yield and content of milk fat: non-protected fat in the diet. In International Dairy Federation Bulletin *Factors Affecting the Yields and Contents of Milk Constituents of Commercial Importance*. IDF Document 125, pp. 88–95

Sutton, J.D. (1984) *Feeding and Fat Production. (Occasional Publication of the British Society of Animal Production)*, **9**, 43–52

Taylor, St. C.S. (1984) Genetic size-scaling rules in animal growth. *Animal Production*, **30**, 161–165

Thomas, C. (1987) Factors affecting substitution rates in dairy cows on silage-based rations. In *Recent Advances in Animal Nutrition – 1987* (ed W. Haresign), Butterworths, London

Van der Honing, Y. (1975) *Intake and Utilisation of Energy of Rations with Pelletted Forages by Dairy Cows.* Wageningen Agricultural Research reports 836, Wagengingen, Pudoc

Webster, A.J.F. (1984) Energy costs of digestion and metabolism in the gut. In *Digestive Physiology and Metabolism in Ruminants* (eds Y. Ruckebush and P. Thivena), MTP Press, Lancaster, pp. 469–484

Whitelaw, R.G., Milne, J.S., Orskov, E.R. and Smith, J.S. (1986) The nitrogen and energy metabolism of lactating cows given abomasal infusions of casein. *British Journal of Nutrition*, **55**, 537–556

Whittemore, C.T. (1985) Nutritional manipulation of carcass quality in pigs. In *Recent Advances in Animal Nutrition – 1985* (eds W. Haresign and D.J.A. Cole), Butterworths, London, pp. 149–156

Wilson, S. (1983) Apparent re-esterification of fatty acids during lipolysis in pregnant ewes. *Proceedings of the Nutritional Society*, **42**, 130A

Wolin, M.J. and Miller, T.L. (1989) Microbe–microbe interaction. In *The Rumen Microbial Ecosystem* (ed P.N. Hobson) Elsevier, London, pp. 343–359

6

PROTEIN EVALUATION IN PIGS AND POULTRY

A.G. LOW
AFRC Institute for Grassland and Environmental Research, Shinfield, Reading, Berks RG2 9AQ

Introduction

The evaluation of feed proteins for pigs and poultry encompasses (a) their gross chemical composition in terms of nitrogen and amino acids, (b) their digestibility, (c) their availability, (d) their content of anti-nutritional factors, (e) their level of inclusion in diets and (f) their interactions with other dietary constituents. Information on each of these aspects is important for practical pig and poultry nutrition: this is then matched with data on the requirements or the responses of animals to proteins or amino acids to provide feeding strategies. In theory, information on requirements of animals is independent of feed characteristics. However, in practice statements of requirements inevitably incorporate, to some degree, factors for the quality of the feeds used in their empirical derivation.

The eventual aim of protein evaluation is to describe feeds as fully as possible in terms of the above attributes, at high speed, with great accuracy and at low cost.

The term apparent digestibility is used here to describe the disappearance of a nutrient anterior to the gut site at which it is measured and is defined as: (amount in diet − amount in digesta/faeces)/amount in diet.

True digestibility of a nutrient is defined in the same way except that the amounts of the nutrient collected after feeding the animal with a diet omitting the nutrient are subtracted from the amounts found in digesta or faeces when the feed contains the nutrient. Real digestibility is analogous to apparent digestibility but it is measured in terms of the difference in concentration of an isotopic tracer (incorporated in the protein) in the feed and in digesta or faeces. The terms apparent and true absorption are synonymous with apparent and true digestibility. The availability of a nutrient is defined as the proportion in a feed which is digested and which is also available for normal metabolic processes.

The aim of this review is to consider the methods currently being used to evaluate proteins in pigs and poultry in the light of present-day knowledge of protein digestion in these species.

Protein digestion

PIGS

Hydrolysis of proteins begins in the stomach through the action of pepsin and hydrochloric acid: at least half of the protein which leaves the stomach is usually peptide in form, i.e. in peptides with 10 or less amino-acids and much of this is soluble in trichloroacetic acid. Further hydrolysis by the pancreatic proteases trypsin and chymotrypsin and the peptidases carboxypeptidases A and B reduces the size of peptides to a chain comprising two or three amino acids. These short peptides are either hydrolysed on the mucosal surface or are absorbed as such, together with free amino acids, into the enterocytes of the small intestine. There is no evidence of active uptake of amino acids in the stomach, and only marginal uptake in the large intestine; the small intestine is therefore the site of virtually all absorption. At the same time it is recognized that there are substantial endogenous inputs of amino acids, peptides and proteins into the digestive tract in the form of, for example, shed epithelial cells, enzymes, plasma proteins, mucin. In addition it is now becoming clear that the mixture of proteins found throughout the digestive tract includes a substantial component within bacteria, especially at the end of the small intestine (Drochner, 1984; Liu *et al.*, 1985), and in the large intestine. Proteins, peptides and amino acids which enter the large intestine are subjected to a wide variety of microbial metabolic processes but no nutritionally significant absorption of peptides or amino acids into the pig occurs in this region (Zebrowska, 1973). The detailed processes of protein digestion in pigs have recently been reviewed by Low and Zebrowska (1989a).

POULTRY

The principles of protein digestion are very similar to those in pigs. Only the proventriculus of the gastric region is secretory and initial proteolysis under acid conditions occurs there and in the gizzard. The large intestine in poultry is simpler and proportionately shorter than in the pig; it is also the site of microbial activity, though less than in the pig. The detailed processes of protein digestion in poultry have also been reviewed by Low and Zebrowska (1989b).

Methods of protein evaluation

A wide range of reviews of different aspects of protein evaluation is available. Five which are particularly relevant concern the concept of amino acid availability (McNab, 1979), feed evaluation and nutritional requirements in pigs and poultry (Sibbald, 1987; Henry, Vogt and Zoiopoulos, 1988; Den Hartog, Verstegen and Huisman, 1989a), and biochemical and methodological principles of chemical, *in vitro* and *in vivo* assays, especially with small animals (Eggum, 1989). The present review will focus on methodology used in pigs and poultry.

ANALYSIS

The measurement of protein quality *in vivo* and *in vitro* is critically dependent on the nitrogen or amino acid analysis of the best feedstuff or diet. Although there are a very large number of published methods, it has to be recognized that there are still technical problems, especially for amino acids. Sample preparation uses acid or alkaline hydrolysis or enzymic hydrolysis under conditions which represent practical compromises (a) between complete hydrolysis and destruction of some quantities of certain amino acids, (b) between the conditions for complete hydrolysis of feedstuffs of varying susceptibility to hydrolysis and (c) between conditions required for feedstuffs and digesta or faeces. The most thorough attempt to provide a single-step hydrolysis procedure was described by Mason, Bech-Andersen and Rudemo (1980); this procedure performed well in a collaborative test (Andersen, Mason and Bech-Andersen, 1984). Many chromatographic methods of measurement of amino acids from hydrolysates have been described. Similarly there are numerous methods for extracting alpha-amino nitrogen and amino acids from tissues such as blood. In all cases there can be major effects on results due to differences (a) between the chemical and physical methods used, (b) between different instruments measuring the same samples and (c) between different operators using the same techniques. There is currently a regrettable lack of comparative tests between laboratories. Comparisons of data within a laboratory are probably more reliable than those between laboratories, at the present time. There is a tendency to express results to a greater degree of apparent accuracy than the sensitivity of the methods allows: for example, percentage apparent digestibility values should probably be quoted only to the nearest whole number. More emphasis is needed on developing reliable methods with low between-operator, instrument and laboratory variation and greater accuracy. Without such improvements, the considerable recent efforts to refine the *in vivo* and *in vitro* assays of proteins are of limited value. It is also important to be aware of new rapid methods for nutrient analysis, such as near infra-red reflectance spectroscopy. This may well revolutionize amino acid analysis in the future, especially if the limiting amino acid in the system being studied can be identified because of its dominant importance in determining the nutritive value of a protein.

IN VIVO AMINO ACID DIGESTIBILITY MEASUREMENTS

Until the 1970s it was generally thought that faecal apparent or true amino acid digestibility measurements were a reliable indicator of protein quality in pigs and poultry. However, the recognition that the gut microflora could metabolize amino acids within the gut lumen, either to other amino acids in bacterial protein, or to non-amino acid compounds such as ammonia or amines (Michel, 1966) led to the view of Payne *et al.* (1968) that digestibility should be measured at the end of the small intestine before the action of the microflora of the large intestine. This has led to a wide variety of methods for measuring ileal digestibility, especially in pigs, after Zebrowska (1973) demonstrated that proteins entering the large intestine have no nutritional value for the host. Subsequently Just, Jørgensen and Fernandez (1981) confirmed this by demonstrating that amino acids infused into the caecum of pigs did not improve overall nitrogen balance. The extent of microbial activity in poultry appears to be much less than in pigs: Fuller and Coates (1983) were of the opinion that it is not necessary to correct digestibility values for microbial activity but Kussaibati, Guillaume and Leclerq (1982) concluded that the evidence for discounting this activity is not yet sufficient. Three methods have therefore been developed in attempts to overcome the problems caused by microbial action in pigs and poultry: (a) germ-free animals, (b) antibacterial compounds and (c) surgical intervention. For routine purposes method (a) is complex and costly and fails to account for the fact that animals normally have a gut microflora. The effects of antibacterial compounds are variable, and complete sterility within the gut is never achieved. This approach was used in pigs by Just *et al.* (1980) and Livingstone, Fowler and McWilliams (1982). Thus much effort has been expended on surgical intervention especially in pigs.

Ileal Digestibility in Pigs

Methods

In the simplest case, pigs can be anaesthetized and samples of the ileal digesta collected, before the animal is killed. The digestibility of nitrogen or of amino acids can be measured, with reference to an indigestible and non-absorbed marker substance (for a detailed review of markers see Kotb and Luckey, 1972) given with the test feedstuff. This method is expensive because carcasses containing anaesthetics cannot be sold. In addition only one measurement is possible per animal.

A variety of surgical methods have been developed to measure ileal digestibility. The desirable conditions for measurements have been defined by Darcy and Laplace (1980) as: (a) avoidance of transsection of the distal ileum, (b) adequate post-operative recovery period, (c) preservation of the functional role of the ileo-caecal valve, (d) an internal diameter of cannulas of 20 mm and (e) minimal passage through and residence time in tubes connected to the cannulas. In addition it is important that the animal has a normal appetite and that it is growing at an acceptable rate.

During the 1960s and 1970s many groups used pigs with re-entrant cannulas in the distal ileum, prepared by the method described in detail by Markowitz, Archibald and Downie (1954). This comprises transsection of the gut and placing rigid or flexible cannulas (Low, 1980) in the two blind ends, followed by exteriorization. The method allows total collection of digesta but it is necessary to mill coarsely ground feeds before they are given to pigs to avoid blockage (Braude, Fulford and Low, 1976). Such problems have led to the development of a re-entrant cannula placed either side of the ileo-caecal valve (Easter and Tanksley, 1973) or posterior to it (Darcy, Laplace and Villiers, 1980): both of these overcame some of the problems with blockage. In all cases sampling must be followed by return of digesta either manually or automatically.

Subsequently several centres began to use simple 'T' shaped cannulas, placed at the end of the terminal ileum or in the caecum (Van Leeuwen *et al.*, 1988). These were of variable and sometimes large size, allowing coarsely ground feed residues to flow through the barrel. Generally speaking such preparations last longer (up to 2 years according to Kesting *et al.*, 1986) than re-entrant cannulas and there are few problems of leakage. Furthermore, the cannulas can in some cases be replaced (for example, Bjornhag and Jonsson, 1984).

Problems with cannulas led Fuller and Livingstone (1982) to prepare pigs with the terminal ileum anastomosed to the rectum, thus allowing digestion to by-pass the large intestine. This technique, now usually called the ileo-rectal shunt procedure, has been used by several groups, with the large intestine either being excluded or opening into the rectum. Details of surgical methods have been given by Picard *et al.* (1984), Souffrant *et al.* (1985), Hennig *et al.* (1986), Green *et al.* (1987), Bengala Friere *et al.* (1988) and Green (1988a). Such preparations last for many months, allow total collection of digesta and can be used with coarsely ground feedstuffs. A detailed study of the physiological effects of ileo-rectal anastomosis by Hennig *et al.* (1986) suggested that they are relatively minor.

The specific effects of different surgical procedures on digestibility measurement are difficult to summarize because their success also depends on many factors including husbandry and surgical experience, together with the type and size of pig used. Each research group has developed its own routine by experience and has become increasingly successful without always being aware of all the reasons for it.

Validity of Cannulation

An important question concerns whether or not the procedure of cannulation disturbs the processes of digestion and absorption. The evidence is either that there is no effect on overall nitrogen digestibility (Furuya, Takahashi and Omori, 1974; Sauer, Stothers and Parker, 1977; Taverner, 1979; Huisman *et al.*, 1984) or that there may be some adverse effects on nutrient digestibility (Laplace and Borgida, 1976; Sauer, Aherne and Thacker, 1979) or growth (Livingstone and McWilliams, 1985). However, the specific effect of ileal cannulation (simple 'T' cannulas were used) on ileal digestibility of amino acids has only been studied by Moughan and Smith (1987) who found no effect.

The validity of digestibility measurements in pigs depends upon the frequency and total period of sampling. In the case of simple ileal or caecal cannulation studies most authors would agree that regular sampling of large amounts of digesta throughout 12 h periods, replicated on several successive days, will provide representative samples, provided that the pigs are fed at regular intervals. In the case of re-entrant cannulation, collections have usually lasted for 24–72 h, and collections can be taken from pigs with ileo-rectal shunts for periods of many days. Circadian rhythms are important in determining sampling frequency from simple cannulas, as described by Livingstone *et al.* (1980) and Graham and Åman (1986).

The validity of digestibility measurements from 'T' cannulas depends centrally upon the representative nature of the samples and the validity of the marker used: as already noted a wide range of markers is used and all have shortcomings which are rarely fully understood. It is generally accepted that comparisons between the digestibility of different feeedstuffs can be made within a trial using the same marker, but it must be recognized that the behaviour of a marker may be influenced by the physico-chemical nature of the materials with which it moves. There is a need for further systematic and comparative studies on the behaviour of markers used for digestibility measurements in pigs; the studies by Pond *et al.* (1986) are an important step in this direction.

Choice of cannulation method

Choice of cannulation method for measuring the digestibility of amino acids at the end of the ileum is not entirely straightforward. Zebrowska *et al.* (1978) concluded that simple and re-entrant ileal cannulas gave very similar results, the former with less difficulty for staff and with fewer problems for the pigs. Darcy, Laplace and Villiers (1980) studied digestibility in pigs with ileo-caecal or post-valvular ileo-colic cannulation and considered that the latter procedure led to fewer motility disturbances and more valid results. The ileo-colic procedure was subsequently compared with ileo-rectal anastomosis by Darcy-Vrillon and Laplace (1985) who found substantially higher amino acid digestibility in pigs given sugar beet pulp with the former than the latter technique, though there were few differences between the techniques for other diets; it was suggested that the ileo-caecal valve which only functions in the former technique could be important by increasing retention time in the small intestine for diets which lead to large volumes of liquid digesta. Recent experience with the post-valvular caecal simple cannula of Van Leeuwen *et al.* (1988) shows that this is robust and simple to use, it is replaceable and allows normal function of the ileo-caecal valve and the colon. Most of the caecum is removed, but otherwise the gut is intact. During collection all of the digesta pass through the cannula because the ileo-caecal valve protrudes into the cannula and thus no digesta go into the colon. Of the current methods available this appears to be the most satisfactory. A recent comparative study of simple ileal cannulas, post-valve caecal simple cannulas and ileo-caecal re-entrant cannulas by Den Hartog *et al.* (1988) showed some differences in values for digestibility of dry matter. These authors confirmed the views expressed by previous groups that further studies are needed before drawing firm conclusions about which procedure should be adopted. At the present time this conclusion seems to be fully justified.

Endogenous Secretions

The classical method of measuring endogenous nitrogen or amino acids in digesta or faeces is to use a protein- or nitrogen-free diet. In recent years the costly procedure of ^{15}N labelling of either the dietary protein or the whole animal has been used by several groups. A further ingenious alternative method is to guanidinate the lysine side chains in dietary protein to form homoarginine. This compound is digested and absorbed but is not used for protein synthesis and therefore does not reappear in endogenous secretions into the gut (unlike ^{15}N labelled amino acids). In addition homoarginine is transformed to lysine, thus preventing lysine deficiency. This approach was used by Hagemeister and Erbersdobler (1985) who found ileal apparent digestibility values of nitrogen of 74–91% and corresponding true digestibility values, measured by homoargine of 98–99%. Subsequently Roos, Hagemeister and Scholtissek (1990) found that true digestibility by the ^{15}N method gave values on average 4.8% lower than by the homoarginine method.

The very substantial contribution of endogenous nitrogen and amino acids (estimates vary between 30% and 90% of the total) to ileal digesta using the ^{15}N labelling or homoarginine labelling techniques has been noted by Zebrowska *et al.* (1982, 1986), Hagemeister and Erbersdobler (1985) and Simon, Bergner and Partridge (1987) among others. The values obtained by these and the classical methods are to some extent a function of the measurement technique used but it is also evident that they are influenced by the type and amount of non-starch-polysaccharides in the diet (Sauer, Stothers and Parker, 1977; Taverner, Hume and Farrell, 1978; de Lange *et al.*, 1989). In addition de Lange, Sauer and Souffrant (1989) observed lower endogenous nitrogen flows in pigs fed orally with protein-free diets when the pigs were intravenously given a balanced mixture of amino acids. Thus the endogenous losses were influenced by the protein status of the pigs.

There are two schools of thought over whether the ileal digestibility of amino acids should be stated as apparent or true values. It can be said that the different amounts of endogenous secretion induced by different feedstuffs are a component of the net value of the feedstuff to the pig: the amount of amino acids leaving the ileum for the large intestine are thus to be subtracted from the amounts eaten, irrespective of origin, because all of them will fail to be used by the pig. On the other hand, it is attractive to think that the amino acids in each feedstuff are digested to an extent that is determined purely by attributes of the feedstuff, and that endogenous secretions should be discounted in statements of digestibility. If such discounting could be done accurately then the nutritive values of individual feedstuffs could, it is argued, be put together additively to predict the digestibility of complete diets. However, the evidence available at present suggests that the regulation of endogenous protein secretions and their subsequent digestion and reabsorption is a highly complex matter dependent on many nutritional factors. It is therefore unlikely that any soundly based constant value for endogenous secretions can be used to correct apparent digestibility values for practical purposes. This view is reinforced by the recent awareness (Drochner, 1984) that up to 60% of the amino acid content of ileal digesta is of bacterial origin; the factors which govern bacterial growth include protein and energy supply. This bacterial population and activity adds a serious further complication to attempts to provide digestibility values for amino acids which are solely an attribute of the feedstuff. It thus seems prudent at present to use apparent rather than true digestibility values.

It is clear from several studies that the apparent or true digestibility of protein (nitrogen × 6.25) is not a reliable predictor of the digestibility of its constitutent amino acids. Furthermore ileal digestibility values do not form a constant proportion of faecal digestibility values, and ranged from 0.73 to 0.98 in the studies by Wunsche, Bock and Meinl (1984).

Factors Influencing Ileal Digestibility

Measurements of ileal digestibility are known to be influenced by numerous nutritional factors such as extrusion cooking (Fadel et al., 1988), starch type (Darcy, Laplace and Villiers, 1981), non-starch polysaccharides (Dierick et al., 1983; Just, Fernandez and Jørgensen, 1983; Partridge, Simon and Bergner, 1986; Den Hartog et al., 1988), particle size (Owsley, Knabe and Tanksley, 1981), tannins (Cousins et al., 1981), level of feeding (Furuya and Takahashi, 1980; Haydon, Knabe and Tanksley, 1984), and associative (i.e. non-additive) effects between feeds (Laplace et al., 1989). Observations such as these underline the complexity of the factors that collectively determine digestibility of proteins.

Predictive Value of Ileal Digestibility Measurements

Despite the considerable practical interest in using ileal apparent digestibility values there has been relatively little validation of the approach in terms of the extent to which the values obtained are accurate predictors of rates of lean tissue deposition. The results of the few published studies on this topic present a conflicting picture. In a comparison of the relative efficacy of two methods of supplementing diets with free amino acids to meet the requirement for ideal protein, Fuller et al. (1981) found no benefit in basing the supplements on ileal apparent digestibility rather than gross amino acid composition, when animal responses were assessed in terms of improvements in nitrogen balance. Such a result could be explained as being due to the inappropriateness either of the composition of ideal protein proposed at that time or of the ileal digestibility procedure used. Similarly, Jagger, Cole and Wiseman (1987) found no performance benefits of formulating diets containing various heated fish meals on the basis of their ileal apparent digestibility values probably because these values did not reveal differences in availability. By contrast, Tanksley and Knabe (1984) concluded that there were important performance benefits in formulating diets on the basis of their apparent ileal digestibility values for amino acids when they contained unusual protein sources, but not for diets based on maize and soya; the latter are products of relatively uniform digestibility and requirements for pigs in the USA have mainly been derived from empirical studies in which they have been used, and in which their digestibility has therefore inevitably been incorporated. Similar conclusions were drawn in a review by Sauer and Ozimek (1986). Better correlations between protein deposited in female pigs and ileal rather than faecal digestible amino acids values were found by Just, Jørgensen and Fernandez (1985). Dierick *et al.* (1987) correlated daily gain and feed conversion in pigs with apparently digested crude protein measured in ileal digesta of faeces and found the former provided a better prediction of performance. Low and Partridge (1984) found a close correlation between ileal apparent digestibility of lysine and lysine deposition in the whole body but only when this amino acid was limiting. This illustrates the important point that it is only information about the first, second or perhaps third limiting amino acid in the diet which is of importance.

Recent studies by E.S. Batterham (unpublished observations) have shown that under some conditions ileal apparent digestibility values of lysine are a considerable over-estimate of its availability as measured in an *in vivo* slope ratio assay. This applied to cottonseed meal, but not to soybean meal which has a much higher availability. The author concluded that ileal digestibility is not a reliable indicator of lysine digestibility.

In conclusion, it is important to recognize that ileal apparent digestibility values do not necessarily provide a satisfactory degree of predictive power for diet formulation and further methodological studies are still merited to resolve the major uncertainties which have been described. Ten years ago there was still extensive debate about whether the digestibility of proteins could be measured in a meaningful way using faecal analysis, even though there were clear indications that microbial activity in the large intestine could greatly modify the composition and amount of amino acids in digesta, to the extent that amino acid composition is remarkably similar for many types of diet and that it resembles the composition of faecal bacteria. It is now almost universally accepted that such an approach is fundamentally unsound and an increasing number of organizations in Europe and North America are using ileal apparent digestible amino acid values in diet formulation for growing pigs. Ileal digestibility values are also becoming available for feeds used in piglets from three weeks of age. However it is now becoming clear that there may be important differences between digestibility and availability and these raise major doubts about the validity of the ileal digestibility approach.

Amino Acid Digestibility Measured in the Small and Large Intestines of Poultry

Ileal Cannulation

By contrast with pigs, amino acid digestibility has been relatively rarely measured in the ileum of poultry because of surgical complexity (Summers, Berzins and

Robblee, 1982; Raharjo and Farrell, 1984a) and because of cannula rejection and impeded digesta flow (Thomas and Crissey, 1983).

Caecectomy

Awareness of microbial activity in the large intestine of poultry has led to experimental use of caecectomized birds, in which this activity is greatly reduced, as reviewed in detail by McNab (1973). There is no evidence of amino acid absorption in the caecae, and some of the ileal effluent may not enter either of these blind sacs, which have an excretory role. Slight increases in faecal amino acid output have been seen in fasted caecectomized birds compared with intact birds (Kessler, Nguyen and Thomas, 1981; Parsons, 1984), while some minor differences have also been seen in fed birds (Parsons, 1984; Raharjo and Farrell, 1984b). Recent comparisons between intact and caecectomized cockerels by Green *et al.* (1987a) for maize, wheat and barley revealed no differences in true digestibility of amino acids. In a similar study (Green *et al.*, 1987b) on soybean, sunflower and coconut meals, small differences were seen for threonine, glycine (probably deaminated) and lysine (probably syhtesized) only. In a third related study by Green (1988b), a variety of levels of dietary fibre were added to a protein-free basal diet; there was no effect of diet on faecal amino acid output, but there was evidence of degradation or retention in the caecae of some amino acids either contained in maize husks or of endogenous origin. As is the case in the pig, it seems that intestinal microbial activity does not contribute significantly to the protein nutrition of poultry (Salter, 1973; Salter, Coates and Hewitt, 1974); it is thus of some importance to assess whether the observed microbial activity is sufficient to reduce the validity of digestibility values obtained with intact or colostomized birds.

Colostomy

Because faeces and urine are voided together through a common cloaca, methods have been developed to separate these components using colostomy (Imbayashi, Kametaka and Hatano, 1955) or exteriorization of the ureters (Dixon and Wilkinson, 1957). Variations on the methods used were reviewed in detail by Sibbald (1982). Colostomy has been the most common procedure and example of measurements of amino acid digestibility have been given by Waring (1969), McNab and Shannon (1971, 1974) and Shannon and McNab (1973). When colostomized and intact birds were compared (Gruhn, 1974) the latter gave slightly lower values because avian urine contains small amounts of amino acids (Gruhn and Raue, 1974). However, these differences were not detected in studies by Skrede, Krogdahl and Austreng (1980) who fed raw cod fillet to colostomized hens, again indicating that surgical intervention may not be essential.

Faecal Excreta Assays

At a practical level there is no clear consensus as to which method of measurement of amino acid digestibility using faecal excreta should be used. However, three principal methods have been used:

1. Feeding complete diets to adapted animals for a period of some days (substitution is used to calculate values for individual ingredients).
2. Rapid assays, using starvation before and after consumption of a known weight of complete diet (substitution is again used to calculate individual ingredient values).
3. Rapid assays relying on tube-feeding of the test material into the crop of starved birds (avoiding the need to feed a complete diet, and substitution calculation) (Sibbald, 1979, 1986, 1987).

Although method (1) has the greatest general acceptance there are experimental difficulties in diet preparation and accurate measurement of intake and collection of excreta. The aim of method (2), as proposed by Farrell (1978), is speed, combined with standardization and substantial intakes of pelleted diets, but in practice the method has not always been completely successful. The development of force-fed or tube-fed assays (3) has been rapid in recent years. Here intakes can be accurately controlled, but because input weights are small sampling errors are of increased importance. In the hands of trained operators feeding is rapid (15–30 seconds) and there is little evidence of stress beyond that in handling the birds. Under these experimental conditions the effects of fermentation are minimal. It is usual in such assays to measure true digestibility because food intake level can markedly alter apparent digestibility; true digestibility measurements are arguably independent of level of intake (Sibbald, 1979). If between-feedstuffs comparisons are to be made constant intakes are essential to avoid a likely systematic bias.

The concepts involved in measuring true digestibility in poultry are the same as those discussed earlier for pigs but it seems that in practice endogenous amino acids form a small proportion of total amounts excreted after feeding most feedstuffs. Practical refinement of assays of this kind still depends on more understanding of the factors that influence endogenous secretions and caecal microbial activity. The age of the birds is also of importance; and in addition it is not clear whether values obtained with adult cockerels may be used for turkeys or ducks. Rapid assays have allowed an enormous increase in the number of feeds that can be assayed at low cost without complication of mixed diets, variable feed intakes or the uncertainties of extrapolation. The degree of precision which these assays bring is impressively high, but their level of accuracy in the full biological sense still requires further rigorous assessment, just as in the case for amino acid digestibility measurement in pigs.

At present there are no comparative studies on digestibility and availability in poultry. It is important to establish whether the marked differences between digestibility and availability of low-quality proteins fed to pigs are also found in poultry. It seems likely that such differences would be seen.

Mobile Bag Measurements of Digestibility in Pigs

The concept of enclosing a small sample of a feedstuff in a permeable bag and allowing it to be introduced into part of the digestive tract and either anchored there, or allowed to move freely to the end of the gut, it not new. Petry and Handlos (1978) developed the method for use in pigs and obtained very high values, possibly because of prolonged gastric retention. Sauer, Jørgensen and Berzins (1983) developed the procedure with an *in vitro* digestion before introduction through a duodenal cannula. The effects of predigestion conditions were further examined by Cherian, Sauer and Thacker (1988) who found that pepsin concentration, the length of time and pH of predigestion were all critical factors: 377 IU pepsin/litre, 4 h predigestion and pH 2.0 gave the best result in terms of similarity of digestibility of nitrogen in the bag and overall nitrogen digestibility. The method was then used for a wide variety of feedstuffs in pigs by Sauer *et al.* (1989) with generally good agreement between mobile bag and conventional *in vivo* overall nitrogen digestibility values. Taverner and Campbell (1985) found that although the mobile bag technique consistently underestimated dry matter digestibility, the values for different materials ranked in the same order as conventionally derived values; furthermore, the type of basal diet consumed by the pigs did not affect the mobile bag digestibility values. Some discrepancies between nitrogen digestibility values obtained with mobile bags and conventional *in vivo* values were noted by Graham *et al.* (1985) and these were discussed in detail. To date there are no published values for ileal digestible amino acids using mobile bags; when available these will be of more relevance to protein evaluation than those studies already published. While there are considerable attractions in the mobile bag approach, it is worth noting that some proteins containing antinutritional factors only induce their secretagogue effects when they are in contact with the gut wall; if these materials were enclosed in a nylon bag such interactions would not occur. The method would, however, seem to offer promise for rapid assessment of large numbers of samples of a given type of feedstuff, at least to provide a measure of their relative protein quality ranking. As in the case of ileal digestibility measurements, this procedure cannot be expected to detect differences in amino acid availability.

Protein Evaluation Using Blood Measurements

Practical assessment of protein quality using measurements of blood amino acid concentrations have not been attractive in recent years because of the realization that the values are influenced not only by the type of diet fed but by circadian rhythms, age, species, physiological state, meal size and feeding frequency of pigs and poultry. The amounts present in the blood represent a balance between inputs from the digestive tract, withdrawal by tissues for new protein synthesis and outputs of surpluses from tissues. Plasma concentrations can, however, be used to identify the limiting amino acid in pig and poultry diets, and the approximate point at which a requirement is met may be demonstrated by an increase in its concentration.

A strong inverse correlation between biological value of proteins and blood urea values was established by Munchow and Bergner (1967) for rats and pigs, and this was further confirmed in extensive studies in young growing rats by Eggum (1970, 1973). The biological value of the 47 protein sources used ranged from 2.1 to 95.5 and this correlated with blood urea values with a coefficient of -0.95. Further work on the accuracy and precision of blood urea concentrations as a means of measuring protein quality in pigs would seem worthwhile, and it is also likely that urinary urea output could be a more useful predictor than plasma urea (Brown and Kline, 1974), but, again, its sensitivity needs to be investigated.

Rérat, Vaugelade and Villiers (1980) described a method for quantitive measurement of amino acids absorbed and transported into the heptic portal vein.

A series of subsequent studies by this group has demonstrated major differences in the qualitative, quantitative and kinetic aspects of the digestion of different dietary protein sources (Rérat, Vaissade and Vaugelade, 1979; Rérat, Jung and Kande, 1988) or endogenous proteins (Rérat, Vaissade and Vangelade, 1988). The methods are, however, very complex and time consuming because a large number of analyses are needed to construct a profile of absorption, and they are not therefore suitable for practical feed evaluation.

IN VITRO ASSAYS TO PREDICT PROTEIN QUALITY

There is considerable appeal in evaluating proteins by means of *in vitro* assays, either with a single enzyme (usually pepsin) or a mixture of enzymes; this mixture can be relatively pure or it may be small intestinal digesta, or faeces, or a microbial culture, or a combination of two or more of these. Such assays are relatively simple, inexpensive and rapid, and high levels of precision are often achieved. In addition many samples can be compared at the same time. Until now virtually all assays have been used to predict the apparent digestibility of nitrogen and are thus rather non-specific. However, these assays do allow meaningful discrimination between feedstuffs of very different digestibility. A frequently used method using laboratory-grade enzymes is the two-step multi-enzyme system described by Pedersen and Eggum (1981). The use of digesta collected from the pig jejunum to provide an enzymic mixture was described by Furuya, Sakamoto and Takahashi (1979) and more recently duodenal and ileal digesta have been used by Lowgren, Graham and Åman (1989) and Graham, Lowgren and Åman (1989). Correlations between these methods and overall ileal nitrogen digestibility are generally quite close. Graham *et al.* (1989) also found good correlation between ileal *in vivo* nitrogen digestibility and *in vitro* digestion using duodenal digesta, though the absolute *in vitro* values were consistently higher. It is nevertheless likely that further refinement of the methods of *in vitro* digestion by enzymes will lead to a closer approximation to absolute values for *in vivo* digestion; in addition it is necessary for such methods to be used with amino acid analysis in order to be of more value for protein evaluation.

Microbiological assays were routinely used to measure amino acid concentrations in protein hydrolysates before the advent of ion-exchange chromatography. Subsequently such assays, all based on microbial growth as the response criterion, have been used to assess the digestibility and availability of proteins and they have been a satisfactory means of quality control, ranking materials of similar composition. It is evident, however, that the availability of amino acids or short peptides to bacteria is not necessarily the same as that to pigs and poultry, and this is thus a weakness of such assays.

In studies on the prediction of ileal protein and amino acid digestibility, Hall *et al.* (1987) showed that inclusion of neutral detergent fibre, neutral detergent fibre nitrogen and nitrogen into a single equation by a stepwise regression technique explained 92–96% of the variation between samples in ileal protein and amino acid digestibility. This implies that amino acids associated with cell walls are those least likely to be digested and also that cell wall material influences endogenous amino acid secretion, as shown in other studies.

Chemical methods for assessing the availability of amino acids have been almost invariably associated with lysine, assessed by the reactivity of its ε-amino group with 1-fluoro-2, 4-dinitrobenzene (FDNB) (Carpenter *et al.*, 1957; Carpenter,

1960; Roach, Sanderson and Williams, 1967) 2,4,6-trinitrobenzenesulphonic acid (TNBS) (Kakade and Liener, 1969) and dye-binding with Acid Orange 12 (Hurrel and Carpenter, 1976). In a series of trials in which chemical estimates of available lysine were compared with growth or carcass assays in pigs by Batterham, Murison and Lewis (1978, 1979), Batterham, Murison and Lowe (1981), Taverner and Farrell (1981) and Batterham, Murison and Andersen (1984) all found marked discrepancies. It is nevertheless the case that the Carpenter (1960), Roach et al. (1967) and Hurrel and Carpenter (1976) assays can provide useful quality control information, especially where heat damage is suspected and where ranking of similar samples rather than absolute values of quality are required.

In a recent discussion of various rapid protein quality assays for poultry, Dudley–Cash and Halloran (1988) described the use of pepsin assays of protein quality and pointed out that the concentration of pepsin used is critical. They noted that assays using the AOAC recommended concentration of 2 g pepsin/litre failed to discriminate between good and heat-damaged fishmeal, whereas those using 2.0 mg or 0.2 mg/litre showed quality differences clearly (Johnston and Coon, 1979; Lovern, 1965).

Most recent in vitro assays have been developed to measure digestibility rather than availability. They are thus open to criticism when these two attributes of a protein differ. In vitro digestibility procedures are intended to represent in vivo measurements: it is important to recognize that they do not mimic apparent digestibility because endogenous secretions are either not included at all or only at low levels, in assays using digesta. Whether such assays correspond with true digestibility in vivo is not clear, although theoretically they should. However, the problems of measuring true digestibility accurately in vivo are considerable. There is thus a need to develop methods which minimize this conflict. Nevertheless, the differences between true and apparent digestibility of a protein may be much smaller than the differences between either of these measures and its availability; this latter difference may be of much greater nutritional significance.

There is considerable scope for critical re-evaluation of existing assays and also for new approaches to measuring the availability of proteins.

Growth Assays

The yardstick by which all the above procedures can and should be compared is the protein growth of pigs and poultry, either in terms of the whole animal, or of their carcass content, or egg output. The most accurate and most costly procedure is to kill and analyse the whole body chemically in a group of animals at the beginning of the trial, and for related animals to be slaughtered and analysed at the end of feeding the test ingredients or diet. For protein evaluation at least the nitrogen content and preferably the amino acid content should be measured. Scrupulous care with all aspects of growth and slaughter studies is essential and there are many published accounts of specific procedures which will not be reviewed further here. It is notable that overall growth responses to diets of differing protein quality or quantity are a sensitive means of evaluation in growing poultry, which have a relatively constant body composition irrespective of any marked dietary composition changes which may be imposed. By contrast, pigs are more sensitive to variations in protein quality and will tend to deposit fat and lean in different proportions. Thus overall growth of poultry corresponds more closely with protein deposition in poultry than in pigs.

Measurement of protein deposition in pigs is often measured indirectly by measuring nitrogen balance *in vivo*, because it is substantially cheaper than comparative slaughter and it is a sensitive tool for discriminating between the availability of proteins if close attention is given to experimental technique. Measures such as collection of urine via a bladder catheter, faecal collection into colostomy bags, ensuring that all feed is actually consumed and adaptation of the animals to the diet and pen or cage arrangement can collectively make this a reliable technique. It is important to recognize that all the sources of experimental error tend to lead to nitrogen balance being over-estimated *in vivo*, but between-feed discrimination can be good. Although nitrogen balance has been used in poultry it is not a widely practised procedure.

An alternative *in vivo* assay for measuring protein quality is the slope-ratio procedure described by Batterham *et al.* (1979), who concluded that carcass rather than total weight gain is the best dependent variable because it avoids variation associated with gut fill; later gain:feed ratio, based on carcass weight was used as the response criterion (Batterham *et al.*, 1984). Subsequently, Leibholz (1986) found that the three regressed responses (weight gain, feed:gain, nitrogen retention) of piglets to various test materials did not lead to a single intercept value, though the differences were relatively small. Each response was related to the response of the piglets to free lysine. Such assays, though expensive and time-consuming, are sensitive tests of protein availability.

Many aspects of growth assays in pigs and poultry have been discussed in detail by McNab (1979) and Sibbald (1987).

BETWEEN-SPECIES COMPARISONS

To a considerable extent the staple feedstuffs used for pigs are also used for poultry. It would therefore be very useful in practice if their nutritive values were demonstrated to be similar in both species. It would also be an advantage if nutritive values measured in small animals such as rats could be applied to pigs and to poultry. This was the background to a study by Slump *et al.*, (1977) who fed three contrasting diets to these three animal species. In general apparent digestibility values for pigs were higher than for cockerels, which tended to be higher than for rats: the ranges for overall faecal total amino acid digestibility values were 83–88%, 72–76% and 57–66% for the three diets: a similar pattern was seen for true digestibility. The differences were such that reliable between-species predictions could not be made. However, as already noted, faecal amino acid digestibility is not a satisfactory index of protein quality. Similarly, Green and Kiener (1989) compared digestibilities in pigs with ileo-rectal anastomoses and caecectomized or intact tube-fed cockerels given diets containing soyabean, sunflower, meat and rapeseed meals, and again found some marked differences between the species.

Furthermore, *in vitro* estimates of protein digestibility with pepsin were insensitive. Comparisons of rat and pig ileal apparent digestibility values by Moughan *et al.* (1987) were more encouraging and indicated agreement for all essential amino acids in barley except methionine. It is possible that at least some of the between-species differences observed were the result of insufficiently similar experimental conditions being applied, rather than intrinsic differences in the way pigs and poultry digest proteins.

Conclusions

Present methods of protein evaluation have shortcomings because of imperfect analytical procedures, frequent use of nitrogen rather than amino acid analysis, the absence of any totally satisfactory point for measurement of digestibility within the digestive tract, variations in endogenous secretions, and over-simplification of attempts to mimic digestibility and availability *in vitro*. However, feed evaluation in its practical sense is and always will be a compromise between the need for simplicity, speed and low cost on the one hand, and the great complexity of feedstuffs and of the living organism on the other. There is a good prospect of improving this compromise through improvement of *in vitro* assay systems, in particular, coupled with more accurate amino acid analysis, and validated by growth studies. More revolutionary development ought to be possible, however, as modern analytical methods such as near infra-red reflectance spectroscopy become more thoroughly understood and can be used not only to quantify total amounts of amino acids in feeds but also to identify the nature of inter-amino acid structure and the relationship between proteins and other components of feedstuffs. Such developments in analytical chemistry may also provide a means of identifying structures resistant to digestion, or which may be unavailable, with far greater accuracy than hitherto; in turn, specific methods of overcoming these resistant structures may be devised, leading to increased protein quality.

Translation of research results into practice can present formidable challenges as demonstrated by Elbers *et al.*, (1989) who found that the organic matter digestibility of the same diet on 17 pig farms differed between 75% and 85% for a grower diet, and between 78% and 86% for a finisher diet. It can be assumed that amino acid digestibility would also have differed to a similar extent on the farms. Clearly management, environment, disease and genetic factors could all play a part in determining the digestibility of these diets and ultimately nutrition must therefore be seen as an interacting constituent of the whole process of animal production.

In summary, there is a lack of comprehensive studies of the relative value of different methods of measuring protein quality; without such studies, supported by improved understanding of nutritional physiology, the fragmentary and somewhat contradictory picture presented here will not change.

References

Andersen, S., Mason, V.C. and Bech-Andersen, S. (1984) EEC collaborative studies on a streamlined hydrolysate preparation method for amino acid determinations in feedstuffs. *Zeitschrift für Tierphysiologie Tierernährung und Futtermittelkünde*, **51**, 113–129

Batterham, E.S., Murison, R.D. and Andersen, L.M. (1984) Availability of lysine in vegetable protein concentrates as determined by the slope-ratio assay with growing pigs and rats and by chemical techniques. *British Journal of Nutrition*, **51**, 85–100

Batterham, E.S., Murison, R.D. and Lewis, C.E. (1978) An evaluation of total lysine as a predictor of lysine status in protein concentrates in growing pigs. *British Journal of Nutrition*, **40**, 23–28

Batterham, E.S., Murison, R.D. and Lewis, C.E. (1979) Availability of lysine in protein concentrates as determined by the slope-ratio assay with growing pigs and rats and by chemical techniques. *British Journal of Nutrition*, **41**, 383–393

Batterham, E.S., Murison, R.D. and Lowe, R.F. (1981) Availability of lysine in vegetable protein concentrates as determined by the slope-ratio assay with growing pigs and rats and by chemical techniques. *British Journal of Nutrition*, **45**, 401–410

Bengala Freire, J., Peiniau, J., Lebreton, Y. and Aumaitre, A. (1988) Determination of ileal digestibility by shunt technique in the early-weaned pig: methodological aspects and utilization of starch-rich diets. *Livestock Production Science*, **20**, 233–247

Bjornhag, G. and Jonsson, E. (1984) Replaceable gastro-intestinal cannulas for small ruminants and pigs. *Livestock Production Science*, **11**, 179–184

Braude, R., Fulford, R.J. and Low, A.G. (1976) Studies on digestion and absorption in the intestines of growing pigs. Measurements of the flow of digesta and pH. *British Journal of Nutrition*, **36** 497–510

Brown, J.A. and Cline, T.R. (1974) Urea excretion in the pig: an indicator of protein quality and amino acid requirements. *Journal of Nutrition*, **104**, 542–545

Carpenter, K.J. (1960) The estimation of available lysine in animal protein feeds. *Biochemical Journal*, **77**, 604–610

Carpenter, K.J., Ellinger, G.M., Munro, M.I. and Rolfe, E.J. (1957) Fish products as protein supplements to cereals. *British Journal of Nutrition*, **11**, 162–172

Cherian, G., Sauer, W.C. and Thacker, P.A. (1988) Effects of predigestion factors on the apparent digestibility of protein for swine determined by the mobile nylon bag technique. *Journal of Animal Science*, **66**, 1963–1968

Cousins, B.W., Tanksley, T.D., Jr., Knabe, D.A. and Zebrowska, T. (1981) Nutrient digestibility and performance of pigs fed sorghums varying in tannin concentration. *Journal of Animal Science*, **53**, 1524–1537

Darcy, B. and Laplace, J.P. (1980) Digestion in the pig small intestine. 1. Definition of digesta collection conditions. *Annales de Zootechnie*, **29**, 137–145

Darcy-Vrillon, B. and Laplace, J.P. (1985) Ileal amino acid digestibility measurement in pigs fed high fiber diets: ileo-rectal anastomosis versus ileo-colic post-valve fistulation. In *Report from National Institute of Animal Science* (eds A. Just and H. Jørgensen), No. 580, pp. 184–187

Darcy, B., Laplace, J.P. and Villiers, P.A. (1980) Collection of digesta flowing into the large intestine after post ileocolic valve fistulation: preliminary results. *Reproduction, Nutrition and Development*, **20**, 1197–1202

Darcy, B., Laplace, J.P. and Villiers, P.A. (1981) Digestion in the pig small intestine. 4. Kinetics of the passage of digesta at the ileo-caeco-colic junction and digestion balances according to the dietary starch and protein. *Annales de Zootechnie*, **30**, 31–62

De Lange, C.F.M., Sauer, W.C. and Souffrant, W. (1989) The effect of protein status of the pig on the recovery and amino acid composition of endogenous protein in digesta collected from the distal ileum. *Journal of Animal Science*, **67**, 755–762

De Lange, C.F.M., Sauer, W.C., Mosenthin, R. and Souffrant, W.B. (1989) The effect of feeding different protein-free diets on the recovery and amino acid composition of endogenous protein collected from the distal ileum and feces in pigs. *Journal of Animal Science*, **67**, 746–754

Den Hartog, L.A., Huisman, J., Thielen, W.J.G., Van Schayk, G.H.A., Boer, H. and Van Weerden, E.J. (1988) The effect of including various structural polysaccharides in pig diets on ileal and faecal digestibility of amino acids and minerals. *Livestock Production Science*, **18**, 157–170

Den Hartog, L.A., Van Leeuwen, P., Huisman, J., Van Heugten, E., Van Ommeren, H.J. and Van Kleef, D. (1988) Comparison of ileal digestibility data obtained from pigs provided with a different type of cannula. In *Digestive Physiology in the Pig* (eds L. Buraczewska, S. Buraczewski, B. Pastuzewska and T. Zebrowska), Polish Academy of Sciences, Jablonna, pp. 275–282

Den Hartog, L.A., Verstegen, M.W.A. and Huisman, J. (1989) Amino acid digestibility in pigs as affected by diet composition. In *Absorption and Utilization of Amino Acids* (ed M. Friedman), C.R.C. Press, Boca Raton, Florida (in press)

Dierick, N., Vervaeke, I., Decuypere, J. and Hendricks, H.K. (1983) Influence of the nature and level of crude fibres on the apparent ileal and fecal digestibility of dry matter, proteins and amino acids and on the nitrogen retention in swine. *Révue de l'Agriculture*, **36**, 1691–1712

Dierick, N.A., Vervaeke, I.J., Decuypere, J.A., Van der Heyde, H. and Hendricks, H.K. (1987) Correlation between ileal and faecal digested protein and organic matter to production performances in growing pigs. In *EAAP Publication 35. Proceedings of the 5th International Symposium on Protein Metabolism and Nutrition*, Vol. 37, Rostock, GDR, pp. 50–51

Dixon, J.M. and Wilkinson, W.S. (1957) Surgical technique for the exteriorization of the ureters of the chicken. *American Journal of Veterinary Research*, **18**, 665–667

Drochner, W. (1984) The influence of changing amounts of crude fibre and pectic components on precaecal and postileal digestive processes in the growing pig. *Advances in Animal Physiology and Animal Nutrition*, **14**, 125

Dudley-Cash, W.A. and Halloran, H.R. (1988) A better test for protein availability may exist. *Feedstuffs*, **60**, 14

Easter, R.A. and Tanksley, T.D. (1973) A technique for re-entrant ileo-caecal cannulation of swine. *Journal of Animal Science*, **36**, 1099–1103

Eggum, B.O. (1970) Blood urea measurement as a technique for measuring protein quality. *British Journal of Nutrition*, **24**, 983–988

Eggum, B.O. (1973) The levels of blood amino acids and blood urea as an indicator of protein quality. In *Proteins in Human Nutirtion* (eds J.W.G. Porter and B.A. Rolls), Academic Press, London, pp. 317–327

Eggum, B.O. (1989) Biochemical and methodological principles. In *Protein Metabolism in Farm Animals* (eds H.D. Bock, B.O. Eggum, A.G. Low, O. Simon and T. Zebrowska), Oxford University Press, Oxford, pp. 1–52

Elbers, A.R.W., Den Hartog, L.A., Verstegen, M.W.A. and Zandstra, T. (1989) Between and within herd variation in the digestibility of feed for growing-fattening pigs. *Livestock Production Science*, **23**, 183–194

Fadel, J.G., Newman, C.W., Newman, R.K. and Graham, H. (1988) Effects of extrusion cooking of barley on ileal and faecal digestibilities of dietary components in pigs. *Canadian Journal of Animal Science*, **68**, 891–898

Farrell, D.J. (1978) Rapid determination of metabolizable energy of foods using cockerels. *British Poultry Science*, **19**, 303–308

Fuller, M.F., Baird, B., Cadenhead, A. and Aitken, R. (1981) An assessment of amino acid digestibility at the terminal ileum as a measure of the nutritive value of proteins for pigs. *Animal Production*, **32**, 396

Fuller, R. and Coates, M.E. (1983) Influence of the intestinal microflora on nutrition. In *Physiology and Biochemistry of the Domestic Fowl*, Vol. 4 (ed B.M. Freeman), Academic Press, London, pp. 55–61

Fuller, M.F. and Livingstone, R.M. (1982) *Annual Report of Studies in Animal Nutrition and Allied Sciences*, Vol. 39, Rowett Research Institute, Aberdeen, p.45

Furuya, S., Takahashi, S. and Omori, S. (1974) The establishment of T-piece cannula fistulas into the small intestine of the pig. *Japanese Journal of Zootechnical Science*, **45**, 42–44

Furuya, S., Sakamoto, K. and Takahashi, S. (1979) A new *in vitro* method for the estimation of digestibility using the intestinal fluid of the pig. *British Journal of Nutrition*, **41**, 511–520

Furuya, S. and Takahashi, S. (1980) Factors influencing digestibility by swine – relationship of level of feeding and retention time in the intestines to digestibility. *Japanese Journal of Zootechnical Science*, **51**, 33–39

Graham, H. and Åman, P. (1986) Circadian rhythm in composition of duodenal and ileal digesta from pigs fitted with T-cannulas. *Animal Production*, **43**, 133–140

Graham, H., Åman, P., Newman, R.K. and Newman, C.W. (1985) Use of a nylon-bag technique for pig feed digestibility studies. *British Journal of Nutrition*, **54**, 719–726

Graham, H., Lowgren, W. and Åman, P. (1989) An *in vitro* method for studying digestion in the pig. 2. Comparison with *in vivo* ileal and faecal digestibilities. *British Journal of Nutrition*, **61**, 689–698

Green, S. (1988a) A note on amino acid digestibility measured in pigs with pre- or post-valve ileo-rectal anastomoses, fed soyabean, pea and meat meals. *Animal Production*, **47**, 317–320

Green, S. (1988b) Effect of dietary fibre and caecectomy on the excretion of endogenous amino acids from adult cockerels. *British Poultry Science*, **29**, 419–429

Green, S., Bertrand, S.L., Duron, M.J.C. and Maillard, R. (1987a) Digestibilities of amino acids in maize, wheat and barley meals, determined with intact and caecectomized cockerels. *British Poultry Science*, **28**, 631–641

Green, S., Bertrand, S.L., Duron, M.J.C. and Maillard, R. (1987b) Digestibilities of amino acids in soyabean, sunflower and groundnut meals, determined with intact and caecectomized cockerels. *British Poultry Science*, **28**, 643–652

Green, S., Bertrand, S.L., Duron, M.J.C. and Maillard, R.A. (1987) Digestibility of amino acids in maize, wheat and barley meal, measured in pigs with ileo-rectal anastomosis and isolation of the large intestine. *Journal of the Science of Food and Agriculture*, **41**, 29–43

Green, S. and Kiener, T. (1989) Digestibilities of nitrogen and amino acids in soya-bean, sunflower, meat and rapeseed meals measured with pigs and poultry. *Animal Production*, **48**, 157–180

Grühn, K. (1974) Excretion of amino acids in urine and faeces of colostomized and intact hens – contribution to the technique of amino acid absorbability. 2. Hydrolysis of faeces and droppings of colostomized and untreated animals. *Archiv für Tierernährung*, **24**, 75–83

Grühn, K. and Raue, B. (1974) Excretion of amino acids in urine and faeces of colostomized and intact hens – contribution to the technique of amino acid absorbability. 1. Dependence of urinary amino acid excretion on amino acid intake. *Archiv für Tierernährung*, **24**, 67–73

Hagemeister, H. and Erbersdobler, H. (1985) Chemical labelling of dietary protein by transformation of lysine to homoarginine: a new technique to follow intestinal digestion and absorption. *Proceeedings of the Nutrition Society*, **44**, 133A

Hall, D.D., Fernandez, J.A., Jørgensen, H. and Boisen, S. (1987) Prediction of ileal and faecal digestibility of crude protein and amino acids in pigs diets by NDF and NDF-nitrogen analyses. In *EAAP Publication No. 35. Proceedings of the 5th International Symposium on Protein Metabolism and Nutrition*, Vol. 37, Rostock, GDR, pp. 54–55

Haydon, K.D., Knabe, D.A. and Tanksley, T.D., Jr. (1984) Effects of level of feed intake on nitrogen amino acid and energy digestibility measured at the end of the small intestine and over the total digestive tract of growing pigs. *Journal of Animal Science*, **59**, 717–724

Hennig, U., Noel, R., Herrmann, U., Wünsche, J. and Mehnert, E. (1986) Nutrition – physiologic studies in pigs with ileo-rectal anastomoses. 1. Operation methods, biochemical and morphological findings. *Archiv für Tierernährung*, **7**, 585–596

Henry, Y., Vogt, H. and Zoiopoulos, P.E. (1988) Pigs and poultry (Feed evaluation and nutritional requirements). *Livestock Production Science*, **19**, 299–354

Huisman, J., Van Weerden, E.J. Van Leeuwen, P., Van Hof, G. and Sauer, W.C. (1984) Effect of the insertion of cannulas on the rate of passage and nutrient digestibilities in pigs. *Journal of Animal Science*, **59** (suppl.), 271

Hurrell, R.F. and Carpenter, K.J. (1976) An approach to the rapid measurement of 'reactive lysine' in foods by dye binding. *Proceedings of the Nutrition Society*, **35**, 23A

Imbayashi, K., Kametaka, M. and Hatano, T. (1955) Studies on digestion in the domestic fowl. 1. 'Artificial anus operation' for the domestic fowl and the passage of the indicator throughout the digestive tract. *Tohoku Journal of Agricultural Research*, **6**, 99–117

Jagger, S., Cole, D.J.A. and Wiseman, J. (1987) Effect of diet formulation using apparent faecal and ileal amino acid digestibility of heated fish meals on pig performance. *Animal Production*, **44**, 486

Johnston, J. and Coon, C.N. (1979) A comparison of six protein quality assays using commercially available protein meals. *Poultry Science*, **58**, 919–927

Just, A., Jørgensen, H. and Fernandez, J.A. (1981) The digestive capacity of the caecum-colon and the value of the nitrogen absorbed from the hind gut for protein synthesis in pigs. *British Journal of Nutrition*, **46**, 209–219

Just, A., Fernandez, J.A. and Jørgensen, H. (1983) The net energy value of diets for growth in pigs in realtion to the fermentative processes in the digestive tract and the site of absorption of the nutrients. *Livestock Production Science*, **10**, 171–186

Just, A., Jørgensen, H. and Fernandez, J.A. (1985) Correlations of protein deposited in growing female pigs to ileal and faecal digestible protein and amino acids. *Livestock Production Science*, **12**, 145–159

Just, A., Sauer, W.C. and Bech-Andersen, S. (1980) The influence of the hind-gut microflora on the digestibility of protein and amino acids in growing pigs elucidated by addition of antibiotics to different fractions of barley. *Zeitschrift für Tierphysiologie Tierernährung und Futtermittelkünde*, **43**, 83–91

Kakade, M.L. and Liener, I.E. (1969) Determination of available lysine in proteins. *Analytical Biochemistry*, **27**, 273–280

Kessler, J.W., Nguyen, T.H. and Thomas, O.P. (1981) The amino acid excretion values in intact and caecectomized negative control roosters used for determining metabolic plus endogenous urinary losses. *Poultry Science*, **60**, 1576–1577

Kesting, U., Klukas, H., Englemann, H. and Bolduan, G. (1986) Methodological studies with pigs of ileal flow measuring by means of big intestinal cannulae. *Archiv für Tierernährung*, **36**, 793–802

Kotb, A.R. and Luckey, T.D. (1972) Markers in nutrition. *Nutrition Abstracts and Reviews*, **42**, 813–845

Kussaibati, R., Guillaume, J. and Leclerq, B. (1982) The effects of the gut microflora on the digestibility of starch and proteins in young chicks. *Annals de Zootechnie*, **31**, 483–488

Laplace, J.P. and Borgida, L.P. (1976) Physiological problems posed by chronic re-entrant fistulation of the ileum in the pig. *Annales de Zootechnie*, **25**, 361–371

Laplace, J.P., Darcy-Vrillon, B., Perez, N.M., Henry, Y., Giger, S. and Sauvant, D. (1989) Associative effects between two fibre sources on ileal and overall digestibility of amino acids, energy and cell-wall components in growing pigs. *British Journal of Nutrition*, **61**, 75–88

Leibholz, J. (1986) The utilization of lysine by young pigs from nine protein concentrates compared with free lysine in young pigs fed *ad lib. British Journal of Nutrition*, **55**, 179–186

Livingstone, R.M., Baird, B.A., Atkinson, T. and Crofts, R.M.J. (1980) Circadian variation in the apparent digestibility of diets measured at the terminal ileum in pigs. *Journal of Agricultural Science, Cambridge*, **94**, 399–405

Livingstone, R.M., Fowler, V.R. and McWilliam, R. (1982) The effect of two antibacterial agents in a pig diet on the digestibility of protein and organic matter measured at the terminal ileum and over the whole gut. *Animal Production*, **34**, 403–404

Livingstone, R.M. and McWilliam, R. (1985) The effect of terminal ileum cannulation on the performance of growing pigs. *British Veterinary Journal*, **141**, 186–191

Lovern, J.A. (1965) Some analytical problems in the analysis of fish and fish products. *Journal of the Association of Official Agricultural Chemists*, **48**, 60–68

Low, A.G. and Partridge, I.G. (1984) The value of *in vivo* and *in vitro* digestibility measurements of amino acids as predictors of performance and carcass composition in pigs. In *Proceedings of VI Symposium on Amino Acids*, Polish Scientific Publishers, Warsaw, pp. 77–87

Low, A.G. and Zebrowska, T. (1989a) Digestion in pigs. In *Protein Metabolism in Farm Animals* (eds H.D. Bock, B.O. Eggum, A.G. Low, O. Simon and T. Zebrowska), Oxford University Press, Oxford, pp. 53–121

Low, A.G. and Zebrowska, T. (1989b) Digestion in poultry. In *Protein Metabolism in Farm Animals* (eds H.D. Bock, B.O. Eggum, A.G. Low, O. Simon and T. Zebrowska), Oxford University Press, Oxford, pp. 122–142

Lowgren, W., Graham, H. and Åman, P. (1989) An *in vitro* method for studying digestion in the pig. 1. Simulating digestion in the different compartments of the intestine. *British Journal of Nutrition*, **61**, 673–687

Markowitz, J., Archibald, J. and Downie, H.G. (1954) *Experimental Surgery*, 3rd edn. London, Baillière, Tindall and Cox

Mason, V.C., Bech-Andersen, S. and Rudemo, M. (1980) Hydrolysate preparation for amino acid determinations in feed constituents. In *Proceedings 3rd EAAP Symposium on Protein Metabolism and Nutrition* (eds H.J. Oslage and K. Rohr), Braunschweig, FRG, pp. 351–355

McNab, J.M. (1973) The avian caeca: a review. *Worlds Poultry Science Journal*, **29**, 251–263

McNab, J.M. (1979) The concept of amino acid availability in farm animals. In *Recent Advances in Animal Nutrition*, (eds W. Haresign and D. Lewis), London, Butterworths, pp. 1–9

McNab, J.M. and Shannon, D.W.F. (1971) Studies on the potential value of maize germ meal using colostomised laying hens. *Journal of the Science of Food and Agriculture*, **22**, 600–601

McNab, J.M. and Shannon, D.W.F. (1974) The nutritive value of barley, maize, oats and wheat for poultry. *British Poultry Science*, **15**, 561–567

Michel, M.C. (1966) Metabolism of the intestinal flora of the pig. 1. Degradation of the L and D forms of amino acids. *Annales de Biologie Animale, Biochimie, Biophysique*, **6**, 33–46

Moughan, P.J. and Smith, W.C. (1987) A note on the effect of cannulation of the terminal ileum of the growing pig on the apparent ileal digestibility of amino acids in ground barley. *Animal Production*, **44**, 319–321

Moughan, P.J., Smith, W.C., Kies, A.K. and James, K.A.C. (1987) Comparison of the ileal digestibility of amino acids in ground barley for the growing rat and pig. *New Zealand Journal of Agricultural Research*, **30**, 59–66

Münchow, H. and Bergner, H. (1967) Examination of techniques for protein evaluation of feedstuffs. Part 2. *Archiv für Tierernährung*, **17**, 141–150

Owsley, W.F., Knabe, D.A. and Tanksley, T.D., Jr. (1981) Effect of sorghum particle size on digestibility of nutrients at the terminal ileum and over the total digestive tract of growing-finishing pigs. *Journal of Animal Science*, **52**, 557–566

Parsons, C.M. (1984) Influence of caecectomy and source of dietary fibre or starch on excretion of endogenous amino acids by laying hens. *British Journal of Nutrition*, **51**, 541–548

Partridge, I.G., Simon, O. and Bergner, H. (1986) The effects of treated straw meal on ileal and faecal digestibility of nutrients in pigs. *Archiv für Tierernährung*, **36**, 351–359

Payne, W.L., Combs, G.F., Kifer, R.R. and Snyder, D.G. (1968) Investigation of protein quality–ileal recovery of amino acids. *Federation Proceedings*, **27**, 1199–1203

Pederson, B. and Eggum, B.O. (1981) Prediction of protein digestibility by *in vitro* procedures based on two multi-enzyme systems. *Zeitschift für Tierphysiologie Tierernährung und Futtermittelkünde*, **45**, 190–200

Petry, H. and Handlos, B.M. (1978) Investigations for determining the digestibility of nutrients and feed energy using the nylon bag technique in swine. *Archiv für Tierernährung*, **28**, 531–543

Picard, M., Bertrand, S., Genin, F. and Maillard, R. (1984) Digestibility of amino acids: interest of the ileo-rectal shunt technique in the pig. *Journées Recherche Porcine en France*, **16**, 355–360

Pond, W.G., Pond, K.R., Ellis, W.C. and Matis, J.H. (1986) Markers for estimating digesta flow in pigs and effects of dietary fiber. *Journal of Animal Science*, **63**, 1140–1149

Raharto, Y. and Farrell, D.J. (1984a) A new biological method for determining amino acid digestibility in poultry feedstuffs using a simple cannula and the influence of dietary fibre on endogenous amino acid output. *Animal Feed Science and Technology*, **12**, 29–45

Raharto, Y.C. and Farrell, D.J. (1984b) Effects of caecectomy and dietary antibiotics on the digestibility of dry matter and amino acids in poultry feeds determined by excreta analysis. *Australian Journal of Experimental Agriculture and Animal Husbandry*, **24**, 516–521

Rérat, A., Jung, J. and Kande, J. (1988) Absorption kinetics of dietary hydrolysis products in conscious pigs given diets with different amounts of fish protein. 2. Individual amino acids. *British Journal of Nutrition*, **60**, 105–120

Rérat, A., Vaissade, P. and Vaugelade, P. (1979) Absorption kinetics of amino acids and reducing sugars during digestion of barley or wheat meals in the pig: preliminary data. *Annales de Biologie Animale, Biochemie, Biophysique*, **19**, 739–747

Rérat, A., Vaissade, P. and Vaugelade, P. (1988) Quantitative measurement of endogenous amino acid absorption in unanesthetized pigs. *Archiv für Tierernährung*, **38**, 463–480

Rérat, A., Vaugelade, P. and Villiers, P. (1980) A new method for measuring the absorption of nutrients in the pig: a critical examination. In *Current Concepts of Digestion and Absorption* (eds A.G. Low and I.G. Partridge), NIRD, Reading, pp. 177–216

Roach, A.G., Sanderson, P. and Williams, D.R. (1967) Comparison of methods for the determination of available lysine value in animal and vegetable protein sources. *Journal of the Science of Food and Agriculture*, **18**, 274–278

Roos, N., Hagemeister, H. and Scholtissek, J. (1990) Protein digestibility measured by ^{15}N and homoarginine. *Proceedings of the Nutrition Society*, (in press)

Salter, D.N. (1973) The influence of gut microorganisms on utilization of dietary proteins. *Proceedings of the Nutrition Society*, **32**, 65–71

Salter, D.N., Coates, M.E. and Hewitt, D. (1974) The utilization of protein and excretion of uric acid in germ-free and conventional chicks. *British Journal of Nutrition*, **31**, 307–318

Sauer, W.C., Aherne, F.X. and Thacker, P. (1979) Comparison of amino acid digestibilities in normal and cannulated pigs. In *University of Alberta 58th Annual Feeders Day Report*, University of Alberta Faculty of Extension, Edmonton, pp. 28–29

Sauer, W.C., Den Hartog, L.A., Huisman, J., Van Leeuwen, P. and De Lange, C.F.M. (1989) The evaluation of the mobile nylon bag technique for determining the apparent protein digestibility in a wide variety of feedstuffs for pigs. *Journal of Animal Science*, **67**, 432–440

Sauer, W.C., Jørgensen, H. and Berzins, R. (1983) A modified nylon bag technique for determining apparent digestibilities of protein in feedstuffs for pigs. *Canadian Journal of Animal Science*, **63**, 233–237

Sauer, W.C. and Ozimek, L. (1986) Digestibility of amino acids in swine: results and their practical application. A review. *Livestock Production Science*, **15**, 367–388

Sauer, W.C., Stothers, S.C. and Parker, R.J. (1977) Apparent and true availabilities of amino acids in wheat and milling by-products for growing pigs. *Canadian Journal of Animal Science*, **57**, 775–784

Sibbald, I.R. (1979) A bioassay for available amino acids and true metabolizable energy in feedstuffs. *Poultry Science*, **58**, 934–939

Sibbald, I.R. (1986) *The T.M.E. System of Feed Evaluation: Methodology, Feed Composition Data and Bibliography*. Technical Bulletin 1986-4E. Research Branch, Agriculture Canada, Ottawa

Sibbald, I.R. (1987) Estimation of bioavailable amino acids in feedingstuffs for poultry and pigs: a review with emphasis on balance experiments. *Canadian Journal of Animal Science*, **67**, 221–300

Simon, O., Bergner, H. and Partridge, I.G. (1987) Estimation of the endogenous N proportions in ileal digesta and faeces in ^{15}N-labelled pigs. *Archiv für Tierernahrung*, **37**, 851–859

Skrede, A., Krogdahl, A. and Austreng, E. (1980) Digestibility of amino acids in raw fish flesh and meat-and-bone meal for the chicken, fox, mink and rainbow trout. *Zeitschrift fur Tierphysiologie Tierernährung und Futtermittelkünde*, **43**, 92–101

Slump, P., Van Beek, L., Janssen, W.M.M.A., Terpestra, K., Lenis, N.P. and Smits, B. (1977) A comparative study with pigs, poultry and rats of the amino acid digestibility of diets containing crude protein with diverging digestibilities. *Zeitschrift für Tierphysiologie Tierernährung und Futtermittelkünde*, **39**, 257–272

Souffrant, W.B., Schumann, B., Matkowitz, R. and Gebhardt, G. (1985) Studies on the absorption of nitrogen and amino acids in the small intestine of growing pigs. 1. Method of the animal experiment, nitrogen content and amino acid composition of the chyme in the small intestine during the feeding of various proteins. *Archiv für Tierernährung*, **35**, 781–789

Summers, D.J., Berzins, R. and Robblee, A.R. (1982) Ileal cannulation of chickens. *Poultry Science*, **61**, 1551–1552

Tanksley, T.D. and Knabe, D.A. (1984) Ileal digestibilities of amino acids in pig feeds and their use in formulating diets. In *Recent Advances in Animal Nutrition 1984* (eds W. Haresign and D.J.A. Cole), Butterworths, London, pp. 75–95

Taverner, M.R. (1979) *Ileal Availability for Pigs of Amino Acids in Cereal Grains*. PhD Thesis, University of New England, Australia

Taverner, M.R. and Campbell, R.G. (1985) Evaluation of the mobile nylon bag technique for measuring the digestibility of pig feeds. In *Digestive Physiology in the Pig*. Report No. 580 of National Institute of Animal Science, Copenhagen, pp. 385–388

Taverner, M.R. and Farrell, D.J. (1981) Availability to pigs of amino acids in cereal grains. 3. A comparison of ileal availability values with faecal, chemical and enzymic estimates. *British Journal of Nutrition*, **46**, 173–180

Taverner, M.R., Hume, I.D. and Farrell, D.J. (1981) Availability to pigs of amino acids in cereal grains. 1. Endogenous levels of amino acids in ileal digesta and faeces of pigs given cereal diets. *British Journal of Nutrition*, **46**, 149–158

Thomas, O.P. and Crissey, S.D. (1983) Recent advances in the field of amino acid bioavailability. *Proceedings of 4th European Symposium on Poultry Nutrition* (ed M. Larbier), Tours, France, pp. 82–90

Van Leeuwen, P., Huisman, J., Verstegen, M.W.A., Baak, M.J., Van Kleef, D.J., Van Weerden, E.J. and Den Hartog, L.A. (1988) A new technique for collection of ileal chyme in pigs. In *Digestive Physiology in the Pig* (eds L. Buraczewska, S. Buraczewski, B. Pastuzewska and T. Zebrowska), Polish Academy of Sciences, Jablonna, pp. 289–296

Waring, J.J. (1969) The nutritive value of fish meal, meat-and-bone meal and field bean meal as measured by digestibility measurements on the adult colostomized fowl. *British Poultry Science*, **10**, 155–163

Wünsche, J., Bock, H.D. and Meinl, M. (1984) Variations between apparent faecal and apparent ileal crude protein and lysine digestibility in pigs. *Archiv für Tierernährung*, **11**, 761–767

Zebrowska, T. (1973) Digestion and absorption of nitrogenous compounds in the large intestine of pigs. *Roczniki Nauk Rolniczych*, **95B3**, 85–90

Zebrowska, T., Buraczewska, L., Pastuszewska, B., Chamberlian, A.G. and Buraczewski, S. Effect of diet and method of collection on amino acid composition of ileal digesta and digestibility of nitrogen and amino acids in pigs. *Roczniki Nauk Rolniczych*, **99B1**, 75–86

Zebrowska, T., Munchmeyer, R., Bergner, H. and Simon, O. (1986) Studies on the secretion of amino acids and of urea into the gastrointestinal tract of pigs. 2. Net secretion of leucine into the small and large intestines. *Archiv für Tierernährung*, **36**, 17–24

Zebrowska, T., Simon, O., Münchmeyer, R., Wolf, E., Bergner, H. and Zebrowska, H. (1982) Flow of endogenous and exogenous amino acids along the gut of pigs. *Archiv für Tierernährung*, **32**, 431–444

7

INFLUENCE OF PALATABILITY ON DIET ASSIMILATION IN NON-RUMINANTS

T.L.J. LAWRENCE
Department of Animal Husbandry, University of Liverpool, Veterinary Field Station, Neston, South Wirral L64 7TE, UK

Introduction

From the other chapters of this book there is an abundance of information to indicate how the nutritive value of complete diets and of those raw materials of which they are compromised may be either determined or predicted. By using such information diets formulated to meet certain specifications are either given restrictedly to animals or are offered *ad libitum*. In either case the anticipation is the same: there will be a certain intake of the diet sufficient to meet nutritional requirements of the animal(s) which have eaten it. If the anticipated intake is not realized, than the pre-determined nutritive value of the diet will have been of limited value relative to meeting the nutritional requirements of the animals for which it was formulated. *In extremis* a diet will have no nutritive value if it is not eaten at all by the animals for which it was specifically designed. Therefore the two interrelated sides of any food evaluation system, that is an assessment of nutrient requirements on the one hand and an assessment of the nutritional value of the food on the other hand, will have as a determinant link the intake of the food in question. Indeed the intake link in the middle is of no less importance than are the two factors that it separates.

If intake is so crucial, then it is essential to describe the factors that affect it. In animals a variety of sensory systems are involved in the selection and ingestion of food. There is likely to be a central integration of these factors (Figure 7.1) but it is important to appreciate at the outset that the factors are not insular in exerting their effects but act together in a concerted fashion. In birds, and to a large extent in mammals, the stages in food ingestion and the sensory systems may be summarized as in Figure 7.2. However, it is likely that the relative importance of the different sensory systems relative to these stages differ between birds and mammals.

The domesticated hen is a nest building (nidifugous) bird and her chick is essentially precocial, that is it can thermoregulate, move about and feed itself almost immediately after hatching, and differs from both altricial birds, which are fed in their nests by their parents for quite long periods of time (e.g. birds of prey and sparrows) and semi-precocial birds which are intermediate in their ability to cope with life so quickly after hatching (e.g. the common and the sooty terns). The important implication in this categorization is that domesticated chicks must have

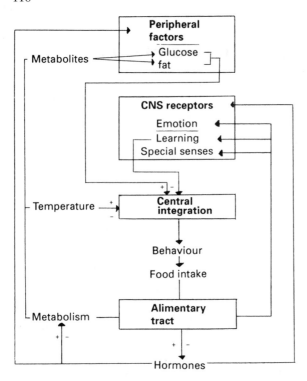

Figure 7.1 Correlation of the control systems that regulate food intake (adapted from Bell, 1976)

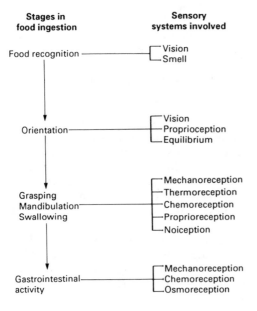

Figure 7.2 Stages in food ingestion (based on Kuenzel, 1983) and sensory systems used (based on Gentle, 1985)

innate behavioural patterns at hatching, which perhaps mitigate against the intake of undesirable substances, but which must be modified later on the basis of the gaining of new knowledge from experiences obtained in indulging in different behavioural patterns copied from their parents. The necessity of learning so early in life appears to be heavily dependent on visual factors to which further, more detailed attention, will be given later. Clearly, however, this overall position differs vastly from that of mammals, where the suckling pattern which becomes established early in life relies on a limited set of instinctive behavioural traits in the first instance although, as time progresses, there will be a progressive learning programme based on the parent's behaviour. As will become evident later, this programme will be far less dependent on visual activity than it is in birds and other senses play a more important part. Also, in the case of the pig, the practice of abrupt weaning at considerably earlier ages than occur in nature necessitates that the learning processes have to be developed earlier and in a different way to that which nature endowed.

In addition to learning programmes, animals appear to have an inherent wisdom which enables them to select a diet which is nutritionally optimal for their physiological status at that particular moment in time. This is particularly so for specific nutrients, for example salt hunger in conditions of adrenal insufficiency in birds.

In the context of these interacting factors and systems and the overall frameworks proposed in Figures 7.1 and 7.2, it is necesary initially to attempt to define palatability.

Palatability

Palatability may be defined as the total perception of a food at the time it is being consumed (Kitchell and Baker, 1972), that is the overall sensory impression received from the food. The weight of food eaten per unit of time is often referred to as the rate of eating and it is of interest in the context of assessment of palatability that rate of eating sometimes declines towards the end of a meal (Auffray and Marcilloux, 1983). Furthermore the total amount of food eaten during a given time, usually 1 day, is often called the voluntary food intake. Palatability is therefore the integrated response of several different sensory systems in the mouth simultaneously (Baldwin, 1976) and Kitchell and Baker (1972) suggest that palatability will reflect flavour, appearance, temperature, size, texture and consistency. Thus it becomes apparent that chiefly the chemosensory receptors of olfaction and taste, but also the receptors of touch and temperature, are involved (Baldwin, 1976). The word taste in the narrowest sense refers to the sensation of taste mediated through the taste buds but its general usage usually includes olfaction, gustation and the common sense of touch whilst taste interpretation in birds, and probably in mammals, is influenced also by temperature, the shape and colour of food and prior experience (Kare, Black and Allison, 1957). Flavour itself is often used to connote the perception of food derived from the input of taste and olfactory receptors plus general chemical receptors from the oral, nasal and pharyngeal cavities but does not generally include the neural responses to the temperature, size, colour, texture and appearance of the food, although all of these factors may affect the flavour of a substance (Kitchell and Baker, 1972). Odour and smell are often used to describe the effect of volatile components of the food on olfactory receptors in the nasal cavity whilst the term 'olfactory stimuli' is used to describe the physico-chemical aspects of odour on the olfactory receptors (Kitchell and Baker, 1972). In addition the phrase 'olfactory impulses' is taken to imply the impulses arising from the stimulation of the olfactory receptors and which run in the olfactory nerves and central neural pathways to the cerebrum.

Sense organs

There is a considerable variation in sense organ structure and function both between and within birds and mammals. It is not only of interest, but also of value in giving an overall understanding, to consider the sensory systems of birds and mammals in the context of how they compare with those of the ancestors from which they were derived and with other wild animals of the same species. In this way a better understanding of palatability in the domesticated fowl and pig, to which the later parts of this paper will address itself, becomes possible.

EYES

In birds the eyes form a greater proportion of the head and are large relative to the size of the brain compared with the situation in mammals. Thus in domesticated fowl the weight ratio of the two eyes to the brain is about 1:1, whilst in man the corresponding ratio is about 1:25 (King–Smith, 1971).

The eyes of birds also differ anatomically from the eyes of mammals. With the exception of penguins most birds are capable of both monocular and binocular vision: in the former case one eye being focused on one subject and in the latter case both eyes being focused. The movement of the two eyes can be independent but becomes coordinated when binocular vision is used. Often birds follow a moving object by moving the entire head and then use either monocular or binocular vision to bring the object into final, sharp focus. The visual field is the angle or field through which the birds can see without moving its head and is determined by the position of the eyes in the head. Therefore birds with flat eyes, such as the pigeon and domesticated fowl, have visual fields that are wider than those of birds that have eyes that are globose or elongated and tubular (as in birds of prey). The shape of the head and the shape and position of the eye in the head determine the proportion of the total visual field that is binocular. Eyes positioned laterally in the head have a smaller binocular field than do eyes that are placed in the front of the head. Comparisons are: owls and hawks with frontal tubular eyes, 60–70° wide and homing pigeons about 24° wide.

The visual acuity of the eye also varies between birds and mammals and between birds themselves. A general concensus is that visual acuity of birds is superior to that of man, although not greatly so but that rate of assimilation of detail is likely to be considerably greater (Pumphrey, 1961). For example, several species of eagle have visual acuities between two and three and a half times greater than man and it appears likely that the superiority in acuity over man and other mammals is a result of the need for quick visual interpretation in birds that are moving in flight quickly. The acuity of the eye depends on the structures and concentrations of rods and cones in the retina. In diurnal birds the sites in the retina known as areae are where cone density is highest and within which there are pit-like depressions or foveas. In birds such as hawks these areas represent the centres of maximal optical resolution and contain in each square millimetre about 1×10^6 cones compared with about 0.4×10^6 in the sparrow. Also, outside the fovea, the hawk retina has about twice as many cones as the human retina. Most mammals have only one area of foveas whereas eagles, hawks and swallows have two foveas in each eye, one central and the other temporal or lateral. In comparison domesticated fowl and turkeys have one central fovea.

All diurnal birds have colour or chromatic vision, whilst nocturnal birds have achromatic vision. In the former case the eye has more cones and less rods: in the latter case vice-versa. The carotenoid proteins in the retina appear to be responsible for its sensitivity to light and colour. These probably control the response of the eye to coloured light. For example, if the human eye becomes adapted to dark conditions, it responds maximally to yellow/green light at a wavelength of about $510 \mu m$ compared with a response at $560 \mu m$ in the light adapted eye. Therefore the eye that is light adapted sees better in light of a longer wavelength than does the dark adapted eye and the shift in response is known as the Purkinje phenomenon (Armington and Thieve, 1956). The Purkinje phenomenon is present in birds, including pigeons and domesticated fowl. Colour sensitivity in birds is also probably linked to, and in the case of some (e.g pigeon) may be totally dependent on, the brightly coloured oil droplets present in most avian retinas (King–Smith, 1969). These are placed in the inner parts of the retina and therefore nearly all light reaching the outer segments of the retina will have passed through oil droplets. The droplets are orange, yellow or red in colour and are suggested as acting as intraocular filters of light by intensifying similar colours but at the same time reducing the discrimination of others, for example blue and violet. Indeed there are some reports indicating a particular insensitivity to the colour blue. On the other hand, some birds such as hawks, woodpeckers and parrots have few or no red oil droplets and probably see blues and violets in the same way as man (Kare and Rogers, 1976) whilst yet others which are strongly nocturnal, such as owls, have only faintly pigmented droplets. Several groups of workers have reported on the spectral limits of domesticated fowl (cited by Kare and Rogers, 1976). The limits appear to be from $700 \mu m$ to $715 \mu m$ at one end of the spectrum to between 395 and $405 \mu m$ at the other end. The maximum sensitivity of the eye of the domesticated chick is at $560 \mu m$ and that of the adult fowl at $580 \mu m$ and Kare and Rogers (1976) suggest that this shift towards the red end of the spectrum is probably a result of an increase in the density of oil droplets with increasing age.

That vision plays an important role in feeding behaviour in many birds is without doubt, so that whilst the owl may represent an exception, hawks demonstrate the supreme linkage of vision to the securing of food. Domesticated fowl appear to respond to both light intensity and colour with evidence of a bi-modal preference in the latter with one peak occurring in the orange region of the spectrum and a second peak in the blue region (references cited by Kare and Rogers, 1976). In contrast it appears that ducklings have a narrower range of preferences with a single peak in the green/yellow region.

Vision relative to the perception of food therefore differs between birds and mammals and between different birds. Interpretation relative to the palatability of food given to the domesticated species, in particular fowl, will be attempted later. In domesticated mammals vision appear likely to play a very much smaller part in food perception as evidenced by the work of Ewbank, Meese and Cox (1974) with pigs.

THE CHEMICAL SENSES

Chemoreception is concerned primarily with the taste buds of the oral cavity and with the olfactory epithelium of the olfactory nasal chamber. There are three main classes of chemical sense: olfaction (smell), gustation (taste) and the common chemical sense. There is often no clear-cut division between the three classes and a single chemical may affect all three senses. The common chemical sense may be regarded as a sense which responds to non-specific stimuli in the environment which are often chemical irritants, for example ammonia, in a number of different areas through stimulation of the free nerve endings of numerous surfaces in, for example, the eyelids, the mouth and the nasal chamber. In birds the nasal cavities are innervated by the trigeminal and olfactory nerves and aquatic birds such as ducks have a greater development of the former compared with the latter. According to Kare and Rogers (1976) this may be a reflection of a functional role in feeding because aquatic birds may be expected to come into contact with chemical irritants more frequently than other birds. This is an interesting hypothesis when considered in the context of the generally accepted primitive nature of the common senses and their likely precursory position relative to the senses of olfaction and taste and the fact that in fishes, representing a lower animal form, taste buds are found extensively over the body surface (Moncrieff, 1967). The major difference between olfaction and gustation is that the former is mediated basically via a telereceptor system which has the capacity to receive airborne chemical stimuli in extreme dilution over great distances whilst the latter usually requires more intimate contact of higher concentrations of the chemical stimuli with the taste receptors which are usually found in the taste buds of the oral cavity (Kare and Rogers, 1976). Olfaction and gustation or taste are large and complex areas and merit separate attention, although both may be closely integrated in the perception of food.

Olfaction

An interesting discourse on many aspects of the evolutionary biology of mammalian olfaction was given by Stoddart (1980). An extensive review of the olfactory systems of birds was written by Kare and Rogers in 1976. In fish there are two sets of sensory apparatus which are concerned with the common chemical sense and in part with olfaction. Taste receptors on barbels occur around the mouth whilst in front of the eyes, and totally unrelated to the respiratory system, is the olfactory sac through which water and dissolved substances flow, passively in most cases. However, there does appear to be a fundamental difference between olfaction in fish and in air breathing tetrapods. This is in spite of the fact that in air breathing tetrapods the peripheral processes which extend from olfactory cells to the surface, where they bear a tuft of several hair like projections which are the actual receptors for the sense of smell, are covered by moist mucus and that, in consequence, materials to be smelled probably go into solution before they can reach the sensory cells (Frandson, 1986).

The olfactory system of fish is very sensitive to minute quantities of substances of very high molecular weight whilst air breathing tetrapods can only perceive substances the vapour pressure of which, under particular environmental conditions, allows them to be volatile. Therefore air breathing animals are restricted to low molecular weight substances with few C atoms because it is only substances such as these which can be taken into the nasal cavities. However, unlike the largely passive situation in fishes, the olfactory perception is not passive and sniffing plays an important part. Stoddart (1980) perceives this basic difference in evolutionary terms in the following way: 'One of the inescapable consequences of man's evolution was that as he raised himself into an upright posture his nose, once close to the earth, came to be held high above it. Vision and hearing became more acute and gradually the newly emergent human being came to rely less and less upon his nose for information about his environment'. If this is accepted, then there is every reason to suppose that there may be differences in olfactory powers both between and within mammals, birds and fishes. Such a supposition is not supported in terms of anatomical differences for the olfactory system is, and in contrast to the visual system, structurally very simple and constant in form throughout not only the five classes of vertebrates but throughout the animal kingdom. There are, however, differences in size of olfactory epithelia (Table 7.1) and these may be related to differences in olfactory acuity to certain substances.

Table 7.1 AREA OF OLFACTORY EPITHELIA AND NEURONE DENSITY IN OLFACTORY EPITHELIA IN DIFFERENT MAMMALS (DODD AND SQUIRRELL, 1980)*

Species	*Total area of olfactory epithelium* (cm^2)	*Density of olfactory neurones* ($\times 10^{-6}$cm^{-2})	*Calculated no. of olfactory neurons* ($\times 10^{-6}$)
Cat	13.9	9.6	133
Dog	150	1.5	225
Frog	c. 0.5	1.5	0.75
Man	2 to 4	3.0	9.0
Rabbit	7.27	9.0	65.5
Squirrel monkey	3.0	–	–

*See original reference for individual references to species.

In birds olfactory systems are more or less developed depending on species and smell is probably of seconday importance in the majority, but not entirely in all, species. According to Kare and Rogers (1976) the sense of smell is well developed in the vulture, in the albatross and in petrels. In comparison it is moderately well developed in the pigeon and in most birds of prey whilst song birds have poorly developed senses of smell. Interesting peculiarities are those of the turkey vulture which exhibits strong olfaction in locating carrion initially but then once in the more immediate area of the carrion relies more on vision, and the Kiwi, which is nocturnal, is the only bird with nostrils at the tip of the beak, has poor vision, feeds largely on earthworms and other hidden food and, unlike other birds, sniffs when foraging for its food. Several aquatic species, including wild geese, have well developed senses of smell. Relative to this background it is difficult to know where domesticated birds of economic importance should be placed. The modern domesticated turkey has an ancestry in the wild turkey of North America (Nixey, 1988) and domesticated fowl have an ancestry in the Jungle Fowl of south east Asia (Wood–Gush, 1959). There is no evidence of where these ancestors fit into any league of olfactory acuity and it seems likely that whilst duck and geese may have more acute senses of smell than either fowl or turkeys, all have relatively poor senses of smell overall and that whilst for wild birds there is no evidence of a pre-eminent role for olfaction amongst their natural functions, in domesticated birds there is a very low response to odours in the environment in general and to food in particular. Vision would appear to be of much greater importance but nevertheless olfaction cannot be totally excluded (Gentle, 1985) even though domesticated fowl have been found to eat grain contaminated with wild onion bulbs which other animals have rejected (Forbes, 1986).

In mammals the data of Table 7.1 leave no doubt that there are considerable differences between species. The acute sense of smell of the dog is evident but there are problems with attempting to fit the pig into any league table of olfactory acuity. It appears, as evidenced by the fact that olfactory bulbectomy has been shown to have no effect on the voluntary food intake of pigs (Baldwin and Cooper, 1979), and that they appear to eat readily a variety of foods with different odours, that olfactory acuity may be low. However, whilst this may be true overall, there can be no doubt that an acute sense of smell for certain aromatic compounds is present as evidenced by the use of pigs to sniff out, from beneath the ground, the truffle fungus to meet the requirements of the human palate. However, the intelligence of the pig is often too great for the ingenuity of man in designing experiments to test its sense of smell because although, as pointed out above, Baldwin and Cooper (1979) found olfactory bulbectomy to be without effect on either rate of intake or on total voluntary food intake, they found it difficult to confirm in absolute terms that the sense of smell had been removed because the pigs learned very quickly to locate hidden food by exploration. Forbes and Blundell (1989) conclude that whilst the odour of a single food does not seem to influence the level of intake, this does not preclude the possibility that smell is used when selecting from a range of available foods. Also, social interactions in older pigs which affect food intake are important and appear more likely to be dependent on sensory systems involving pheromones rather than on sight (Ewbank, Meese and Cox, 1974).

Taste

'The function of taste is to encourage the ingestion of nutrients, to discriminate among foods that are available and, possibly, to avoid those that are toxic. The taste system in a particular species can be expected to complement digestion, metabolism and the dietary requirements of that species' (Kare and Rogers, 1976). Also 'taste appears to be an important factor in the ability of an animal to select food containing elements or a factor in which the animal is deficient' (Frandson, 1986). The neural mechanisms involved in taste are described by many (e.g. Kitchell and Baker, 1972) and the receptors in domesticated animals are located in the taste buds on the tongue and in those on the palate, in the pharynx and on the epiglottis. Underlying each taste bud is a subepithelial plexus of small myelinated and unmyelinated fibres and this plexus is the entry point for taste fibres to infiltrate the taste bud. The opening of taste buds on to the surface of the tongue is via a taste pore and the apical tips of receptors cells extend through this pore. Microvilli, about $2\,\mu m$ long and between 0.1 and $0.2\,\mu m$ wide extend through this pore and it is thought that these play an important part in the transduction process of the taste chemoreceptor. Again there are differences between birds and mammals and between species of both.

The taste buds of birds are intermediate in shape between those of fishes and mammals (resembling those of reptiles) and are also innervated by the glossopharyngeal nerve. Other sensory organs and free nerve endings which may be concerned with pressure sensations (see later) are found in the hard palate and on the beak (Hill, 1971). Numbers of taste buds in animals are given in Table 7.2. It will be observed from this table that birds have relatively few taste buds (with most occurring in the parrot) compared with other animal classes and that in domesticated fowl there is a slight increase with age. It should also be noted that anatomists find some difficulties in deciding precisely what is a taste bud so that some give much higher numbers for domesticated fowl (e.g. Saito, 1966). The increase with age in domesticated fowl contrasts to the position in other birds such as the bullfinch where numbers decrease with increasing age (Kare and Rogers, 1976).

Table 7.2 TASTE BUD NUMBERS IN DIFFERENT ANIMALS*

Animal	Numbers
Chicken†‡	24
Pigeon	37
Bullfinch	46
Starling	200
Duck	200
Parrot	350
Japanese quail	62
Snake	0
Kitten	473
Bat	800
Squirrel	6000
Human and Hare	9000
Sheep and Marsupials	10000
Pig and Goat	15000
Rabbit	17000
Calf	25000
Ox	35000
Catfish	100000

*As cited by Kare and Rogers (1976) from a modification of the data provided by Kare and Ficken (1963) except for ox, squirrel, sheep and marsupials, and hare (from Moncrieff, 1967)
†This number would relate to a fowl of about 3 months of age. In the young chicken the number would be about 12 (Hill, 1971) but as low as 5 with a mean of 8 (Lindenmaier and Kare, 1959).
‡Size (μm): width 30, length 70.

In birds taste buds are found at the base of the tongue, caudal to the the row of large papillae which cross its base and in the form of a letter 'V' with its apex pointing rostrad and on the floor of the pharynx and in the posterior part of the oral cavity. In domesticated fowl the distribution of taste is such that 54% are in the palate, 42% are in the floor of the oral cavity and only 4% in the tongue whilst in duck the taste buds are in the areas of the mouth where there is a prolonged contact with food and would thus allow a gustatory dissemination (Gentle, 1985). The anterior part of the tongue is highly cornified and is devoid of taste buds. Most buds are closely associated with salivary ducts and histochemical studies show that alkaline and acid phosphatase together with 3- and 5- nucleotidase enzymes are present in the mucosal layers. The fact that in birds taste buds are so few in number compared with other animals suggests that chemoreceptors may be important and according to the enzyme inhibition hypothesis a flavoured substance inactivates one or more of the enzymes of the mouth and the resulting temporary imbalance of the normal state gives rise to a characteristic sensation. However, and unlike most mammals, there is no chewing of food to accompany the secretion of saliva and in the fowl, and in most other birds, the food is retained in the mouth for a very short period of time and the saliva secreted per unit weight of food consumed is less than in mammals, particularly man (Dukes, 1955).

In mammals the tongue is very mobile and is covered by papillae. This contrasts to the situation in birds where a horny covering of skin is very prominent and where the tongue is very poorly supplied with muscles and does not, therefore, have the same mobility. It is interesting to note in passing that the tongue of reptiles is very highly developed, for example, the forked tongue of snakes and lizards and the telescopic tongue of chameleons, whilst in fishes it is merely a swelling on the floor of the mouth although in some, notably the pike, teeth are developed on it. In man the taste receptors occur in the tongue and the soft palate, in the epiglottis and at the beginning of the gullet. When adult no taste receptors are found on the underside of the tongue, in the middle of the upperside of the tongue, on the inside of the cheeks, on the lips, on the nostrils or on the uvula. In the child, taste buds occur all over the upper surface of the tongue and on the inside of the cheeks (Moncrieff, 1967). The upper surface of the tongue is densely populated with papillae which give rise to the rough texture which is evident to the human eye. There are four types of papillae and a difference in distribution compared with the pig is apparent (Kitchell and Baker, 1972).

With the anatomical differences described above it is pertinent to consider next how such variation is reflected in acuity of taste and in the detection of different flavours, with particular reference to the domesticated fowl and to man and pigs. In the case of birds there is an opinion expressed in some quarters to the effect that birds do not need a sense of taste because they swallow their food almost as soon as it is siezed by their beaks. However, as Lindenmaier and Kare (1959) point out, this does not preclude the possibility that when birds are faced with a strange food, they will first eat very cautiously. Notwithstanding this qualification the evidence available does not appear to support the opinion expressed above. On the evidence available up to the mid-1950s and based on their own experiments Kare, Black and Allison (1957) concluded that the chick has a sense of taste which is rather stronger than 'rudimentary'. Later reviews of published work in this area (e.g. Gentle, 1972 and 1985) imply that the sense of taste is fairly acute. However, compared with man, for certain, and probably compared with pigs as well, it appears that this sense of taste is both less acute and less sophisticated. The basis for this statement is considered below.

In man there are basically four taste modalities: sweet, salt, bitter and sour (acid). The quality of a stimulus, whether sweet, bitter sour or salty may depend on which neurons respond and on the extent of their response. Also in addition to the four modalities there is the so called 'water response'. In this the response of taste receptors to water occurs when the tongue is rinsed with a low concentration salt solution (0.1 M) followed by rinsing with tap or distilled water. Such responses have been recorded for the cat, the dog and the pig by Liljestrand and Zotterman (1954), for the cat and the pig by Zotterman (1956) and also for domesticated fowl by

Kitchell, Ström and Zotterman (1959) and by Halpern (1963) and for pigeons by Kitchell *et al.* (1959). A similar response appears to be absent in man, in sheep, in goats and in cattle. In the context of the sweet modality Kare and Rogers (1976) conclude that most avian species, including domesticated fowl but excluding parrots, budgerigars and some humming birds, find sugar solutions attractive but will not avidly select them in preference to water if given the chance.

This lack of interest tends to be in marked contrast to the preference exhibited by a number of domesticated mammals, but not all (for example, the cat and the armadillo), for some compounds which are sweet to the human palate (Kare and Ficken, 1963) and details pertaining to the pig will be presented later. Kare and Rogers (1976) point out however that synthetic sweetners, such as saccharin, are nearly always rejected by birds and that a number of factors other than taste may or may not be involved in response to sugar solutions, for example osmotic pressure, viscosity, melting point, nutritive value, toxicity and optical characteristics. Nevertheless, there are differences between different sugars. Gentle (1972) indicates that domesticated fowl do not perceive glucose in solution until it reaches a concentration of 25 g/kg water. Above this concentration they exhibit a significant reduction in intake of glucose solution relative to water. Fructose induces very different and very complex responses whilst there is an indifference to graded levels of sucrose up to 250 g/kg solution, thereby indicating that sweetness as perceived by man has no meaning for domesticated fowl (Kare and Medway, 1959).

In this latter context viscosity also appears to be a factor of little importance in that sugar solutions of different viscosity were without effect on either selection or intake. In these studies the diet was nutritionally adequate but when energy is restricted, domesticated fowl will select a sucrose solution to which they are normally indifferent and will increase their fluid intake of the sucrose solution to counteract the deficiency. As Kare and Rogers (1976) point out the function of taste in nutrition is very much an engima and possibly no physical or chemical quality can be used to predict reliably, under all circumstances, how domesticated fowl will respond to different sugars in solution. Nevertheless, in terms of nutritive value of sugars it is interesting that Kare and Medway (1959) found that xylose was clearly rejected, in common with arabinose and mannose which have been investigated in other work.

Baldwin (1976) reviewed much of the published work pertaining to the choice of sugars by pigs which had been carried out up to that time and from this and from

Table 7.3 PIGLET PREFERENCE FOR RATIONS WITH SUCROSE IN FREE CHOICE TESTS WITH SEVERAL RATIONS (AUMAITRE, 1980)

Sucrose % in feed	0	0.62	3	5	10	20	30
Proportion of feed intake free choice*							
Piglets 10–30 days†	–	–	18	24	58	–	–
Piglets 14–56 days †	1	–	–	–	6	25	68
Piglets 21–56 days	11	89	–	–	–	–	–

*Expressed as % of total of the average intake in each experience.
†Piglets 10–30 days, suckling piglets (Salmon-Legagneur and Fevrier, 1956), piglets 14–56 days, weaned piglets (Combs and Wallace, 1959) and piglets 21–56 days, weaned piglets (Grinstead, Speer and Hays, 1961).

later reported work (Aumaitre, 1980) sucrose appears to be very attractive to young pigs (Table 7.3). Relative to the data of Table 7.3 it is important to note in the work of Combs and Wallace (1959) that food with 300 g/kg sucrose was preferred to all others whereas in the work of Grinstead, Speer and Hays (1961) the food with 6.2 g/kg of sucrose was preferred almost to the exclusion of the diet containing no sucrose. Undoubtedly, and as will be discussed later in general terms, the method of assessment of preference may play a big part in influencing the responses obtained. For example, Wahlstrom, Hauser and Libal (1974) found that young pigs consumed more of a diet containing sucrose than of one in which no sucrose had been included when both diets were available at all times but not if only one of the diets was available singly. Kare, Pond and Campbell (1965), using concentrations varying from 5 to 40 g/l found that glucose, lactose and sucrose were all preferred to water, particularly at the higher concentrations, but that sucrose was the most attractive of the three sugars. Kennedy and Baldwin (1972), using long- and short-term preference tests, investigated sucrose, glucose, sodium saccharin, sodium cyclamate and sodium chloride. Using the crtieria of Goatcher and Church (1970a, b, c, d) where percentage preference = (volume of test solution consumed/total fluid volume consumed) × 100, and where 60–80% indicates a moderate or weak preference, and 80–100% indicates a strong preference, their findings showed big differences between the various sugars.

A preference threshold (concentration at which at least 60% of total fluid intake was in the form of a sucrose solution) was found to be between 0.005 and 0.01 mol/l. At concentrations above 0.03 mol/l a strong preference was shown with little variation between pigs. There were no differences between the long- and short-term tests. In comparison with sucrose, glucose was found to be a less preferred sugar with a threshold preference of between 0.01 and 0.03 mol/l and a strong preference was evident for all concentrations above 0.12 mol/l up to 0.6 mol/l. As for sucrose, long and short-termed tests produced similar responses. For sodium saccharine, in both long and short-termed tests, there was considerable variation between individual pigs and the preference threshold was between 0.005 and 0.01 mol/l but with the most preferred concentration the percentage preference rose to 90% only. Above concentrations of 0.01 mol/l the preference rapidly fell. These results are at variance with those of Kare, Pond and Campbell (1965), where sodium cyclamate and the sugars gave different responses and where there was an indifference to any concentration up to 0.01 mol/l and a rejection of 0.1 M solutions.

Of the other taste modalities in birds, there is an apparent wide range of tolerance to pH on either side of neutrality in drinking water (although, as pointed out later, there is an awareness of changes in pH), and, therefore, a wide tolerance in taste to substances that are to varying degrees sour. In the bitter modality many chemicals have been shown to be offensive to taste at low concentrations. For example, dimethyl anthranilate, used in flavouring human food, has been found to be offensive to growing domesticated fowl and to turkey poults and to depress food intake but, as pointed out later, the reduction in intake may be very dependent on the experimental design used to make the assessment (Kare and Rogers, 1976). Domesticated fowl appear to have distinct taste tolerance thresholds to salt in solution where it has been found that they are willing to indulge in a pattern of delayed drinking for extended periods to avoid consuming a salt solution of a greater concentration than the ability of the kidneys to cope with it. Acceptance of salt up to concentrations in solution of 9 g/kg (0.15 M) only is evident but many birds are indifferent to various concentrations up to the point at which rejection occurs. In addition to these responses Kare and Rogers (1976) pointed out the importance of temperature of water in influencing taste. It appears that domesticated fowl are acutely sensitive to water temperature and that acceptability decreases as water temperature increases above ambient temperature.

Conversely, at the other extreme, water is readily accepted at temperatures down to freezing and the hypothesis is that temperature takes precedence over all chemical stimulants. Finally, there is considerable variation between individuals in taste to a variety of chemicals. There is a distribution of thresholds which is continuous with reactions between birds to a single concentration of one chemical varying from preference to rejection. In this work Kare and Rogers (1976) pointed out that chemical specificity must play a part because an individual that can taste a specific chloride at either very low or very high concentrations will often respond in an average manner only to other concentrations.

In comparison with birds and domesticated fowl, in the pig the thresholds of sour (acid), salt and bitter have been investigated less and not a great deal is known. Unlike birds, pigs appear to tolerate, and even to like, high levels of salt as evidenced by their willingness to eat domestic food waste (swill) with salt levels as high as 50 g/kg, particularly if a plentiful supply of water is readily available. The tolerance to bitter substances is basically unknown whilst sour or acid tolerances appear to be very high as evidenced by the willingness of pigs to eat cereals preserved with very high levels of organic, particularly propionic, acid (e.g Lawrence, 1976a). Lactic acid also appears to be tolerated well (Lawrence, 1973) whilst mixtures of volatile fatty acids at 40 g/kg of the diet have been tolerated well and have been without effect on food intake (Bowland, Young and Milligan, 1971). Responses to mineral acids appear to be undetermined.

Factors interacting with and affecting sensory perception and the design of preference experiments

Some of the more important factors that interact with sensory perception and which dictate that a very careful approach must be made to the design of preference study experiments must next be considered before the practical side of the palatability of poultry and pig foods can be brought into focus. Some aspects of experimental design have been alluded to briefly in consideration of taste acuities.

In the first place there is evidence in domesticated fowl, by using electro-encephalographic (EEG) techniques, that an awareness of differences between dilute solutions of hydrochloric and acetic acids (0.05–0.1 N) and water may not be reflected in any preferential intakes until concentrations reach 0.2 N or above, where there will be an increasing rejection of both acids (Gentle, 1972). Therefore, on this basis, the ability to discriminate chemical solutions from water is better than can be ascertained by studying behavioural responses which will be reflected ultimately in differences in intake. This, and the fact that domesticated fowl in the first 3–5 days of any trial attempting either to assess response to concentrations of any one flavour, or to different flavours of the same concentration, exhibit greater variation than in the periods which follow, suggests strongly than an initial period may be one of learning.

In consequence the time period over which responses are measured must be carefully chosen if confidence is to be placed in results on which conclusions may be based. Some are of the opinion that it is possible to manipulate the design of experiments to produce any type of response (Kare, Black and Allison, 1957) and there is a considerable amount of truth in this. There is evidence that in domesticated fowl the unfamiliar is initially rejected and the ability to taste is not uniformly present in all but that age is of little consequence (Kare and Pick, 1960). In taste experiments with birds, and more specifically with domesticated fowl, water has been used more as a medium than has dry food and there is considerable evidence to suggest that this permits a much greater degree of flavour discrimination (e.g Kare and Pick, 1960). This could be related to the comparatively limited saliva production in birds limiting the conveying of stimuli to taste receptors. With mammals, and more specifically with pigs, there is a dearth of information on which to base a decision on which medium will give the most acute discrimination. Irrespective of this, however, the behavioural studies of taste, in particular, and the measurement of any sensation or perception in general, possibly generates four types of question in the animal (Galanter, 1962):

1. Is anything there?, indicating a detection problem.
2. What is it?, indicating a problem of recognition.
3. How much of it is there?, the scaling problem.
4. Is this different to that?, leading to a discrimination problem.

With food preferences a fifth question is raised: does the animal (do you) like the food substance or does the animal (do you) prefer the food substance to another food substance? Kitchell and Baker (1972) proposed that questions one to four above are directed as 'does the animal' (or 'can you') detect differences, whilst question five is directed as 'does the animal' (or 'do you') prefer. The degree of preference could then be classified as in Table 7.4.

Table 7.4 CLASSIFICATION OF DEGREE OF PREFERENCE (KARE, BLACK AND ALLISON, 1957)

Marked
Moderate
Terminal
Acceptance
Terminal acceptance
Moderate rejection
Marked rejection
Rejection
Unclassified

Most studies of detection and preference in animals have been designed to answer the questions above, and those in Table 7.4, by studying the animals' behaviour using the so called 'preference method'. Kitchell and Baker (1972) describe the important aspects of the preference (free-choice) method in the following manner. They state that the acceptance threshold is the maximum concentration of a noxious substance consumed during the taste period which is equal in quantity to the consumption of a neutral substance such as water or the minimum concentration of an acceptable substance reliably preferred (consumed in equal quantities) to a neutral substance such as water. In contrast the rejection threshold is the minimum concentration of a substance consistently rejected in favour of neutral substances such as water. The rejection and acceptance thresholds are 'taste preference' thresholds and are not usually the same as 'taste detection' thresholds. The vast majority of preference experiments have been used as tools for determining taste thresholds and to analyse responses to substances which fall into one of the four major taste modalities and for determining preferences of substances and individual foods one (some) over the other(s). There are distinct limitations to the meaning of results thus obtained. In the first place an animal presented with a substance or food without choice (no-choice situation) may well exhibit very different responses, both in time scale and ultimate consumption (either in total or per unit of time) compared with that under a free-choice situation. For example the substance used for flavouring human food, dimethyl anthranilate, and referred to previously, has been shown to need a ten-fold increase in a no-choice situation compared with a free-choice situation to give the same reduction in intake in domesticated chicks and turkey poults (Kare and Rogers, 1976). In the situation where exhibition of preference is possible the interaction of post-ingestional factors, mediated perhaps via the feedback from the various receptors present in the gut (see Figure 7.2 and Forbes, 1986), with taste, and their effect on intake of food, has led to the development of preference tasting procedures in which these effects can be minimized; the so-called 'operant flavour testing procedure' as opposed to measuring preference responses based on the intakes of the animals.

In this approach the animals' responses are measured by their learning to press a lever to release food of the type that they have built up a preference for. The procedure has found most use with mammals, dogs and pigs particularly, but still causes problems in interpretation. Both in this type of assessment and in the context of the free-choice and non-choice assessments there is always the possibility that animals become accustomed to a particular taste and/or odour to the point where they develop an aversion to it and then decrease their intake, although with dogs, in operant testing conditions, this has been shown to be preventable by giving a 'reward' for consistently pressing the correct lever. On the other hand animals may take a considerable period of time to become accustomed to a particular taste and/or odour in a food but may eventually be willing to eat as much or more of it compared with a food which they have eaten more avidly more quickly. Clearly then, the effects of the basic physiology of the animal and experimental design can have a profound influence on the interpretation of data obtained from experiments in which assessments of preferences for substances and food have been investigated. In many experiments many of the factors discussed as having the potential to influence palatability have been confounded.

Genetic selection of animals and palatability

In domesticated animals breeding programmes have concentrated strongly on economic factors, particularly growth rate, efficiency of food conversion and attributes of carcass quality. The senses of taste and olfaction, and of visual acuity in the case of birds, have received apparently no attention. In the case of the fowl and the rat domestication has changed the anatomical, physiological and behavioural characteristics. In the latter case the laboratory rat has been found to consume greater quantities of a variety of sugar solutions than the wild Norway rat

(Maller and Kare, 1965), thereby hinting that the function of taste may have been altered in its relationship to the selection and ingestion of food. In the case of domesticated fowl the same workers (Kare and Maller, 1967) were unsure if the relative indifference to sugar solutions, discussed earlier, had any physiological and/or evolutionary significance. The distant history of domesticated fowl shows that initial selection was probably for fighting ability, with a later change in emphasis to selection related to religious practices (Wood–Gush, 1959).

What happened after this is not certain but jungle fowl, from which it is likely that domesticated fowl originated, have been shown to exhibit preference differences for sucrose solutions compared with their domesticated derivatives, differences of preference behaviour more complementary to their nutritional needs than in their domesticated counterpart (Kare and Ficken, 1963). The jungle fowl corrected for diet dilution more effectively and reduced food intake after diet enrichment. Therefore domestication has apparently produced an animal with diminished sensitivity to energy intake regulation; the wild animal being more responsive to the nutritional consequences of its food and liquid intake. The sluggishness in response of domesticated fowl in this case has also found parallels in comparisons of different domesticated breeds. For example, Englemann (1950) found that heavy breeds (for example, Rhode Island Red × Red-barred Plymouth Rock) were more sluggish in their taste reactions than lighter breeds. Also, it has been suggested that the marked variability often evident between individual domesticated fowl in taste preferences may be due to common selection procedures used in practice where there is no selection for taste sensitivity (Gentle, 1972).

There is less evidence of changes of a similar nature to those above having been induced in mammals, though clearly in mammals such as dogs and pigs, the domesticated animal has to perceive a very different range of tastes and smells, and food of different shapes, sizes and textures, compared with its wild counterpart. The changes in the animal which have occurred under domestication to deal with this are largely unknown. In the pig future selection pressures will be to breed an animal with a large appetite but which can convert the additional food into lean meat with high efficiency (Webb, 1989). Selection pressures in these directions will presumably be unable to embrace considerations of the taste and/or olfactory senses. These possibly competing demands are further complicated by the fact that the test station environment designed to obtain accurate food intakes by feeding pigs individually, contrasts strongly to the environment of the commercial farm where pigs are group-fed. The different environments, one with competition at the feeding trough and the other without, may give very different food intake responses to diets which have either different or similar taste and olfactory properties, and different physical forms.

Factors affecting palatability and intake of foods with particular reference to domesticated pigs and fowl

GENERAL

Notwithstanding the many limitations inherent in assessments of palatability, discussed above, the second part of this chapter will concern itself with how various facets of diet, dietary regime and environment and management may alone, or in concert with each other, affect palatability of the diet and consequent intake in domesticated pigs and fowl. Whenever possible the experimental method used to assess palatability will be stated. In most cases total intakes have been measured over a period of time. There are few experiments where rates of intake of diets have been objectively measured. This latter point might be particularly important in situations where animals have to compete with each other at the feeding trough and where individuality in eating rate may affect food intake because of differences in speed of response of the perceptive faculties. On the other hand, and as pointed out already, adjustment over a period of time might nullify initial differences and bring, overall, apparent differences initially very much closer together after a period of time.

It would appear feasible to divide the factors affecting palatability and intake of foods, with particular reference to domesticated pigs and fowl, into four main areas: physical form and appearance of the diet, raw materials with particular reference to anti-nutritional factors, flavouring additives and feeding methods. It is conceded at the start that this is the broadest of approaches and without too much finesse. On the other hand a careful consideration leads to the conclusion that to find a more sophisticated approach that has any practical significance is almost impossible.

PHYSICAL FORM AND APPEARANCE OF DIET

The physical form of the diet can vary according to its component parts and to the manner in which these have been processed, with the way in which the overall diet has been processed and with the method of presentation to the animal (Table 7.5).

Table 7.5 PHYSICAL FORM AND PRESENTATION VARIATION FOR PIGS AND POULTRY

Physical form (component parts and/or whole diet)	*Presentation variation*
Whole cereal grains	Dry feeding
Hammer milled cereal grain and other dietary components	Feeding diet mixed with water (including soaking)
Rolled cereal grains	Feeding components of diets separately (un-mixed)
Various heat treatments of cereal grains and other dietary components	
Pelleting of cereals and diets	

It will only be possible to consider some of these factors in any detail to give some idea of the way in which the majority of experimental results have pointed. In view of the fact that for both pigs and poultry, cereals invariably form the highest proportions of diets, most of the discussion will relate to the manner in which their physical forms are likely to affect intake.

Cold processing methods

Cereals with less than about 160 g moisture per kg, and other dietary components, may be ground through hammer mill screens of varying sizes to give products with different sized particles. In addition cereals can be cold rolled or, alternatively, left whole. There is evidence that the palatability of such materials varies, both within and between the avian and porcine species and that the rate of intake per unit of time and the total intake ultimately achieved, may be affected. In growing pigs rate of intake, as measured by the time taken to eat a given quantity of food, appears to have been subjectively, but not objectively, assessed in several experiments. In cases where subjective assessments have been used it appears that whole cereal grains are eaten much more slowly than cereal grains that have been ground or rolled and that pigs have some difficulty in the prehension of whole grains (e.g. Lawrence, 1970). Also it appears that palatability problems may result from material that has been ground too finely, particularly if the cereal is wheat, because of a cloying, pasty mass produced in the mouth. In terms of voluntary intakes on *ad libitum* feeding in no-choice situations the data of Table 7.6 indicate preferences for certain particle sizes for barley, maize and oats with coarse grinding and whole grains apparently being preferred less than finer ground grains. For barley and wheat the unsubstantiated opinion of many is that in free-choice situations, rolling to give 'flattened discs' from each grain and not coarsely angled particles, is likely to be the preferred form of the pig.

Table 7.6 VOLUNTARY INTAKES OF DIFFERENTLY PROCESSED DRIED CEREALS FOR GROWING PIGS*

Processing	Cracked	Grinding			Rolling	Whole
		Coarse	*Medium*	*Fine*		
Barley†	–	–	–	109	116	100
	–	–	–	143	113	100
Oats‡	–	100	106	116	–	–
Maize§	100	–	108	–	–	–

*Within each set of data lowest intake taken as 100.
†Haugse *et al.* (1966) (two experiments).
‡Crampton and Bell (1946).
§Maxwell *et al.* (1970).

With high moisture (>16 g/kg and up to about 40 g/kg) cereals, however, and where grinding is not possible, some evidence suggests that the softer texture of the grain, due to the higher moisture content, may give very small differences in intake between offering the cereal whole or in rolled forms (Table 7.7). It is interesting in this context that the pigs in the experiment cited in Table 7.7, where *ad libitum* free-choice feeding was used, exhibited no apparent trend to adapt in favour of either one form or another over a period of 70 days. Equally, however, it is clear that the same maize after drying and grinding, and with the same amount of propionic acid added, was preferred almost to the exclusion of the rolled and whole maize when all three forms were available together. The experiment was

Table 7.7 DAILY INTAKES (kg/day) OF GROWING PIGS* OFFERED IN A FREE-CHOICE SITUATION DIETS BASED ON DRIED GROUND MAIZE OR HIGH MOISTURE MAIZE EITHER IN ROLLED OR WHOLE FORMS (LAWRENCE, 1976b)

Form of maize	Dried ground†	High moisture rolled‡	High moisture whole‡
0–14 days	0.78	0.02	0.07
14–35 days	0.89	0.05	0.07
35–49 days	–§	0.46	0.47
49–70 days	–§	0.72	0.62

*Overall mean live weight at start was 19.2 kg and at finish after 70 days was 53.6 kg.
†Moisture content 113 g per kg with 26 g propionic acid per kg added.
‡Moisture content 374 g per kg with 26 g propionic acid per kg added.
§All dried ground maize consumed, none avilable for feeding.

deliberately not of factorial design and so a valid comparison of the dried material in rolled and whole forms, with the wet material, is not possible. If it were, then a fuller clarification of moisture level × processsing interactions would be possible, notwithstanding the fact that there could be a moisture level × propionic acid inclusion level of importance as well, important in the context of its possible effects on the olfactory and taste receptors of the pig. There would appear to be a dearth of published work with very young pigs and with breeding pigs from which conclusions may be drawn.

In poultry the story is very different and there is considerable evidence that shape, size and colour of food particles probably play a more important part in determining intake, and therefore in reflecting palatability, than do the chemical senses. That visual acuity is of relatively greater importance than the chemical senses has been referred to previously in a general discourse and here domesticated fowl in particular, will receive attention.

In adult fowl (White Leghorns) Hurnick *et al.* (1971) investigated in free-choice situations preferences for both the colour of the food and the trough into which the food was placed. They found that red was the most preferred trough colour but the least preferred food colour with a decreasing significant order between other food colours from blue to yellow and green. The newly hatched chick and turkey poult have been found to exhibit innate colour preferences for food in free-choice situations. From the particular survey of the literature made by Gentle (1985) the following points emerge:

1. For newly hatched chicks and turkey poults a preference of green over red may exist though some work has shown, and as mentioned earlier, that a bi-modal preference may exist with peaks in the orange and blue regions of the spectrum.
2. In chicks reared in the dark an initial preference for red and blue may be evident but can be eradicated by giving chicks previous experiences with achromatic light.
3. The attractiveness of an innately non-preferred colour can be increased by simple exposure to it.
4. The strong bias to use colour in learning situations allows, on the basis of colour selection, an avoidance of substances which after ingestion produce ill-effects.
5. Novelty may be an important component in learened illness-induced aversions and in discerning unpalatable substances.

Therefore light intensity and colour appear to be important factors in affecting perception of food by the young chick (and turkey poults as far as the more limited evidence suggests). However, in most modern production systems, where the newly hatched chick cannot learn from studying the behaviour of the broody hen, but only from its equally un-initiated contemporaries within its immediate vicinity, initial and subsequent food intake depend on a learning process of differentiating between food and non-food particles and substances which give ultimately a satisfactory intake to meet requirements. An association has to be formed between the visual, tactile and gustatory stimuli from the food and the positive long-term effects of the food ingestion (Gentle, 1985). In this particular context, whilst there is some evidence of olfactory involvement, it seems that the actual grasping, mandibulation and swallowing of the food is likely to be of the greatest importance. If this is so, then feedback from receptors in the oral cavity, in the pharynx and in the upper oesophagus must play an important part, particularly in relation to particle size and shape of the food itself.

Preference for round, rather than angular shaped, particles has been shown by the newly hatched chick, although such a preference may be modified if angular shaped particles are offered alone for a few days before round objects are re-introduced (Frantz, 1957). In addition, however, the preference for round objects may be associated with their form, that is whether they are solid (ball like preferred) or two-dimensional like a coin (less preferred) (Dawkins, 1968). Relative to this discernment of shape it is common knowledge that adult fowl will readily eat whole cereal grains and experimental results confirm this and often show a preference for whole grain compared with the same material which has been ground. For example, Summers and Leeson (1979) offered (no-choice or free-choice not specified, but probably the latter) laying hens simialr diets but with the 600 g maize and the 400 g barley per kg diet either ground to a meal (mash) or the former coarsely cracked and mixed with whole barley grain. Over seven 28-day periods the daily intakes (g) per bird were 114.5 and 102.0 for the whole (and cracked) based diets and the ground (mash) diet respectively. The adult fowl therefore differs markedly from the pig in prehending whole cereal grains and in achieving high intakes from them. In the young chick the story may be different and the willingness initially to eat whole grains of the major cereals may be limited ultimately by negative feedback mechanisms because of the difficulties of swallowing down a small gullet the relatively large particles. The corollary is that small cereal grains (e.g. millett, which often form a high proportion of cage-bird diets) may be eaten more readily and more consistently.

Many diets given to fowl are mashes in which particle sizes vary for a variety of reasons. Early work on particle sizes in mash diets for broilers (Eley and Hoffman, 1949) indicated a preference for larger sized particles and work with young chicks (Davis *et al.*, 1951) showed that a dislike for finely ground food could be carried to the extreme and be the cause of death because of self-induced starvation. Recently published work with laying hens (Portelle, Caston and Leeson, 1988a) and with broiler chicks (Portella, Caston and Leeson, 1988b) has shown how these two different classes of fowl may respond to different dietary particle sizes within a given diet. In the work with laying hens a diet containing maize (669 g/kg) and soyabean meal (230 g/kg plus vitamins and minerals) was given to groups of four birds per group as either mash, regular crumbles, small crumbles or large crumbles (two groups of birds, one with a 15-day adaption period – as for the first three mentioned forms above – the other without). Crumbles, formed by applying heat and pressure to the mash diet, may legitimately be regarded as a variant in examining particle size differences as much work has shown that changes in nutrient availability *per se* from the process have no or very little effect on intake. 200 g of food were offered at 05.00 h and at each hour subsequently until 20.00 h when food remaining was removed from the trough, sieved, weighed, remixed and returned to the feeders. The crumbled diet contained 323 g/kg of particles greater than 2.36 mm with a nutrient composition similar to that of the overall diet. In contrast the mash contained only 30 g/kg of particles greater than 2.36 mm and these contained a marginal nutrient composition (e.g. less protein, calcium and phosphorus).

Table 7.8 PARTICLE DISAPPEARANCE, EXPRESSED AS PERCENTAGES AND GRAMS OF FOOD, RELATIVE TO THE INITIAL CONCENTRATION IN EACH SEGREGATION IN LAYING HENS GIVEN A DIET OF SIMILAR COMPOSITION IN EITHER MEAL (MASH) OR CRUMBLE FORM (PORTELLA, CASTON AND LEESON, 1988a)

Particle size (mm)	Experiment*									
	1		2		3		4		5	
	%	g	%	g	%	g	%	g	%	g
>2.36	75.2	48.6	68.0	4.0	100.0	105.8	100.0	81.0	–	–
>1.18	55.2	42.7	59.3	33.8	–	–	–	–	73.4	91.7
>0.85	42.0	8.7	60.3	24.3	–	–	–	–	53.0	15.4
>0.71	42.3	3.6	68.3	27.0	–	–	–	–	33.3	4.0
>0.60	31.6	2.4	53.6	9.8	–	–	–	–	30.1	2.6
>0.60	32.0	6.4	38.0	15.8	–	–	–	–	38.8	9.9
Total intake		112.4		114.7		105.8		81.0		123.6

*1. Crumbles offered for a 15-day adaptation period. Then 200 g offered at 05.00 h, food removed at hourly intervals subsequently, sieved, re-mixed and returned.
2. As for 1 using a meal (mash).
3. Crumbles as in 1 but sieved to leave particles >2.36 mm only which were then offered for a 15-day adaptation period after which same procedure as in 1 adopted.
4. Abrupt change from normal crumbles to large (>2.36 mm) crumble particles (i.e. no adaptation period).
5. Crumbles sieved to provide only small particles (<2.36 mm) with a 15-day adaptation period.

Particle disappearance is shown in Table 7.8. The results showed that there was a marked disappearance of larger particles when the birds were given regular crumbles although smaller particles disappeared as the concentration of the larger particles decreased throughout the day. The birds appeared unwilling to eat particles smaller than 0.60 mm but when small particles only (<2.36 mm) were offered, food intake increased (experiment 5, Table 7.8). Also, and unlike some other findings, the birds appeared not to discriminate between regular crumbles (in essence 'large' particles) and mash but to decreases their intakes dramatically when abruptly changed from small to large particles (treatment 4, Table 7.8). Therefore it appears that laying fowl are responsive to different dietary particle sizes and that they will increase or decrease their intakes according to various factors. In the experiments with broiler chicks a maize, soyabean, wheat and fat based diet was offered in either a crumble or pellet form. The major findings from this work may be summarized thus:

1. At all ages particles >1.18 mm were eaten more quickly than were smaller particles.
2. Between 8 and 16 days of age disappearance of particles between >1.18 mm and <2.36 mm was most pronounced.
3. As birds became older their apparent preference for particles >2.36 mm was more marked and possibly suggests that particle preference may be related to beak dimensions.
4. Smaller particles were eaten at the ends of periods of study when no other food remained.
5. Changing particle size abruptly from crumbles to pellets did not adversely affect overall food consumption.

Hot processing methods

Hot processing methods include steam flaking, micronizing, expanding jetsploding and pelleting cereals and other raw materials and the application of various heat and pressure processes to complete diets. Extensive reviews of these fields of study have been published for pigs (Lawrence, 1972; 1976a and 1978; 1982; Tait and Beames, 1988) and for poultry (Vohra, 1972; Wilson and McNab, 1976 and McNab, 1982). In all cases improvements in nutritive value have been established to a greater or lesser extent for all of the processes, whilst for poultry there is a more extensive literature on the effects on intake than there is for pigs.

With pigs the small amount of published work which has investigated the effects of various heat and pressure processes, other than pelleting, on intake, indicates small and equivocal results. In some cases much may depend on the method of feeding used. Thus, whilst a relatively small amount of a flaked cereal may enhance the palatability of a diet for a young pig by virtue of 'opening up' that diet and giving it texture, the same inclusion level, or more particularly a higher level, may have the opposite effect if the diet were to be soaked in water before feeding in that it would become 'soggy'. With other heat and/or pressure processes it is also conceivable that if the temperature and/or pressure are too high, with the result that the gelatinization of cereal starches is excessive, then such material may be too glutenous for the animal to deal with readily. Pelleting, however, has been studied in much greater detail. Vanschoubroek, Coucke and Van Spaendonck (1971) examined the results of a large number of experiments which had investigated the effects of pelleting processes of varying types on a variety of different diets. The main conclusions from their examination were:

1. In free-choice situations young pigs (up to about 60 days of age) prefer pellets to meals (by factors which varied from × 1.5 to 12.0 in individual experiments).
2. In no-choice situations young pigs may eat smaller quantities of pellets than meal but give as good or slightly better performance responses.
3. In growing pigs, in no-choice situations and on unrestricted intakes, the same results as under (2) above may apply.

In their survey of work with young pigs they could reach no conclusions on the effects of intake of sugar inclusion (sucrose mostly, but lactose and molasses as well), of pellet size (where Aumaitre and Salmon–Legagneur (1961) had found a preference for 2.5 mm diameter, cf. 5 or 7 mm diameter, pellets) or the physical consistency of the pellet (hardness). With growing pigs there appeared to be an optimum pellet diameter of between 4 and 8 mm with, in some cases, young growing pigs exhibiting initial difficulties in coping with 13 mm pellets. Published work since has not given any reason to alter views on these conclusions and one of the main problems in this area is that measurements of voluntary intake of meal, compared with pellets, will almost always be confounded by differences in wastage between the two forms: more wastage for meal, less for pellets.

The application of various moderate heat treatments to dietary ingredients for poultry has been shown to improve nutritive value (for example the review of McNab, 1982) but it is impossible to form an unequivocal opinion on the effects on intake although in many cases increases have been found (e.g. Allred, Jensen and McGinnis, 1957). Pelleting (and crumbling – referred to previously), usually with steam, is a widely used process for complete diets and to a lesser extent for dietary ingredients, but again it is difficult to draw unequivocal conclusions from published work in which a very large number of variables have been present. There is evidence that feeding behaviour is changed when a pelleted, compared with a mash, diet is offered. Jensen *et al.* (1962) found that chicks and turkey poults offered mash or pellets spent the following percentages of the day actually feeding: chicks (turkey poults) 14.3 (18.8) and 4.7 (2.2) for mash and pellets respectively. In this context it is interesting to reflect upon the differences in dietary density that result from pelleting. These are likely to be greatest when inclusion rates of bulky, fibrous materials are relatively high but typical density differences are those of 0.57 and 0.71 weights per unit of volume, for mash and pellets respectively, cited by Hussar and Robblee (1962). Such differences might be important in determining both total and rate of intake and Fujita (1974) found that at similar total food intakes the daily eating times for mash and pellets were over 500 min and less than 200 min respectively. Extrapolation of results such as these suggest tentatively that pelleting, because of a decrease in activity and in the energy used to consume food, may improve dramatically, perhaps by as much as 30%, the net availability of the metabolizable energy intake (Jensen *et al.*, 1962; Reddy *et al.*, 1961). However, a resultant disadvantage from being less occupied could be an increases in vice, (Bearse *et al.*, 1949) and, possibly at least in part because of this, in broilers an increased downgrading because of feather loss (Merritt *et al.*, 1960).

RAW MATERIALS AND ANTI-NUTRITIONAL FACTORS

To achieve pre-determined nutritional compositions in diets, raw materials substitutions are made frequently and the effect on overall palatability of diets of individual raw materials is of no less importance than is a consideration of their contribution to the overall nutrient content. Changes in dietary composition by substitutions of different raw materials, because of cost and/or availability, present the animal often with a new taste or smell to which it is unaccustomed. This is the case also at certain points in life and production cycles where diets are changed to meet the changing nutrient requirements of the animal. The earlier part of this paper suggested that olfaction and taste were likely to be of greater importance in controlling food intake in mammals than in birds. In consequence this section will concentrate on some of the more important aspects of this whole area in pigs.

In many cases the relative acceptability of a raw material is inextricably linked with anti-nutritional factors contained within it. In other cases there are inherent characteristics of the raw material, either in smell, or probably more particularly in taste, which determine its attractiveness. In the latter category effects on intakes of young pigs are probably more important than on growing and breeding pigs.

The earlier part of this paper presented data which showed the attractiveness of sucrose in diets for young pigs but at the same time the problems inherent in assessing an optimal level of inclusion (Aumaitre, 1980). There is a general consensus of opinion based on a large body of published work which supports this. Also, after weaning young pigs appear to prefer diets based mainly on constituents derived from cows' milk rather than those based on cereal and soyabean meal (English *et al.*, 1978). Of the major cereal grains, from published work it would appear that in free-choice situations rolled oats are very palatable with a decreasing order after this from wheat to barley to maize (Salmon–Legagneur and Fevrier, 1959). However, there are considerable difficulties in extrapolating from results obtained from free-choice experiments, to the practical situation in which a single diet is offered. This has been shown to be the case with various raw materials (e.g. whey, with and without added salt – (Wahlstrom, Hauser and Libal, 1974) – and sucrose and glucose – (Aumaitre, 1980). Offering of a single feed in practice, found to be unpalatable in a free-choice situation, is likely to give higher intakes than in the free-choice situation and Aumaitre (1980) suggests that whilst preference or aversion for a product is generally established during the first week of a test, and that in the situation where preference is strong and the quantity chosen represents 80% of the total quantity, precocity and continuity of that preference are necessary to give maximum palatability for that product, particularly in the practical situation.

There are a number of raw materials which can cause palatability problems and therefore which can affect intake because of anti-nutritional factors which are present. Interpreted in the widest possible context, amongst the more important of these are the glucosinolates (and their associated compounds) and the tannins of rapeseed meals, the tannins of some cereals and other raw materials, the oxidative degradation products of fats, the products of putrefaction of fish and meat meals, the lectins of legumes, the proteins and glycoproteins of soyabeans and the metabolites of the fungi which may contaminate all raw materials but particularly cereals. Fats present an enigma in that although they are already widely used, and in many circumstances could be used more than is the practice currently, there is insufficient evidence to know the concentrations of certain fatty acids, particularly stearic and palmitic, and their balance with medium-chain and unsaturated fatty acids, which may affect palatability and intake deleteriously. Fowler and Gill (1989) suggest that at present the most simple guideline is to ensure that the melting point of the fat complement is about 5°C below the pig's body temperature of about 37°C. It is impossible to deal here, in other than the most cursory form, with the effects on the palatability of diets of fungal metabolites which are known collectively as mycotoxins. Clearly, however, the review papers of Norred (1986 – aflatoxins) and of Osweiler (1986 – tricothecenes and zearalenone) indicate that the concentrations of these fungal metabolites that will deleteriously affect intake will be considerably above those that will cause both reductions in performance and which will be reflected in both sub-clinical and clinical manifestations of ill-health, even though the latter may cause a decrease in intake *per se*. The dominance of the oilseed rape crop in Canada and in Northern Europe and the desire to use, in particular, extracted meals from it in pig feeding, perhaps places it at the moment in a pre-eminent position for consideration. Although it is likely that the tannins of the seed coat are very bitter and cause problems with palatability, there can be no

Table 7.9 VOLUNTARY INTAKES (kg/per day) AND INTAKES (g) RELATIVE TO METABOLIC BODY WEIGHT ($W^{0.75}$) OF DIETS CONTAINING 200 g/kg OF EITHER SOYABEAN MEAL (SMB) OR THE LOW GLUCOSINOLATE RAPESEED MEALS TOWER (TRSM) OR ERGLU (ERSM) (SINGAM AND LAWRENCE, 1979)

Week of experiment	Liveweights (kg)			Voluntary intake (kg per day)			Daily intake relative to metabolic body weight (g/kg $W^{0.75}$)		
	SBM	*TRSM*	*ERSM*	*SBM*	*TRSM*	*ERSM*	*SBM*	*TRSM*	*ERSM*
1	26.7	25.6	24.8	1.59	1.47	1.38	135	129	124
2	32.6	30.7	29.0	1.90	1.81	1.61	139	138	130
3	37.9	35.5	33.3	2.07	2.03	1.89	135	140	137
4	43.2	40.2	37.7	2.37	2.34	2.17	140	146	143
5	49.6	45.6	42.8	2.47	2.54	2.37	132	145	142
6	55.2	51.5	48.0	2.65	2.71	2.52	131	141	138
7	60.9	57.2	53.1	2.86	2.94	2.74	131	141	139
8	67.3	63.0	58.7	3.12	3.12	2.89	133	139	136

doubt that the 'hot' taste of the glucosinolates are also of extreme importance (Hill and Lee, 1980; Lee, Pittman and Hill, 1980). For young pigs even low glucosinolate meals (<5 μmols/g) substituted for soyabean meal in the diet cause a depression in intake (Baidoo and Aherne, 1987). From this work, regression analyses of intake data for pigs weaned at 3 weeks (6 kg) and at 5 weeks of age (9–10 kg), and given diets with either 193 g soyabean meal per kg diet or 270 g low glucosinolate meal per kg diet at the extremes, or one of three diets between these extremes in which the ratios of soyabean meal to rapeseed meal were 75:25, 50:50, 25:75, showed daily reductions in intake of between 3.5 g and 4.6 g (from different experiments) for each 10 g addition of rapeseed meal per kg diet. Growing pigs of 20 kg, also in no-choice situations, may show an initial reluctance to eat diets containing 200 g/kg of soyabean meal or low glucosinolate rapeseed meals offered *ad libitum* but may, after a period of adjustment, compensate and equal or slightly exceed those offered soyabean meal based diets (Table 7.9 – Singam and Lawrence, 1979). On restricted, but nevertheless generous scales of feeding, low glucosinolate meal at concentrations up to 330 g/kg of diet have been shown to induce minor adaptation problems only for an initial period of a few days when offered to pigs of about 25 kg live weight (Rowan and Lawrence, 1986). In common with work conducted at many other centres, work (unpublished) at this centre suggests that high glucosinolate rapeseed meals (>50 μmol/g) are eaten less readily and that meals containing in excess of 80 μmol/g may be totally refused at dietary concentrations of about 150 g/kg or above.

FLAVOURING ADDITIVES

Flavouring agents have been added to diets for pigs, but much less frequently to diets for poultry, in attempts to improve palatability and to increase intake. With young pigs additions have been primarily to induce the eating of dry food at as young an age as possible; for older pigs usually the intention has been to mask undesirable flavours present in raw materials. The apparent attractiveness of sucrose for young pigs has been discussed previously. Other sweetening agents have been used in attempts to increase intakes of dry food. For example, Aumaitre (1980) investigated high fructose corn syrup in a free-choice situation. This substance had a high sweetening power and was a glucose syrup rich in fructose, the fructose having been produced by glucose isomerization in an industrial process. Although a similar monosaccharide to sucrose, young pigs showed little preference for it compared with sucrose. Various non-sweetening agents have also been tried and the results reported in a very large number of published papers. Once again the use of preference tests within groups of pigs makes any extrapolation to no-choice situations very difficult. One of the few published papers to yield an unequivocal result in a no-choice situation is that which reported extensive investigations by McLaughlin *et al.* (1983). In this work over 129 flavours were examined and a compound with a cheesy aroma was found to increase food intake after weaning. Unpublished work at this centre, which has attempted to mask the undesirable taste of high glucosinolate rapeseed meals using a variety of flavouring agents, has given negative results and a review of the literature does not allow even the most tentative of conclusions to be drawn in the very wide area of flavouring additions and intake.

FEEDING METHODS

Feeding methods may alter the palatability of a diet for an animal. Quite apart from the obvious consideration that free water must be readily available if dry food intakes are to be optimal, feeding the diet dry, or with varying proportions of water, may alter palatability *per se* as well as affecting intake *via* the effects of feed-back mechanisms stimulated by gut fill. This may be particularly so in the case of poultry where, as previously pointed out, acuity of taste is apparently greater for some substances in water compared with offering the same substances in a dry form. However, in terms of the different ingredients present in the diet, poultry appear to be able to balance their nutrient intakes reasonably well from components, of vastly different taste (and physical form), or if presented separately (for example whole cereal grains on the one hand and protein/vitamin/mineral components on the other). This feat pigs appear unable to match with even a low degree of efficiency. There are many other aspects of feeding methods which could be considered as possibly influencing palatability but space precludes other than the most cursory of examinations of the two factors mentioned above.

With pigs, the addition of water to dry food may enhance or detract from flavours present and/or may improve palatability because of an amelioration of the inhibiting effects of some physical forms. For example, a finely ground diet may be unpalatable when offered in a dry form, may be even more unpalatable when a small quantity of water is added to form a thick porridge-like mass (particularly if the cereal component is mostly wheat with a high gluten content) and may only improve when a large quantity of water is added so that a thinner consistency is induced in the mixture. Similarly, work with pigs (Danish National Committee, 1986) and with dogs (Kitchell and Baker, 1972) has shown a preference for pellets fed wet (or semi-moist) compared with offering them dry: a daily increases of about 12% in the former case and, in the latter case over a period of about 6 months in a free-choice situation, 91.5% and 8.5% of the total intake were for the pellets offered semi-moist and dry respectively. It is conceivable that flavours in foods may be altered by water additions but there is a dearth of information on which any

Table 7.10 FOOD AND NUTRIENT INTAKES OF LAYING FOWL OFFERED COMPLETE DIETS OR SPLIT DIETS SIMULTANEOUSLY (KARUNAJEEWA, 1978)

Intake	Age (weeks)	Wheat		Barley		Significance	
		Cereal component separate	Complete diet	Cereal component separate	Complete diet	Wheat	Barley
Food (kg per bird)	25–49	18.70	19.69	19.18	22.17	NS	*
	49–73	18.96	20.82	19.83	23.84	NS	*
Protein (g per hen day)	25–49	17.90	19.0	15.50	19.40	NS	*
	49–73	17.90	19.0	17.30	19.40	NS	*
ME (MJ per hen day)	25–49	1.30	1.34	1.21	1.38	NS	*
	49–73	1.31	1.42	1.24	1.48	NS	*
Calcium (g per hen day)	25–49	3.22	3.21	3.09	3.54	NS	*
	49–73	3.56	3.89	3.65	3.81	*	NS

conclusions may be drawn. With poultry the effects of water temperature on intake were discussed previously but it is not possible to ascertain from the literature the relative preferences likely to be shown for dry or wet mashes and pellets. Soaking in water before feeding has been shown recently to improve the nutritive value and intake of rye based diets for chickens (Ward and Marquardt, 1988). In this context it was thought likely that the anti-nutritional water-soluble pentosans, which are highly viscous, were destroyed in the soaking process by enzymatic processes. If this were the case then the improved intakes would have emanated from a decreased viscosity.

The ability of birds to meet specific nutrient requirements has been used in attempts to feed poultry different components of a diet separately but simultaneously and thereby avoid the process of mixing in the first instance. In such approaches it is clear that the different components of the diet must have considerably different make-ups and must, therefore, present to the bird materials of very different taste, shape and odour. One example of experimental work in which choice-feeding was investigated is that of Karunajeewa (1978) who offered laying hens wheat or barley based diets either complete or with the cereal (in whole grain form) and protein/vitamin/mineral components offered simultaneously but separately. The intake data of Table 7.10 indicate the ability of fowl in these circumstances to balance their food and daily energy intake reasonably well from wheat-based diets but their inability to do so to the same extent from barley-based diets (rate of lay and egg weight were not significantly affected except in the 49–73 week period for the wheat-based diets where egg weight was less for the complete diet). In this experiment the hens were presented with materials of different shapes, sizes, texture, taste, colour and, perhaps, smell, but clearly showed an ability in some cases to meet their needs by adjusting their intakes, albeit imperfectly.

References

Allred, J.B., Jensen, L.S. and McGinnis, J. (1957) Factors affecting the response of chicks and poults to pellet feeding. *Poultry Science*, **36**, 517–523

Armington, J.C. and Thieve, F.C. (1956) Electroretinal demonstration of a Purkinje shift in the chicken eye. *American Journal of Physiology*, **186**, 258

Auffray, P. and Marcilloux, J.C. (1983) Etude de la séquence alimentaire du porc adulte (Study of the feeding pattern in adult pigs). *Reproduction, Nutrition, Developement*, **23**, 517–524

Aumaitre, A. (1980) Palatability of piglet feeds: trial methods and practical results. In *First International Symposium on Palatability and Flavour Use in Animal Feeds, October 1978, Zurich* pp. 86–95.

Aumaitre, A. and Salmon–Legagneur, E. (1961) Les préférences alimentaires du porcelet. 5. Comparison de divers modes de distribution de l'aliment (Feed preference in young in young pigs. 5. Comparison of different ways of distributing the feed). *Annales de Zootechnie*, **10**, 197–203

Baidoo, S.K. and Aherne, F.X. (1987) Canola meal as a protein supplement for growing – finishing pigs. In *66th Annual Feeders' Day Report, University of Alberta*, Faculty of Extension: Alberta, pp. 4–6

Baldwin, B.A. (1976) Quantitative studies on taste preferences in pigs. *Proceedings of the Nutrition Society*, **35**, 69–73

Baldwin, B.A. and Cooper, T.R.(1979) The effects of olfactory bulbectomy on feeding behaviour in pigs. *Applied Animal Ethology*, **5**, 153–159

Bearse, G.E., Berg, L.R., McClary, C.F. and Miller, V.L. (1949) The effect of pelleting chicken rations on the incidence of cannibalism. *Poultry Science*, **28**, 756

Bell, F.R. (1976) Regulation of food intake. *Proceedings of the Nutrition Society*, **35**, 63–67

Bowland, J.P., Young, B.A. and Milligan, L.P. (1971) Influence of dietary volatile fatty acid mixtures on performance and on fat composition of growing pigs. *Canadian Journal of Animal Science*, **51**, 89–94

Combs, G.E. and Wallace, H.D. (1959) Palatable creep feeds for pigs. *Florida Agricultural Experimental Station Bulletin*, **610**, 1–12

Crampton, E.W. and Bell, J.M. (1946) The effect of fineness of grinding on the utilization of oats by market hogs. *Journal of Animal Science*, **5**, 200–210

Danish National Committee for Pig Breeding and Production 1984 (1986). *Svineaval og – Produktion I Danmark*, 1986

Davis, R.L., Hill, E.G., Sloan, H.J. and Briggs, G.M. (1951) Detrimental effect of corn of coarse particle size in rations for chicks. *Poultry Science*, **30**, 325–328

Dawkins, R. (1968) The ontogeny of a pecking preference in domestic chicks. *Zeitschrift für Tierpsychologie*, **25**, 170–186

Dodd, G.H. and Squirrell, D.J. (1980) Structure and mechanism in the mammalian olfactory system. In *Olfaction in Mammals* (ed D.M. Stoddart) Academic Press, London, pp. 35–56

Dukes, H.H. (1955) *The Physiology of Domestic Animals*, 7th edn. Cornell University Press, Ithaca

Eley, C.P. and Hoffman, E. (1949) Feed particle size as a factor in water consumption and elimination. *Poultry Science*, **28**, 215–222

Engelmann, C.(1950) Über den Geschmackassin des Huhnes. IX. *Zeitschrift für Tierphysiologie*, **7**, 84–121

English, P.R., Deligeorgis, S.G., Davidson, F.M., Dias, M.F.M., Smith, J.M. and Fowler, V.R. (1978) Evaluation of alternative diets and feeding systems for early-weaned pigs. *Animal Production*, **26**, 398 (Abstr.)

Ewbank, R., Meese, G.B. and Cox, J.E. (1974) Individual recognition and the dominance hierarchy in the domesticated pig. The role of sight. *Animal Behaviour*, **22**, 473–480

Forbes, J.M. (1986) *The Voluntary Food Intake of Farm Animals*, Butterworths, London

Forbes, J.M. and Blundell, J.E. (1989) Central nervous control of voluntary food intake. *British Society of Animal Production Occasional Publication*, **13**, 7–26

Fowler, V.R. and Gill, B.P. (1989) Voluntary food intake in the young pig. *British Society of Animal Production Occasional Publication*, **13**, 51–60

Frandson, D.R. (1986) *Anatomy and Physiology of Farm Animals*, 4th edn. Lea and Febiger, Philadelphia

Frantz, R.L. (1957) Form preferences in newly hatched chicks. *Journal of Comparative and Physiological Psychology*, **50**, 422–430

Fujita, H. (1974) Quantitative studies on the variations in feeding activity of chickens. 3. Effect of pelleting the feed on the eating patterns and the rate of feed passage through the digestive tract in chicks. *Japanese Poultry Science*, **11**, 210–216

Galanter, E. (1962) Contemporary psychophysics. In *New Directions in Psychology* (eds R. Brown, E. Galanter, E.H. Hess and G. Mandler), Holt, Rinehart and Winston, Inc. New York, pp. 87–156

Gentle, M.J. (1972) Taste preferences in the chicken (Gallus Domesticus). *British Poultry Science*, **13**, 141–155

Gentle, M.J. (1985) Sensory involvement in the control of food intake in poultry. *Proceedings of the Nutrition Society*, **44**, 313–321

Goatcher, W.D. and Church, D.C. (1970a) Taste responses in ruminants. I. Reactions of sheep to sugars, saccharin, ethanol and salts. *Journal of Animal Science*, **30**, 777–783

Goatcher, W.D. and Church, D.C. (1970b) Taste responses in ruminants. II. Reactions of sheep to acids, quinine, urea and sodium hydroxide. *Journal of Animal Science*, **30**, 784–790

Goatcher, W.D. and Church, D.C. (1970c) Taste responses in ruminants. III. Reactions of pygmy goats, normal goats, sheep and cattle to sucrose and sodium chloride. *Journal of Animal Science*, **31**, 364–372

Goatcher, W.D. and Church, D.C. (1970d) Taste responses in ruminants. IV. Reaction of pygmy goats, sheep and cattle to acetic acid and quinine hydrochloride. *Journal of Animal Science*, **31**, 373–382

Grinstead, L.E., Speer, V.C. and Hays, V.W. (1961) Sucrose for baby pigs. *Journal of Animal Science*, **20**, 934

Halpern, B.P. (1963) Gustatory nerve responses in the chicken. *American Journal of Physiology*, **203**, 541–544

Haugse, C.N., Dinusson, W.E., Erickson, D.O. and Bolin, D.W. (1966) Effect of the physical form of barley in rations for fattening pigs. *North Dakota Agricultural Experimental Station Bulletin*, **17**

Hill, K.J. (1971) The structure of the alimentary tract. In *Physiology and Biochemistry of the Domestic Fowl*: Vol. 1, (eds D.J. Bell and B.M. Freeman), Academic Press, London, pp. 1–22

Hill, R. and Lee, P. (1980) The voluntary food intake of young growing pigs given diets containing a high proportion of rapeseed meal. *Proceedings of the Nutrition Society*, **39**, 75A

Hurnik, J.F., Jerome, F.M., Reinhart, B.S. and Summers, J.D. (1971) Color as a stimulus for food consumption. *Poultry Science*, **50**, 944–949

Hussar, N. and Robblee, A.R. (1962) Effects of pelleting on the utilization of feed by the growing chicken. *Poultry Science*, **41**, 1489–1493

Jensen, L.S., Merrill, L.H., Reddy, C.V. and McGinnis, J. (1962) Observations on eating patterns and rate of food passage of birds fed pelleted and unpelleted diets. *Poultry Science*, **41**, 1414–1419

Kare, M.R., Black, R. and Allison, E.G. (1957) The sense of taste in the fowl. *Poultry Science*, **36**, 129–138

Kare, M.R. and Ficken, M.S. (1963) Comparative studies on the sense of taste. In *Olfaction and Taste* Vol. 1, (ed Y. Zotterman), Pergamon Press, London, pp. 285–298

Kare, M.R. and Maller, O. (1967) Taste and food intake in domesticated and jungle fowl. *Journal of Nutrition*, **92**, 191–196

Kare, M.R. and Medway, W. (1959) Discrimination between carbohydrates by the fowl. *Poultry Science*, **38**, 1119–1127

Kare, M.R. and Pick, H.L. (1960) The influence of the sense of taste on feed and fluid consumption. *Poultry Science*, **39**, 697–706

Kare, M.R., Pond, W.C. and Campbell, J. (1965) Observations on the taste reactions in pigs. *Animal Behaviour*, **13**, 265–269

Kare, M.R. and Rogers, J.G. (1976) Sense organs. In *Avian Physiology*, (ed P.D. Sturkie), Springer-Verlag, New York, pp. 30–52

Karunajeewa, H. (1978) The performance of cross-bred hens given free choice feeding of whole grains and a concentrate mixture and the influence of source of xanthophylls on yolk colour. *British Poultry Science*, **19**, 699–708

Kennedy, J.M. and Baldwin, B.A. (1972) Taste preferences in pigs for nutritive and non-nutritive sweet solutions. *Animal Behaviour*, **20**, 706–718

Kitchell, R.L. and Baker, G.G. (1972) Taste preference studies in domestic animals. In *Nutrition Conference for Feed Manufacturers : 6* (eds Swan, Henry and Lewis), Churchill Livingstone, Edinburgh, pp. 158–202

Kitchell, R.L., Ström, L. and Zotterman, Y. (1959) Electrophysiological studies of thermal and taste reception in chickens and pigeons. *Acta Physiologica Scandanavica*, **46**, 133–151

King–Smith, P.E. (1969) Absorption spectra and function of the coloured oil drops in the pigeon retina. *Vision Research*, **9**, 1391–1399

King–Smith, P.E. (1971). Special senses. In *Physiology and Biochemistry of the Domestic Fowl*, Vol. 2, (eds D.J. Bell and B.M. Freeman), Academic Press, London, pp. 1039–1083

Kuenzel, W.J. (1983) Behavioural sequence of food and water intake: its significance for elucidating neural mechanisms controlling feeding in birds. *Bird Behaviour*, **5**, 2–15

Lawrence, T.L.J. (1970) Some effects of including differently processed barley in the diet of the growing pig. 1. Growth rate, food conversion efficiency, digestibility and rate of passage through the gut. *Animal Production*, **12**, 139–150

Lawrence, T.L.J. (1972) Developments in cereal processing and digestion – growing pigs. In *Cereal Processing and Digestion*, U.S. Feed Grains Council, London, pp. 77–106

Lawrence, T.L.J. (1973) Volatile fatty acids as sources of energy in the diet of the growing pig. *International Research Communication System* (73–3) 45–6–1

Lawrence, T.L.J. (1976a) Some effects of processing on the nutritive value of feedstuffs for growing pigs. *Proceedings of the Nutrition Society*, **35**, 237–243

Lawrence, T.L.J. (1976b) High moisture maize grain for growing pigs – some effects on acceptability, digestibility, nitrogen retention and tocopherol supplementation. *Journal of Agricultural Science, Cambridge*, **86**, 315–324

Lawrence, T.L.J. (1978) Processing and preparation of cereals for pig diets. In *Recent Advances in Animal Nutrition* (eds Haresign, William and Lewis), Butterworths, London, pp. 83–98

Lawrence, T.L.J. (1982) Effect of processing on nutritive value of diets for pigs. In *Handbook of Nutritive Value of Processed Foods, Volume II Animal Foodstuffs* (ed Rechcigl, M. Jr.) CRC Press, Boca Raton, Fl., pp. 389–401

Lee, P., Pittam, S. and Hill, R. (1980) The effect of treatments of rapeseed meal and of extracts of the meal on the voluntary food intake of young growing pigs. *Proceedings of the Nutrition Society*, **39**, 76A

Liljestrand, G. and Zotterman, Y. (1954) The water taste in mammals. *Acta Physiologica Scandinavica*, **32**, 291–303

Lindenmaier, P. and Kare, M.R. (1959) The taste end-organs of the chicken. *Poultry Science*, **38**, 545–550

Maller, O. and Kare, M.R. (1965) Selection and intake of carbohydrates by wild and domesticated rats. *Proceedings of the Society for Experimental Biology and Medicine*, **119**, 199–203

Maxwell, C.V., Reimann, E.M., Hoekstra, W.G., Kowalczyk, T., Benevenga, N.J. and Grummer, R.H. (1970) Effect of dietary particle size on lesion development and on the contents of various regions of the swine stomach. *Journal of Animal Science*, **30**, 911–922

McLoughlin, C.L., Baile, C.A., Buckholtz, L.L. and Freeman, S.K. (1983) Preferred flavors and performance of weanling pigs. *Journal of Animal Science*, **56**, 1287–1293

McNab, J.M. (1982) Effects of processing on nutritive value of diets for poultry. In *Handbook of Nutritive Value of Processed Foods, Volume II Animal Foodstuffs* (ed Rechcigl, M. Jr.) CRC Press, Boca Raton, Fl., pp. 403–424

Merritt, E.S., Downs, J.H., Bordeleau, R. and Tinney, B.F. (1960) Growth, variability of growth and market quality of broilers on mash and pellets. *Canadian Journal of Animal Science*, **40**, 7–14

Moncrieff, R.W. (1967) *The Chemical Senses*, 3rd edn, Leonard Hill, London

Nixey, C. (1988) The turkey. *Biologist*, **35**, 35–49

Norred, W.P. (1986) Occurrence and clinical manifestations of aflatoxicosis. In *Diagnosis of Mycotoxicoses* (eds Richard, J.L and Thurston, J.R.) Dordrecht, Martinus Nijhoff, pp. 11–29

Osweiler, G.D. (1986) Occurrence and clinical manifestations of tricothecene toxicoses and zearalenone toxicoses. In *Diagnosis of Mycotoxicoses* (eds Richard, J.L. and Thurston, J.R.) Dordrecht, Martinus Nijhoff, pp. 31–42

Portella, F.J., Caston, L.J. and Leeson, S. (1988a) Apparent feed particle size preference by laying birds. *Canadian Journal of Animal Science*, **68**, 915–922

Portella, F.J., Caston, L.J. and Leeson, S. (1988b) Apparent feed particle size preference by broilers. *Canadian Journal of Animal Science*, **68**, 923–930

Pumphrey, R.J. (1961) The sensory organs: hearing in birds. In *Biology and Comparative Physiology of Birds*, Vol. 2, (ed Marshall, A.J.), Academic Press, New York, pp. 55–86

Reddy, C.V., Jensen, L.S., Merrill, L.H. and McGinnis, J. (1961) Influence of pelleting on metabolizable and productive energy content of a complete diet for chicks. *Poultry Science*, **40**, 1446

Rowan, T.J. and Lawrence, T.L.J. (1986) Ileal apparent digestibilities of amino acids, growth and tissue deposition in growing pigs fed low glucosinolate rapeseed meals. *Journal of Agricultural Science, Cambridge,* **107**, 493–504

Saito, I. (1966) Comparative anatomical studies of the oral organs of the poultry. V. Structures and distribution of taste buds of the fowl. *Bulletin of the Faculty of Agriculture, Miyazahi University,* **13**, 95–102

Salmon–Legagneur, E. and Fevrier, R. (1956) Feed preferences in young pigs. 2. Sugar in rations for young pigs. *Annales de Zootechnie,* **5**, 73–79

Salmon–Legagneur, E. and Fevrier, R. (1959) Feed preference in young pigs. 3. Appetite for some cereals. *Annales de Zootechnie,* **8**, 87–93

Singam, D.R., The Late and Lawrence, T.L.J. (1979) Studies on the acceptibility and nitrogen utilization of Tower and Erglu rapeseed meals by the growing pig. *Journal of the Science of Food and Agriculture,* **30**, 21–26

Stoddart, D.M. (1980) Aspects of the evolutionary biology of mammalian olfaction. In *Olfaction in Mammals* (ed Stoddart, D.M.), Academic Press, London, pp. 1–13

Summers, J.D. and Leeson, S. (1979) Diet presentation and feeding. In *Food Intake Regulation in Poultry* (eds Boorman, K.N. and Freeman, B.M.), British Poultry Science, Edinburgh, pp. 445–469

Tait, R.M. and Beames, R.M. (1988) Processing and preservation of cereals and protein concentrates. In *World Animal Science, B4 Disciplinary Approach,* (ed Ørskov, E.R.) Elsevier, Amsterdam, pp. 151–176

Vanschoubroek, F., Coucke, L. and Van Spaendonck, R. (1971) The quantitative effect of pelleting feed on the performance of piglets and fattening pigs. *Nutrition Abstracts and Reviews,* **41**, 1–9

Vohra, P. (1972) Developments in cereal processing and digestion – poultry. In *Cereal Processing and Digestion,* U.S. Feed Grains Council, London, pp. 53–66

Wahlstrom, R.C., Hauser, L.A. and Libal, G.W. (1974) Effects of low lactose whey, skim milk and sugar on diet palatability and performance of early weaned pigs. *Journal of Animal Science,* **38**, 1267–1271

Ward, A.T. and Marquardt, R.R. (1988) Effect of various treatments on the nutritional value of rye or rye fractions. *British Poultry Science,* **29**, 709–720

Webb, A.J. (1989) Genetics of food intake in the pig. *British Society of Animal Production Occasional Publication,* **13**, 41–50

Wilson, B.J. and McNab, J.M. (1976) The effect of pretreatment on the nutritive value of diets for poultry. *Proceedings of the Nutrition Society,* **35**, 231–236

Wood–Gush, D.G.M. (1959) A history of the domestic chicken from antiquity to the 19th century. *Poultry Science,* **38**, 321–326

Zotterman, Y. (1956) Species differences in the water taste. *Acta Physiologica Scandinavica,* **37**, 60–70

8

THE IMPORTANCE OF INTAKE IN FEED EVALUATION

P.C. GARNSWORTHY and D.J.A. COLE
University of Nottingham School of Agriculture, Sutton Bonington, Loughborough, Leics LE12 5RD, UK

Introduction

The traditional concept of feed evaluation relates to the nutritional and energy-yielding properties of a unit mass of the material. Subsequently, through a total system involving animal requirements or responses, these are related to productivity. However, the nature of the feedstuff may be such that it influences productivity in other ways. The increasing tendency to feed farm animals *ad libitum*, rather than at a restricted level, means that intake, and how it is influenced by diet composition, is an important aspect of feed evaluation. This applies to single-ingredient diets, to mixed diets and also to the associative effects that a dietary ingredient may have on intake of other ingredients in the diet. Conversely, if intake is restricted, the nutritive value per unit of the feedstuff may be influenced by the level at which intake is set at both the evaluation and subsequent feeding stages. It is these aspects of feed evaluation that this paper addresses.

Dietary factors affecting voluntary feed intake

Voluntary feed intake is influenced by many factors. At its simplest, an animal can be considered to have a target level of intake which is that level of intake that exactly supplies the nutrients required for its genetically determined level of production. The ability of the animal to achieve this target will be influenced by the environment, by the physical capacity of the intestines and the bulkiness of the diet, and by the need to avoid imbalances or excesses of particular nutrients. Environmental influences on feed intake are physiological control mechanisms largely independent of diet composition, although different dietary energy concentrations may affect the ability of animals to cope with environmental stress. Physical capacity and bulkiness of the diet are linked. A particular ingredient may, because of its bulkiness, restrict feed intake if the amount of food that an animal should eat to meet its requirements is more than its physical capacity will allow. Further, a complete diet with a nutrient imbalance may inhibit or stimulate intake through the homoeostatic mechanisms of the animal. This situation occurs primarily with diets where intake is not limited by physical capacity.

DIETARY ENERGY CONCENTRATION

The most important factor determining production in farm animals is energy supply and meeting energy requirements is normally regarded as the primary goal in the control of voluntary feed intake. Cole, Hardy and Lewis (1972) proposed a model (Figure 8.1) which suggests that there is a range of diet quality over which the pig eats more of a low energy diet to achieve a constant daily intake of digestible energy. However, a point is reached (A) where the pig is not able to eat more due to the physical limitation of gut capacity. It was further suggested that there might be a minimum amount (B) of food required to avoid gastric hunger sensation. Complete compensation within the range of physiological control has recently been questioned (Cole and Chadd, 1989). They reported complete compensation in the case of boars and incomplete compensation with gilts (Figure 8.2) but did not suggest a general sex relationship. However, they stated that incomplete compensation was a common occurrence with modern pigs and can be used as a means of increasing energy intake.

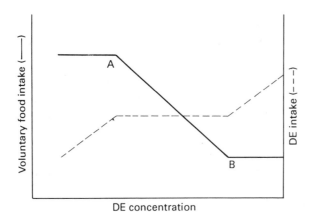

Figure 8.1   Effect of dietary energy concentration on voluntary intake of food and digestible energy (DE) in pigs. A = point where physical capacity becomes limiting, B = minimum food intake required to satisfy hunger (after Cole, Hardy and Lewis, 1972)

Ruminants are seldom fed on diets with a sufficiently high energy concentration to observe the effects of lack of gut fill, but physical capacity is certainly of major importance in the regulation of feed intake with the majority of ruminant diets. Montgomery and Baumgardt (1965) proposed a model for the relationships between nutrient concentration and feed intake in ruminants (Figure 8.3) which is similar to the response at the lower energy concentrations in the model of Cole, Hardy and Lewis (1972). When feeds of low nutritive value are offered, intake is limited by physical capacity of the rumen. This is primarily a control of meal size so that an animal stops eating when the rumen is full. It will only start eating again when a certain amount of food has left the rumen either because it has been digested or has a particle size small enough to allow passage through the reticulo-omasal orifice. As dietary energy concentration or digestibility increase, feed intake increases but, at some undefined point, feed intake starts to decline

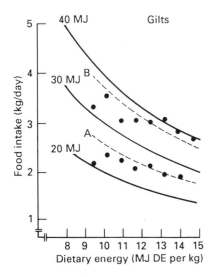 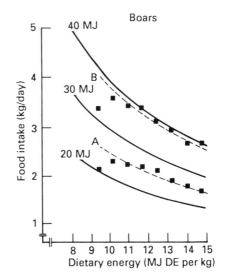

Figure 8.2 Incomplete food intake compensation for dietary energy concentration in gilts and complete compensation in boars. A = 25 to 60 kg live weight, B = 60 to 120 kg weight. The 20, 30 and 40 MJ lines represent intake required for complete compensation (Cole and Chadd, 1989)

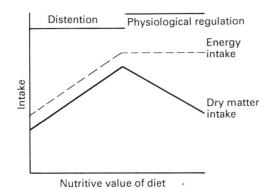

Figure 8.3 Effect of dietary energy concentration on voluntary intake of food in ruminants (after Montgomery and Baumgardt, 1965)

with increasing energy concentration as physiological control mechanisms attempt to maintain a constant energy intake.

If the inclusion of a particular ingredient causes a physical limitation on intake, this influences its value and the effect of inclusion level needs to be identified. It is bulky foods which are likely to be of importance in this context. For example, the use of diets containing oat husks was examined in pigs by Cole, Duckworth, Holmes and Cuthbertson (1968). They reported a physical limitation when 33% oat feed was incorporated into a diet containing 8.8 MJ DE/kg and offered to pigs from 59 to 91 kg live weight. The average dry matter intake of the group that had the physical limitation was 4.05 kg/day with a faecal output of 1.9 kg/day.

Such a characteristic will only hold good for a certain situation. For example, the problems of bulkiness will vary with the maturity of a non-ruminant animal. Owen and Ridgman (1968) offered diets ranging from 10.3 to 14.1 MJ/kg to pigs over four periods as they grew from 27 to 118 kg live weight. At live weights above 50 kg, pigs were able to compensate for low energy concentration and consumed similar levels of energy but below 50 kg, compensation was not complete and energy intake was reduced with low energy diets.

PHYSICAL FORM

A means by which the limitations of bulky feeds on intake may be overcome is by changing their physical form. In ruminants, when a forage is chopped or ground there is usually an increase in voluntary feed intake due to a faster rate of passage of material from the rumen, but this can also result in a slight decrease in digestibility because the feed is subjected to the action of micro-organisms in the rumen for a shorter period of time. For example, Wilkins, Lonsdale, Tetlow and Forrest (1972) mixed dried grass that had either been chopped to approximately 3 cm length or ground through a 2 mm screen in different ratios to give a range of particle size distributions and then formed the material into wafers. When fed to cattle or sheep, particle size was found to be negatively related to voluntary intake, but positively related to digestibility of organic matter (Table 8.1). In a subsequent experiment, they found that as particle size decreased, rate of passage through the digestive tract increased.

Table 8.1   EFFECT OF PARTICLE SIZE OF DRIED GRASS ON INTAKE AND DIGESTIBILITY OF ORGANIC MATTER IN CATTLE (C) AND SHEEP (S) (WILKINS *et al.*, 1972)

		Milled material in grass mixture					
		0	*0.25*	*0.5*	*0.75*	*1.0*	*s.e.*
Intake of organic matter (g/kg$^{0.75}$)	C	83.6	83.1	88.6	92.0	89.8	1.78
	S	65.7	68.7	75.4	75.8	75.8	4.23
Organic matter digestibility (%)	C	68.9	66.8	67.1	65.8	61.7	0.87
	S	68.3	68.6	68.2	64.5	63.8	1.00

PROTEIN DEFICIENCY

Over the range of protein concentrations normally found in diets for farm animals there is no evidence for control of protein intake. However, when diets with a very low protein content are offered, intake is depressed. In pigs, the critical protein concentration is 60 g/kg (Robinson, Holmes and Bayley, 1974). Ruminants are capable of utilizing non-protein nitrogen which is used by rumen micro-organisms to synthesize true protein. They are also able to compensate to some extent for low dietary concentrations of protein by recycling nitrogen, in the form of urea, via saliva. Nevertheless, intake is reduced with diets of low nitrogen content. This is mainly due to a reduction in the activity of rumen micro-organisms and hence cellulose digestibility. Milford and Minson (1966) found that intake by sheep was reduced with diets containing less than 70 g/kg protein (Figure 8.4). However, there is evidence that protein supplementation of diets with low protein concentrations can also have an effect beyond the rumen. Egan and Moir (1964) infused casein or urea into the duodenum of sheep and found that intake of a low protein forage increased immediately with casein infusion but at a much slower rate with urea infusion. Casein did not alter rumen activity but urea increased the rate of digestion in the rumen. It can be concluded that casein infusion increased the ability of the sheep to utilize energy by removing a protein deficiency and therefore stimulated intake by metabolic means whereas urea infusion probably increased the supply of nitrogen to the rumen via the saliva.

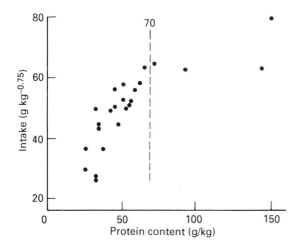

Figure 8.4 Relationship between crude protein content and intake of tropical forages by sheep (after Milford and Minson, 1966)

PROTEIN SOURCE

In ruminants offered forage based diets, differences in intake have been observed when supplements containing protein from different sources are used. Belete–Adinew and Garnsworthy (1989) supplemented grass silage with isonit-rogenous compounds containing either rapeseed and urea (H) or fishmeal and soya (L) as protein sources. When fed to steers, nitrogen in supplement L was degraded more slowly in the rumen. This was accompanied by a faster rate of digestion of silage, higher proportions of acetic acid and lower ammonia levels in the rumen. Consequently, dry matter intake and live-weight gain were higher with supplement L (Table 8.2). It was suggested that the slower rate of nitrogen release from supplement L more closely matched the requirements of the cellulolytic bacteria in the rumen and resulted in a more efficient fermentation. In other words, the relative timing (time from feeding) of the release of nitrogen and energy-yielding substrates in the rumen is important. Micro-organisms have specific requirements for both nitrogen and energy and, unless their supply is synchronized, maximal

Table 8.2   FERMENTATION CHARACTERISTICS OF SUPPLEMENTS FOR SILAGE CONTAINING DIFFERENT PROTEIN SOURCES AND THEIR EFFECT ON DIGESTION OF SILAGE IN THE RUMEN AND DRY MATTER INTAKE WHEN FED TO BEEF STEERS (BELETE-ADINEW AND GARNSWORTHY, 1989)

	Protein source	
	Rape + urea	*Fishmeal + soya*
Nitrogen degradability	0.75	0.66
Immediate nitrogen release	0.37	0.21
Rate of nitrogen release	0.09	0.08
Silage digestion rate in the rumen	0.04	0.08**
Dry matter intake	7.28	7.56

microbial growth, and hence rate of digestion of feed, will not be achieved. If the rate of digestion of feed in the rumen is decreased, intake will be decreased as a result.

AMINO ACID BALANCE

Imbalance of amino acids has been shown to depress intake in various species (Harper and Kumata, 1970). Work at Nottingham by G.M. Sparkes, D.J.A. Cole and D. Lewis (unpublished, cited by Cole and Chadd, 1989) examined the effect of dietary lysine content on food intake in pigs. Lysine was included at ten rates between 6.3 and 30 g/kg diet, with other essential amino acids being balanced to provide the ideal protein and the diets were isoenergetic. Food intake increased as lysine concentration was increased to 10.3 g/kg. It then fell with subsequent increases in lysine until a plateau was reached at 14 g/kg (Figure 8.5). This, and data from subsequent experiments, led to the model proposed by Cole and Chadd (1989) in which maximal intake occurs when an individual amino acid is supplied in the proportion required for the ideal protein (Figure 8.6).

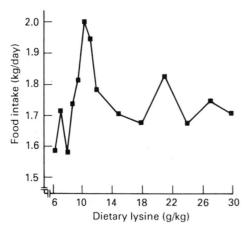

Figure 8.5   Food intake response to dietary lysine concentration in boars (from 25 to 55 kg live weight (from Cole and Chadd, 1989)

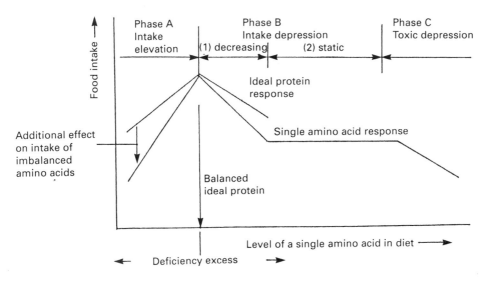

Figure 8.6 Schematic representation of voluntary food intake under different protein nutrition (Cole and Chadd, 1989)

The effect of feeding level on the nutritional value of a feedstuff

When evaluating feedstuffs, it is usual to feed animals at a restricted level because of problems in accounting for feed refusals. Often the amount of feed offered is calculated to supply only sufficient energy and protein to meet the maintenance requirements of the animals, especially with longer term experiments where changes in live weight through the course of the experiment may affect the results. Even if a feedstuff is evaluated under *ad libitum* conditions, the actual intakes will be specific to those animals under the particular conditions of the trial and other animals may eat more or less of the same diet. It is well established that the level of feeding can affect the ability of animals to extract nutrients from the diet and this needs to be taken into account in feedstuff evaluation.

DIGESTIBILITY

Animals usually digest a greater proportion of the nutrients in their feed when fed at a restricted level than when fed *ad libitum* (Schneider and Flatt, 1975). However, with diets that had a low crude fibre content (approximately 50 g/kg DM), Peers, Taylor and Whittemore (1977) found that levels of feeding for pigs from maintenance to three times maintenance had no effect on the apparent digestibility of energy and nitrogen (Table 8.3). Cunningham, Friend and Nicholson (1962) found a considerable effect of level of feeding on digestibility in pigs given a diet with a crude fibre content of 200 g/kg, but little effect with a diet containing 43 g/kg (Table 8.4). Generally, it could be concluded that there is little influence over the normal range of intakes, with diets that do not contain high levels of fibre.

Table 8.3 APPARENT DIGESTIBILITY COEFFICIENTS OF CEREAL BASED DIETS FOR PIGS FED AT TWO LEVELS OF INTAKE (PEERS *ET AL.*, 1977)

Digestibility of	*Level of feeding*	
	Maintenance	*3× Maintenance*
Dry matter	0.822	0.827
Energy	0.833	0.835
Nitrogen	0.834	0.839

Table 8.4 EFFECT OF FEEDING LEVEL ON APPARENT DIGESTIBILITY IN PIGS OF TWO DIETS CONTAINING DIFFERENT LEVELS OF CRUDE FIBRE (DATA FROM CUNNINGHAM, FRIEND AND NICHOLSON, 1962)

Crude fibre (g/kg)	*43.4*		*200.4*	
Feeding level (× maintenance):	*1*	*2.5*	*1*	*2.5*
Digestibility of:				
Dry matter	0.798	0.796	0.631	0.573
Crude protein	0.852	0.838	0.731	0.676
Crude fibre	0.308	0.260	0.302	0.154
Energy	0.818	0.811	0.633	0.584

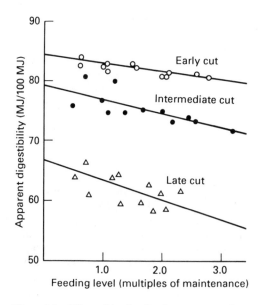

Figure 8.7 Effect of feeding level on apparent digestibility in sheep of ryegrass harvested at three stages of maturity (after Blaxter, 1969a)

Ruminants are usually fed on diets that have a much higher crude fibre content than diets for pigs. Blaxter (1969a) found that the digestibility of ryegrass decreased with increasing feeding level, and that the decrease depended on the stage of maturity at which the grass was harvested (Figure 8.7). Regression analysis showed that digestibility decreased by 1.5, 2.4 and 3.3 percentage units for the young material, older material and mature material respectively with each multiple of maintenance by which feeding level was increased. Blaxter (1969b) analysed the results of 53 experiments with cattle and sheep and again found that apparent digestibility decreased with increasing level of feeding but the absolute depression depended on the apparent digestibility determined at maintenance, according to the expression:

$$\Delta A = 0.017 - 0.113 \, A_m$$

where ΔA = decrease in apparent digestibility per unit rise in feeding level.
A_m = apparent digestibility of feed determined at maintenance.

The cause of the decrease in digestibility with increasing feeding level is the increase in rate of passage through the rumen when feed intake is increased, as found when material is ground. The reason for the greater depression in digestibility with feeds of lower quality is that the quantity of feed containing the nutrients for maintenance is greater with poorer quality feeds so an increase from the maintenance to twice maintenance level of feeding will involve a proportionately greater increase in total feed intake.

It can be concluded that, with diets or feedstuffs of high fibre content, digestibility values determined under conditions of restricted feeding will over-estimate the amount of digestible nutrients that can be extracxted from the diet by animals fed at higher levels of feeding or *ad libitum*.

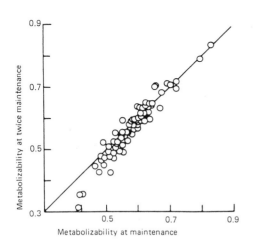

Figure 8.8   Effect of feeding level on metabolizability of dietary energy (ARC, 1980)

METABOLIZABILITY

Some compensation for the effects of feeding level on digestibility is found when evaluation systems based on metabolizable energy are used for ruminants since energy losses in methane and urine generally decrease with increased feeding level. Blaxter and Clapperton (1965) found that the loss of energy in methane decreased by about 15 kJ/MJ food gross energy and Blaxter, Clapperton and Martin (1966) found that the loss of energy in urine decreased by an average of 7.3 kJ/MJ food gross energy on raising the feeding level from maintenance to twice maintenance. Blaxter (1969b) derived the equation:

$$q_L = q_m + (L - 1) (0.20[q_m - 0.623])$$

where q_L is metabolizability of the gross energy at feeding level L and q_m is that determined at maintenance. The results of 72 experiments were analysed for ARC (1980) and this confirmed that metabolizability is only decreased when feeding level increases with diets where q_m is less than 0.62 (Figure 8.8). Therefore, unless diets of low quality are to be evaluated and fed, intake level has little effect on the metabolizable energy content of a feedstuff.

ENERGY UTILIZATION

As energy intake is increased, energy retention increases but the relationship is not linear and varies with the type of feed given (Figure 8.9). With high quality feeds, the increases in energy retention per unit increase in feeding level is less affected by

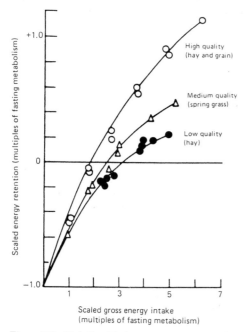

Figure 8.9   Relationship between energy retention and energy intake for feeds of different quality (ARC, 1980)

actual feeding level than with low quality feeds. ARC (1980) states that the reasons for curvilinearity are not completely understood but are thought to involve changes in rumen fermentation and changes in the efficiency of utilization of metabolizable energy due to differences in the partitioning of energy to body fat and protein as energy intake is increased. Therefore, it appears that feeding level needs to be taken into account with evaluation systems based on Net energy. However, the curvilinearity of the response has recently been questioned by Emmans and Oldham (see Chapter 5).

PROTEIN DEGRADABILITY

The effect of intake on protein degradability has been mentioned in Chapter 4. As intake is increased, the retention time of material in the rumen is decreased so that less protein is degraded by micro-organisms and protein degradability is lower. It is therefore necessary to estimate the rumen outflow rate in order to predict the effective degradability of proteins under different feeding conditions and it has been shown that the effect of outflow rate on effective degradability varies markedly with different protein sources (Figure 8.10). There is an added problem

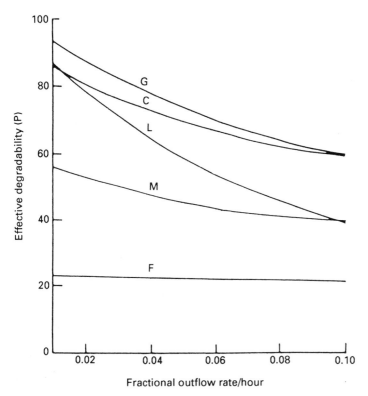

Figure 8.10 Effect of rumen outflow rate on effective degradability (P) for different protein supplements. G = ground nut meal, C = cottonseed meal, L = linseed meal, M = meat and bone meal, F = fish meal (after Ørskov, 1982)

when evaluating protein sources for dairy cows using other classes of livestock since it is difficult to get sheep or growing cattle to eat at the same level as dairy cows. At present, the standard method is to determine outflow rate in the test animals (usually, as a coefficient, between 0.02 and 0.06/hr) and assume an outflow rate of 0.08/hr when predicting effective degradability for high-yielding dairy cows (ARC, 1984).

This is not the complete answer since it often involves extrapolation beyond the range of outflow rates possible in test animals. Even when two feeding levels are employed with mature steers and outflow rates both above and below 0.08/hr are achieved, it is difficult to predict protein degradability for dairy cows, as illustrated in Table 8.5 (P.C. Garnsworthy and G.P. Jones, unpublished observations). The protein degradability of two compound feeds with high (H) or low (L) energy concentration was determined. The effective degradability of the two compounds was similar at the twice maintenance level of feeding, but not at the maintenance level. When adjusted to an outflow rate of 0.08/hr, predicted degradability values are very similar using degradability curves determined at twice maintenance, but are markedly different when the curves determined at maintenance are used. This emphasizes the fact that it is inadvisable to extrapolate from figures determined at low feeding levels to the dairy cow that is fed at a high level.

Table 8.5   EFFECT OF FEEDING LEVEL ON PROTEIN DEGRADABILITY OF TWO COMPOUNDS DETERMINED IN STEERS AND PREDICTED FOR DAIRY COWS (P.C. GARNSWORTHY AND G.P. JONES, UNPUBLISHED OBSERVATIONS)

			Degradability	
Compound	*Feeding level**	*Outflow rate/h*	*Actual†*	*Predicted dairy cow‡*
High	M	0.05	0.79	0.74
energy	2M	0.09	0.75	0.77
Low	M	0.05	0.88	0.84
energy	2M	0.09	0.76	0.78

*M = maintenance, 2M = twice maintenance.
†Determined from *in situ* measurements.
‡Predicted at outflow rate of 0.08/h.

ASSOCIATIVE EFFECTS ON INTAKE

Certain feed ingredients, because of their nature, may affect the intake of other ingredients when included in a diet. This could be considered as qualitive intake effects on nutritive value rather than the quantitive effects that have previously been discussed. Feed ingredients which produce unpalatable flavours when they deteriorate in storage may reduce intake (see Chapter 7). It is difficult to allow for such effects in feed evaluation. Other ingredients, particularly legumes, may contain anti-nutritive factors that interface with the digestive process of the animal (Wiseman and Cole, 1988). Certain classes of feed ingredients, such as materials with a high content of starch or fat, may affect intake in ruminants when included at high levels in mixed diets, through their influence on rumen fermentation. When starch is added to roughage diets for ruminants, particularly those with a low nitrogen content, soluble nitrogen is depleted due to the rapid growth of amylolytic organisms. The growth rate of cellulolytic organisms is reduced because of the shortage of nitrogen and cellulose digestibility is decreased. High concentrations of fat in roughage diets decrease the digestibility of fibre in the rumen. Devendra and Lewis (1974) suggested that this effect may be due to physical coating of the fibre by fat, toxic effects of fat on rumen microbes, surfactant effects of fatty acids on cell membranes or the formation of insoluble cation soaps.

Associative effects are difficult to allow for in feed evaluation since their magnitude depends upon the other dietary ingredients. The usual approach is to limit the inclusion level of certain ingredients at the diet formulation stage.

Conclusion

The effects of feed ingredients on intake are very important and the intake modifying characteristics of feedstuffs should be included in their evaluation. Some of these characteristics are features of individual ingredients and others become apparent when the inclusion of certain feeds results in imbalanced diets.

When evaluating feedstuffs, any effect of feeding level needs to be taken into account and evaluation programmes should ideally include more than one feeding level. This could be a particular problem when evaluating feeds for dairy cows where intake level can be considerably higher than in animals fed at maintenance. Intake should be considered in qualitative as well as quantitative terms to allow for associative effects.

References

Agricultural Research Council (1980) *The Nutrient Requirements of Ruminant Livestock*, Commonwealth Agricultural Bureaux, Slough

Agricultural Research Council (1984) *The Nutrient Requirements of Ruminant Livestock*, Suppl. 1. Commonwealth Agricultural Bureaux, Slough

Belete-Adinew, A. and Garnsworthy, P.C. (1989) The effects of rate of protein degradation in silage supplements on performance of beef cattle. *Animal Production*, **48**, 638–639

Blaxter, K.L. (1969a) *The Energy Metabolism of Ruminants*, Hutchinson Scientific and Technical, London

Blaxter, K.L. (1969b) The efficiency of energy transformations in ruminants. In *Energy Metabolism of Farm Animals* (eds K.L. Blaxter, J. Kielanowski and G. Thorbek), European Association for Animal Production Publication No. 12, Oriel Press Limited, Newcastle upon Tyne, pp. 21–28

Blaxter, K.L. and Clapperton, J.L. (1965) Prediction of the amount of methane produced by ruminants. *British Journal of Nutrition*, **19**, 511–522

Blaxter, K.L., Clapperton, J.L. and Martin, A.K. (1966) The heat of combustion of the urine of cattle in relation to its chemical composition and to diet. *British Journal of Nutrition*, **20**, 449–460

Cole, D.J.A. and Chadd, S.A. (1989) Voluntary food intake of growing pigs. In *Voluntary Food Intake of Pigs* (eds J.M. Forbes, M.A. Varley and T.L.J. Lawrence), Occasional Publication 13, British Society of Animal Production, Edinburgh

Cole, D.J.A., Duckworth, J.E., Holmes, W. and Cuthbertson, A. (1968) Factors affecting voluntary feed intake in pigs. 3. The effect of a period of feed restriction, nutrient density of the diet and sex on intake, performance and carcass characteristics. *Animal Production*, **10**, 345–357

Cole, D.J.A., Hardy, B. and Lewis, D. (1972) Nutrient density of pig diets. In *Pig Production* (ed D.J.A. Cole), Butterworths, London

Cunningham, H.M., Friend, D.W. and Nicholson, J.W.G. (1962) The effect of age, bodyweight, feed intake and adaptability of pigs on the digestibility and nutritive value of cellulose. *Canadian Journal of Animal Science*, **42**, 167–175

Devendra, C. and Lewis, D. (1974) The interaction between dietary lipids and fibre in sheep. 2. Digestibility studies. *Animal Production*, **19**, 67–76

Egan, A.R. and Moir, R.J. (1964) Nutritional status and intake regulation in sheep. 1. Effects of duodenally infused single doses of casein, urea and propionate upon voluntary intake of a low protein roughage. *Australian Journal of Agricultural Research*, **16**, 437–449

Harper, A.E. and Kumata, U.S. (1970) Effects of ingestion of disproportionate amounts of amino acids. *Physiological Reviews*, **50**, 428–558

Milford, R. and Minson, D.J. (1966) Intake of tropical pasture species. *Proceedings of the 9th International Grassland Congress*, Departmento da Produção Animal da Secretaria da Agricultura do Estado de São Paulo, São Paulo, pp. 815–822

Montgomery, M.J. and Baumgardt, B.R. (1965) Regulation of food intake in ruminants. 1. Pelleted rations varying in energy concentration. *Journal of Dairy Science*, **48**, 569–574

Ørskov, E.R. (1982) *Protein Nutrition in Ruminants*. Academic Press, London

Owen, J.B. and Ridgman, W.J. (1968) Further studies of the effect of dietary energy content on the voluntary intake of pigs. *Animal Production*, **10**, 85–91

Peers, D.G., Taylor, A.G. and Whittemore, C.T. (1977) The influence of feeding level and level of dietary inclusion on the digestibility of barley meal in the pigs. *Animal Feed Science and Technology*, **2**, 41–47

Robinson, D.W., Holmes, J.H.G. and Bayley, H.S. (1974) Food intake regulation in pigs. 1. The relationship between dietary protein concentration, food intake and plasma amino acids. *British Veterinary Journal*, **130**, 707–715

Schneider, B.H. and Flatt, W.P. (1975) *The Evaluation of Feeds Through Digestibility Experiments*, The University of Georgia Press, Athens

Wilkins, R.J., Lonsdale, C.R., Tetlow, R.M. and Forrest, T.J. (1972) The voluntary intake and digestibility by cattle and sheep of dried grass wafers containing particles of different size. *Animal Production*, **14**, 177–188

Wiseman, J. and Cole, D.J.A. (1988) European legumes in diets for non-ruminants. In *Recent Advances in Animal Nutrition – 1988* (eds W. Haresign and D.J.A. Cole) Butterworths, London

9

CHEMICAL EVALUATION OF POLYSACCHARIDES IN ANIMAL FEEDS

P. ÅMAN and H. GRAHAM

Swedish University of Agricultural Sciences, S-750 07 Uppsala, Sweden

Introduction

Polysaccharides in animal feeds are a complex group of components differing widely in physical properties and physiological activity. The recent upsurge of interest in polysaccharides has resulted from the development of more sophisticated and accurate methods for analysis and the realization that polysaccharides are not only energy-yielding compounds but also can control both the rate and extent of assimilation of other dietary components. Nevertheless it should be borne in mind that the relationship between structure and physiological activity is as yet poorly understood.

Polysaccharides may be classified as cell-wall polysaccharides or storage polysaccharides, including starch, fructans and cell-wall storage polysaccharides. Small amounts of other types of polysaccharides, including glycoconjugates, may also be present in animal feeds. In this chapter the structure, occurrence and methods for the chemical evaluation of polysaccharides in higher plants will be presented, although, in some cases, polysaccharides from other sources may also be included in feeds.

Structural analysis

The primary structure of a polysaccharide is known only when the following characteristics have been elucidated:

1. The glycosyl residue composition, including the identity and ratio of the residues.
2. The absolute configuration of each glycosyl residue.
3. The glycosy linkage composition of each residue.
4. The ring form of each glycosyl residue.
5. The anomeric configuration of the glycosidic linkage of each residue.
6. The sequence of the glycosyl residues.
7. The identity, points of attachment and stereochemistry, if appropriate, of any non-carbohydrate moieties.

Many standard methods are described for the elucidation of the glycosyl residue composition (e.g. Sawardeker, Sloneker and Jeanes, 1965), absolute configuration of glycosyl residues (e.g. Leontein, Lindberg and Lönngren, 1978) and glycosyl linkage composition (e.g. Björndal *et al.*, 1970) while more advanced techniques are often necessary for the determination of the other characteristics. Recently a scheme was outlined for the complete determination of the primary structure of polysaccharides (McNeil *et al.*, 1982). The main steps in this scheme are as follows:

1. Purify the polysaccharide and determine its glycosyl composition. Determine the absolute configuration of the glycosyl residues. Determine, if possible, the anomeric configuration and non-carbohydrate substituents by NMR on the intact polysaccharide.
2. Prereduce uronic acid residues, if appropriate or possible, and methylate the polysaccharide. Determine the glycosyl linkage composition on an aliquot of the methylated polysaccharide. Information about the ring form may be obtained. Carboxyl groups of uronic acid residues in the methylated polysaccharide may be reduced.
3. Partially hydrolyse the methylated polysaccharide after optimizing the conditions, and reduce and ethylate the resulting oligosaccharides.
4. Fractionate the mixture of peralkylated oligosaccharide alditols by HPLC and detect the alditol by refractive index or HPLC-MS. Isolated peralkylated di- tri- or tetraalditols may be analysed by GLC-MS and larger alditols by direct probe MS. Determine the anomeric configurations of the glycosyl linkages of the fractionated alditols by ^{1}H-NMR.
5. Fully hydrolyse, reduce, and acetylate the fractionated oligosaccharide alditols and characterize the partially alkylated, partially acetylated alditols obtained by GLC and GLC-MS. Deduce the glycosyl sequence of the fractionated peralkylated oligosaccharide alditols and determine the ring form of the glycosyl residues.
6. Determine the primary structure of the polysaccharide from its glycosyl linkage composition, the structure of the oligosaccharides derived from it and the content of any labile non-carbohydrate component.

As an example, a partially methylated trisaccharide alditol (Figure 9.1) was obtained in a mixture by methylation, reduction of uronosyl residue carboxyl groups with lithium aluminium deuteride (LiAID$_4$), partial hydrolysis, reduction of the resulting partially methylated oligosaccharide alditols with sodium boro deuteride (NaBD$_4$), and ethylation of the newly formed hydroxyl groups (Robertsen *et al.*, 1981). The trisaccharide alditol was purified by high pressure liquid chromatography and the location of deuterium and O-ethyl groups were ascertained during subsequent analysis and were used for the identification of the uronic acid residues and in addition, where other glycosyl residues were linked to the trisaccharide alditol.

Knowledge about the molecular weight and conformation is also of vital importance for an understanding of the properties of the polysaccharide. Since plant polysaccharides, unlike bacterial polysaccharides, often lack true repeating units and are frequently interconnected in macromolecular complexes such as the plant cell wall, purification and fractionation often becomes a delicate and time-consuming matter. In many cases it is possible only to isolate a small fragment of a pure polysaccharide from the macromolecular complex, and this fragment may not always be representative of the intact polymer.

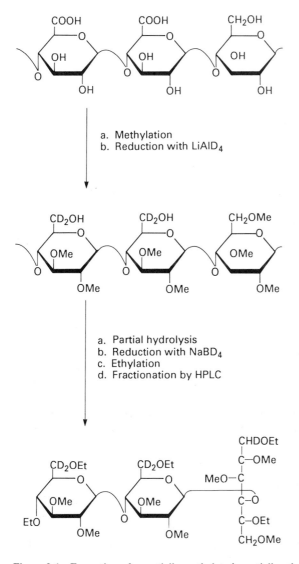

Figure 9.1   Formation of a partially methylated, partially ethylated trisaccharide alditol from the acidic polysaccharide secreted by Rhizobium trifolii NA 30 (From Robertsen *et al.*, 1981)

The nutritional utilization of cell wall components is different for various plant species and stages of development, as well as for different types of cell walls within the same plant (Chesson *et al.*, 1986; Nordkvist, 1987). Information about the composition and organization of plant cell walls is therefore of vital importance in order to understand the mechanisms of digestion and to optimize pretreatment and utilization of feeds. Relationships between glycosyl linkage composition of cell walls and digestion has recently been studied by Gordon, Lomax and Cheeson, 1983; Cheeson, Gordon and Lomax, 1985; and Cheeson *et al.*, 1986. The importance of surface chemistry in relation to cell-wall digestion has also been stressed by this group.

Storage polysaccharides

GENERAL

Storage or reserve polysaccharides are usually formed during periods of intense photosynthetic activity and are utilized by the plant as carbohydrate monomers. Starch is probably the only polysaccharide formed in plastids, and plastids are the only cellular compartments which form starch in higher plants, while the non-starch polysaccharides are stored either in vacuoles or outside the plasmalemma (Meier and Reid, 1982). Non-starch reserve polysaccharides are found mainly in seeds, roots, rhizomes, tubers, bulbs, shoot axes and, to a lesser extent, in leaves. The number of reports about non-starch reserve polysaccharides in plants has increased in recent years, and they are probably more widespread than previously thought.

STARCH

Starch is found in all parts of higher plants but the content may vary from less than 1% to over 80% (Greenwood, 1970). The reserve carbohydrates of many algae, fungi, bacteria and protozoa also resemble plant starches or animal glycogen in overall structure (Greenwood, 1970; Whistler, BeMiller and Parschall, 1984).

Starch is laid down in the form of insoluble particles – starch granules – with a size and shape characteristic of the botanical source. The main components of the granules are amylose and amylopectin but minor components such as proteins, lipids and bound phosphorus are also present in varying amounts. The molecular weight and fine structure of both amylose and amylopectin as well as the relative ratio of these components varies with botanical origin and stage of maturity of the plant. Recent studies have indicated heterogenicities within amylose and it now appears very likely that in most cases amylose possesses a limited degree of branching (Whistler, BeMiller and Parshall, 1984). The polymers in the granule are arranged in an ordered radial manner, as shown by the birefrigent properties, but the exact organization of the starch granule is not fully understood. The heterogenous nature of starch has made structural studies as well as quantitive analysis of this major feed component a difficult task.

Many methods have been developed using hydrolytic or non-hydrolytic procedures for starch analysis (Lyne, 1976; Hassid and Neufeld, 1964), but most are often limited in being applicable only to a certain type of material. At present enzymatic methods using amylases prevail (Thivend, Mercier and Guilbot, 1972; Batey, 1982; Åman and Hesselman, 1984; Åman, Westerlund and Theander, 1989). These methods include gelatinization and solubilization of starch, enzymic hydrolysis to glucose by amyloglucosidase, alone or in combination with α-amylase, with subsequent determination of glucose released. The major problems in these methods are incomplete dissolution of starch and limited accessibility to enzymes caused by, for example, the formation of retrograded amylose (Batey, 1982), starch-lipid complexes (Holm *et al.*, 1983), resistant starch in heat-processed products (Englyst, Andersson and Cummings, 1983) and encapsulation of starch in protein or fibre structures (Batey, 1982; Holm *et al.*, 1983). The use of a thermostable α-amylase during the gelatinization step has been shown to reduce or eliminate many of these problems.

A procedure for starch analysis including the thermostable α-amylase, Termamyl, is as follows (Åman, Westerlund and Theander, 1989):

1. Grind a representative 50 g samples of grain in a Tecator Cyclone Sample Mill to pass a 0.5 mm screen.
2. Weigh duplicate 50.0 mg samples into 35 ml thick-walled Pyrex glasstubes with screw caps containing tightly-fitting Teflon washers.
3. Extract low-molecular weight sugars with 15 ml 80% ethanol in a boiling water-bath for 30 min.
4. Gelatinize and degrade starch using a thermostable α-amylase (Termamyl) in a boiling waterbath and further to glucose using an amyloglucosidase.
5. Calculate the starch content from the glucose content in the supernatant as measured by the conventional glucose-oxidase method.

The method has a high precision and the coefficients of variation for different cereals have been shown to vary between 0.5% and 1.5% (Åman and Hesselman, 1984). With the described method, analysis of at least 52 samples may be started each day and no advanced equipment is necessary. The method has been used in our laboratories for several years on many different types of material and proved to be very reliable.

Amylase in saliva and pancreatic juice hydrolyses starch to low molecular-weight carbohydrates. The final hydrolysis to free glucose is carried out by disaccharidases in the brush border membrane of the small-intestinal enterocytes. It is known that the availability of starch for amylases varies with both plant source and heat treatment (Asp *et al.*, 1987). Factors of importance besides granule size and structure may include:

1. Feed architecture, i.e. large particle size, intact cell structures and encapsulation in matrix structures may reduce the availability.
2. Degree of gelatinization where ungelatinized starch generally has a low availability.
3. Covalent bonds resistant to amylase. These bonds are introduced in starch derivatives and may be formed during heat treatment.
4. Resistant starch, which consists in all probability of strongly retrograded amylose having low accessibility to amylases unless first solubilized in alkali or dimethylsulfoxide. Retrogradation of amylopectin, which causes staling of bread during storage, may also reduce amylolysis.
5. Starch–lipid complexes which are formed between polar lipids and mainly amylose. In spite of their relative resistance to amylases *in vitro*, they have a relatively high availability *in vivo*.
6. A number of other factors have also been shown or suggested to impair starch digestion. These include starch–protein interactions, amylase inhibitors and antinutritional factors such as phytic acid and tannins.

In order to study the significance and basic mechanisms of these factors on starch digestion, a wide range of methods are necessary. These may include chemical, physical, microscopic and biological approaches. It is outside the scope of this chapter to discuss these methods further.

FRUCTANS

The most important non-starch storage polysaccharide in animal feeds is fructans. They are stored in a dissolved or colloidal state in the vacuoles of vegetative tissues of both mono- and dicotyledons (Meire and Reid, 1982). Fructans were first extracted from rhizomes of Jerusalem artichoke and these types of fructans were later named inulins. In monocotyledons other types of linear fructans, denoted phleins or levans, and branched fructans are also found. Fructans are non-reducing carbohydrates with sucrose at the reducing end (Figure 9.2). In addition inulins contain unbranched (2→1)-linked fructofuranosyl residues, phleins unbranched (2→6)-linked fructofuranosyl residues and branched fructans either an inulin or a phlein type backbone with side chains of β-D-fructofuranosidic residues. Characteristic properties of fructans are high solubility in aqueous solutions and high sensitivity to acid hydrolysis due to the fructofuranosidic linkages. Values between 20 and 40 have been reported for the average degree of polymerization of inulins from Compositae, and phleins have a degree of polymerization varying from 10 to 300 (Meirer and Reid, 1982). Low molecular-weight homologues of fructan also occur frequently in plant materials (Nilsson, 1988).

In plant material the fructans content may vary from less than 1% to as much as 65% or more of the dry matter, as in the edible parts of onions (Darbyshire and

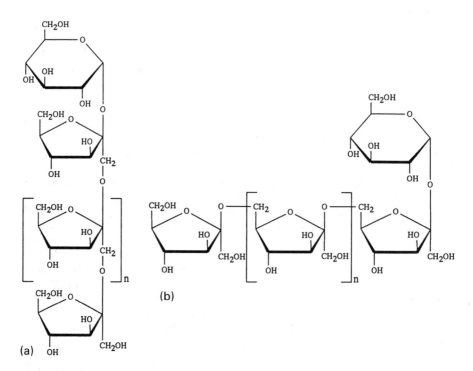

Figure 9.2   General structure of fructans. (*a*) Inulin type with (2→)-linked β-D-fructofuranosidic residues. (*b*) Phlein type with (2→6)-linked β-D-fructofuranosidic residues. Branched fructans with inulin or phlein type backbone are also found

Henry, 1978; Nilsson, 1988). In temperate forage grasses high amounts of fructans can be found, although the content varies with growing conditions, crop, plant part and stage of development. Probably the most thorough investigation on the distribution of fructans in different parts of plants has been conducted by Smith (1967, 1973) in brome grass and in timothy (Table 9.1).

Table 9.1 CONTENT OF FRUCTANS IN VARIOUS PARTS OF THE FLOWERING TILLERS OF TIMOTHY AND BROME GRASS AT EARLY ANTHESIS (ADAPTED FROM SMITH 1967, 1973)

Plant		Timothy fructan % DM	Brome grass fructan % DM
Inflorescence		1.2	4.2
Internode	1	2.0	2.1
	2	4.4	7.9
	5 (stembase)	30.5	16.3
Leaf sheath	2	1.0	5.7
	4	2.7	9.4
Leaf blade	2	1.5	4.6
	4	1.0	7.1
Total tiller		8.7	7.5

Fructans are not hydrolysed by endogenous enzymes of man or animals and only limited, if any, acid hydrolysis takes place in the stomach of rats (Nilsson, 1988). On the other hand fructans are easily fermented by microorganisms in the gastro-intestinal tract (Graham and Åman, 1986; Nilsson, 1988). Fructans may therefore be included in the dietary fibre complex, at least if a chemical definition, i.e. non-starch polysaccharides and lignin, is used.

Fructose-containing carbohydrates are generally quantitatively determined using, more or less specific, colorimetric (Dische and Borenfreund, 1951; Dische, 1962) or enzymatic (Larsson and Bengtsson, 1983) methods. An enzymatic method in which fructose and glucose residues in fructo-oligosaccharides and fructans are released by selective acid hydrolysis and determined by enzymatic methods is as follows:

1. Extract fructo-oligosaccharides, fructans and other soluble components with 0.05 M acetate buffer (pH 5.0) at 65°C.
2. Analyse the content of glucose, fructose and sucrose in the extract using essentially the conventional enzymatic methods by Boehringer and Mannheim.
3. Hydrolyse specifically the acid labile fructofuranosidic linkages using 0.032 M H_2SO_4 for 70 min at 80°C.
4. Analyse the free fructose and glucose in the extract by the enzymatic method of Boehringer and Mannheim.
5. Calculate the content of fructo-oligosaccharides and fructans as the sum of anhydroglucose and anhydrofructose released by the weak acid hydrolysis.

This method has a high precision when significant amounts of fructose-containing carbohydrates are present in the material analysed.

CELL-WALL STORAGE POLYSACCHARIDES

Cell-wall storage polysaccharides occur widely in seeds. They are stored outside the plasmalemma, sometimes in large quantities, and have a chemical structure similar to the other cell-wall polysaccharides present (Meier and Reid, 1982). Those which constitute large substrate reserves can be subdivided into three groups, mannans, xyloglucans and galactans, but in all likelihood other structures frequently occur in smaller amounts.

The mannan-type of cell-wall reserves are all based on a linear $(1\rightarrow4)$-linked β-D-mannopyranosidic backbone and seem to occur only in endosperms. They may be subdivided into 'pure' mannans, glucomannans in which some of the mannose residues are replaced with $(1\rightarrow4)$-linked β-D-glucopyranosidic residues, and galactomannans in which the backbone carries $(1\rightarrow6)$-linked sidechains of α-D-galactopyranosidic residues (Figure 9.3; Dea and Morrison, 1975; Meier and Reid, 1982). In animal feeds 'pure' mannans occur mainly in palm seeds and galactomannans in leguminous seeds.

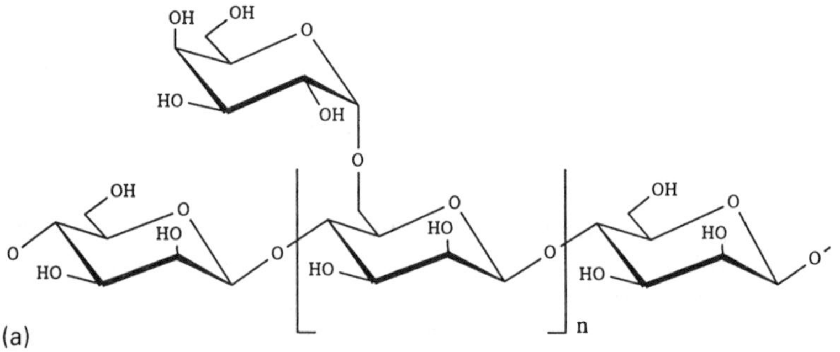

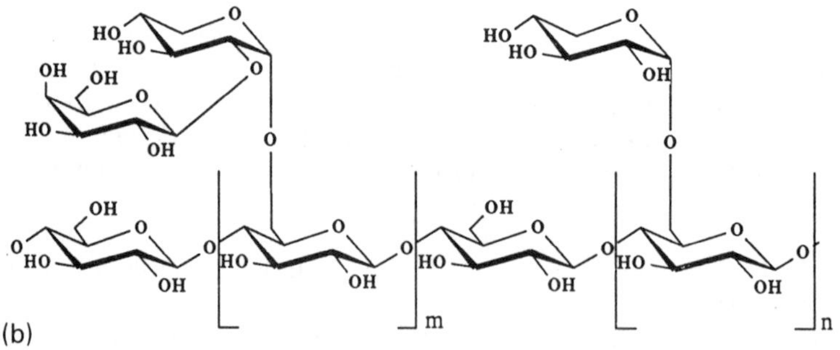

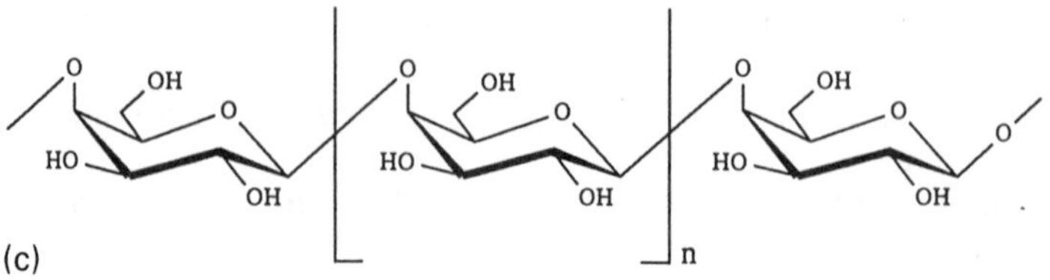

Figure 9.3   General structure of some cell-wall storage polysacchardies. (*a*) galactomannan, (*b*) xyloglucan and (*c*) lupin galactan

Storage xyloglucans are present mainly in dicotyledonous seeds and are based on a linear backbone of (1→4)-linked β-D-glucopyranosidic residues to which are attached short side chains containing D-xylopyranosidic and D-galactopyranosidic residues (Figure 9.3; Kooiman, 1960). Little is known about the importance of storage xyloglucans in animal feeding.

Cotyledon cells of *Lupinus* seeds have long been known to have unusually thickened walls and a high content of galactans. Structural studies of the galactans from white lupin have revealed (1→4) linked β-D-galactopyranosyl residues (Figure 9.3) to be the major component, although arabinose, uronic acid and branched galactose units also were present (Jones and Tanaka, 1965; Åman and Cheeson, unpublished observations). Lupins are sometimes used in animal feeds and a high content of storage galactans will undoubtedly have significant effects on its feeding properties.

No specific analytical methods are developed for the quantitive analysis of storage cell-wall polysaccharides and therefore general carbohydrate methods have to be used.

Cell-wall polysaccharides

GENERAL

The plant cell wall consists predominantly of carbohydrates, glycoprotein and lignin. These components are organized in a complex three-dimensional structure which is neither uniform nor completely described for different plants or plant fractions. Over one hundred different monosaccharide residues have been described in nature. Fortunately, only ten of these dominate quantitatively as building blocks of cell walls of higher plants. These are – arabinose and xylose (pentoses); glucose, galactose and mannose (hexoses); rhamnose and fucose (6-deoxy hexoses) and galacturonic, glucuronic and 4-O-Me-glucuronic acid (Lexuronic acids) (Figure 9.4).

Plant cell-wall polysaccharides associated with cellulose are commonly isolated using selective extraction procedures. Gums are extracted with water, pectic substances with hot aqueous solutions containing chelating agents for calcium, and hemicelluloses with alkali. However, the extracts obtained generally contain mixtures of different polysaccharides and often polysaccharides with only small differences in chemical structure may be found in all fractions. It is therefore important to use a more precise classification based on structure, which allows plant cell-wall polysaccharides to be divided into five groups (Aspinall, 1981). These are glucans, including cellulose, callose, mixed-linked (1→3, 1→4)-β-glucans and xyloglucans, rhamnogalacturonans and associated arabinans and arabinogalactans, mannans including glucomannans and galactoglucomannans, xylans and glucuronomannans. Details on the primary structure of these polysaccharides are generally available.

During plant development, the primary cell wall is formed first and is connected to the middle lamella, an intracellular layer rich in pectin. This pectin has a high content of methyl-esterified galacturonic acid residues and a low content of essentially unbranched rhamnose residues compared to the pectin in the primary wall (Selvendran, Stevens and DuPont, 1987). The primary wall has a skeleton of cellulose microfibrils which are embedded in a matrix of polysaccharides, lignin and

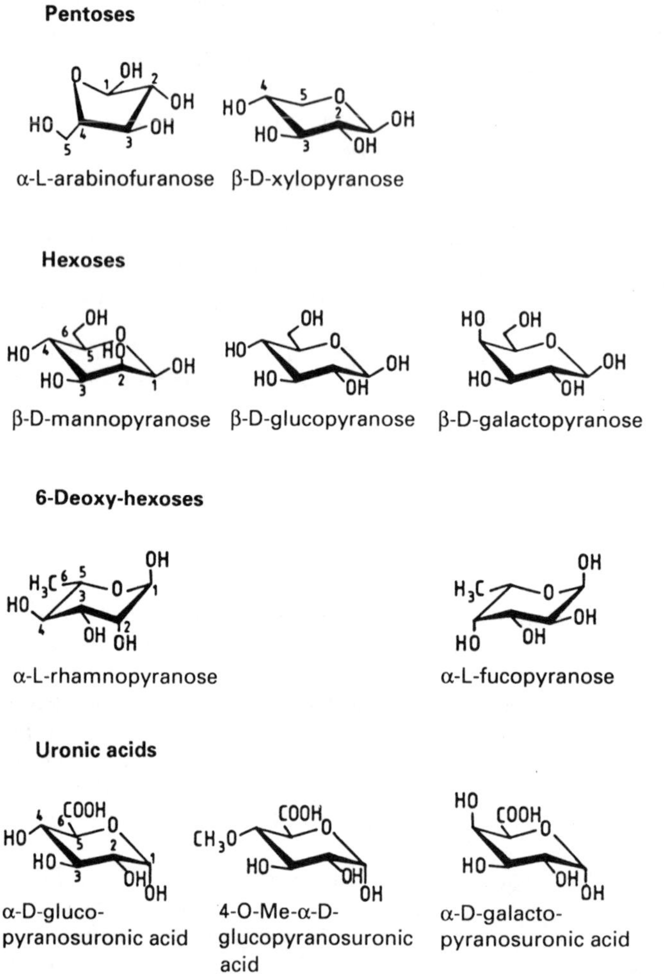

Pentoses

α-L-arabinofuranose   β-D-xylopyranose

Hexoses

β-D-mannopyranose   β-D-glucopyranose   β-D-galactopyranose

6-Deoxy-hexoses

α-L-rhamnopyranose   α-L-fucopyranose

Uronic acids

α-D-gluco-   4-O-Me-α-D-   α-D-galacto-
pyranosuronic acid   glucopyranosuronic   pyranosuronic acid
                     acid

Figure 9.4 Main cell-wall polysaccharide constituents presented in their preferred conformation

other non-carbohydrate constituents such as phenolic components and acetyl groups. The matrix polysaccharides of primary walls in dicotyledons are predominantly composed of xyloglucans, galactans, arabinogalactans and rhamno-galacturonans or pectin (Albersheim, 1976) while primary walls in monocotyledons are mainly composed of arabinoxylans, glucuronoarabinoxylans and mixed-linked β-glucans (Cheeson, Gordon and Lomax, 1985). Models for the primary cell wall have been proposed in which covalent linkages are found between most of the polysaccharide and glycoprotein components but not, however, between cellulose and the other components (McNeil *et al.*, 1984).

The secondary wall is deposited inside the primary wall and is much thicker and more rigid. It is more or less lignified and may reach a completely dominant position among the cell constituents, contributing up to 90% of the cell dry matter. Cellulose and xylans are often important constituents of secondary plant cell walls. For more details about the composition and structure of the plant cell walls, reviews by Tolbert (1980) and Colvin (1982) are recommended.

ANALYSIS OF CELL-WALL POLYSACCHARIDES

The growing interest over the last decades in the physiological effects of plant cell wall polysaccharides has raised many questions concerning the definition and analysis of fibre. Originally, dietary fibre was defined as 'the remnants of plant cell walls that are resistant to digestive enzymes' (Trowell, 1972). This definition was later expanded to include, for example, indigestible storage polysaccharides, and the current definition of dietary fibre, to which most people adhere, is indigestible polysaccharides (or rather non-starch polysaccharides) plus lignin (Trowell *et al.*, 1976; Theander and Åman, 1979). This classification is more narrow in the sense that it excludes, for example, indigestible cell wall protein, cutins and waxes, but expanded in that it may also include polysaccharide additives. However, in spite of a generally adopted definition there is still controvsery with respect to what to include, and it has recently been suggested that lignin, which may be a major

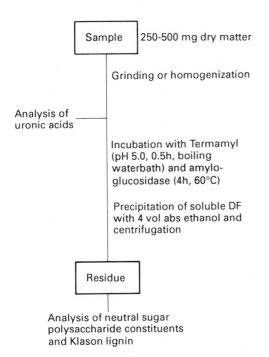

Figure 9.5   Scheme for the analysis of dietary fibre by the Uppsala procedure (Theander and Åman, 1979; Theander and Westerlund, 1986)

cell-wall constituent, should be excluded. Resistant starch and unavailable oligosaccharides are also considered by some as dietary fibre.

A major problem in research on plant cell-wall polysaccharides or dietary fibre has been the evaluation of suitable analytical methods. It was apparent relatively early that dietary fibre could not be analysed as crude fibre. Currently two different principles for the measurement of dietary fibre have been developed. These are the enzymatic-gravimetric methods which attempt to determine the part of the feed that resists hydrolysis by starch- and protein-degrading enzymes (e.g. Prosky *et al.*, 1985) and the enzymatic-chemical methods which, after removal of starch and low-molecular-weight carbohydrates, determine non-starch polysaccharides and lignin (e.g. Theander and Åman, 1979; Theander and Westerlund, 1986; Englyst and Cummings, 1988).

The Uppsala methodology (Figure 9.5) for the accurate analysis and chemical characterization of total dietary fibre has been in use for more than ten years (Theander and Åman, 1979). Here a thermostable α-amylase (Termamyl) in combination with amyloglucosidase is used for starch removal, modern GLC-methodology, including correction factors, for the analysis of neutral non-starch polysaccharide residues, a stoichiometric decarboxylation method for uronic acid residues, and a gravimetric method for Klason lignin. The method has recently been improved by using ethanol precipitation of soluble fibres and 1-methylimidazole as a catalyst during the derivatization of neutral non-starch polysaccharide residues (Theander and Westerlund, 1986). The method has been applied to a number of divergent samples, including foods, feeds, digesta and faeces, and has proven to be robust, reproducible, adaptable and accurate. For a skilled analyst it is possible to run up to forty samples per week and it can easily be adapted to analyse separately soluble and insoluble fibre components. Total and insoluble fibre will include resistant starch determined as glucose residues. If desired, 'resistant' starch can be determined by a direct one-tube method on the original sample after initial removal of starch. In this procedure resistant starch is solubilized in 2 M aqueous potassium hydroxide, hydrolysed with amyloglucosidase and glucose released determined by the glucose oxidase method (Westerlund *et al.*, 1989):

1. Grind representative samples in a Tecator Cyclone Sample Mill to pass a 0.5 mm screen.
2. Weigh duplicate samples, containing 0.1–2.0 mg resistant starch, into 35 ml thick-walled Pyrex glasstubes with screw caps containing tightly fitting Teflon washers.
3. Gelatinize and degrade starch by two incubations with thermostable α-amylase (Termamyl) in a boiling waterbath and one incubation with amyloglucosidase at 60°C.
4. Wash the insoluble residue with water and dissolve resistant starch in 2M KOH at room temperature.
5. After neutralization and centrifugation, hydrolyse the dissolved starch with amyloglucosidase and calculate the content of resistant starch from the released glucose as measured by the conventional glucose-oxidase method.

In many samples like cereal grains total fibre can be estimated by subtracting the content of free sugars, starch, crude protein, crude fat using acid hydrolysis and ash from the dry matter of the sample (Åman and Hesselman, 1984). Such a 'by difference' method will include small amounts of components other than non-starch polysaccharides and Klason lignin, but has been shown for cereals to correlate

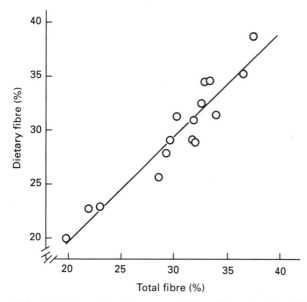

Figure 9.6 Correlation between total fibre (determined by difference) and dietary fibre (determined as non-starch polysaccharides and Klason lignin) in oats (Åman, 1987)

closely to dietary fibre (Figure 9.6). The determination of total fibre by difference is not as accurate or informative as determination by enzymatic – chemical methods but in certain cases it seems more appropriate to determine the available nutrients with high precision and calculate the fibre content with a somewhat lower precision.

Currently crude fibre and detergent methods (Van Soest, 1982) are frequently used for fibre analysis. However, these analytical methods determine only a fraction of the fibres present, with no close relationship between dietary fibre, neutral detergent fibre (NDF), acid detergent fibre (ADF) and crude fibre for different feedstuffs or feeds (Table 9.2). Several factors, such as available

Table 9.2 THE DIETARY FIBRE, NEUTRAL DETERGENT FIBRE (NDF), ACID DETERGENT FIBRE (ADF), AND CRUDE FIBRE CONTENTS OF SOME FEEDSTUFFS (FROM GRAHAM, 1988)

Feed	Content (g/kg DM) of:				As % of dietary fibre	
	Dietary fibre	NDF	ADF	Crude fibre	NDF	Crude fibre
Maize	94	82	22	20	91	21
Wheat	108	98	32	23	91	21
Barley	188	152	50	42	81	22
Feed peas	163	104	71	52	64	32
Soybean meal	241	151	88	76	63	32
Pig grower feed	203	134	53	42	66	21
Wheat straw	798	830	522	447	104	56

equipment, amount of time, accuracy, precision and amount of information required, are important for the choice of analytical procedure for fibre.

SOLUBILITY OF CELL-WALL POLYSACCHARIDES

In recent years there has been a growing interest in soluble or viscous fibres since these components are known to give specific metabolic effects. A division of fibre into a soluble and an insoluble component will, however, depend very much on the analytical procedure (Graham, Grön Rydberg and Åman, 1988) and since the fibre complex is continuously modified during its passage through the gastro-intestinal tract it is impossible to simulate this process with a simple chemical analysis of the feed (Theander and Åman, 1982).

Several factors such as the fine structure of the polysaccharides, interrelationships between cell-wall constituents, activity of endogenous hydrolytic enzymes, pretreatment of the sample and extraction conditions will influence the yield as well as the chemical and physical properties of both the soluble and insoluble fibres. In a recent experiment, water-soluble fibres were extracted from wheat, rye, barley and oats using two different methods; (A) pH 5.0 acetate buffer at 96°C for 4 h during starch degradation and (B) water at 38°C for 2 h (Graham, Grön Rydberg and Åman, 1988). For wheat, rye and barley the higher temperature of method A resulted in higher yields of soluble fibre while method B gave higher yield for the oats sample. This difference was due presumably to a high endogenous hydrolytic activity in the oat sample solubilizing fibres at 38°C.

ANALYSIS OF SPECIFIC DIETARY FIBRE COMPONENTS

Few methods have been developed for the analysis of specific dietary fibre components. However the detergent system developed by Van Soest (1982) includes the determination of cellulose and lignin by hydrolysis or oxidation of isolated ADF, and specific analytical methods for mixed-linked β-glucans in cereals have been developed (e.g. McCleary and Glennie–Holmes, 1985; Åman and Graham, 1987).

In most cases, however, classical methods for carbohydrate analysis have to be used for the determination of specific components. It is also important to remember that many cell-wall polysaccharides are linked covalently or in other ways to the other macromolecules in the plant cell walls and that in these cases it is the macromolecular complex and not the specific components which will give a physiological response.

Conclusions

During recent years there has been a tremendous development in methods for the analysis of polysaccharides and associated components. However, only a few of these methods have been generally used for the characterization of animal feeds. It is important to realize that polysaccharides in feeds are a complex group of components, often associated to each other or to other components, with different properties and physiological activities. The chemical structure of these polymers or complexes can affect both the rate and extent of digestion and can be used to control and optimize nutrient assimilation in different parts of the gastro-intestinal tract. The use and development of better analytical methods and animal techniques for the evaluation of polysaccharides in feeds will therefore be a prerequisite for a greater understanding of the role of polysaccharides in animal feeds.

References

Albersheim, P. (1976) The primary cell wall. In *Plant Biochemistry* (eds Bonner, J. and Varner, J.E.) Academic Press, New York, pp. 225–272

Åman, P. (1987) The variation in chemical composition of Swedish oats. *Acta Agriculture Scandinavica*, **37**, 347–352

Åman, P. and Graham, H. (1987) Analysis of total and insoluble mixed-linked β-(1→3), (1→4)-D-glucans in barley and oats. *Journal of Agricultural Food and Chemistry*, **35**, 704–709

Åman, P. and Hesselman, K. (1984) Analysis of starch and other main constituents of cereal grains. *Swedish Journal of Agricultural Research*, **14**, 135–139

Åman, P., Westerlund, E. and Theander, O. (1989) Determination of starch using a thermostable α-amylase. *Methods in Carbohydrate Chemistry* (in press)

Asp, N.G., Björck, I., Holm, J., Nyman, M. and Siljeström, M. (1987) Enzyme resistant starch fractions and dietary fibre. *Scandinavian Journal of Gastroenterology*, **22**, Suppl. 129, 29–32

Aspinall, G.O. (1981) Constitution of plant cell wall polysaccharides. In *Plant Carbohydrates II, Extracellular Carbohydrates* (eds Tanner, W. and Loewus, F.A.), Springer – Verlag, Berlin, pp. 3–8

Batey, I.L. (1982) Starch analysis using thermostable alpha-amylases. *Starch*, **34**, 125–128

Björndal, H., Hellerqvist, C.G., Lindberg, B. and Svensson, S. (1970) Gas liquid chromatography and mass spectrometry of methylated polysaccharides. *Angewante Chemie International Edition*, **9**, 610–619

Cheeson, A., Gordon, A.H. and Lomax, J.A. (1985) Methylation analysis of mesophyll, epidermis and fibre cell-walls isolated from the leaves of perennial and Italian ryegrass. *Carbohydrate Research*, **141**, 137–147

Cheeson, A., Stewart, C.S., Dalgarno, K. and King, T.P. (1986) Degradation of isolated grass mesophyll, epidermis and fibre cell walls in the rumen and by rumen bacteria in axenic culture. *Journal of Applied Bacteriology*, **60**, 327–336

Colvin, J.R. (1981) Ultrastructure of the plant cell wall: Biophysical viewpoint. In *Plant Carbohydrates II, Extracellular Carbohydrates* (eds Tanner, W. and Loewus, F.A.) Springer – Verlag, Berlin, pp. 9–27

Darbyshire, B. and Henry, R.J. (1978) The distribution of fructans in onions. *New Phytology*, **81**, 29–34

Dea, I.C.M. and Morrison, A. (1975) Chemistry and interactions of seed galactomannans. *Advances in Carbohydrate Chemistry and Biochemistry*, **31**, 241–312

Dische, Z. (1962) Color reactions of hexoses. *Methods in Carbohydrate Chemistry*, **I**, 488–494

Dische, Z. and Borenfreund, E.J. (1951) A new spectrophotometric method for the detection and determination of keto sugars and trioses. *Journal of Biological Chemistry*, **192**, 583–587

Englyst, H., Andersson, V. and Cummings, J.H. (1983) Starch and non-starch polysaccharides in some cereal foods. *Journal of the Science of Food and Agriculture*, **34**, 1434–1440

Englyst, H. and Cummings, J.H. (1988) Improved method for measurement of dietary fiber as non-starch polysaccharides in plant foods. *Journal of the Association of Official Analytical Chemists*, **71**, 808–814

Gordon, A.H., Lomax, J.A,. and Chesson, A. (1983) Glycosidic linkage composition of legume, grass and cereal straw cell walls before and after extensive digestion by rumen micro-organisms. *Journal of the Science of Food and Agriculture*, **34**, 1341–1350

Graham, H. (1988) Dietary fibre concentration and assimilation in swine. *ISI Atlas of Science, Animal and Plant Sciences*, **1**, 78–80

Graham, H. and Åman, P. (1986) Composition and digestion in the pig gastrointestinal tract of Jerusalem artichoke tubers. *Food Chemistry*, **22**, 67–76

Graham, H., Grön Rydberg, M. -B. and Åman, P. (1988). The extraction of soluble fiber. *Journal of Agricultural and Food Chemistry*, **36**, 494–497

Greenwood, C.T. (1970) Starch and glycogen. In *The Carbohydrates, Chemistry and Biochemistry*, Vol. II B. (eds Pigman, W. and Horton, D.) Academic Press, New York, pp. 471–514

Hassid, W.Z and Neufeld, E.F. (1964) Quantitative determination of starch in plant tissues. *Methods in Carbohydrate Chemistry*, **IV**, 33–36

Holm, J., Björck, I., Ostrowska, S., Eliasson, A.-C., Asp, N.G., Larsson, K. and Lundkvist, I. (1983) Digestibility of amylose-lipid complexes *in vitro* and *in vivo*. *Starch*, **35**, 294–297

Jones, J.K.N. and Tanaka, Y. (1965) Galactan – Isolation from lupinus seed. *Methods in Carbohydrate Chemistry*, **V**, 132–134

Kooiman, P. (1960) On the occurrence of amyloids in plant seeds. *Acta Botanica Neerlandica*, **9**, 208–219

Larsson, K. and Bengtsson, S. (1983) *Bestämning av lätt tillgängliga kolhydrater i växtmaterial*. Metodbeskrivning 22, SLL, Uppsala, Sweden

Leontein, K., Lindberg, B. and Lönngren, J. (1978) Assignment of absolute configuration of sugars by g.l.c. of their acetylated glycosides from chial alcohols. *Carbohydrate Research*, **62**, 359–362

Lyne, F.A. (1976) Determination of starch in various products. In *Examination and Analysis of Starch and Starch Products* (ed Radley, J.A.), Science Publishers, London, pp. 167–188

McCleary, B.V. and Glennie–Holmes, M. (1985) Enzymatic quantification of $(1\rightarrow 3)$, $(1\rightarrow 4)$-β-D-glucan in barley and malt. *Journal of the Institute of Brewing*, **91**, 285–295

McNeil, M., Darvill, A.G., Åman, P., Franzén, L.-E. and Albersheim, P. (1982) Structural analysis of complex carbohydrates using high-performance liquid chromatography, gas chromatography, and mass spectrometry. *Methods in Enzymology*, **83**, 3–45

McNeil, M., Darvill, A.G., Fry, S.C. and Albersheim, P. (1984) Structure and function of the primary cell walls of plants. *Annual Reviews in Biochemistry*, **53**, 625–663

Meier, H. and Reid, J.S.G. (1982) Reserve polysaccharides other than starch in higher plants. In *Plant Carbohydrates I, Intracellular Carbohydrates* (eds Loewus, F.A. and Tanner, W.), Springer – Verlag, Berlin, pp. 418–471

Nilsson, U. (1988) *Cereal Fructans – Preparation, Characterization, Fermentation and Bioavailability*. PhD thesis, Department of Applied Nutrition, University of Lund, S-221 00 Lund, Sweden.

Nordkvist, E. (1987) *Composition and Degradation of Cell Walls in Red Clover, Lucerne and Cereal Straw*, PhD thesis, Department of Animal Nutrition and Management, The Swedish University of Agricultural Sciences, S-750 07 Uppsala, Sweden

Prosky, L., Asp, N.-G., Furda, I., DeVries, J.W., Schweizer, T. F. and Harland, B. (1985) Determination of total dietary fibre in foods and food products: Collaborative study. *Journal of the Association of Official Analytical Chemists*, **68**, 677–679

Robertsen, B., Åman, P., Darvill, A., McNeil, M. and Albersheim, P. (1981) Host-symbiotic interactions, V. The structure of acidic extracellular polysaccharides secreted by *Rhizobioum leguminosarum* and *Rhizobium trifolii*. *Plant Physiology*, **67**, 389–400

Sawardeker, J.S., Sloneker, J.H. and Jeanes, A.R. (1965) Quantitative determination of monosaccharides as their alditol acetates by gas liquid chromatography. *Analytical Chemistry*, **37**, 1602–1604

Selvendran, R.R., Stevens, B.J.H. and DuPont, M.S. (1987) Dietary fiber: Chemistry, analysis, and properties. *Advances in Food Research*, **31**, 117–209

Smith, D. (1967) Carbohydrates in grasses. II. Sugar and fructosan composition of the stem bases of bromegrass and timothy at several growth stages and in different plant parts at anthesis. *Crop Science*, **7**, 62–67

Smith, D. (1973) The nonstructural carbohydrates. In *Chemistry and Biochemistry of Herbage* (eds Butler, G.W. and Bailey, R.W.), Academic Press, London, pp. 105–155

Theander, O. and Åman, P. (1979) Studies on dietary fibre. 1. Analysis and chemical characterization of water-soluble and water-insoluble dietary fibre. *Swedish Journal of Agricultural Research*, **9**, 97–106

Theander, O. and Åman, P. (1982) Studies on dietary fibre. A method for the analysis and chemical characterization of total dietary fibre. *Journal of the Science of Food and Agriculture*, **33**, 340–344

Theander, O. and Westerlund, E. (1986) Studies on dietary fiber. 3. Improved procedures for analysis of dietary fiber. *Journal of Agricultural and Food Chemistry*, **34**, 330–336

Thivend, P., Mercier, C. and Guilbot, A. (1972) Determination of starch with glucoamylase. *Methods in Carbohydrate Chemistry*, **VI**, 100–105

Tolbert, N.E. (1980) *The Biochemistry of Plants. Vol. 1; The Plant Cell*, Academic Press, New York

Trowell, H. (1972) Ischemic heart disease and dietary fiber. *American Journal of Chemical Nutrition*, **25**, 926–932

Trowell, H., Southgate, D.T.A., Wolever, T.M.S., Leeds, A.R., Gassull, M.A. and Jenkins, D.A. (1976) Dietary fibre redefined. *Lancet*, **i**, 967

Van Soest, P. (1982) *Nutritional Ecology of the Ruminant*, O & B Books, Corvallis, Oregon

Westerlund, E., Theander, O., Andersson, R. and Åman, P. (1989) Breadmaking, 2. Effects of baking on polysaccharides in white bread fractions. *Journal of Cereal Science*, **10**, 149–156

Whistler, R.L., BeMiller, J.N. and Parschall, E.F. (1984) *Starch: Chemistry and Technology*, 2nd edn, Academic Press, Orlando

10

NUTRITIONAL SIGNIFICANCE AND NUTRITIVE VALUE OF PLANT POLYSACCHARIDES

A. CHESSON
Rowett Research Institute, Bucksburn, Aberdeen AB2 9SB, UK

Introduction

A distinction can be conveniently made between the 'nutritional significance' of plant polysaccharides and their potential 'nutritive value'. The former term relates to the influence which intact polysaccharides, either in isolation or in combination with other polymers, exert on the utilization of other dietary components, for example by influencing the rate and extent of nutrient uptake or by promoting the removal of potentially toxic compounds, while the latter relates to the potential of ingested polysaccharides, after hydrolysis, to contribute directly to the nutrition of livestock. The nutritive value of a polysaccharide is thus dependent on the extent to which its component monosaccharides are released and the manner of their subsequent metabolism. As the assessment of the nutritive value of ingested polysaccharides is the prime concern of the evaluation chemist, reference to the nutritional significance of intact polysaccharides is restricted to those circumstances where they have a demonstrable effect on nutritive value of other diet components.

Predicting the nutritive value of ingested polysaccharides poses a number of major problems for the evaluation chemist. Looking to the future it will be no longer adequate to consider polysaccharides simply as a source of dietary energy treated in isolation from other dietary components. The factors determining the rate of nutrient supply to livestock are multiple and involve a dynamic interplay between digestible and indigestible dietary components, the host's digestive secretions and the action of the gut microflora. The limitations associated with the present Metabolizable Energy (ME) and the ARC (1980, 1984) schemes for separately assessing the energy and protein requirements of ruminants are already well recognized (MacRae, 1986; MacRae *et al.*, 1988). For example, when the level of propionate is insufficient to meet the requirements for glycogenic intermediates, glucogenic amino acids substitute for propionate, reducing the amount of amino acids available for protein synthesis and deposition. Although most acutely seen in ruminants, estimates of the supply of energy-yielding substrates cannot be divorced from estimates of the supply of amino acids in any system which hopes to predict the productive response of livestock. The challenge to all concerned with feed evaluation and diet formulation is to devise assessment methods capable of predicting the supply of metabolizable substrates (rather than metabolizable energy), whether in the form of monosaccharides or microbial fermentation products, available for tissue metabolism. From such data the likely metabolic fate of absorbed nutrients and ultimate productive benefit to the animal can be assessed.

Release of metabolizable substrates from polysaccharides

Although plant polysaccharides are a heterogeneous group of compounds showing a wide range of physical and chemical properties they share a number of common features when fed to livestock. All are dependent on enzymatic hydrolysis to release their components monosaccharides and all are potentially fully hydrolysable. However most fail to reach total hydrolysis largely because of their association with other dietary components.

Most evaluation schemes distinguish between the intracellular storage polysaccharide, starch and the structural polysaccharides which form the major part of the plant cell wall (often loosely referred to as plant fibre or roughage). Starch forms the bulk of most concentrates fed and is the only plant polysaccharide which both mammals and birds are capable of degrading through the action of their endogenous enzyme systems. Both poultry and non-ruminant animals are capable of efficiently degrading starch and further degrading the breakdown products to glucose which is absorbed and used directly for tissue metabolism. In contrast, the hydrolysis of the plant structural polysaccharides is wholly dependent on the action of the gut microflora and any degradation products serve as a carbon source for the growth and multiplication of the gut micro-organisms and are not directly available to the host. The supply of nutrients which derive from the structural polysaccharides, indirectly supplied to the host as microbial fermentation products, is intimately linked to the supply of other nutrients necessary to support microbial growth and to the metabolic status and composition of the resident microbial population.

The site of polysaccharide degradation within the digestive tract has little bearing on the mechanism of hydrolysis other than to determine whether starch hydrolysis is a product of host or microbial derived amylolytic activities. There is little evidence to suggest that the mechanism of microbial colonization and secretion of cell wall (fibre) degrading enzymes differs in principle between the rumen or, for example, the hindgut of non-ruminant species. The factors determining the rate and extent of cell wall breakdown are predominantly substrate led and operate equally in both sites. However, while the mechanism of polysaccharide hydrolysis can be considered independently of the site of enzyme action, the site of hydrolysis does have a major influence on the nature of the end products presented to the host animal. The end product of starch hydrolysis in the small intestine is glucose while in the rumen or hundgut, starch hydrolysis occurs through the action of the microflora and the end products available to the the host animal are volatile fatty acids (VFA).

In ruminants polysaccharide hydrolysis is intimately linked to the capture of nitrogen in the form of microbial protein and to the supply of amino acids to the animal. The nutritive value of the diet as a whole is much more dependent on matching the supply of all nutrients to the rumen microflora and thus the rate of polysaccharide breakdown can be as important as the extent of degradation. For this reason, it is more practical to consider separately the factors influencing nutrient supply in ruminant and non-ruminant livestock.

Polysaccharide degradation and nutrient supply in ruminants

Carbohydrates enter the rumen in three distinct forms:

1. Water-soluble carbohydrates consisting largely of simple sugars and some soluble polysaccharides such as fructosans.
2. Starch, an α-glucan based storage polysaccharide.
3. Structural polysaccharides contributing to the plant cell wall.

In forage fed animals, the bulk of the polysaccharide ingested is in the form of structural polysaccharides or fibre with a lesser contribution from some water-soluble sugars. Starch is present in small amounts only as leaf starch. It is well recognized that the fibre fraction is rarely, if ever, fully degraded and many evaluation schemes distinguish, directly or indirectly, between digestible and non-digestible fibre fractions. This rather simplistic distinction has proved valid when used in those mathematical models of ruminant digestive function concerned with the prediction of intake and fibre digestion in forage fed animals (Mertens and Ely, 1979; Robinson *et al.*, 1986). Although such models are capable of dealing with a dynamic system in which both rate of passage and rate of digestion are included, they do not include any concept of microbial metabolism and cannot be used to predict the nature and rate of production of the products of feed degradation. These models do, however, indicate that the proportion of potentially digestible fibre present is more important that the rate of digestion in affecting digestibility and intake (Mertens and Ely, 1982).

STOICHIOMETRY OF RUMEN FERMENTATION

Relating fermentation products and microbial cell synthesis to diet composition requires knowledge of the biochemical pathways involved in the microbial metabolism of polysaccharides and other diet components. Substrate and product balance equations can be generated for individual dietary components and iterative programming used to develop models based on as many balance equations as are found necessary to produce a desired output. Although balance equations relate to steady-state kinetics they can also be linked to other mathematical descriptions describing the rate of degradation of diet components such as fibre. The stoichiometric parameters for the fermentation of carbohydrate required by models whose output attempts to describe nutrient supply have been generated from both *in vivo* observations, such as the data compiled by Murphy *et al.* (1982), or from *in vitro* observations made with rumen microorganisms (Reichl and Baldwin, 1976). Any evaluation scheme based on nutrient supply to animal tissues defines the outputs necessary and, in theory, the modelling process itself should identify the necessary inputs and therefore the future demands likely to be made of evaluation chemists. However at present we have a chicken and egg situation where stoichiometric parameters are defined in terms of known dietary inputs determined by existing evaluation methods. Such methods have proved inadequate for the stoichiometric modelling of the majority of experimental diets (Hobson and Jouany, 1989).

A major factor determining the overall stoichiometry of fermentation is the nature of the microbial population involved. The proportion of soluble carbohydrate, starch, hemicellulose and cellulose converted to acetate, propionate and butyrate is substantially affected by whether the carbohydrate components are derived from a concentrate or roughage based diet (Table 10.1; Murphy *et al.*, 1982). Such results are not unexpected because of the major shift in the rumen population known to occur during adaption from concentrate to roughage diet or vice-versa. Although work with gnotobiotic animals has shown that limited number of bacterial species, notably *Streptococcus bovis* and *Bacteroides amylophilus*, are predominantly responsible for starch hydrolysis in the rumen, the shift in VFA stoichiometry cannot be ascribed simply to variations in the size of the amylolytic population. The increases in the proportion of starch fermented to propionate found in high concentrate diets has been observed to relate to an adaptive balance between amylolytic- and lactate-utilizing bacteria (Mackie and Gilchrist, 1979). Thus changes in diet can markedly affect organisms not primarily responsible for polysaccharide metabolism and stoichiometric parameters established *in vitro* solely with amylolytic bacteria are unlikely to be those operating *in vivo* in the rumen.

Table 10.1   PROPORTION OF FERMENTED SUBSTRATE CONVERTED TO VFA IN ROUGHAGE (R) AND CONCENTRATE (C) FED ANIMALS. DATA TAKEN FROM MURPHY *ET AL*. (1982)

Carbohydrate substrate	Group	Acetate	Propionate	Butyrate
WSC	R	0.69	0.21	0.11
	C	0.45	0.21	0.30
Starch	R	0.59	0.14	0.21
	C	0.40	0.30	0.20
Hemicellulose	R	0.57	0.18	0.21
	C	0.56	0.26	0.11
Cellulose	R	0.66	0.09	0.23
	C	0.79	0.06	0.07

CELL WALL FERMENTATION

Establishing the stoichiometry of cell wall degradation is fraught with difficulties, not least because of the large number of different organisms apparently involved, directly or indirectly, in end product formation. Attempts to establish a viable and defined fibre-degrading flora in gnotobiotic animals have shown that 40–70 or more strains have to be introduced before the defined population becomes stable and fibre degradation approaches that of the conventional animal (Hobson and Jouany, 1989). It can be argued that models of the digestive process which most closely represent the biology of the system are likely to prove more robust in practice than models which deal only with aspects of the system. If this is the case then treatment of the fibre fraction of the feed in terms of digestible and non-digestible components or in terms of polysaccharide classes (cellulose, hemicellulose) is inadequate since it does not accurately reflect the mechanism of cell wall degradation by gut microorganisms. Ingested plant particles contain a variety of cell types each of which posses cell walls of different composition and degradability. Microscopic examinations of plant tissues undergoing degradation have shown that cells with primary walls are the first to be degraded, usually to completion, while cells possessing secondary-thickened (lignified walls) are much more slowly attacked. In an attempt to quantify this process, ryegrass leaf tissue was mechanically fractionated to provide samples of each cell type forming the leaf tissue. Cell walls were then prepared from each of the homogeneous cell samples, analysed and their degradability measured (Gordon *et al.*, 1985; Chesson *et al.*, 1985, 1986). As Table 10.2 shows, from a biological point of view the structural polysaccharides found in grass leaf consist of three fractions, a fully digestible fibre fraction provided by the primary cell walls, a more slowly digested fraction of the secondary-thickened walls and a non-digestible residue derived solely from the secondary walls.

Table 10.2 DISTRIBUTION OF CARBOHYDRATE IN RYEGRASS LEAF

	% Leaf dry matter				
	WSC*	Starch	Primary cells†	Secondary cells‡	Total
Cell wall content	–	–	35.1	18.2	53.3
Carbohydrate (CHO)	4.3	0.3	29.9	15.8	50.3
Fraction of total CHO	0.09	0.01	0.59	0.31	1.00
Digestibility (%) 12 h	100	100	100	46	–

*WSC = Water-soluble carbohydrate.
†Primary cells = leaf mesophyll and epidermis.
‡Secondary cells = vascular bundles and sclerenchyma.

The ratio of primary walls or highly digestible fibre to that of partially digestible fibre found in secondary walls is approximately 2:1. It is interesting to note that attempts to maximize the degradability of poorly used fibre sources such as cereal straws are most successful when fed with a highly digestible fibre source such as sugar beet pulp or turnip and when the content of straw is limited to approximately 30% of total dry matter intake (Silva *et al.*, 1989). Formulating a diet on this basis would appear to recreate a forage-equivalent in which the soluble sugars present in turnip or sugar beet substitute for the water soluble carbohydrate (WSC) and starch in the forage diet; turnip or sugarbeet cell walls provide the highly digestible primary cell walls and the straw provides the source of secondary wall material. Many feeding problems arise only when the diet fed deviates from this general forage model. Increasing the proportion of WSC or readily fermented starch may give rise to acidosis, while the rate of carbohydrate release from diets with extreme levels of secondary cell walls is too low to support growth.

The inclusion of separate values for WSC and starch as inputs into models of rumen fermentation seems fully justified on both biological and pragmatic grounds. It would also appear justifiable to treat the pectic substances as WSC when dicotyledenous plants form a large part of the diet. Legume pectic substances are degraded considerably faster than other legume cell wall polysaccharides (Chesson and Monro, 1982; Barry *et al.*, 1984; Ulyatt *et al.*, 1984). The remaining carbohydrates, the cell wall fraction of the diet, present the greatest difficulty since they can be variously described (Table 10.3). Selection of the appropriate description is thus crucial to the development of models capable of predicting nutrient supply.

Table 10.3 OPTIONS FOR DESCRIBING THE CELL WALL (FIBRE) CONTENT OF
RUMINANT FEEDS

Cell wall content:

Single component model	–	Total cell wall content – NDF
Two-component model	–	Digestible and indigestible fractions
Three-component model	–	Primary cell wall content – fully digestible
		Digestible fraction of the secondary wall
		Indigestible fraction of the secondary wall

Cell wall polysaccharide:

Single-component model	–	Total cell wall carbohydrate – NSP
Two-component model	–	Cellulose and hemicellulose
Multiple component model	–	Individual cell wall polymers, e.g. xylan

---

NDF = Neutral detergent fibre, NSP = Non-starch polysaccharide

Murphy *et al.* (1982) chose to distinguish between two classes of fibre polysaccharides, cellulose and hemicellulose, and found significant differences in the proportion of VFA produced (see Table 10.2). These differences could have arisen for a number of interrelated reasons. Hemicellulose is predominantly composed of pentose sugars while cellulose is a hexose polymer and the two forms of sugars follow different metabolic pathways in rumen microorganisms. However the bulk of pentose appears to be converted to hexose and triose and is channelled towards pyruvate thus minimizing any likely differences in end-product formation (Baldwin and Allison, 1983). If monosaccharide composition is the major factor determining the nature of end product formation, then future evaluation schemes would require inputs which, in some way, describe the monosaccharide content of the feed. This seems unlikely however since it is well recognized that microbial metabolism of the β-linked glucan, cellulose, produces higher proportions of acetate than metabolism of the α-glucan, starch (see Table 10.1). In this case differences in VFA proportions arise because of the radically different amylolytic and cellulolytic flora and not because of the nature of the sugar fermented. This could be taken to imply that different microbial populations are involved in the breakdown and fermentation of cellulose and hemicellulose. However, unlike starch, cellulose and hemicellulose do not exist independently of one another or of the other polymers found in the cell wall.

All of the evidence obtained to date suggests that there is no selective degradation of cell wall polysaccharides, but that all components of the wall are hydrolysed at a common rate and that the organisms primarily responsible for the initial hydrolysis of both cellulose and hemicellulose are the 'cellulolytic' bacteria (Chesson *et al.*, 1986). A more likely explanation of the differences in VFA proportions described by Murphy *et al.* (1982) is that the authors were indirectly distinguishing between the breakdown and fermentation of primary (cellulose-rich) and secondary-thickened (hemicellulose-rich) cell walls (see xylose content, Figure 10.1). There is far more evidence to suggest population differences in the flora associated with the degradation of the cell walls of different cell types than with different cell wall polysaccharides. *Bacteroides succinogenes*, for example, is found in greater numbers associated with secondary plant walls. Since strains of this organism are capable of hydrolysing xylans but are unable to utilize the products of xylan hydrolysis (Morris and Van Gylswyk, 1980), the pentose sugars released are available to support the metabolic activities of other organisms. Similarly the

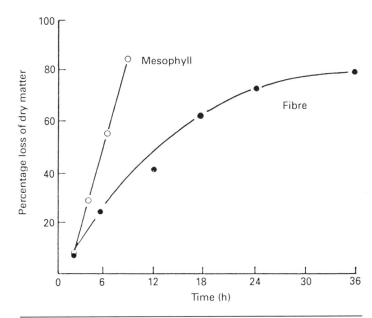

	Cell wall dry matter (%)	
	Mesophyll	Fibre
Monosaccharide residues		
Rhmanose	0.8	0.2
Fucose	0.1	0.1
Arabinose	6.4	3.8
Xylose	11.0	27.4
Mannose	0.3	0.1
Galactose	2.5	0.8
Glucose	59.4	50.9
Uronic acid	4.7	3.8
Total carbohydrate	85.2	87.1

Figure 10.1 Carbohydrate composition and rumen degradability of primary (mesophyll) and secondary-thickened (sclerenchyma) cell walls prepared from leaves of perennial ryegrass (Data from Chesson, 1986)

rumen fungi are predominantly found associated with tissues rich in secondary walls while the protozoa are generally found associated with primary tissue (Akin, 1988).

A further factor which may influence the stoichiometry of VFA production is the rate of polysaccharide hydrolysis which can be shown to be far higher in primary than in secondary cell walls (Figure 10.1). Both the rate of supply of metabolizable nutrients to the micro-organism and the balance between catabolic and anabolic activities are likely to influence microbial metabolic status. A clear example of this is *Selenomonas ruminantium* which is capable of adjusting its metabolic pathways in response to nutrient supply. When growth rates are low because of nutrient limitation, intracellular pyruvate concentration is low, the enzyme lactate dehydrogenase remains inactive and acetate and propionate are the major end products. With increasing nutrient supply and growth rate, lactate dehydrogenase is activated and lactate is released as an end product with a consequence reduction in the levels of acetate and propionate formed (Wallace, 1978). Unfortunately there is little similar work available which has attempted to relate growth rates of individual 'cellulolytic' bacteria and the formation of end products other than cell yield.

The distinction made between the WSC (including pectic substances from dicotyledenous plants) and starch contents of diets appears justified in biological terms since it recognizes the differences both in availability and hence rate of fermentation and in the microbial flora responsible. Similar arguments applied to the fibre fraction suggest that the remaining ingested polysaccharides are best considered in terms of primary and secondary cell walls rather than in terms of their component polysaccharides. Although values for the primary cell wall content and the digestible fraction of the secondary cell walls can be combined to give a single figure describing the total available cell wall polysaccharides, this fails to provide important information on the rate of nutrient supply to the microorganisms and hence on the rate of formation of fermentation products; it also fails to take account of the differences in the associated microbial flora and the effect this apparently has on the stoichiometry of fermentation.

ESTIMATION OF PRIMARY AND SECONDARY CELL WALL CONTENT

Evaluation schemes for forage-based diets do not distinguish between the various cell wall types present in plant tissues. Although this could be done by microscopic methods coupled with image analysis to provide data in a numerical form, such a process would be both slow and laborious. Chemical methods, preferably instrumental and capable of adaption to automated analysis, would be required for routine evaluation purposes. The development of such a scheme presupposes that there are suitable marker compounds, capable of being measured and of being used to quantify the differences between primary and secondary wall contents.

In the *Gramineae*, phenolic acids could serve as suitable candidates for marker compounds. It is well recognized that the degradability of grass samples is inversely correlated with the *p*-coumaric acid content and with the ratio *p*-coumaric : ferulic acid (Chaves *et al.*, 1982; Burritt *et al.*, 1984). Work in this laboratory has shown that ferulic acid is the only phenolic acid found in more than trace amounts in the primary walls of ryegrass and immature barley. In contrast, *p*-coumaric acid appears to be associated with the process of secondary wall formation and is characteristic of lignified tissue. The ratio of the two phenolic acids could thus be used to apportion the total cell wall content (determined as neutral detergent fibre NDF), or the cell wall polysaccharide content (determined as NDF-acid detergent lignin ADL) between primary and secondary sources. In addition, estimation of the digestible residues (e.g. by NCD, neutral detergent cellulose digestibility) would provide a value which could be subtracted from the secondary wall content to provide an estimate of the digestible secondary wall content of the feed. Clearly work would have to be done to establish the reliability of this measure and to determine whether ratios could be applied across species and to samples taken at various stages of maturity.

Unfortunately the phenolic acids have a restricted distribution (Harris and Hartley, 1980) and are not found in the dicotyledenous species used as animal feed. A component of arabinoxylan is a possible alternative since this polysaccharide is not found in the primary wall of dicotyledenous species but is the major hemicellulosic polysaccharide formed during secondary thickening. Xylose residues themselves would not be appropriate since xylose is found as a component of primary wall xyloglucan. Were it not for the difficulty in analysis, glucuronic acid would be ideal since its distribution is restricted to secondary tissues. Again work would be required to establish whether glucuronic acid or some other marker compound could be used with confidence to quantify differences in the amounts of primary and secondary wall material in feeds.

Polysaccharide degradation and nutrient supply in non-ruminants

Most diets formulated for pigs and poultry are based on cereal grains with wheat used by feed compounders to the virtual exclusion of other cereals in the United Kingdom. Starch is thus the major source of polysaccharide ingested and other forms of carbohydrate contribute relatively little to the dry matter content of the diet (Table 10.4). However, the presence of non-starch polysaccharide may significantly influence the extent of starch hydrolysis and depress the uptake of both starch and protein degradation products in the foregut. In this respect the nutritional significance, as opposed to nutritive value, of ingested polysaccharide is greater in non-ruminant species.

Table 10.4 WATER-SOLUBLE CARBOHYDRATE (WSC), STARCH AND NON-STARCH POLYSACCHARDE (NSP) CONTENT OF CEREAL GRAINS AND OTHER INGREDIENTS OF PIG DIETS. DATA FROM ENGLYST *ET AL*. (1983), HENRY (1985) AND THIS LABORATORY

Source	*% Dry matter*		
	WSC	*Starch*	*NSP*
Barley	1.4	72.1	11.8
Wheat grain	1.4	64.6	9.6
Rye grain	2.6	66.7	13.2
Wheatfeed	5.5	20.4	36.6
Maize gluten	1.4	23.8	31.3
Soyabean meal	10.7	0.7	19.6
Lupin seed meal	9.0	trace	50.2
Sunflower seed meal	5.4	N.D.	32.4

NUTRITIVE VALUE OF POLYSACCHARIDES FED TO PIGS AND POULTRY

Problems of estimating nutrient supply to the host tissues in pigs and poultry predominantly fed carbohydrate in the form of soluble sugars and cereal starch are, in practice, less acute than those posed by ruminants. Since the non-ruminant directly absorbs simple sugars and the products of starch and protein degradation, estimation of the bulk of substrates available for tissue metabolism can be directly determined provided the extent of digestion in the foregut can be predicted or measured.

Starch hydrolysis

The extent of starch breakdown is dependent on a number of factors; some intrinsic to the plant, such as the chemical nature of the starch, its physical form, the presence of phospholipid and protein coating the starch granule and the distribution of starch in relation to the cell wall content of the feed; some extrinsic, such as the method of feed processing. These factors have been widely considered elsewhere (Dreher *et al.*, 1984; Williams and Chesson, 1989) and, in general, it is sufficient to note that, provided grains are partially disrupted, uncooked cereal starch is digested with high efficiency by both poultry (Riesenfeld *et al.*, 1980) and pigs (Keys and DeBarthe, 1974; Kidder, 1982). Total starch content can be readily measured by physical methods or by enzymic methods based on the use of thermostable amylase preparations (Hesslemen and Aman, 1985). *In vitro* measures of apparent digestibility using porcine amylase generally give values lower than those obtained *in vivo* which remains the method of choice (Dreher *et al.*, 1984). Simple *in vitro* systems, however, can be useful when comparing starch from different sources or following different methods of feed preparations.

VFA production from polysaccharide

Starch escaping digestion in the upper foregut of the pig and the non-starch polysaccharide component of the diet are potentially available to the gut microbial flora as substrates for fermentation. Conventionally the hindgut is considered as the site of microbial fermentation but there is increasing evidence that a significant amount of microbial activity is found in the lower part of the ileum (Millard and Chesson, 1984; Graham *et al.*, 1986). Although it is of little nutritive consequence whether VFA production occurs in the fore- or hindgut since the nutritive value of the VFA remains constant, the site of fermentation may be of nutritional significance. A well established foregut flora can compete directly with the host for potentially absorbable substrates if the absorption rates are reduced by the presence of other nutrients, notably gel-forming polysaccharides. Foregut fermentation can also complicate determinations of VFA production from the NSP component of the feed. Measurement of the disappearance of carbohydrate beyond the terminal ileum makes no allowance for foregut fermentation and estimations of VFA production derived by this method are likely to be an underestimate.

In diets based on wheat grain and soyabean the contribution of absorbed VFA to tissue metabolism is relatively small compared to the contribution made by absorbed glucose. In this respect, attempts to estimate VFA production could be seen as of academic rather than practical interest. However the NSP content of diets can be substantially increased when diet ingredients other than wheat and soya are incorporated. Many alternative protein and energy sources such as sunflower and lupin meals, wheatfeed and maize gluten, have a relatively high cell wall (NSP) content (Table 10.4). Since the cell walls in such ingredients are largely primary, they are readily degraded by the gut microflora and VFA production is likely to make a greater contribution to the nutrition of the animal. Metz *et al.* (1985) found that 14–39% of organic matter digestion occurred beyond the terminal ileum of pigs depending on the nature of the diet fed.

In some respects estimating VFA production in pigs appears simpler than in ruminants since microbial growth is rarely limited by nitrogen supply. Although the microflora found in the pig gut have the same requirement for nitrogen as the rumen microbes, maximizing nitrogen capture and microbial cell yield is not a prime concern. Thus it is not necessary to match closely the rates of nitrogen and carbohydrate supply. Adequate levels of nitrogen appear to be provided by residual feed protein escaping foregut digestion, endogenous nitrogen in the form of mucus and epithelial cells and urea recycled into the hindgut, almost regardless of the diet fed. Similarly, since the proportion of the various VFA produced and available for tissue metabolism is probably not a critical factor, it is unnecessary to distinguish between the various forms of carbohydrate supplied. Inputs into any model of hindgut fermentation can thus be restricted to a single description of carbohydrate available for fermentation, although account may have to be taken of transit time through the hindgut compartment and any limitation this may impose on the time available for fermentation.

While the mechanism of polysaccharide degradation and the structural factors determining the extent of hydrolysis are independent of the host species and of the site within the gastro-intestinal tract, the microbial population directly or indirectly involved in polysaccharide fermentation can show considerable variation. There is increasing evidence to suggest that the 'rumen model' cannot be directly applied to fermentation in non-ruminant animals and that stoichiometric parameters determined from data obtained from ruminants cannot be applied with any confidence to non-ruminants. Calculation of the dry matter fermented in growing pigs fed sugarbeet pulp made from observations of the molar proportion of individual VFA present in the hindgut and the rate of methane production and using rumen stoichiometry gave a value less than 20% of that observed. Similarly comparisons between measured and predicted methane production gave predicted values approximately 3.7 times greater than those actually observed. Free hydrogen was detected, but in amounts too low to account for the discrepancy (Zhu *et al.*, 1988). As the authors conclude, hydrogen sinks other than methane and propionate appear to be significant in the pig. It is possible to speculate that aerobic metabolism may be much more extensive in the pig hindgut than in the rumen. The greater surface area of the gut relative to its volume in the pig compared with the rumen would tend to promote gaseous interchange. If this is the case then part of the 'potentially fermentable polysaccharide' could be fully metabolized to CO_2 with oxygen acting as the terminal acceptor. This would have obvious implications for the estimation of VFA production.

The similarity in TMEn (true metabolizable energy corrected to zero nitrogen retention) measurements made using intact and caecectomized cockerels suggest that hindgut fermentation contributes very little to the nutrition of poultry. Any differences detected invariably fall within the experimental error and are rarely significant (J. McNab, unpublished observations). However, while non-starch polysaccharides may have a negligible nutritive value when fed to poultry, they can have marked nutritional significance (Janssen and Carre, 1985).

NUTRITIONAL SIGNIFICANCE OF POLYSACCHARIDES FED TO PIGS AND POULTRY

Physiological responses to the cell wall ('dietary fibre') content of diets is seen throughout the digestive tract of pigs and poultry although the nature of the responses varies with different sites in the digestive tract and with the composition of the diet fed. Water-soluble polysaccharide leached from some primary cell walls ('soluble dietary fibre') primarily induces effects in the foregut but, in pigs at least, has little effect elsewhere since it is rapidly degraded by the microflora. Insoluble fibre, in contrast, influences the digestive process throughout the gut.

Effect of dietary polysaccharides in the foregut

Endogenous secretions into the foregut are increased considerably in the presence of fibre. Gastric, biliary and pancreatic secretions were virtually doubled in pigs fed a barley-based diet compared to the rates in the same pigs fed a semi-purified low fibre diet (Zebrowska *et al.*, 1983). Similarly the general secretion of nitrogen into the pig foregut lumen appears to be increased in the presence of soluble fibre (Low and Rainbird, 1984). Some cell wall polysaccharides may be partially solubilized during passage through the digestive tract, a process encouraged by the acidic conditions found in the stomach of the pig and the proventriculus/gizzard of the bird. These include pectic substances from dicotyledenous plants, mixed-linked glucans from the endosperm walls of barley grain and arabinoxylans from rye grains. The presence of soluble polysaccharide may substantially increase the viscosity of digesta, delaying passage of the aqueous phase from the stomach to the duodenum and reducing the diffusion of released glucose and amino acids from the

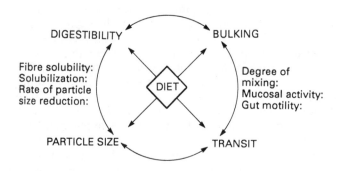

Figure 10.2 Interactions between fibre composition, digestibility, bulking and transit time in the hindgut of the pig

gut lumen to the epithalial cells of the intestinal villi. The rate of glucose absorption can be considerably reduced in the presence of soluble fibre and the appearance of glucose and amino-nitrogen in blood plasma delayed and peak concentrations reduced. This phenomenon has been extensively studies in humans because of its implications for the control of some forms of diabetes (Anderson, 1980). In general, the apparent digestibility of organic matter and nitrogen measured at the terminal ileum is inversely related to the fibre content of the diet (Low, 1985). Impaired absorption due to the presence of soluble polysaccharide cannot wholly explain this observation which has been made on many occasions. Reduced apparent digestibility can occur in diets with a low soluble fibre content. Similarly the absorption of free lysine was found to be unaffected by the presence of soluble polysaccharide added to pig diets while the apparent digestibility of protein was reduced (Murray *et al.*, 1977). The cell wall content of the diet clearly produces multiple effects in the foregut, not all of which are currently recognized.

Soluble polysaccharide can survive intact in poultry to be voided with the excreta where they may give rise to the problems of 'sticky droppings'. This, and the associated problems of depression of nutritive value, is why there is a reluctance to use barley and rye grains in diets for broiler chickens. Hydrolysis of the gel-forming polysaccharide by addition of an appropriate enzyme activity can overcome this problem, albeit at some financial cost (Hessleman and Aman, 1986; GrootWassink *et al.*, 1989).

Transit time and hindgut effects

The effect of fibre on tranist time of digesta in the pig, as in ruminants, is of some nutritional significance since an increased rate of passage through the hindgut can influence the nature of the microbial flora present, the time available for microbial fermentation and, hence, the nutritive value of ingested fibre. The mechanism underlying the effect of fibre on transit time and colonic motility is not fully resolved but clearly relates to a number of highly interrelated factors including the chemical composition of the fibre source, its water holding capacity and its digestibility (Figure 10.2). In general fibre sources rich in primary cell walls (e.g. vegetables) have a high water holding capacity (WHC). However, because of their inherent digestibility, structural integrity is rapidly lost in the hindgut, mean particle size is reduced and the fibre loses its WHC and contributes little to digesta of faecal bulk. In contrast, fibre sources containing lignified cell walls (cereal fibre, bran) are resistant to microbial degradation and retain much of their cellular integrity during passage. Although the intrinsic WHC of such fibre sources are only a fraction of the WHC of most vegetables fibres, WHC is retained during passage through the gut and the fibre contributes far more to the bulking of digesta and faeces. As a result, lignified fibre sources such as bran, tend to shorten the transit time through the hindgut compartment (Robertson *et al.*, 1986; Robertson, 1988; Low, 1988).

Dietary NSP has both nutritional significance and nutritive value for pigs and poultry neither of which, as Low (1988) concludes, can be satisfactorily assessed by conventional evaluation methods. Equations for the prediction of digestible energy in pig diets which make some allowance for the fate of NSP can be improved if values for the crude fibre content of diets are replaced by NDF values (Morgan, 1987). This may be sufficient for diets heavily weighted towards cereal grains.

However any move towards the inclusion in diets of a greater range and level of by-products with a high NSP content is likely to highlight the inadequacy of this approach to feed evaluation.

Acknowledgements

I would like to thank my colleagues, Drs J.C. MacRae, M.F. Fuller and C.S. Stewart for their helpful comments during the preparation of this paper.

References

Agricultural Research Council (1980) *The Nutrient Requirements of Farm Livestock No 2. Ruminants.* Commonwealth Agricultural Bureaux, Farnham Common

Agricultural Research Council (1984) *Report of the Protein Group of the Agricultural Research Council Working Party on the Nutrient Requirements of Ruminants*, supplementary report to Chapter 4 (ARC, 1980) Commonwealth Agricultural Bureaux, Farnham Common

Akin, D.E. (1988) Biological structure of lignocellulose and its degradation in the rumen. *Animal Feed Science and Technology*, **21**, 295–310

Anderson, J. W. (1980) Dietary fiber and diabetes. In *Medical Aspects of Dietary Fibre* (eds G.A. Spiller and R. McPherson Kay) Plenum Medical Book Company, New York, pp. 193–221

Baldwin, R.L. and Allison, M.J. (1983) Rumen metabolism. *Journal of Animal Science*, **57**, Suppl.2, 461–477

Barry, T.N., Manley, T.R. and Duncan, S.J. (1984) Quantitative digestion by sheep of carbohydrates, nitrogen and S-methyl-L-cysteine sulphoxide in diets of fresh kale (*Brassica oleracea*). *Journal of Agricultural Science*, **102**, 479–486

Burritt, E.A., Bittner, A.S., Street, J.C. and Anderson, M.J. (1984) Correlation of phenolic acids and xylose content of cell walls with *in vitro* dry matter digestibility of three maturing grasses. *Journal of Dairy Science*, **67**, 1209–1213

Chaves, C.M., Moore, J.E., Moye, H.A. and Ocumpaugh, W.R. (1982) Separation, identification and quantification of lignin saponification products extracted from digitgrass and their relation to forage quality. *Journal of Animal Science*, **54**, 196–203

Chesson, A. (1986) The evaluation of dietary fibre. In *Feedingstuffs Evaluation, Modern Aspects, Problem, Future Trends* (ed R.M. Livingstone), Rowett Research Institute, Aberdeen, pp. 18–25

Chesson, A. and Monro, J.A. (1982) Legume pectic substances and their degradation in the ovine rumen. *Journal of the Science of Food and Agriculture*, **33**, 852–859

Chesson, A., Stewart, C.S., Dalagarno, K. and King, T.P. (1986) Degradation of isolated grass mesophyll, epidermis and fibre cell walls in the rumen and by cellulolytic rumen bacteria in axenic culture. *Journal of Applied Bacteriology*, **60**, 327–336

Dreher, M.L., Dreher, C.J. and Berry, J.W. (1984) Starch digestibility of foods: a nutritional perspective. *CRC Critical Reviews in Food Science and Nutirion*, **20**, 47–71

Englyst, H.N., Anderson, V. and Cummings, J.H. (1983) Starch and non-starch polysaccharides in some cereal foods. *Journal of the Science of Food and Agriculture*, **34**, 1434–1440

Gordon, A.H., Lomax, J.A., Dalagarno, K. and Chesson, A. (1985) Preparation and composition of mesophyll, epidermis and fibre cell walls from leaves of perennial ryegrass. (*Lolium perenne*) and Italian ryegrass (*Lolium multiflorum*). *Journal of the Science of Food and Agriculture*, **36**, 509–519

Graham, H., Hesselman, K. and Aman, P. (1986) The influence of wheat bran and sugar-beet pulp on the digestibility of dietary components in a cereal based pig diet. *Journal of Nutrition*, **116**, 242–251

GrootWassink, J.W.D., Campbell, G.L. and Classen, H.L. (1989) Fractionation of crude pentosanase (arabinoxylanase) for improvement of the nutritional value of rye diets for broiler chickens. *Journal of the Science of Food and Agriculture*, **46**, 289–300

Harris, P.J. and Hartley, R.D. (1980) Phenolic constituents of the cell walls of monocotyledons. *Biochemical Systematics and Ecology*, **8**, 153–160

Henry, R.J. (1985) A comparison of the non-starch carbohydrates in cereal grains. *Journal of the Science of Food and Agriculture*, **36**, 1243–1253

Hesselman, K. and Aman, P. (1985) Analysis of cereal grains and the influence of β-glucans on digestion in broilers. In *New Approaches to Research on Cereal Carbohydrates* (eds R.D. Hill and L. Munck), Elsevier Science Publishers, Amsterdam, pp. 363–372

Hesselman, K. and Aman, P. (1986) The effect of β-glucanase on the utilisation of starch and nitrogen by broiler chickens fed in barley of low- or high-viscosity. *Animal Feed Science and Technology*, **15**, 83–93

Hobson, P.N. and Jouany, J.-P. (1989) Models, mathematical and biological, of the rumen function. In *The Rumen Microbial Ecosystem* (ed P.N. Hobson), Elsevier Applied Science, London, pp. 461–511

Janssen, W.M.M.A. and Carré, B. (1985) Influence of fibre on digestibility of poultry feeds. In *Recent Advances in Animal Nutrition – 1985* (eds W. Haresign and D.J.A. Cole), Butterworths, London, pp. 71–86

Keys, J.E. and DeBarthe, J.V. (1974) Site and extent of carbohydrate, dry matter, energy and protein digestion and rate of passage of grain diets in swine. *Journal of Animal Science*, **39**, 57–62

Kidder, D.E. (1982) Nutrition of the early weaned pig compared with the sow-reared pig. *Pig News and Information*, **3**, 25–28

Low, A.G. Role of dietary fibre in pig diets. In *Recent Advances in Animal Nutrition – 1985* (eds W. Haresign and D.J.A. Cole) Butterworths, London, pp. 87–112

Low, A.G. (1988) Gut transit and carbohydrate uptake. *Proceedings of the Nutrition Society*, **47**, 153–159

Low, A.G. and Rainbird, A.L. (1984) Effect of guar gum on nitrogen secretion into isolated loops of jejunum in conscious growing pigs. *British Journal of Nutrition*, **52**, 499–505

Mackie, R.I. and Gilchrist, F.M.C. (1979) Changes in lactate-producing and lactate-utilizing bacteria in relation to pH in the rumen of sheep during stepwise adaption of a high concentrate diet. *Applied and Environmental Microbiology*, **38**, 422–430

MacRae, J.C. (1986) An appraisal of current systems for the evaluation of the energy and protein needs of ruminants. In *Feedingstuffs Evaluation, Modern*

Aspects, Problems, Future Trends (ed R.M. Livingstone), Rowett Research Institute, Aberdeen, pp. 11–17

MacRae, J.C., Buttery, P.J. and Beever, D.E. (1988) Nutrient interactions in the dairy cow. In *Nutrition and Lactation in the Dairy Cow* (ed P.C. Garnsworthy) Butterworths, London, pp. 55–75

Mertens, D.R. and Ely, L.O. (1979) A dynamic model of fibre digestion and passage in the ruminant for evaluating forage quality. *Journal of Animal Science*, **49**, 1085–1095

Mertens, D.R. and Ely, L.O. (1982) Relationship of rate and extent of digestion to forage utilization – a dynamic model evaluation. *Journal of Animal Science*, **54**, 895–905

Metz, S.H.M., Dekker, R.A. and Everts, H. (1985) Effect of dietary composition on the contribution of large intestine to total digestion in the growing pig. *Beretning fra Statens Husdrybrugsforsog*, **580**, 227–230

Millard, P. and Chesson, A. (1984) Modifications to swede (*Brassica napus* L.) anterior to the terminal ileum of pigs: some implications for the analysis of dietary fibre. *British Journal of Nutrition*, **52**, 583–594

Morgan, C.A. (1987) Chemical components of fibre and their nutritional significance. *Journal of the Science of Food and Agriculture*, **40**, 28–29

Morris, E.J. and Van Gylswyk, N.P. (1980) Comparison of the action of rumen bacteria on cell walls of *Eragrostis tef*. *Journal of Agricultural Science*, **95**, 313–323

Murphy, M.R., Baldwin, R.L. and Koong, L.J. (1982) Estimation of stoichiometric parameters for rumen fermentation of roughage and concentrate diets. *Journal of Animal Science*, **55**, 411–421

Murray, A.G., Fuller, M.F. and Pirie, A.R. (1977) The effect of fibre in the form of various polysaccharides on the apparent digestibility of protein in the pig. *Animal Production*, **24**, 139

Reichl, J.R. and Baldwin, R.L. (1976) A rumen linear programming model for evaluation of concepts of rumen microbial function. *Journal of Dairy Science*, **58**, 879–890

Riesenfeld, G., Skelan, D., Bar, A., Eisner, U. and Hurwitz, S. (1980) Glucose absorption and starch digestion in the intestine of the chicken. *Journal of Nutrition*, **110**, 117–121

Robertson, J.A. (1988) Physicochemical characteristics of food and the digestion of starch and dietary fibre during gut transit. *Proceedings of the Nutrition Society*, **47**, 143–152

Robertson, J.A., Murison, S.D. and Chesson, A. (1986) Loss of selected water-insoluble polysaccharides and component neutral sugars from swede (*Brassica napus* (cv. Danestone)) and cereal bran measured during digestion in the pig caecum. *Journal of the Science of Food and Agriculture*, **37**, 359–365

Robinson, P.H., Fadel, J.G. and Tamminga, S. (1986) Evaluation of mathematical models to describe neutral detergent residue in terms of its susceptibility to degradation in the rumen. *Animal Feed Science and Technology*, **15**, 249–271

Silva, A.T., Greenhalgh, J.F.D. and Ørskov, E.R. (1989) Influence of ammonia treatment and supplementation on the intake, digestibility and weight gain of sheep and cattle on barley straw diets. *Animal Production*, **48**, 99–108

Ulyatt, M.J., Waghorn, G.C., John, A., Reid, C.W.S. and Monro, J. (1984) Effect of intake and feeding frequency on feeding behaviour and quantitative aspects of digestion in sheep fed lucerne hay. *Journal of Agricultural Science*, **102**, 645–657

Wallace, R.J. (1978) Control of lactate production by *Selenomonas ruminantium:* homotrophic activation of lactate dehydrogenase by pyruvate. *Journal of General Microbiology*, **107**, 45–52

Williams, P.E.V. and Chesson, A. (1989) Cereal raw materials and animal production. In *Cereal Science and Technology* (ed G.H. Palmer), Aberdeen University Press, Aberdeen, pp. 413–442

Zebrowska, T., Low, A.G. and Zebrowska, H. (1983) Studies on gastric digestion of protein and carbohydrate, gastric secretion and exocrine pancreatic secretion in the growing pig. *British Journal of Nutrition*, **49**, 401–410

Zhu, J.-Q., Fowler, V.R. and Fuller, M.F. (1988) The production of methane and volatile fatty acids resulting from the fermentation of sugar beet pulp in the gut of growing pigs. *Proceedings of the 4th Symposium on Digestive Physiology of the Pig* (eds L. Burczewska, S. Burczewska, B. Pastuszewska and T. Zebrowska). Polish Academy of Sciences, Institute of Animal Physiology and Nutrition, Jablonna, Poland

11

CHEMICAL ANALYSIS OF LIPID FRACTIONS

B.K. EDMUNDS
Intermol, King George Dock, Hedon Road, Hull HU9 5PR, UK

Introduction

With the advent of high nutrient concentration diets for farm animals, the inclusion of fat into compound feed has steadily risen over the last 15–20 years. Currently, some high performance broiler diets can contain up to 110 g/kg total fat of which 60–70 g is added. Limitations to higher inclusions in other diets can be a function of milling technology rather than nutritional constraints. Taking the energy value of a blended fat to be 36 MJ/kg, 70 g/kg of added fat represents approximately 20% of the total energy value of a modern broiler diet. Consequently, assessment of the total amount of fat in a diet and predicting its nutritive value from chemical analyses is of considerable importance in feed formulation and animal production.

Due to the major influence that added fat can have on energy value of the diet, the present paper will concentrate on the chemical assessment of feed fats and their raw materials while highlighting aspects of methodology and nutritional relevance.

Determination of total oil in feedstuffs

Before discussing the various lipid fractions, it is relevant to consider the determination of total oil in feedstuffs. Since 1986, two official European Economic Community (EEC) methods (Procedures A and B) have been in use for the analysis of oils and fats (The Feedingstuffs Regulation, 1985). As discussed by Sanderson (1986), Procedure A involves the extraction of oil and fat into light petroleum and weighing the dried residue whereas Procedure B includes a preliminary acid hydrolysis step followed by extraction. The latter method was introduced to harmonize oil analysis throughout the EEC and to enable protected fats to be included in the total oil declaration of a finished feed. Protected fats used in ruminant diets are usually calcium or magnesium salts of fatty acids and, consequently, are water soluble and will not be extracted by light petroleum. The acid hydrolysis step converts any soaps present in the feedstuff to fatty acids, rendering them soluble in the solvent. As pointed out by Cooke (1986), the analysis of raw materials and compound feed by Procedure B always produces a higher oil value even when protected fats are not included in the feedingstuff. As yet, the 'additional' fat extracted has not been fractionated but it is assumed to be composed of poorly digested waxes.

A major problem with Procedure B as detailed in the Feedingstuffs Regulation (1985), is of incomplete recovery of oil from the sample which may be caused by losses during filtration. Following acid hydrolysis, excess acid is removed by washing with up to 1000 ml of water.

Although the analyst is warned to examine the filtrate for oil, in practice this presents difficulty. The method specifies a sample weight of 2,5 g and a loss, for example, of 10% oil on filtration from a sample with 60 g/kg total oil would involve the detection of 15 mg of oil in up to 1000 ml of water. The higher the oil concentration, the greater the possibility for loss although there is little practical evidence to suggest contamination of the filtrate. Pre-extraction of the sample with light petroleum before acid hydrolysis and subsequent extraction eliminates the loss but has implications in time and cost for routine analysis (Sanderson, 1986). Cooke (1986) calculated that differences between Procedures A and B would result in a 0.28 MJ/kg drop in the energy value of a broiler feed if a Procedure A result was utilized in the current EEC poultry Metabolizable Energy (ME) equation (The Feedingstuffs Regulation, 1988) which is based upon Procedure B. Therefore, it is not only important to distinguish between Procedures A and B but also to clarify if pre-extraction has been carried out with the latter method.

Lipid fractions found in added fat

The remainder of this paper will concentrate on the lipid fractions found in fat added to animal diets due to the major influence that added fat can have an energy value of the diet (Huyghebaert, De Munter and De Groote, 1988; Janssen, 1988). Ideally, account should be taken of the fatty acid profile of all constituent raw materials in a feed but it is the quantity and quality of added fat that has received most attention (e.g. Wiseman *et al.*, 1986). Thus, the chemical assessment of blended feed fats and their raw materials will be addressed, highlighting aspects of methodology and nutritional interpretation.

A simplified breakdown of the lipid fractions in feed fat is given in Figure 11.1 based on those components with and without nutritional worth. The classification cannot be rigid as, for example, the unsaponifiable matter will contain useful fat

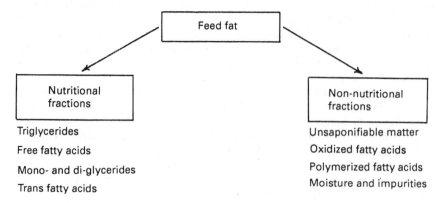

Figure 11.1   A simplified fractionation of the chemical constituents of feed fat of nutritional significance

soluble vitamins but the majority will be of low nutritional value. The classification is not comprehensive in that for example, it does not include contaminants (e.g. pesticide residues). However, the emphasis is upon those fractions that have relevance to nutritional performance.

Nutritional fractions

TRIGLYCERIDES

In feed formulation, the composition of the triglyceride fraction of fat has received little attention even though established methods are available using both packed and capillary column gas-liquid chromatography (GLC, Pocklington and Hautfenne, 1985). Freeman, Holme and Annison (1968) have shown that, apart from poorly digested fats, the configuration of constituent triglyceridres has little effect upon overall fat digestibility. However, Davies and Lewis (1969) compared the digestibility of fatty acids to tallow and lard with that of interesterified lard. In the latter, palmitic acid was randomly distributed between the 1, 2 and 3 positions in the triglyceride and its digestibility fell compared to lard in which the favoured 2-position for pancreatic lipase predominates. Freeman (1984) suggested that the configuration of the constituent triglycerides can have an influence on the rate of fat digestion. In practical feed fat blends, it would be difficult to address any requirement for specifying the triglyceride fraction.

FREE FATTY ACIDS (FFA)

The conventional assessment of the free fatty acid content of lipid has been titration with sodium hydroxide (BS 684, 1976). The latter specifies agreement between duplicates of 0.2% for FFA levels above 100 g/kg and repeatability within a laboratory using a blended feed fat is shown in Table 11.1. The FFA concentration in feed fat raw materials varies over a wide range (Table 11.2) and the high variability found reflects the difference between raw material sources. FFA concentrations in feed fat are not only relevant due to the possibility of corrosion but also due to the potential for soap formation in the animal (Atteh and Leeson, 1983). Wiseman and Cole (1983) demonstrated a decrease in pig digestible

Table 11.1 REPEATABILITY OF FREE FATTY ACID (FFA), SLIP POINT (SP), UNSAPONIFIABLE MATTER (US), OXIDIZED FATTY ACIDS (OFA) AND MOISTURE DETERMINATIONS ON A SAMPLE OF BLENDED FEED FAT WITHIN A LABORATORY. RESULTS ARE IN g/kg . FOR METHODS SEE BS 684 (1976) (WITH KIND PERMISSION OF PAULS AGRICULTURE.)

	FFA	*SP (°C)*	*US*	*OFA*	*Moisture*
No. repeats	10	10	10	10	5
Mean	364	268	19.7	8.6	6.2
SD	2.5	5.9	2.4	1.5	6
CV (%)	7	22	122	174	95

Table 11.2   MEAN (±SD) FREE FATTY ACID CONCENTRATION IN FEED FAT RAW MATERIALS. ALL VALUES ARE BASED UPON THE MOLECULAR WEIGHT OF OLEIC ACID

Material	Free fatty acids (g/kg)	
	Mean	No. of samples
Crude soyabean oil	7 ± 3	28
Recovered vegetable oil	44 ± 44	164
Tallow (feed grade)	114 ± 52	108
Soya/sunflower acid oil	428 ± 134	31
Mixed soft acid oil	495 ± 131	59
Fish acid oil	611 ± 133	12
Palm acid oil	716 ± 131	17
Palm fatty acid distillate	889 ± 32	12

Table 11.3   COMPARISON OF THE FATTY ACID PROFILE OF THE FREE FATTY ACID FRACTION OF FEED FATS WITH THE TOTAL PROFILE. METHYL ESTERS OF FATTY ACIDS WERE PREPARED EITHER WITH (TOTAL FATTY ACIDS, TFA) OR WITHOUT (FREE FATTY ACIDS, FFA) SAPONIFICATION

Fatty acid g/kg	PFAD/tallow (50:50 mixture)		CSBO	
	TFA	FFA	TFA	FFA
C14:0	20	17	1	3
C14:1	1	1	–	–
C16:0	380	414	109	159
C16:1	13	9	–	–
C18:0	107	78	38	46
C18:1	370	360	215	191
C18:2	64	82	551	501
C18:3	6	8	78	76
C20 and above	2	7	5	7

PFAD = palm fatty acid distillate; CSBO = crude soyabean oil

energy with increasing FFA concentration, although Vernon and Perry (1981) found no significant effect of increasing FFA in tallow or soya oil on the production response of broilers. Table 11.3 suggests that the fatty acid profile of the free fatty acid fraction broadly follows that of the total fat.

TRANS FATTY ACIDS

Trans fatty acids occur naturally in small amounts except in specific instances. Tung oil contains up to 800 g/kg trans fatty acids but the major natural source arises from the biohydrogenation of polyunsaturated fatty acids in the rumen.

Consequently, milk can contain significant amounts of trans fatty acids (up to 210 g/kg of the total monounsaturated acids; Gurr, 1986). The processing of highly unsaturated oils (e.g., fish oils) by the food industry to produce hardened oils by hydrogenation is a major source of trans fatty acids (Enig *et al.*, 1983). The nutritional consequences arising from the human consumption of trans fatty acids has been linked to a possible increase in athersclerosis but it would appear that conclusive evidence is lacking (Gurr, 1986). The digestibility of trans fatty acids in farm species is not well documented although Huyghebaert, De Munter and De Groote (1988) have found the concentration of trans C18:1 in rendered fat used in Dutch feed fat to range from 20 to 220 g/kg of the total oleic acid content.

Total trans fatty acids in lipid are usually estimated by infra-red analysis as isolated trans bonds exhibit a specific absorption band arising from the C-H bond deformation (AOCS, 1978). However, modifications have been developed for methyl esters (Paquot, 1979) and for particular applications (e.g., milk, Deman and Deman, 1983).

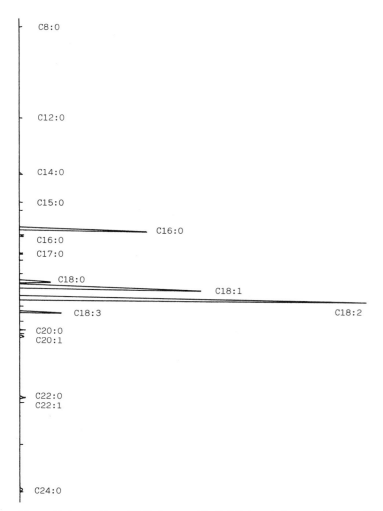

Figure 11.2   Capillary GLC of soya acid oil. 0.5µl methyl esters of fatty acids injected; column: BP20 (15 m); Perkin Elmer 8500; run time: 16 minutes

FATTY ACID PROFILE

Determination of the component fatty acids in lipid by GLC has been developed into a standard technique which is widely used (AOCS, 1978). Recent work has centred on fine tuning the method, for example, examination in detail of relative response factors (Craske and Bannon, 1987). The development of capillary columns has significantly improved resolution and analysis times (Figure 11.2; D'A Alonzo, Kozarek and Wharton, 1981). The fatty acid profile can provide extensive information about the fat sample, including the saturated to unsaturated fatty acid ratio, the concentration of essential fatty acids and the concentration of long and short chain fatty acids. Analysis times by capillary GLC of 16 minutes (Figure 11.2) have resulted in fatty acid profile determinations being able to replace iodine value and produce more valuable information. Modern GLC has also improved the determination of non-elutable material (NEM) in feed fats which will be discussed below.

Non-nutritional fractions

Concern about the non-nutritional fractions in feed fat stems from the chemical changes that may have occurred to feed fat raw materials (Veen, 1985). The majority of raw materials used in current feed fat blends are by-products of the refining of fats and oils for human and industrial consumption. The financial penalty involved in the use of crude, (except in some instances crude soyabean oil) and refined oils precludes their inclusion. However, the quality of raw materials available to the fat blending industry can vary widely as data presented below will demonstrate. Consequently, it is necessary to be able to utilize relevant chemical evaluation to assess nutritional worth.

UNSAPONIFIABLE MATTER

The unsaponifiable matter (UM) fraction is composed of a range of compounds which can be specific to particular oils. Consequently, in some instances, UM can be useful as the adulteration of refined oils can be detected by a detailed examination of the compounds of the UM fraction (Fedeli, 1966). The UM includes sterols (including cholesterol), squalene and tocopherols but will also contain contaminants such as polyethylene and pesticide residues. The UM fraction in crude vegetable oils is normally less than 20 g/kg (Gutfinger and Letan, 1974) but it can be considerably higher in feed fat raw materials (Table 11.4). The UM fraction will vary within a specific raw material and also from one source (Table 11.5). The higher concentrations of UM in vegetable acid oils arise from the inclusion of materials removed at the deodorizing stage in the chemical refining of vegetable oils. The greater the efficiency of the refining plant, the higher the concentration of UM. Although the fraction will contain some tocopherols (Gutfinger and Letan, 1974), their stability, availability and usefulness as an antioxidant is not known. The nutritional value of the UM is considered to be low due to its poor digestibility (Cooke, 1982).

The analysis of fats for their UM content is well established (BS 684, 1976) which involves the conversion of triglycerides and FFA into soaps by heating with strong

Table 11.4 MEAN (±SD) UNSAPONIFIABLE MATTER FOUND IN COMMON FEED FAT RAW MATERIALS USING BS 684 (1976) (NUMBERS IN PARENTHESES ARE NUMBER OF SAMPLES)

Material	Unsaponifiable matter (g/kg)			
Crude soyabean oil	5	±	2	(6)
Recovered vegetable oil	7	±	4	(36)
Palm acid oil	15	±	4	(6)
Tallow	2	±	8	(20)
Fish acid oil	21	±	5	(4)
Palm fatty acid distillate	23	±	11	(5)
Mixed soft acid oil	3	±	14	(21)
Soya/sunflower acid oil	64	±	27	(25)

Table 11.5 VARIATION IN THE UNSAPONIFIABLE MATTER OF CONSECUTIVE DELIVERIES OF A VEGETABLE ACID OIL FROM A SINGLE SOURCE

Delivery	Unsaponifiable content (g/kg)
1	75
2	89
3	16
4	38
5	59
6	33
7	46
8	93
Mean = 56 ± 28	

alkali. Maxwell and Schwartz (1979) have developed a dry saponification technique followed by grinding with Celite powder and elution of the UM in a glass column with dichloromethane. Schwartz (1988) has further refined the technique to correct for contaminating soap and establish optimum time temperature conditions for saponification. The technique shows good agreement with the official AOAC method (1984). The UM fraction may also contain some highly polymerized fatty acids which cannot be saponified (Schwartz, 1988) which complicates the precise fractionation of fat. Determination of the UM in finished feed will not produce a result that can be reliably used to indicate the quality of the added fat due to the wide and variable amounts of UM in feed raw materials.

The unsaponifiable fraction will also contain any polyethylene and pesticide residues present in the fat. The latter subject is treated elsewhere in these proceedings so will not be commented upon. Contamination of feed fat with polyethylene arises from the inclusion of packaging materials in the raw material used in rendering plants. Some renderers filter hot tallow to remove polyethylene and comply with the 200 g/kg maximum stated in commercial contracts. However, severly contaminated loads are produced and polyethylene has also been detected in recovered vegetable oil. A rapid, semi-quantitative assessment of the polyethylene content of fats can be made using the flocculation test where dissolved polyethylene precipitates in a mixture of acetone and methanol. Quantification can be achieved using the lengthy official procedure (BS 684, 1976), although data in Table 11.6 would suggest that low recoveries can be obtained by this procedure.

Table 11.6   COMPARISON OF THE STANDARD METHOD OF DETERMINATION OF POLYETHYLENE (mg/kg) IN FATS (BS 684, 1976) WITH A MODIFIED PROCEDURE USING A LONGER PRECIPITATION PERIOD AND EXTRACTION INTO CARBON TETRACHLORIDE (C. WEBSTER, UNPUBLISHED OBSERVATIONS)

Sample		Modified	
	BS 684 (A)	*Procedure (B)*	*A/B (%)*
1	80	235	34
2	25	90	28
3	50	160	31
4	120	430	28
5	95	260	37
6	30	170	18

Table 11.7   MEAN (±SD) VARIATION IN THE MOISTURE CONTENT OF FEED FAT RAW MATERIALS DETERMINED BY DISTILLATION UNDER TOLUENE (BS 684, 1976) (RESULTS IN PARENTHESES ARE NUMBERS OF SAMPLES)

Material	*Moisture* (g/kg)		
Palm acid oil	2.8	2.3	(5)
Crude soyabean oil	1.4	0.7	(5)
Soya/sunflower acid oil	6.8	5.4	(9)
Tallow	5.7	5.8	(23)
Mixed soft acid oil	18.2	14.4	(31)
Recovered vegetable oil	13.6	14.8	(39)
Fish acid oil	11.7	8.8	(6)

MOISTURE

A variable amount of moisture is always present in feed fats (Table 11.7) and large amounts are to be avoided due to the dilution effect on energy value and the possibility of residual acidity resulting from the acidulation of soapstocks. Moisture is routinely determined by distillation under toluene (BS 684, 1976) although rapid results can be obtained using Karl–Fisher titration. Oven drying will produce erroneous results as fatty acids become volatile above 60°C.

IMPURITIES

Impurities in fat are defined as the amount of material that will not dissolve in light petroleum (BS 684, 1976). However, instances can arise where fine insoluble material is present that will not be retained on filter paper and, consequently, is not included in the impurity fraction. Such material contributes to the sludge that can accumulate in fat storage tanks.

RANCIDITY

Conventionally, peroxide value has been used as an indicator of rancidity (BS 684, 1976) and is a valid technique if the history of the oil or fat is known. Peroxide value may be relevant to refined oils but the test has little application to the evaluation of feed fats. Due to the processing that feed fat raw materials may have undergone, considerable oxidative changes may have occurred which make the interpretation of peroxide values difficult.

Other indicators of subsequent stages of lipid oxidation, notably the thiobarbaturic acid assay, suffers from similar problems and is thought not to be highly specific (Ware, 1985).

OXIDIZED AND POLYMERIZED FATTY ACIDS

Without doubt, the oxidized and polymerized fatty acid (OPFA) fraction in feed fat raw materials has a major effect on energy value (Veen, 1985; Janssen, 1988), but it is the most difficult to define and quantify accurately. Interest in the OPFA fraction stemmed from investigations into the effects of human consumption of food fried in overheated oils. Many of the assays developed to assess the OPFA fraction are directed at used frying fats and their direct application to the evaluation of feed fats must be viewed with care.

When fats and oils are heated, both thermal and oxidative changes occur. The type and amount of compounds produced is a function of the original oil or fat and the heating and processing conditions used. Some 400 decomposition products, both volatile and non-volatile, have been identified as a result of the changes that can occur during deep fat frying (Gere, 1982). Determination and quantification of specific compounds in all the wide range of raw materials and sources used in feed fat is impractical and consequently, most methods attempt to assess the group as a whole.

The oxidized fatty acid assay is still used to indicate the extent of OPFA and is quoted in the specifications of most commercial feed fat blends within the UK. The assay determines the proportion of saponifiable fatty acids that will not dissolve in petroleum ether but will dissolve in a mixture of diethyl-ether and acetone due to its higher polarity (BS 684, 1976). The separation is based on solubility and is non-specific. A range of compounds may be present that are sufficiently non-polar that solubility in petroleum ether is possible resulting in an underestimated value. Thermally polymerized fats may fall into this category. Therefore, it is not surprising that Billek, Guhr and Waibel (1978) found that, in used frying fats, the oxidized fatty acid test considerably underestimated the total amount of OPFA determined by gel-permeation chromatography (GPC). Although Billek's data (Figure 11.3) would suggest that 10 g/kg oxidized fatty acids equated to 150 g/kg total OPFA, the data only apply to the used frying oils studied. Feed fat blends and raw materials would be expected to demonstrate specific relationships dependent upon the type of fat or oil and the processing conditions used. Consequently, the oxidized fatty acid test can only give an indication of the total OPFA present.

Furthermore, Billek, Guhr and Waibel (1978) proposed the use of a silica column to separate the polar and non-polar fractions of used frying oils. The method is now officially recognized (AOAC, 1984) and has been used in feed fat evaluation studies (Huyghebaert, De Munter and De Groote, 1988). The method has an advantage in time compared to the oxidized fatty acids test and is simple to carry out. However, the assay was developed for use with used frying oils and its applicability to other feed fat raw materials is limited.

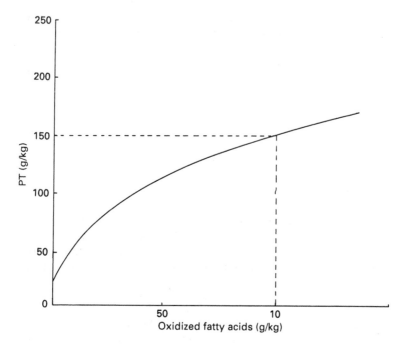

Figure 11.3 Relationship between polymerized triglycerides (PT) and oxidized fatty acids. After Billek *et al.* (1978)

The polar fraction in an oil is retained on the silica column while the non-polar fraction (i.e., unaltered triglycerides) is eluted, weighed and the polar fraction quantified by difference. Unfortunately, the polar fraction will contain not only OPFA but also FFA, mono- and diglycerides and other non-oxidized but polar fats. Feed fat raw materials have a wide range of FFA contents (see Table 11.2) and estimation, for example, of the polar fraction in palm fatty acid distillate (FFA 900 g/kg) would be highly erroneous.

GPC has been used in attempt to quantify high molecular weight polymerized fats (Harris, Crowell and Burnett, 1973; Perkins, Taubold and Hsieh, 1973). The column packing contains pores within which lower molecular weight compounds are retained while higher molecular weight compounds are eluted. The problems associated with GPC are that fatty acids which have undergone cyclysation have similar molecular weights to the unchanged monomer and that dimer and trimer fatty acids have similar molecular weights to di- and triglycerides (Veen, 1985).

High pressure liquid chromatography (HPLC) has also been proposed as a technique to assay OPFA semi-quantitatively using ultra-violet detection (Sortishos, Ho and Chang, 1986) and for the assay of dimers using flame ionization detection (Veasey, 1986). The value of such assays in relation to the cost of equipment required must be considered with reference to the requirements of laboratories evaluating feedstuffs.

All the methods mentioned are either non-specific or, as specificity increases, the techniques and equipment required become inhibitory. Of equal significance is that extensive evaluation of the methods would be required to establish their relevance to feed fat raw materials. For such reasons, the non-elutable material (NEM) assay suggested by Waltking, Seery and Bleffert (1975), listed by the AOAC (1984) and proposed by Howard (1984) for use in feed fats has received attention within the UK feed industry.

NON-ELUTABLE MATERIAL

The basis of the technique is shown in Figure 11.4. An internal marker (usually heptadecanoic acid, C17:0) is added to the test fat and the sample is methylated using methanolic borontrifluoride for separation of the methyl ester by GLC. Calculation of the ratio of marker to fatty acids eluted from the column in comparison to the ratio in the test sample, enables the percentage of material which has not eluted from the column (i.e., non-elutable material, NEM) to be derived (Waltking, Seary and Bleffert, 1975). The use of the fatty acid C17:0 as the internal marker rather than the triglyceride form (triheptadecanoic acid) as given in the AOAC method (AOAC, 1984), enables any glycerol formed on saponification to be included in the NEM value. The NEM fraction will also include moisture, impurities, unsaponifiable matter (UM) and the total oxidized and polymerized fatty acid fraction (OPFA). Consequently, with qualification, NEM can be considered to reflect the total amount of non-nutritional material in feed fat. Typical NEM values for feed fat raw materials are shown in Table 11.8.

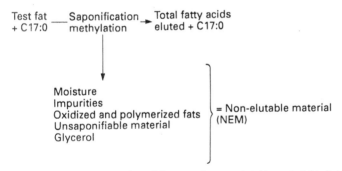

Figure 11.4 Representation of the assay for non-elutable material in fat (based upon AOAC, 1984)

Table 11.8 MEAN (±SD) NON-ELUTABLE MATERIAL (NEM) VALUES FOUND IN FEED FAT RAW MATERIALS. SAMPLES WERE ASSAYED USING C17:0 (FREE ACID) AS AN INTERNAL MARKER. ALL DATA ARE CORRECTED FOR ANY BACKGROUND C17:0 (NUMBERS IN PARENTHESES ARE NUMBER OF SAMPLES.)

Material	*NEM (g/kg)*			
Palm acid oil	40	±	12	(9)
Palm fatty acid distilate	50	±	14	(5)
Crude soyabean oil	52	±	1	(3)
Tallow	70	±	13	(59)
Recovered vegetable oil	86	±	29	(91)
Fish acid oil	102	±	41	(6)
Mixed soft acid oil	105	±	32	(37)
Soya/sunflower acid oil	124	±	28	(22)

In a collaborative study using heated vegetable oils, Waltking (1975) found the coefficient of variation to range from less than 5% for samples with more than 25% NEM to greater than 100% when less than 2% NEM was found. Since publication of these data, the introduction of capillary GLC has greatly improved resolution enabling the coefficient of variation to be reduced, certainly within a laboratory (Table 11.9). However, the AOAC (1984) method does not specify that account has to be taken of the presence of any C17:0 already existing in the sample. The higher the concentration of background C17:0, the greater the error in the calculated NEM value (Table 11.10). Consequently, it is necessary to produce two GLC profiles of the test fat, with and without added internal marker. Using capillary GLC equipment, good resolution can be achieved in 16 minutes (see Figure 11.2) which facilitates duplicate determinations without incurring excessive analytical time. The amount of background C17:0 determined in the test sample is used to correct the amount of C17:0 in the spiked sample. As demonstrated in Table 11.10, small amounts of background C17:0 have a marked effect upon the calculated NEM value so adequate resolution and integration of the C17:0 peak is essential to reduce errors.

Table 11.9   REPEATABILITY OF NON-ELUTABLE MATERIAL (NEM) DETERMINATION IN A FEED FAT RAW MATERIAL

Material	NEM (g/kg)	SD	CV (%)
Sunflower acid oil	136	35	25 (n = 5)

Table 11.10   EFFECT OF CORRECTING FOR BACKGROUND C17:0 ON THE VALUE OF NON-ELUTABLE MATERIAL (NEM) IN FEED FAT RAW MATERIALS. RESULTS ARE GIVEN IN g/kg

Material	C17:0	NEM corrected for C17:0	NEM not corrected for C17:0	Difference
Palm fatty acid distillate	1.8	23	34	11
Sunflower acid oil	2.5	80	95	15
Mixed soft acid oil	5.4	109	139	30
Recovered vegetable oil	5.2	82	118	36
Fish acid oil	9.1	90	138	48
Tallow	15.7	58	143	85

Although the NEM assay may be repeatable within a laboratory (see Table 11.9), inter-laboratory comparisons can show poor reproducibility due to the lack of an accepted and defined analytical procedure. Factors that will affect reproducibility are:

1. GLC column type (packed or capillary).
2. Integration procedures (the use of the total area of the chromatogram or only defined fatty acids peaks).
3. Response factors (the use of 1.0 for all fatty acids or determined response factors).

4. The type of internal standard (triheptacdecanoic acid may be suitable for used frying oils with low FFA but its use in feed fats creates erroneously high NEM values).
5. Method of calculation of the result.
6. The weights of sample and internal standard used.

As already noted, NEM can be considered as the total amount of non-nutritional material in feed fat. However, the assay as described above will produce a NEM value which will include glycerol derived from glyceride in any mono-, di- and triglycerides present in the fat. Glycerol should not be considered as part of the non-nutritive fraction of feed fat. Overall it may be expected than an inverse relationship exists between the FFA and glycerol content of a feed fat. Due to the variation in FFA concentration in feed fats (see Table 11.2), a fat assessed for quality by the NEM assay with a low FFA concentration would appear poor in comparison to a fat of the same nutritional value but with a higher FFA (i.e. the former sample will contain more glyceride resulting in a higher NEM value). Ignoring at this stage the possibility of the presence of mono- and diglycerides in feed fat, at 150 g/kg FFA in a feed fat the glycerol content can be calculated as 38 g/kg (taking the glyceride content of a triglyceride to be 45 g/kg). The same feed fat but with 450 g/kg FFA would contain 25 g/kg glycerol which would result in a difference of 13 g/kg absolute in the NEM value. Therefore, it is proposed that NEM values should be corrected for glyceride content (NEM−G). However, the assumption that mono- and diglycerides are not present in feed fats in significant amounts requires validation. Assay of the actual glyceride content of feed fats (BS 684, 1976) and correlation with FFA concentration may provide a simple means of predicting glyceride concentration.

NEM-G values would then indicate the moisture, impurities, UM and OPFA fractions in fat. Therefore, an indirect means of quantifying the OPFA fraction would be subtract the moisture, impuritites and UM from the NEM-G content. The former 3 tests are all established techniques (BS 684, 1976) and the repeatability within a laboratory using feed fat is acceptable (see Table 11.1). However, it would not be expected that the oxidized fatty acid test would correlate with OPFA calculated from NEM-G due to the non-specificity of the former assay (Billek, Guhr and Waibel, 1987). Some values for feed fat raw materials calculated as proposed are shown in Table 11.11. Although the variation is high (data are means of separate deliveries from several sources), the mean values would appear to agree

Table 11.11 THE TOTAL OXIDIZED AND POLYMERIZED FATTY ACID FRACTION (OPFA) IN FEED FAT RAW MATERIALS CALCULATED FROM GLYCEROL CORRECTED NON-ELUTABLE MATERIAL VALUES (NEM-G). RESULTS ARE GIVEN IN g/kg (NUMBERS IN PARENTHESES ARE NUMBER OF SAMPLES)

Material	OPFA (g/kg)*	SD	
Tallow	16	14	(9)
Recovered vegetable oil	28	15	(8)
Fish acid oil	53	7	(6)
Mixed soft acid oil	53	10	(4)

*Calculated as (NEM-G) − (M + I + U) where M = moisture, I = impurities and U = unsaponifiable material.

with expectations. Tallow and recovered vegetable oil values are relatively low compared to a highly oxidizable fat such as fish acid oil. However, the comparison of soya/sunflower acid oil with mixed soft acid oil (mainly rape acid oil) suggests that further data are required.

The ability to quantify the OPFA fraction and the other lipid fractions discussed would enable the energy evaluation of feed fat raw materials to be correlated with chemical parameters (e.g. Huyghebaert, De Munter and De Groote, 1988). A prerequisite for the accurate description of a feed fat is that the energy contribution from the nutritional and non-nutritional fractions is identified and quantified. The fatty acid profile will have a major influence on the overall energy value of a fat to a particular animal at a given age (Wiseman and Salvador, 1989). Ideally changes in the energy value of a fat due to variation in the fatty acid composition must be separated from those due to the non-nutritional fraction if reliable predictive data from chemical analysis are to be achieved.

Table 11.12 shows the effect of variation in the quality of sunflower oil on its energy value to poultry. The trial compared refined sunflower oil with a sample of poor quality sunflower acid oil. The objective was to maintain the fatty acid profile constant while varying the non-nutritional fraction. Both fats were fed 0, 40, 80 and 120 g/kg inclusion into a basal diet to broilers at 10d and 40d of age according to the procedure of Wiseman (see Chapter 12).

Table 11.12   EFFECT OF THE NON-NUTRITIONAL FRACTION OF SUNFLOWER ACID OIL (SAO) ON THE APPARENT METABOLIZABLE ENERGY (AME) VALUE DETERMINED IN POULTRY. REFINED SUNFLOWER OIL (RSO) ACTED AS A CONTROL. RESULTS ARE GIVEN AS g/kg

	RSO	*SAO*
C16:0	69	82
C18:0	45	48
C18:1	196	191
C18:2	651	620
C20 +	16	25
NEM	41	137
NEM-G	0	108
FFA	0	388
Moisture	2	8
Unsaponifiable matter	1	82
Oxidized fatty acids	0	15
Impurities	0	0
AME (MJ/kg)	37.5	30.6

Table 11.12 clearly demonstrates the effect that poor quality sunflower oil had on energy value. J. Wiseman (unpublished observations) has estimated that 1.5–2.0 MJ/kg of the decrease in energy value can be attributed to differences in FFA concentration. Consequently, an estimated 5 MJ/kg can be related to the non-nutritional fraction (NEM-G = 10.8%). These data would indicate than an increase of 11% in NEM-G resulted in a reduction of around 15% in the energy value of refined sunflower oil. With more extensive data, it would be possible to attribute energy values to the NEM-G fractions (i.e., moisture, impurities, UM and OPFA).

Summary

The established methods for the assessment of lipid fractions in fats and oils have mainly been developed for use with used frying oils. Their application to feed fat has not been validated and in some instances, their usefulness is limited. Reliable chemical assays that can be correlated to determine energy values are necessary, particularly to enable the energy contribution of the non-nutritional fractions to be accurately assessed. Feed fats are a blend of several raw materials which can be variable in quality and nutritive value. Identification and assessment of fat raw materials is necessary to ensure that cost effective feed fats are utilized in diets. The present paper has concentrated on those lipid fractions that have a significant effect on the energy value of fat which is the primary purpose of its inclusion into feeds. Although not dealt with here, the physical properties of fats (e.g. melting point) should also be considered when evaluating fats. Some proposals for the determination of the non-nutritional fraction in feed fats have been made, with the suggestion that glycerol corrected non-eluted material should be evaluated as a useful indicator of fat quality.

References

AOAC (1984) *Official Methods of Analysis of the Association of Official Analytical Chemists* 14th edn. (ed Williams, S.) Virginia: Association of Official Analytical Chemists

AOCS (1978) *Official and Tentative Methods of the American Oil Chemists' Society*, 3rd edn. Virginia: Association of Official Analytical Chemists' Society

Atteh, J.O. and Leeson, S. (1983) Effects of dietary fatty acids and calcium levels on performance and mineral metabolism of broiler chickens. *Poultry Science*, **62**, 2412–2419

Billek, G. Guhr, G. and Waibel, J. (1978) Quality assessment of used frying oils: a comparison of four methods. *Journal of the American Oil Chemists' Society*, **55**, 728–733

BS 684 (1976) *British Standard Methods of Analysis of Fats and Oils*. Milton Keynes: British Standards Institution

Cooke, B.C. (1982) Call for greater fat and blends definition. *Milling Feed and Fertiliser*, **March**, 40–43

Cooke, B. (1986) The implications to research and the feed compounder of the new oils and fats determination. In *Recent Advances in Animal Nutrition* (eds W. Haresign and D.J.A. Cole) London: Butterworths, pp. 83–86

Crashe, J.D. and Bannon, C.D. (1987) Gas liquid chromatography analysis of the fatty acid composition of fats and oils in a total system for high accuracy. *Journal of the American Oil Chemists' Society*, **64**, 1413–1417

D'Alonza, R.P., Kozarek, W.J. and Wharton, H.W. (1981) Analysis of processed soy oil by gas chromatography. *Journal of the American Oil Chemists' Society*, **58**, 215–227

Davis and Lewis, D. (1969) The digestibility of fats of differing glyceride structure and their effects on growth performance and carcass composition of bacon pigs. *Journal of Agricultural Science*, **72**, 217–222

Deman, L. and Deman, J.M. (1983) Trans fatty acids in milkfat. *Journal of the American Oil Chemists' Society*, **60**, 1095–1099

Enig, M.G., Pallansch, L.A., Sampugna, J. and Keeney, M. (1983) Fatty acid composition of the fat in selected food items with emphasis on trans components. *Journal of the American Oil Chemists' Society*, **60**, 1788–1795

Fedeli, E., Langani, A., Capella, P. and Jacini (1966) *Journal of the American Oil Chemists' Society*, **43**, 472

Freeman, C.P. (1984) Digestion, absorption and transport of fats – non-ruminants. In *Fats in Animal Nutrition* (ed J. Wiseman), London: Butterworths

Freeman, C.P., Holme, D.W. and Annison, G.F. (1986) The determination of the true digestibility of interesterified fats in young pigs. *British Journal of Nutrition*, **22**, 651–660

Gere, A. (1982) Studies of the changes in edible fats during heating and frying. *Die Nabrung*, **26**, 923–932

Gurr, M.I. (1986) Trans fatty acids – metabolic and nutritional significance. *British Nutrition Foundation Bulletin*, **11**, 105–122

Gutfinger, T. and Letan, A. (1974) Studies of unsaponifiables in several vegetable oils. *Lipids*, **9**, 659–663

Harris, W.C., Grawell, E.P. and Burnett, B.B. (1973) Quantitative analysis of polymerised fatty acids using gel permeation chromatography. *Journal of the American Oil Chemists' Society*, **50**, 537–539

Huyghebaert, G., De Munter, G. and De Groote, G. (1988) The metabolisable energy (AMEn) of fats for broilers in relation to their chemical composition. *Animal Feed Science and Technology*, **20**, 45–58

Janssen, W.M.M.A. (1988) Polymer levels in commercial fats: their effect on broiler production. In *International Livestock Feed Symposium*, London: National Renderers Association

Maxwell, R.J. and Schwartz, O.P. (1979) A rapid, quantitiative procedure for measuring the unsaponifiable matter from animals, marine and plant oils. *Journal of the American Oil Chemists' Society*, **56**, 634–636

Paquat, C. (1979) *Standard Methods for the Analysis of Oils, Fats and Derivatives*, 6th edn. Oxford: Pergamon Press

Perkins, E.G., Taubold, R. and Hsiek, A. (1973) Gel permeation of heated fats. *Journal of the American Oil Chemists' Society*, **50**, 223–225

Pocklington, W.D. and Hautfenne, A. (1985) Determination of triglycerides in fats and oils. *Pure and Applied Chemistry*, **57**, 1515–1522

Sanderson, P. (1986) A new method of analysis for crude oils and fats in feedingstuffs. In *Recent Advances in Animal Nutrition* (eds W. Haresign and D.J.A. Cole), London: Butterworths, pp. 77–82

Sotirhas, N., Ho, Chi–Tang and Chang, S.S. (1986) HPLC analysis of oxidative and polymerised decomposition products in commercial vegetable oils and heated fats. *Fette Seifen Anstrichmittel*, **88**, 45–48

Schwartz, D.P. (1988) Improved method for quantifying and obtaining the unsaponifiable matter of fats and oils. *Journal of the American Oil Chemists' Society*, **65**, 246–251

The Feedingstuffs (Sampling and Analysis) (Amendment) Regulations (1985) Statutory Instrument No. 1119. London: HMSO

The Feedingstuffs Regulations (1988) Statutory Instrument No. 396, London: HMSO

Veasey, R.L. (1986) Rapid analysis of dimer acid by HPLC/FID. *Journal of the American Oil Chemists' Society*, **63**, 1043–1046

Veen, W.A.G. (1985) Polymerised fat, feeding value and quantitative analysis. In *Increasing Efficiency with Animal Fat and Protein*, London: National Renderers Association, pp. 25–34

Vernon, B.G. and Perry, F.G. (1981) Fats in energy evaluating systems. In *3rd European Symposium on Poultry Nutrition, Edinburgh* (eds D.W.F. Shannon and I.E. Wallace), World Poultry Science Association

Waltking, A.E., Seery, W.E. and Bleffert, G.W. (1975) Chemical analysis of polymerisation products in abused fats and oils. *Journal of the American Oil Chemists' Society*, **52**, 96–100

Waltking, A.E. (1975) Evaluation of methods for the determination of polymers and oxidation products of heated vegetable oils: collaborative study of the gas-liquid chromatographic method for non-elution materials. *Journal of the Association of Analytical Chemists*, **58**, 898–905

Ward, D.D. (1985) The TBA assay and lipid oxidation: an overview of the relevant literature. *Milchwissenschaft*, **40**, 583–588

Wiseman, J. and Cole, D.J.A. (1983) Interaction between dietary fat, fatty acids and calcium in growing pigs. In *Proceedings of 5th World Conference on Animal Production, Tokyo*, pp. 423–424

Wiseman, J. and Salvador, F. (1989) Dietary energy values of fats for poultry as influenced by age and rate of inclusion. *British Poultry Science*, **28**, 631–634

Wiseman, J., Cole, D.J.A., Perry, F.G., Vernon, B.G. and Cooke, B.C. (1988) Apparent metabolisable energy values for broiler chicks. *British Poultry Science*, **27**, 561–576

12

VARIABILITY IN THE NUTRITIVE VALUE OF FATS FOR NON-RUMINANTS

JULIAN WISEMAN
University of Nottingham, School of Agriculture, Nottingham, UK

Introduction

It is unquestioned that fats destined for inclusion into diets for non-ruminants are diverse entities and that this variability has a pronounced effect upon their subsequent nutritional value. This topic has been the subject of a number of previous reviews which have considered both the physiological basis for the digestion and absorption of fats, and the factors which are responsible for the large differences in their subsequent dietary energy values (e.g. Freeman, 1976, 1984; Wiseman, 1984; Krogdahl, 1985).

Thus it is well established that factors including chemical structure of fats, age of animal or bird to which they are fed and the methodology employed in their evaluation are important variables. The principles surrounding these differences are accepted, and numerous individual experiments have confirmed their importance. However, it is difficult to analyse the existing available but rather disparate information in anything other than a somewhat qualitative and superficial fashion, especially when it is realized that the methodologies employed in the evaluation of fats are so diverse and that, frequently, descriptions of the products evaluated are confined to names and origins with no accompanying chemical characterization of a more precise nature.

It would seem that systematic studies on the quantitative contribution of the variables involved in the dietary energy value of fats appear limited and, accordingly, such a programme was initiated. The objective of this review is to draw together the major findings of the comprehensive programme of fat evaluation that has been conducted principally at the University of Nottingham but also at the Station de Recherches Avicoles, Tours, France. Studies have been undertaken with both pigs and poultry, but those based upon the latter species have progressed further due to the relatively lengthier and more complex experimental procedures associated with trials with pigs. Accordingly, the major species to be considered in this discussion is poultry, although preliminary observation with pigs indicate that the two species respond similarly.

Methodology

RATE OF INCLUSION

It is customary in most countries to express the dietary energy value of compound diets or raw materials for poultry in terms of their metabolizable energy (ME) value. The current debate concerning the relative value of True and Apparent ME has been considered in Chapter 3. A principal feature of TME is that the response of dietary energy input to excreta energy output, fundamental to the estimation of endogenous losses, is linear. This assumption is the basis of the claim that TME is independent of intake and, accordingly, is likely to be less variable than AME. In addition, such an assumption allows the estimation of TME by regression rather than by trials involving crop intubation.

There is evidence, however, that fats are utilized more efficiently at lower rates of intake (e.g. Wiseman *et al.* 1986), a situation which is more pronounced the poorer the nutritive value the fat and the younger the bird in question (Wiseman and Lessire, 1987). A more detailed consideration of this observation would reveal that, with a diet containing fat, the response of excreta energy output to dietary input would in fact be curvilinear. The consequence of this response would be that TME would be higher at lower rates of intake. Evidence that this in fact is the case was supplied by Kussaibati *et al.* (1982) and Figure 12.1 present data emerging from that study. It is of course accepted that AME itself is not independent of intake, but the fact remains, nevertheless, that TME may not remove variability in data associated with fluctuations in intake.

Fats, by virtue of their physical nature, may not be evaluated in isolation of any other dietary component and it is conventional to include them into a basal diet and

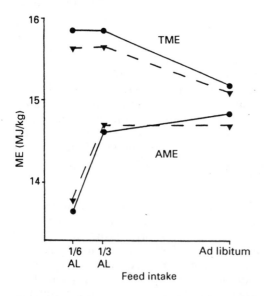

Figure 12.1 Influence of rate of intake below *ad libitum* of diets containing 150 g/kg fat (maize oil, ●——●, animal, fat, ▲ – – ▲), on the true (TME) and apparent (AME) metabolizable energy value of the diets. Data calculated from Kussabaiti *et al.*, 1982

to determine their dietary energy value by the difference between the value obtained for the basal and that obtained for the basal plus fat. However, a further consequence of the possibility, described above, that fats may be more effectively utilized at lower levels of intake is that their rate of inclusion into the basal diet may be an important experimental variable in their evaluation. It would seem important when evaluating fats, accordingly, that experimental design should allow an assessment of whether an effect of rate of inclusion is apparent. Non-linearity in the response of dietary energy value of experimental diets to increasing rates of inclusion of fats would be evidence for such an effect.

In selecting rate of inclusion as the independent variable in studies of this nature, variability in feed intake between treatments may confound the responses obtained. Whilst this poses problems for AME evaluation generally, it is potentially of greater importrance in the evaluation of fats due to the possibility that their utilization is higher at lower intakes. Accordingly, it may be necessary to ensure that feed intakes are equalized across treatments. In practice, however, experience has shown that effects of treatment on feed intake are very limited. Although this may seem surprising if the range in rates of fat are considered (which, it would have been thought, would lead to some form of compensatory mechanism as dietary energy levels alter) the relatively short time over which metabolism experiments are conducted would, in all likelihood, preclude compensation in feed intake (assuming, even, that the broiler chicken in fact responds in this manner). Finally, although a relatively minor issue, in trials employing replicates consisting of a cage of more than one bird (a fairly common occurrence, especially if chicks are used), the loss of a bird during the experimental programme will obviously alter the pattern of intake which may confound interpretation of data if intake is the criterion for assessment but not if rate of inclusion is employed.

In an experiment of fixed size, three rates of inclusion of fat are sufficient to estimate this effect. The analysis of variance performed upon the experimental data generated can then include a non-linear term for rate of inclusion. Estimation of the dietary energy value of fat is achieved by interpolation of the function derived. The choice of which non-linear model most adequately describes the responses obtained is of some considerable importance, and this has been the subject of many previous reports particularly those considering the response of populations of birds to nutrient supply. However, evaluating the change in dietary energy value of diets based upon incremental increases in the concentration of fats is merely an attempt to characterize the role of the fat itself and its decline in dietary energy value, if any, at higher rates of inclusion. Accordingly, simple quadratic functions have been found to be the most appropriate both from the point of view of statistics (i.e. goodness of fit) and of biology because, with their use, is the acceptance that the dietary energy value of a fat declines with increasing rate of inclusion. With respect to this latter consideration, higher level polynomial functions (third or fourth power, for example) are considered conceptually unsound (even though they may, under certain circumstances, explain more of the variation) as their use requires an explanation as to why dietary energy values of fats increase at higher rates of inclusion. The use of inverse polynomial functions may be considered more appropriate, but their use introduces a degree of complexity into evaluation of fats that is probably not justified. Such an appraoch would in addition require more than three rates of inclusion of fat and would therefore be associated with a decrease in the efficiency of utilization of resouces if a programme of work is being undertaken.

What is of fundamental importance, however, is the need to fit the same model to data considering different fats to allow valid comparisons between these fats. Experiments that fail to do this, and rely solely upon goodness of fit in selecting a range of functions, will generate data that may certainly indicate differences between fats. However it is not possible to separate differences attributable to fats themselves from those arising simply from variability between the models employed.

In studies on the effect of rate of inclusion, the general conclusions to be drawn are that non-linear responses are evident with poorer quality fats evaluated with younger birds. Such responses are less apparent with older birds (Figure 12.2). Age is in fact of considerable importance in the characterization of the nutritive value of

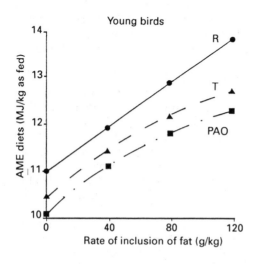

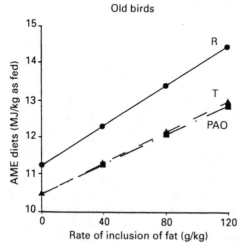

Figure 12.2   Influence of rate of inclusion of fat (R, rape oil; T, tallow; PAO, palm, oil – g/kg diet) on the apparent metabolizable energy (AME, MJ/kg as fed) of diets. Data are derived from different studies, hence lack of common intercept. From Wiseman and Lessire, 1987; Salvador and Wiseman, 1989

fats. The physiological basis for this would appear to be a poor development in the capacity for young poultry to utilize fats (e.g. Carew *et al.*, 1972), a situation that is more pronounced the more saturated the fat in question (e.g. Fedde *et al.*, 1960). It was concluded by Wiseman and Salvador (1989) that there is little improvement in utilization of fats beyond 7.5 weeks of age and, accordingly, two ages (being 1.5 and 7.5 weeks) are adequate in the evaluation of fats.

Further conclusions on the effect of rate of inclusion would appear to indicate that it is, in fact, of relatively minor importance (as indicated by Freeman, 1976). However, the use of multi-rate assays in the evaluation of fats is considered a more appropriate procedure because, in addition to allowing the estimation of any non-linear trends, they permit the derivation of dietary energy values of fats through regression analysis. This approach is associated with considerably lower standard errors than that based simply upon a 'by difference' procedure. In this approach, the dietary energy value of a fat is estimated from the value obtained from a basal diet and the basal diet containing a known proportion of fat. However the, necessarily, low proportions of fat that are used in experiments of this nature generate standard errors that are invariably too high to allow meaningful comparisons between fats to be drawn.

Finally, if multi-rate assays and subsequent regression analyses are employed to compare fats, with the analysis producing a common intercept value for all those fats evaluated in one experiment, then it is the relative slope of the responses that is the basis for comparison. Accordingly, the debate relating to the applicability of AME and TME may not be relevant.

APPARENT AVAILABILITY OF FAT

The methodology described above is based upon the direct determination of dietary energy values of diets containing added fats. These values are estimated with reference to gross energy consumed and gross energy voided. Dietary energy values of fats may in addition, however, be obtained from the product of their apparent availability and their gross energy as suggested by, for example, Renner and Hill, 1960; Artman, 1964; Whitehead and Fisher, 1975; Mateos and Sell, 1981; Lessire *et al.*, 1982). A principal advantage of this approach is that it removes any confounding effects that might occur as a consequence of interactions between the added fat and the non-fat component of the basal as have been reported by, for example, Mateos and Sell (1980, 1981). It is of course important to state that fat content of excreta needs to evaluated with acid hydrolysis prior to solvent extraction.

The approach to the evaluation of fats through this method is precisely the same as that employed for the direct determination of AME. Thus the response of apparent available fat content of experimental diets (expressed as g/kg diet) to incremental increases in rate of inclusion of fat is extrapolated (if the response is linear) or interpolated (if a non-linear response is obtained). Solutions of these functions will generate data for apparent available fat of the added fat itself which is then multiplied by its gross energy to produce AME figures.

Both methods have been employed during the fat evaluation programme. With the information thus generated, a more detailed comparison between the two methods is possible. Figure 12.3 presents information comparing AME of fats derived either from energy balance (*y*), or as the product of apparent availability of

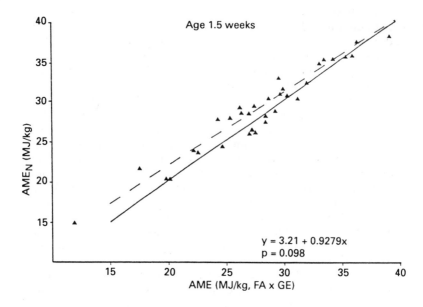

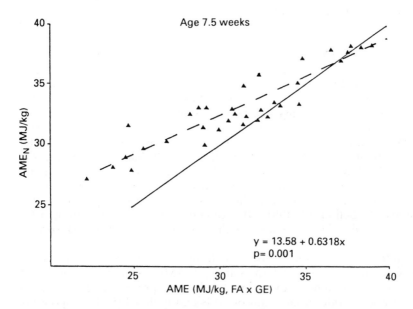

Figure 12.3 Relationship between apparent metabolizable energy corrected to zero nitrogen retention (AMEn, MJ/kg) determined through energy balance and apparent metabolizable energy (AME, MJ/kg) determined as the product of apparent fat availability and gross energy of fats. Data derived from Wiseman and Lessire, 1987, and unpublished data from the University of Nottingham

fats and their respective gross energies (x). The regression of y on x generated the functions 1 and 2 respectively for birds aged 1.5 and 7.5 weeks of age:

$$y = 3.21 + 0.9279x, \text{rsd} = 1.38, \text{r} = 0.938 \dots\dots\dots\dots\dots\dots 1$$
$$\pm 1.20 \pm 0.0423$$

$$y = 13.58 + 0.6318x, \text{rsd} = 1.21, \text{r} = 0.848 \dots\dots\dots\dots\dots\dots 2$$
$$\pm 1.49 \pm 0.0472$$

Comparison of the slopes obtained with the 'ideal' (i.e. that arising if the two sets of data were numerically equal) revealed that there was a significant difference for the older birds ($P<0.001$) and a trend for the younger birds ($P=0.098$). It is evident that AME determined directly generates higher figures than that obtained as the product of fat availability and gross energy, particularly with older birds. In addition, it is apparent that the differences between the two sets of data are more pronounced with lower figures (i.e. with fats of lower nutritive value). From this it follows that under some circumstances the relative rate of loss of fat in the excreta is greater than that for gross energy, a phenomenon which may require further investigation.

In a further examination of the two procedures with a small number of fats (Wiseman and Lessire, 1987) it was concluded that that based upon fat availability was associated with lower standard errors and subsequent studies have confirmed this. As such it is the preferred approach, and data considered subsequently have been obtained with this method.

Chemical structure of fat

Many reviews describing numerous individual experiments have confirmed that the chemical structure of a fat is a major determinant of its subsequent dietary energy value. In this context, the degree of saturation and free fatty acid content are of particular importance although other factors, including the degree of contamination with 'non-nutritive' compounds, which will not be considered here, are relevant (see Wiseman, 1986; Edmunds, 1990).

DEGREE OF SATURATION

In addition to the undenied superiority of unsaturated over saturated fats, the phenomenon of 'synergism' is a major issue in any consideration of degree of saturation and its influence upon the nutritive value of fats. The physiological basis for 'synergism' is well established (e.g. Freeman, 1984) and it refers, basically, to the ability of an unsaturated fatty acid to promote the absorption of one that is saturated. Thus the nutritive value of the latter, if this is defined in terms of the quantity of fat absorbed, is elevated in the presence of the former.

In contrast to the amphiphilic index describing individual fatty acids developed by Freeman (1969) whereby this interaction proceeded in a predictable manner, the term 'synergism' has become associated with an elevation in the dietary energy value of a mixture of a relatively saturated and a relatively unsaturated fat over and above that which would be predicted from values for the two individual fats (e.g.

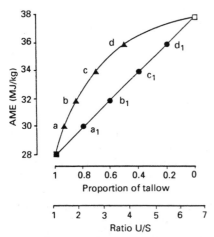

Figure 12.4   Theoretical consideration of the influence of the proportion of a relatively saturated tallow in a mixture with a relatively unsaturated oil, giving blends with an increasing ratio of unsaturated to saturated fatty acids (U/S). The responses in terms of AME of fats shown are those which would be predicted from knowledge of the AME values of the two fats, there being no 'synergism' assumed. a, b, c, d, represent the same mixtures as a1, b1, c1, and d1 respectively. Data obtained from Table 12.1

Lewis and Payne, 1966). Indeed, this would appear to have become the basis for the formulation of fat blends designed for inclusion into diets for non-ruminants.

The interaction may be viewed in terms of the mixture of two fats A, being relatively unsaturated, and B, being more saturated. Arbitrarily, they have been given AME values of 38 and 28 MJ/kg respectively. Mathematically, any blend of the two will result in mixtures with predictable AME values. Thus, for example, a blend of 20A:80B will have a calculated (i.e. predicted) AME value of 30 MJ/kg. The calculated values for any mixture of the two are indicated by the linear response presented in Figure 12.4. The use of descriptive rather than numeric terms for the two fats in question is, however, extremely limited (they could for example be labelled vegetable oil and animal fat respectively for fats A and B) and a more appropriate means of characterization would be that based upon some measurement that is of nutritional relevance. As it is the degree of saturation that is of fundamental importance, then a more sophisticated approach would be to use this as the basis for comparison. Accordingly, the ratio of unsaturated to saturated fatty acids (U/S) has been proposed.

The fatty acid profiles and corresponding U/S ratios of the two fats under consideration are presented in Table 12.1(a). It is crucial to appreciate that the U/S ratio of mixtures of the two fats do not increase in a linear fashion. Accordingly the linear increase in AME with increasing proportions of fat A in the mixture is equivalent to the curvilinear increase in AME when the basis for assessment is the ratio U/S. Thus in Figure 12.4, a, b, c, and d correspond precisely to a1, b1, c1 and d1 respectively.

Evidence for 'synergism', when applied to AME values, would suggest values for mixtures of fats over and above the curvilinear predicted range of values if the ratio U/S was used (or the linear range of values if the relative proportions of A and B

Table 12.1a CHARACTERIZATION OF TWO FATS USED TO DEMONSTRATE EFFECTS OF DEGREE OF SATURATION (SEE FIGURE 12.4 FOR RESPONSES)

Fatty acids*	Fatty acid profile (g fatty acid/kg recoverable fatty acids)	
	A	B
C14 : 0	30	0
C16 : 0	300	20
C18 : 0	165	110
C16 : 1	50	55
C18 : 1	435	245
C18 : 2	20	525
C18 : 3	0	45
Ratio unsaturated:saturated fatty acids (U/S)	1.020	6.692
Apparent metabolizable energy (AME, MJ/kg)	28	38
Mixtures of A and B	US†	AME†
A80 : B20	1.370	30
A60 : B40	1.865	32
A40 : B60	2.623	34
A20 : B80	3.926	36

*Carbon chain length followed by number of double bonds.
†Calculated from the data presented for the two fats together with the appropriate proportions of A and B indicated.

Table 12.1b CALCULATED COEFFICIENT OF AVAILABILITY OF SATURATED FATTY ACID FRACTION OF FATS CHARACTERIZED IN TABLE 12.1a ASSUMING SPECIFIC COEFFICIENTS OF AVAILABILITY FOR THE UNSATURATED FATTY ACID FRACTION. ASSUMED GROSS ENERGY FOR BOTH FRACTIONS IS 39 MJ/kg

Fats	Assumed coefficient of availability of unsaturated fraction				
	0.96	0.97	0.98	0.99	1.00
A	0.471	0.461	0.451	0.440	0.430
80A : 20B	0.508	0.494	0.481	0.467	0.435
60A : 40B	0.560	0.542	0.523	0.504	0.486
40A : 60B	0.640	0.614	0.588	0.562	0.535
20A : 80B	0.778	0.739	0.700	0.660	0.621
B	1.070	1.004	0.937	0.870	0.803

are employed). Figures 12.5(a) and 12.5(b) present the results of two experiments designed to evaluate the contribution of 'synergism' by obtaining data for two fats differing in the ratio U/S, and then comparing predicted values for mixtures with those actually determined. The characterization of the fats evaluated is presented in Table 12.2. In neither case was there any evidence for 'synergism' obtained with older birds in that there was no significant difference between the two responses obtained (determined and predicted), although there did appear to be an increase

Table 12.2   FATTY ACID PROFILES FOR THE FATS EMPLOYED IN THE STUDY OF 'SYNERGISM' (SEE FIGURES 12.5(a) AND 12.5(b) FOR RESPONSES)

*Fatty acids**	*Fatty acid profile*	
	(g fatty acid/kg recoverable fatty acids)	
	Rape oil	Tallow
	A	B
C12 : 0	0	0
C14 : 0	1	37
C16 : 0	54	287
C18 : 0	15	181
C16 : 1	1	48
C18 : 1	624	413
C18 : 2	220	28
C18 : 3	85	6
Ratio unsaturated:saturated fatty acids (U/S)	13.286	0.980

Mixtures of A and B:	5A : 95B	10A : 90B	20 : 80B
Ratio U/S	1.069	1.167	1.392

	Rape oil	Tallow
C12 : 0	0	3
C14 : 0	2	41
C16 : 0	66	186
C18 : 0	32	249
C16 : 1	5	60
C18 : 1	496	368
C18 : 2	255	67
C18 : 3	144	26
Ratio unsaturated:saturated fatty acids (U/S)	9.000	1.088

Mixtures of A and B:	20A : 80B	40A : 60B	60 : 40B
Ratio U/S	1.480	2.054	2.975

*Carbon chain length followed by number of double bonds.

in the determined over the predicted AME value for younger birds (discounting the seemingly anomalous value for fat C) in trial 1 (Figure 12.5(a)) which would appear to support the phenomenon of 'synergism' when applied to AME values.

However, a closer examination for the conceptual basis of this elevation ('synergism') reveals that there are fundamental flaws. Thus its existence requires an explanation as to why a mixture with a specific U/S ratio has a higher determined AME than the predicted AME of the mixture with precisely the same U/S ratio. The phenomenon whereby unsaturated fatty acids improve the utilization of relatively saturated fatty acids is in all probability not the reason as this interaction is already responsible for the existing and accepted increase in AME with increasing U/S ratio. This may be examined in more detail if the two fats considered in Table 12.1(a) are investigated more closely. If the two fractions (unsaturated and saturated fatty acids) are ascribed the same gross energy (in this case 39.00 MJ/kg – figures in practice will not deviate significantly from this) then, with assumed coefficients of availability of the unsaturated fractions contained within the

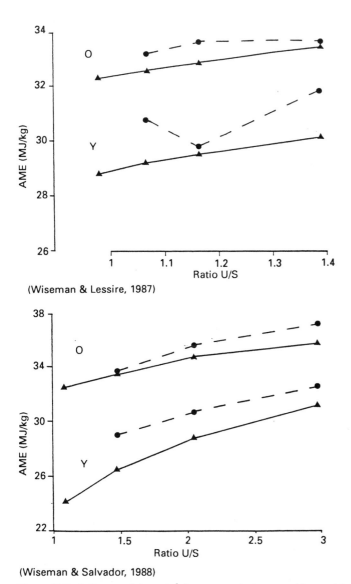

(Wiseman & Lessire, 1987)

(Wiseman & Salvador, 1988)

Figure 12.5 Influence of the ratio of unsaturated to saturated fatty acids (U/S) on the apparent metabolizable energy (AME, MJ/kg) of fats. The response ▲—▲ indicates that which would be predicated from knowledge of the AME values of the two fats employed in formulating the mixtures, and the response ●- -● is that actually determined for the mixtures. Due to scaling the point indicating the AME value of the unsaturated oil is not shown

mixtures, it is possible to calculate those coefficients associated with the saturated fraction assuming the AME values given. The results of such an exercise are presented in Table 12.1(b), with assumed coefficients of availability of the unsaturated fraction of 0.96, 0.97, 0.98, 0.99 and 1.00. These values are well within the range obtained in practice.

It is evident that, with these values for the unsaturated fraction, the coefficients for the saturated fraction must increase with increasing degree of unsaturation of the mixture if the eventual AME values are to be realized (Figure 12.6). It is of course important to appreciate that the unsaturated oil itself contains a proportion of saturated fatty acids. In the example selected, a coefficient of utilization of at least 0.97 must be assumed for the unsaturated fraction in order for the coefficient to be less than 1 for the saturated fraction in the oil itself. There can, therefore, be no plateau in the coefficient of utilization of saturated fatty acids with increasing unsaturation of the fat mixture.

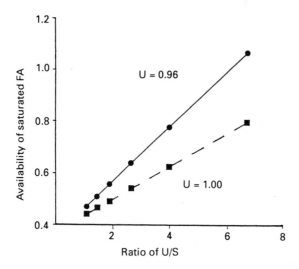

Figure 12.6   Influence of the ratio of unsaturated to saturated fatty acids (U/S) on the coefficient of availability of the saturated fraction. Assumed coefficient of avilability of the unsaturated fraction is 0.96 and 1.00. See Table 12.1(a) for AME and U/S data and Table 12.1(b) for coefficient of availability of fatty acid fractions

The theoretical exercise described above is, for the sake of simplicity, comparatively limited in its approach. Thus for example it is likely that those fatty acids collectively referred to as the saturated fatty acid fraction would, individually, have different coefficients of avilability. However, in describing the possible responses, an attempt was made to employ relatively realistic data and, as such, the conclusions drawn are valid. Practical evidence to support this argument is limited, but that provided by Wiseman and Lessire (1987b) for the availability of palmitic and stearic acids over an, admittedly small, range in ratios of U/S do confirm the trends described above.

It is suggested, in fact, that evidence for 'synergism' in the two trials reported above is due not to any unpredicted interaction between the fats in question, but merely a consequence of limitations in the characterization of fats, which relied solely upon the ratio U/S. In this context, it is probable that free fatty acid content (described below) will modify the responses of fats, in terms of their determined dietary energy value, of differeing U/S ratio.

Previous experiments claiming to have detected 'synergism' in AME values of mixtures of fats have rarely characterized fats in any way other than purely descriptive (which does not allow any rigorous examination of the possible differences of the ratio U/S and the confounding effects of the interaction between the ratio U/S and free fatty acid content). Furthermore, evidence for 'synergism' has been claimed with differences as little as 1.54 MJ/kg between determined and predicted AME values (Sibbald, 1978). Standard errors associated with mean AME data generated from 'by difference' experiments of fat evaluation frequently approach and are, occasionally, higher than this (the more so the lower the the rate of inclusion of fat – Wiseman *et al.*, 1986; Wiseman and Lessire, 1987).

Finally, it is sometimes not appreciated that, as the increase in the ratio U/S with regular progressive increases in the proportions of a relatively unsaturated fat is non-linear, then the rate of increase in AME of mixtures thus produced will itself be non-linear. The curvilinearity of this response has itself, erroneously, been regarded as evidence for 'synergism'. In conclusion it seems likely that evidence for 'synergism' in fat AME values is attributable more to insensitivity in experimental design or interpretation of data than to any unpredictable interaction between two fats.

FREE FATTY ACID CONTENT

Free Fatty Acid (FFA) content is a major variable in chemical composition of fats included into diets for non-ruminants, particularly as there is widespread use of a variety of high FFA raw materials (e.g. acid oils, soapstocks) in the manufacture of fat blends. On a physiological basis, it has been established that the greater the proportion of FFA in a fat, the lower the efficiency with which it is absorbed (e.g. Sklan, 1979), and the relative superiority in terms of AME values of triglyceride as compared to hydrolysed fat is well established (e.g. Young, 1961). This trend appears to be more pronounced the more saturated the fat in question, although the effect has not been extensively studied. Increasing the proportion of FFA would appear to be associated with a linear reduction in fat digestibility (Freeman, 1976), although this topic does not appear to have received much attention. Finally, although Shannon (1971) examined the rate of intake of a fat of varying FFA content, the influence of rate of inclusion has not been studied extensively.

The investigation of the effects of a specific variable should proceed under circumstances where possible effects of interactions with other variables, if these are not recorded, are minimized. Accordingly studies of the influence of FFA content were carried out with specific fats that had been hydrolysed. In view of the possible effect of degree of saturation, each fat was evaluated in conjunction with its hydrolysed product. In addition, three intermediary fats were blended using variable proportions of the two extremes. In this way the response of AME of fat to increasing FFA content could be assessed. The three fats selected, and their hydrolysed products, were tallow and tallow acid oil (FFA content of 138.4 and 952.1 g/kg fat respectively), palm oil and palm acid oil (FFA content of 57.5 and 917.5 g/kg fat respectively) and soya oil and soya acid oil (FFA content of 14.4 and 683.4 g/kg fat respectively). Finally, to investigate the interaction between FFA content and degree of saturation, soya oil was mixed with tallow acid oil and tallow with soya acid oil each to give three blends of varying FFA content (Table 12.3). Evaluation of all fats was achieved with broiler chicks aged 1.5 and 7.5 weeks.

The responses obtained with the three fats and their respective acid oils indicated that there was a progressive, and seemingly linear, decline in AME of fats with

Table 12.3   FREE FATTY ACID CONTENT (FFA g/kg FAT) AND RATIO OF SATURATED TO UNSATURATED FATTY ACIDS (U/S) OF FATS EMPLOYED IN STUDIES ON THE EFFECT OF FFA CONTENT ON APPARENT METABOLIZABLE ENERGY VALUE OF FATS

	FFA (g/kg fat)	*U/S*
Tallow (T)	138.4	1.083
Tallow acid Oil (TAO)	952.1	0.802
75T : 25TAO	341.8	1.005
50T : 50TAO	545.3	0.932
25T : 75TAO	748.7	0.865
Palm (P)	57.5	0.786
Palm Acid Oil (PAO)	917.5	0.848
75P : 25PAO	272.5	0.801
50P : 50PAO	487.5	0.817
25P : 75PAO	702.5	0.832
Soya Oil (S)	14.4	3.444
Soya Acid Oil (SAO)	683.4	9.526
25S : 25SAO	181.7	4.195
50S : 50SAO	348.9	5.250
25S : 75SAO	516.2	6.843
75S : 25TAO	248.8	2.252
50S : 50TAO	483.3	1.564
25S : 75TAO	717.7	1.116
75T : 25SAO	272.7	1.606
50T : 50SAO	410.9	2.478
25T : 75SAO	547.2	4.229

increasing FFA content which was more pronounced the younger the bird (Figures 12.7(a) and 12.7(b) respectively for younger and older birds). Data presented are derived from the linear responses obtained for rate of inclusion of fat. Non-linear responses were obtained but these tended to be confined to younger birds with the poorer quality fats.

The observations reported would appear to be difficult to reconcile with the accepted sequence of events involved with fat digestion and absorption. Thus, basically, hydrolysis under the influence of lipase generally releases two molecules of free fatty acid from each tryglyceride (e.g. Freeman, 1984), from which it could be concluded that FFA contents of the order of 600 g/kg fat would not be accompanied by any marked reduction in the digestibility of fat. However, the current series of trials in using blends of fat and FFA, would exacerbate the problem of total FFA content of digesta because the fat present as triglyceride would itself by hydrolysed.

In addition hydrolysis *in vitro*, which was the basis for the production of the high FFA fats, is indiscriminate whereas that *in vivo* is selective for fatty acids at the 1 and 3 position. The amphiphilic index described by Freeman (1969) indicated the relative superiority of 2-monoglycerides based upon an unsaturated fatty acid over the respective free fatty acid. If proportionally more of the already low levels of unsaturated fatty acids in the relatively saturated fats studied were present as FFA following *in vitro* hydrolysis then this would also reduce overall digestibility.

Further examination of the trends revealed an apparent effect of degree of saturation, with that of palm and palm acid oil being more pronounced than that for

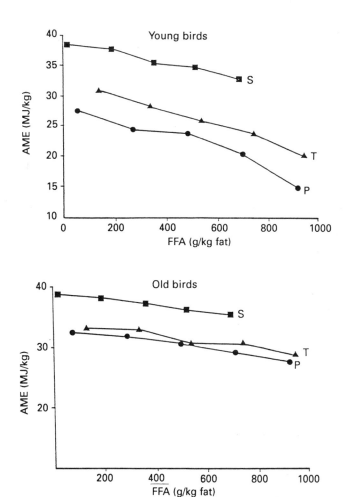

Figure 12.7 Influence of free fatty acid content (FFA, g/kg fat) on the apparent metabolizable energy value (AME, MJ/kg) of soya oil (S), tallow (T) and Palm (P). Data are pooled for rate of inclusion of fats. University of Nottingham unpublished information

tallow and tallow acid oil, which itself showed a greater response than that obtained for soya and soya acid oil. These responses may be explained in terms of the accepted relative efficiencies of absorption of the products of digestion of fat.

This interaction between degree of saturation of fat and its FFA content was further studied in the evaluation of the mixtures of soya and tallow acid oil, and of tallow and soya acid oil. Results for these responses are presented in Figures 12.8(a) and 12.8(b) respectively for young and old birds. For completeness, the results for the four fats determined previously (Figure 12.7(a) and 12.17(b)) are repeated. It is evident that the reduction in AME with increasing FFA content is more pronounced with increasing degree of saturation (e.g. the response of the mixture of soya oil and tallow acid oil compared with that with soya acid oil) and that the worsening in AME of tallow with increasing FFA content is not evident

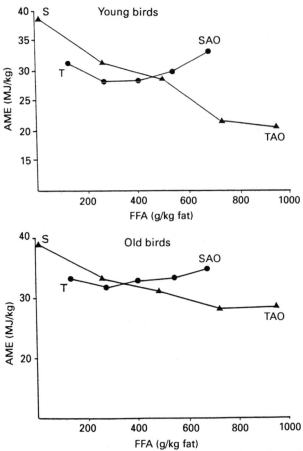

Figure 12.8 Interaction between mixtures of soya oil and tallow acid oil (S and TAO respectively) and between tallow and soya acid oil (T and SAO respectively) Responses shown are apparent metabolizable energy values (AME, MJ/kg) of mixtures as a function of free fatty acid (FFA, g/kg fat) content. Data for S, T, SAO and TAO are taken from Figure 12.6. University of Nottingham, unpublished data

when the acid oil used is soya acid oil. In fact soya acid oil had a greater AME than tallow. As before the responses obtained with younger birds were more pronounced.

In conclusion it is evident that free fatty acid content is an important chemical variable which influences the subsequent nutritive value of a fat. Its effect is, however, influenced considerably by the degree of saturation of the fat in question, as measured by, for example, the ratio of unsaturated to fatty acids.

Prediction of the AME of fats

Prediction of biological responses in an animal from some form of characterization of the food it is fed has for many years been an important aspect of feedstuff evaluation, and the subject has been considered elsewhere in this volume. At its simplest, prediction of the dietary energy value of raw materials has involved the derivation of regression equations, either linear or multiple, with dietary energy value as the dependent variable and chemical measurements of the material(s) under consideration as independent variables either singly or in combination. Whilst the chemical measurements selected do, to a greater or lesser extent, have some biological relevance, this empirical approach is necessarily limited in its value, particularly if reliance upon 'goodness of fit' is the sole criteria for selecting suitable functions. Accordingly, more sophisticated approaches have been developed in which biological considerations assume fundamental importance and models are built around how the animal responds directly to specific inputs. Of course crucial to this approach is the assumption that the animal is responding to the input(s) in question and that account is taken of interactions, if any, between inputs.

In this context, utilization of fats has received some attention. The importance of the degree of saturation or the contents of specific unsaturated and saturated fatty acids has led to their use in predicting dietary energy value or overall fatty acid availability (e.g. Young and Garrett, 1963; Sibbald and Kramer, 1977; Halloran and Sibbald, 1979). However, these studies reported limited success in the prediction of biological responses due, in all probability, to the empirical nature of the approaches adopted. Reliance upon such measurements as iodine value is only of marginal value because, although it does provide an assessment of the degree of unsaturation of constituent fatty acids, it has no direct biological relevance and, perhaps more importantly, will confuse the issue by providing different data for commodities that will in all probability have similar degrees of utilization (for example linoleic and linolenic acids).

The approach adopted at the University of Nottingham, which attempted to identify those variables likely to be of most importance in influencing the dietary energy values of fats and then proceeded to design experiments which would assess their relative role, generated data which could then be incorporated into models describing the variability in the derived dietary energy values. The data selected for treatment were those emerging from the studies on the ratio of unsaturated to saturated fatty acids (U/S) and 'synergism' described above (and taken from Wiseman and Lessire, 1986; Wiseman and Salvador, 1988) together with those considering the effect of free fatty acid content (see Free Fatty Acid Content above) giving a total of 31 data points. These two variables together with age of bird and rate of inclusion of fat were the four considered for study.

Analysis of the effect of these variables revealed that, throughout, the influence of rate of inclusion did not markedly alter numerical values for the components within functions nor did it contribute to an improvement in the accuracy of prediction and its presence in functions did not seem to be appropriate.

The response of AME to U/S was hyperbolic and, accordingly, a suitable function was fitted to the data (Table 12.4(a)). This function was considered adequate in describing the biological response of AME to increasing U/S. The proportion of the variation accounted for by this simple function was 0.636 and 0.868 respectively for the young and old birds. Subsequently, the added influence of FFA content was incorporated, generating the functions described in Table 12.4(b). This approach accounted for considerably more of the variation in AME values (0.816 and 0.925 respectively for young and old birds). Its use, however, fails to account for the possible interaction detected between U/S and FFA (Figure

Table 12.4   PREDICTION OF THE DIETARY ENERGY VALUES OF FATS AND THE PROPORTION OF THE VARIATION (PV) ACCOUNTED FOR

(A) Ratio of unsaturated to saturated fatty acids (U/S)

   AME = A + B × e (C × U/S)

Age 1		PV	Age 2		PV
A	34.832	0.636	A	37.046	0.868
	± 1.433			± 0.791	
B	−25.091		B	−11.994	
	± 8.914			± 2.480	
C	− 0.932		C	− 0.675	
	± 0.421			± 0.242	

(B) U/S and free fatty acids (FFA, g/kg fat) additively

   AME = A + B × FFA + C × e(D × U/S)

Age 1		PV	Age 2		PV
A	38.112	0.816	A	39.025	0.925
	± 1.418			± 0.557	
B	− 0.009		B	− 0.006	
	± 0.002			± 0.001	
C	−15.337		C	− 8.505	
	± 2.636			± 0.746	
D	− 0.509		D	− 0.403	
	± 0.186			± 0.088	

(C) Based upon U/S and FFA interactively

   AME = (A + B × FFA + C × U/S × FFA) × 1 − E × e (D × U/S)

Age 1		PV	Age 2		PV
A	37.262	0.823	A	38.934	0.927
	± 1.573			± 0.629	
B	− 0.014		B	− 0.007	
	± 0.003			± 0.001	
C	0.00096		C	0.0002	
	± 0.0006			± 0.0002	
D	− 0.616		D	− 0.446	
	± 0.260			± 0.099	
E	− 0.379		E	0.217	
	± 0.010			±0.025	

12.8). Accordingly a third function was derived and is presented in Table 12.4(c). This accounted for 0.823 and 0.927 of the variability in AME values respectively for young and old birds.

   The functions described in Table 12.4(c) are, however, not without their problems. Detailed analysis of them reveals that those for different levels of FFA converge at very low and very high U/S (particularly with younger birds) which, in all probability, would not be the case. A likely explanation for this anomaly is that the data set employed did not have points at the extremes of U/S. The overall analysis thus represents a good example of the dangers of extrapolating derived functions outside the range of the data set employed in their derivation. In addition, on a wider issue, it may not be appropriate to consider applying functions derived to those fats containing saturated fatty acids of chain length shorter than those of the current study as these may be associated with improved availability. In this context, the content of saturated fatty acids used to calculate the ratio could be replaced by the sum of palmitic and stearic acids. The data set employed did, however, cover the range of U/S likely to be found in practice and, as such, the analysis should prove valuable in estimating the AME values of fats for poultry.

Acknowledgements

I would like to acknowledge the support of Federico Salvador and those involved in the technical support of this programme together with Jim Craigon who advised on modelling of the responses.

References

Artman, N.R. (1964) Interactions of fats and fatty acids as energy sources for the chick. *Poultry Science*, **43**, 994–1004
Carew, L.B., Machemer, R.H., Sharp, R.W. and Foss, D.C. (1972) Fat absorption by the very young chick. *Poultry Science*, **51**, 738–742
Fedde, M.R., Waibel, P.E. and Burger, R.E. (1960) Factors affecting the absorbability of certain dietary fats in the chick. *Journal of Nutrition*, **70**, 447–452
Freeman, C.P. (1969) Properties of fatty acids in dispersions of emulsified lipid and bile salt and the significance of these properties in fat absorption in the pig and the sheep. *British Journal of Nutrition*, **23**, 249–263
Freeman, C.P. (1976) Digestion and absorption of fat. In *Digestion and Absorption in the Fowl* (eds K.N. Boorman and B.M. Freeman), British Poultry Science, Edinburgh, pp. 117–142
Freeman, C.P. (1984) The digestion, absorption and transport of fat – Non-ruminants. In *Fats in Animal Nutrition* (ed J. Wiseman), Butterworths, London, pp. 105–122
Halloran, H.R. and Sibbald, I.R. (1979) Metabolisable energy values of fats measured by several procedures. *Poultry Science*, **58**, 1299–1307
Krogdahl, A. (1985) Digestion and absorption of lipids in poultry. *Journal of Nutrition*, **115**, 675–685
Kussaibati, R., Guillaume, J. and Leclercq, B. (1982) The effects of age, dietary fat and bile salts and feeding rate on apparent and true metabolisable energy values in chickens. *British Poultry Science*, **23**, 393–403
Lessire, M., Leclercq, B. and Conan, L. (1982) Metabolisable energy value of fats in chicks and adult cockerels. *Animal Feed Science and Technology*, **7**, 365–452
Lewis, D. and Payne, C.G. (1966) Fats and amino acids in broiler rations. 6. Synergistic relationships in fatty acid utilisation. *British Poultry Science*, **7**, 209–218
Mateos, G.G. and Sell, J.L. (1981) Influence of graded levels of fat on utilisation of pure carbohydrate by the laying hen. *Journal of Nutrition*, **110**, 1894–1903
Mateos, G.G. and Sell, J.L. (1980) Nature of the extra-metabolic effect of supplemental fat used in semi-purified diets for laying hens. *Poultry Science*, **60**, 1925–1930
Renner, R. and Hill, F.W. (1960) The utilisation of corn oil, lard and tallow by chickens of various ages. *Poultry Science*, **39**, 849–854
Shannon, D.W.F. (1971) The effect of level of intake and free fatty acid content on

the metabolisable energy and net absorption of tallow by the laying hen. *Journal of Agricultural Science (Cambs.)*, **76**, 217–221

Sibbald, I.R. (1978) The true metabolisable energy values of mixtures of tallow with either soyabean oil or lard. *Poultry Science*, **57**, 473–477

Sibbald, I.R. and Kramer, J.K.S. (1977) The true metaolisable energy values of fats and fat mixtures. *Poultry Science*, **56**, 2079–2086

Sklan, D. (1979) Digestion and absorption of lipids in chicks fed triglycerides or free fatty acids. Synthesis of monoglycerides in the intestine. *Poultry Science*, **58**, 885–889

Whitehead, C.C. and Fisher, C. (1975) The utilisation of various fats by turkeys of different ages. *British Poultry Science*, **16**, 481–485

Wiseman, J. (1984) Assessment of the digestible and metabolisable energy value of fats for non-ruminants. In *Fats in Animal Nutrition* (ed J. Wiseman), Butterworths, London, pp. 277–297

Wiseman, J., Cole, D.J.A., Perry, F.G., Vernon, B.G. and Cooke, B.C. (1986) Apparent metabolisable energy values of fats for broiler chicks. *British Poultry Science*, **27**, 561–576

Wiseman, J. and Lessire, M. (1987a) Interactions between fats of differing chemical content. Apparent metabolisable energy values and apparent fat availability. *British Poultry Science*, **28**, 663–676

Wiseman, J. and Lessire, M. (1987b) Interactions between fats of differing chemical content. Apparent availability of fatty acids. *British Poultry Science*, **28**, 677–691

Wiseman, J. and Salvador, F. (1988) Influence of degree of saturation and interactions between fats on poultry AMEn values. In *Proceedings VI World Association of Animal Production, Helsinki*, p. 323

Wiseman, J. and Salvador, F. (1989) The influence of age, chemical composition and rate of inclusion on the apparent metabolisable energy of fats, to broiler chicks. *British Poultry Science*, **00**, 00–00

Young, R.J. (1961) The energy value of fats and fatty acids for chicks. 1. Metabolisable energy. *Poultry Science*, **40**, 1225–1233

Young, R.J. and Garrett, R.L. (1963) Effect of oleic and linoleic acids on the absorption of saturated fatty acids by the chick. *Journal of Nutrition*, **81**, 321–329

13

THE EVALUATION OF MINERALS IN THE DIETS OF FARM ANIMALS

J.K. THOMPSON and V.R. FOWLER
School of Agriculture, 581 King Street, Aberdeen, UK

Introduction

'Behold now, I have taken it upon me to speak. . . . who am but dust and ashes.'

Abraham; Genesis 18:27

The diversity of elements found in animal tissue reflects the historic availability of such elements as the biochemical pathways developed. The oceans, virtually by definition, contain all the essential elements in dilute solution. However, the distribution of the essential elements on land is much more variable and may well have affected the spread and relative success of terrestrial species. Indeed, in the

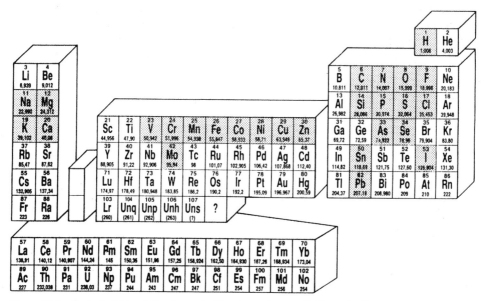

Figure 13.1 Periodic table of the elements. The elements essential for animal life are in the dark boxes

235

restricted nutritional environment of domesticated species, the abundance of a mineral in the feed and its availability to the animal may place severe and unseen limits on its productivity.

At least 26 of the 90 naturally occurring elements are known to be essential for animal life (Figure 13.1). The macroelements are important constituents of tissues with structural and metabolic roles. The trace elements are present in small amounts in animal tissues. For some trace elements information is still insufficient for them to be taken into account in the diets of farm animals. Most, in addition to being essential, are also toxic at high levels of intake. It is clearly important to feed animals quantities of minerals in an acceptable range between these extremes of deficiency and toxicity. In the interests of economy and ecology, the quantities in diets should be close to requirements to minimize waste.

Requirements for minerals have not been easy to define and have become increasingly difficult as the biochemical role of minerals has been better identified. Mertz (1985) stressed the need to define the desirable status for any specific mineral with greater certainty. Desirable status means more than survival of the animal and must provide sufficient minerals to prevent a variety of pathological changes which can develop in the absence of external signs of deficiency. Skeletal mineralization should provide a reserve as well as the minimum mineralization necessary to avoid regular fractures. The wider role of minerals in preventing infectious diseases would need to be accommodated by the quantities supplied by the diet. The minerals in ruminant diets should meet the needs of rumen microbes as well as of the host animal.

The quantities of minerals to meet the requirements of the animal are affected by their availability in the diets and by other dietary constituents which may affect their absorption or metabolism. Mills (1985) has summarized the principal interactions leading to deficiency or toxicity of the trace elements, and Table 13.1 indicates a number of the factors which need to be taken into account when considering whether the dietary concentration of a mineral is adequate.

The mineral nutrition of farm livestock is not well regulated and it has been considered safer to overfeed by appreciable margins. Most foods vary in mineral content and, apart from experimental situations, the availabilities of the minerals is

Table 13.1   DIETARY AND OTHER VARIABLES INCEASING SUSCEPTIBILITY TO DEFICIENCY OR TO TOXICITY

To deficiency		*To toxicity*
Cu	High dietary MO/S or Fe	Stress, with high Cu status; L/SLD*, Low Zn diets
Se	Low vitamin E, high PUFA and high Cu diets	
Co	High concentrate diets	
Zn	Rapid growth, phytate rich feeds (non-ruminants)	Late fetal development, Low Cu/Fe diets; L/SLD
I	Goitrogen-rich diets	
Pb	–	Low Ca/P/S diets; L/SLD
Cd	–	Low Cu status; L/SLD
F	–	Low Ca diets

*L/SLD = liquid/semi-liquid diets (Courtesy of Mills, 1985)

unknown or known only in very broad terms. For these reasons, substantial safety margins have been added to estimates of requirements and this has led to new problems when high concentrations of certain minerals interfere in the absorption and utilization of others, and high rates of excretion result in slurry and manure which pollute drainage water or which affect mineral metabolism in plants. In the United Kingdom, intensive animal production systems have tended to result in the excessive application of potassium to grassland. Concern has been expressed in the Netherlands at the high levels of phosphorus in waterways originating from diets fed to fattening pigs and legislation was proposed in 1987 to limit the amount of phosphorus in manure which can be applied to land.

The threshold between requirement and toxicity is not well defined and is continually being revised. Chase and Sniffen (1982) presented their estimates of the limits within which minerals should be provided to the lactating dairy cow (Table 13.2). These values will not be the same for other species; the maximum safe level for copper is much lower with sheep than with cattle and higher with non-ruminants than with ruminants.

Table 13.2 MINERAL TOLERANCE AND TOXICITY LEVELS FOR LACTATING DAIRY CATTLE (AS % PPM DRY MATTER)

Mineral	Minimum requirement	Maximum safe level	Factor (max ÷ min)
Calcium, %	0.4–0.6	2*	4
Phosphorus, %	0.3–0.4	1*	3
Magnesium, %	0.2	0.5*	2.5
Potassium, %	0.8	3*	3.7
Sodium, %	0.18	1.6*	9
Chloride, %	0.3	2.4*	8
Sulphur, %	0.2	0.4	2
Iron, ppm	50	1000	20
Cobalt, ppm	0.1	20	200
Copper, ppm	10	80	8
Manganese, ppm	40	1000	25
Zinc, ppm	40	1000	25
Iodine, ppm	0.5	50	100
Selenium, ppm	0.1	5	50
Fluorine, ppm	?	30	?
Molybdenum, ppm	1*	6	6

*Estimate
(Courtesy of Chase and Sniffen, 1982)

Analytical techniques for minerals

MINERAL ANALYSES

Mineral elements are determined in plant and animal tissues by a wide range of chemical and physical procedures and, although this is much too large a topic to discuss here, it is appropriate to comment on some of the more recent developments.

ATOMIC ABSORPTION SPECTROSCOPY (AAS)

Atomic absorption has provided the most notable advances in mineral analyses in recent years. It can be applied to more than 60 elements and offers the advantages of excellent sensitivity, speed and convenience. The methods are well suited for routine measurements by relatively unskilled operators, although trace element analysis requires considerable skill and expertise for some elements such as selenium. In atomic absorption at its simplest, an acid solution of the elements is dispersed into small droplets which are then mixed with a gaseous fuel and oxidant and swept into a burner. In the hottest part of the flame, atoms and elementary ions are formed. The vaporized atoms absorb energy to bring about transitions from the ground state to higher energy levels. The source of the radiation is usually a hollow-cathode lamp containing the element under study. These lamps emit radiation with wavelengths characteristic of the element and the absorbed radiation from them can be related to the concentration of the element in the sample.

INDUCTIVELY COUPLED PLASMA EMISSION SPECTROSCOPY (ICP)

This technique has become an established routine analytical tool in the past 10 years. Plasma sources offer several advantages over both flame emission and atomic absorption procedures. Argon ions and electrons form a conducting gaseous mixture or plasma in the high temperatures reached in ICP instruments. Argon ions are capable of absorbing sufficient power from an external source to maintain the temperature at a level at which further ionization sustains the plasma indefinitely and temperatures as great as 10,000°C are encountered. Of the various power sources available, radio frequency or inductively coupled plasma source (ICP) appears to be most sensitive and free from interference. Electromagnetic radiation emitted from atoms and ions of interest within the plasma can be measured simultaneously or sequentially depending on the design of the instrument. Multi-element assays can readily be carried out; a typical rate for a sequential instrument would be 8–12 elements in approximately three minutes.

The determination of major minerals – calcium, magnesium, phosphorus, potassium, sodium – and trace elements – copper, zinc, iron, manganese cobalt, molybdenum – in plant or animal tissue presents no major problems. Relative to AAS, ICP is less sensitive for the alkali metals but has a greater sensitivity for relatively inert elements, such as molybdenum. The detection limit for selenium by hydride generation is noticeably higher than the equivalent system by AAS and the latter would be the preferred technique.

NEUTRON ACTIVATION ANALYSIS (NAA)

Neutron activation analysis techniques have been developed for application in human medicine and their use in animal science has been secondary to this major function. The technique depends on a source of neutrons which may be a radioactive source (for example ^{241}Am-Be) or a 14 MeV sealed tube neutron generator. Following irradiation of the sample and thermal neutron capture a number of elements can be determined from their characteristic gamma emissions.

At the East Kilbride Centre, East *et al.* (1984) have demonstrated the application of NAA in assessing various elements in animal tissues. These have included nitrogen, calcium, phosphorus, sodium and chlorine. Although much of their work has been done with small amounts of homogenized tissue, data for large animals up to 80 kg could be obtained from the existing equipment although live animals would clearly need caging and anaesthetizing. Since the procedures are non-invasive and could be carried out in living animals they offer the attraction of sequential studies on the same animals in investigations of bone materials or electrolyte balance.

Whineray *et al.* (1980) described the application of NAA to the phosphorus nutrition of sheep. Their equipment was designed to irradiate a hind leg of the sheep which was anaesthetized with the leg located in polythene blocks. The systems was transportable and suitable for use in field surveys. Serial changes in leg bone phosphorus could be determined with a precision of 13% but this precision could be improved with additional detectors.

CHOICE OF EQUIPMENT

Investment in analytical equipment must be considered in relation to the scale and complexity of the mineral problem being investigated. All modern laboratories will utilize atomic absorption equipment in addition to other more varied equipment and methods. ICP equipment will cost about twice that of AAS and will increase the range and rate of mineral analyses. NAA is essentially a research tool and its application to animals is usually dependent on equipment established for investigations of mineral problems in man.

BODY COMPOSITION MEASUREMENTS

Changes in the mineral content of the whole body or of specific tissues are frequently required in studies where the retention of minerals is being measured following dietary changes.

The NAA technique already described is an approach which is of value for certain elements but which is not generally available. By far the most common procedure is serial slaughter followed by analyses of the whole body or appropriate tissue. Some tissues can, however, be sampled in living animals. The most successful of these in mineral studies are liver and bone biopsies. Rogers and Poole (1977) used liver biopsies to assess the responses of copper deficient calves and young cattle to copper injections. Little (1972) and Cohen (1975) have discussed bone biopsy in the study of phosphorus status in cattle and sheep. With cattle, rib bone was drilled with a trephine to provide a sample of 700–1000 mg fresh bone. With sheep a section of the entire rib was resected usiung bone cutting forceps. Serial sampling of the same individuals was possible and the technique was claimed to be a sensitive experimental tool for the assessment of skeletal mineral reserves.

Availability

For many years, the concept of nutrient availability has been extended to include many different types of measurement of nutrient utilization and there is a tendency to use different terms such as digestibility, availability and absorbability interchangeably.

Ideally, according to Southon *et al.* (1988), a full assessment of mineral availability from a diet or food should include the measurement of total content, an estimate of the proportion of the total that is in an absorbable form (i.e. maximum theoretical), the actual amount absorbed and finally the proportion of the total which is utilized by the body. The different terms have been defined by Thompson (1964).

Table 13.3 SOME COMMONLY EMPLOYED METHODS OF EXPRESSING PERCENTAGE DIETARY MINERAL INTAKE 'AVAILABLE' TO THE BODY

Term used in the present text	*Equation* $\times$ *I/100*	*Equations expressed in terms of net retention (R)* $\times$ *I/100*
% Apparent digestibility	$= I - F$ (1)	$= R + U$ (5)
% True digestibility	$= I - (F - Fe)$ (2)	$= R + U + Fe$ (6)
% Net retention (R)	$= I - (F + U)$ (3)	
% Availability	$= I - (F - Fe) - (U - Ue)$ (4)	$= R + Fe + Ue$ (7)

where I = intake of mineral
   F = total excretion of mineral in faeces
   U = total excretion of mineral in urine
   Fe = minimum faecal endogenous excretion
   Ue = excretion in urine at zero net retention

(After Thompson, 1964)

The term which describes the maximum absorbable proportion is true digestibility (Equations 2 and 6 in Table 13.3) measured in conditions where intake is below requirement and this is the most useful term in describing the degree to which an element can be extracted from different dietary sources. Although this is the term preferred in this paper much of the pig and poultry data on minerals is concerned with the proportions of the mineral utilized within the body and best described by Equations 4 and 7 in Table 13.3. This approach recognizes the efficiencies of utilizing minerals after absorption and shifts the emphasis from food to animal factors. Availabilities determined by repletion studies (discussed below) compare the proportions of a mineral utilized from different diets or under different physiological circumstances and usually measure retention and biological availability relative to a standard reference source of the mineral.

The measurement of true digestibility or availability depends on an estimate of faecal endogenous loss.

The endogenous excretion of minerals adds much to the complexity of the measurement of true digestibility or availability since it represents the flow of minerals from the animal back into the lumen of the gut and minerals secreted in this fashion cannot be separated chemically from undigested mineral.

Minimum endogenous excretion of minerals should not be confused with the faecal endogenous loss and the former fraction, which includes the urinary endogenous excretion, gives a measure of the net requirement for maintenance. As with most nutrients minimum endogenous excretion has commonly been related to liveweight or metabolic liveweight. It is now accepted that the endogenous excretion of some minerals can be affected by the quantity and nature of the diet

(Braithwaite, 1982; TCORN, 1990), and sometimes by the amount of minerals consumed. Endogenous faecal phosphorus, for example, increases in proportion to the dietary intake of phosphorus. For elements such as phosphorus the view of the ARC (1980) was to measure the minimum endogenous losses by extrapolating to zero intake. There is some debate as to whether this will lead to a satisfactory estimate of the faecal endogenous component of maintenance.

Availability is a major stepping stone towards establishing dietary requirements for a mineral and these can be determined by dividing the net requirement for the mineral by its availability:

$$\text{Dietary requirement} = \frac{\text{Net requirement for tissue} + \text{endogenous losses}}{\text{Availability}}$$

The measurement of availability

The methods used to determine availability of minerals in animal feeds can be divided into two broad classes – balance methods and repletion methods. Within each class there are numerous variations. Suttle (1983) has reviewed these procedures in detail.

BALANCE PROCEDURES

Balance methods include digestibility and retention trials with the endogenous losses estimated by various isotopic labelling techniques.

The term (I-F), that is intake minus faecal loss, is common to all four equations in Table 13.3 and, if the other components are minor or can be accurately predicted, then availability can be calculated from relatively simple digestibility or retention data. In most cases endogenous losses need to be directly measured to allow a confident estimation of availability and this is frequently made by the isotope dilution technique. In this procedure an isotope of the mineral being studied is administered parenterally and the specific radioactivity is measured in the plasma and in the faeces over an appropriate period (usually 8 days). The ratio of the specific activities in faeces to that in plasma is determined by integrating the areas under the two specific radioactivity curves plotted against time (Figure 13.2) and expressing one as a proportion of the other:

$$\text{Faecal endogenous mineral} = \frac{\text{Faecal SR} \times \text{Total faecal mineral}}{\text{Plasma SR}}$$

In variations of the isotope dilution technique, stable isotopes have been used and these have the advantage over radioisotopes with very short half-lives in that they do not decay and that they are generally safer to use. However, they are technically more complex to assess and susceptible to many of the same problems as with radioisotopic procedures (Buckley, 1988). Balance methods have been reviewed by Duncan (1958), by Hegstead (1976) and by Mertz (1987) and they point out the tendency of balance trials to be biased towards falsely high retention. It is difficult to believe that discrepancies between retentions measured by balance trials and

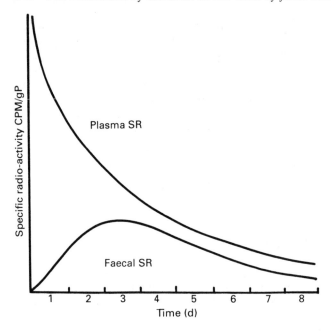

Figure 13.2   The estimation of faecal endogenous loss of phosphorus in sheep.
The ratio of faecal SR to plasma SR is determined by measuring the areas under
the curves (After Scott, 1989)

those from slaughter trials could arise from error alone although the inevitable
small errors in balance trials (spilled feed, incomplete collection of excreta) lead to
an overestimation of retention. However, the order of difference is often much
greater than can be explained by these reasons. Mertz (1987) attributed some of the
defects of balance trials to rapid changes in the body pool size of certain elements
which directly affects the rate of excretion of the element.

Nielsen (1972) performed balance experiments using pigs from 20 to 90 kg
liveweight and analysed the same pigs for their mineral content at the end of the
experiment. He found the retentions from the balance experiment were 39 and
15% higher respectively for phosphorus and calcium than from comparative
slaughter. Similar discrepancies in calcium and phosphorus have been found with
growing lambs between retentions estimated by a balance trial and those from
comparative slaughter (Wan Zahari, 1988).

There is an obvious need for independent methods to estimate retention data and
results obtained by balance procedures must be viewed with some caution.

In spite of these difficulties, balance data are widely used to estimate
availabilities and requirements of minerals, and in defence of these procedures,
there is evidence that requirements for major minerals based on balance trials and
isotope dilution studies are in reasonable accord with estimates from practical trials
(Thompson and Scott, 1990).

Balance methods are recognized to be unsuitable where retentions are very small
since the analytical errors in determining intake and faecal loss although perhaps
small in themselves may be a very large proportion of the retained quantity. For
these minerals, repletion techniques are the preferred methods for estimating
availability.

REPLETION TECHNIQUES

In repletion techniques increments of the mineral under investigation are added to the diet of deficient or depleted animals and their response is measured by a recognized clinical reponse or the concentration in a suitable tissue. This technique will provide a relative measure of availability and if the treatments include a standard form of the element (usually a soluble inorganic salt) the bioavailability of different sources of the element can be compared to a standard. The technique can also be used to assess requirement. Dewar (1986) describes a comprehensive trial with turkey poults fed from hatching to 3 weeks of age with diets supplemented with zinc oxide (Figure 13.3).

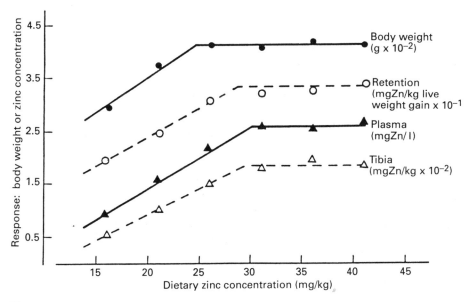

Figure 13.3 Responses of body weight, zinc retention and zinc concentration in deproteinized plasma and dried defatted tibia to increasing concentrations of dietary zinc in turkey poults (After Dewar and Downie, 1984)

The response criteria were body weight, which is commonly used to assess trace element responses in growing poultry, net retention of zinc and the concentrations of zinc in plasma and in the tibia.

The inflection points in the dose-response curves were at 24 mg Zn/kg for maximal body weight, and 29 mg/kg for net retention. Dewar (1986) inferred that the turkey poult has little ability to store zinc once the requirement for maximum body weight has been attained. In support of this the maximal values for tibial and plasma zinc were similar to those for net retention. Suttle (1983) emphasized that repletion should always be measured in pools which have a functional significance and stated that non-functional pools will give low estimates of availability.

The technique has a number of difficulties in its implementation. The responses to repletion are often sigmoid rather than linear reflecting various pools of the element for homeostatic responses in the animal. It is important that the unknown and standard slopes should be linear and have a common intercept for a slopes-ratio assay to be valid. This can sometimes be achieved by restricting the comparison to a linear segment of the curve or by using log transformations and other statistical strategies. Finally, it is necessary that nutrients other than the element being studied should not be limiting animal performance.

The repletion technique has been used to assess the biological availability of iron in calf diets (Bremner and Dalgarno, 1973), and in poultry diets (Davis, Norris and Kratzer, 1968). In both species, the haemoglobin response was measured in depleted animals. Suttle (1974) has found the plasma response of hypocupraemic ewes to dietary copper to be sensitive to differences in copper availability resulting from antagonism from molybdenum and sulphur.

CHEMICAL AND *IN VITRO* METHODS OF DETERMINING AVAILABILITY

Because of the high cost and difficulties of animal experiments there have been many attempts to determine the availabilities of minerals using chemical methods or methods simulating some part of the digestive process.

The chemical methods have measured the solubility of different sources of minerals in water or in dilute acids such as citric and hydrochloric. Although solubility can be an important limitation in the absorption of a mineral it is not necessarily a good indicator of availability. Calcium pyrophosphate for example is soluble in 0.4 per cent HC1 but its availability to pigs and chicks is low. Ferric carbonate is virtually unavailable to the chick or rat compared to ferric ammonium citrate (Fritz *et al.*, 1970). It is apparent that chemical procedures need to be seen to give the same results as those from animal trials and many reports have neglected to make this comparison. Yoshida *et al.* (1979) measured the solubility of phosphate sources in 0.5 per cent citric acid and found a high correlation between solubility and biological availability to chicks. Gueguen (1977) used a method to measure availability of phosphorus to sheep in which the solubility in water was determined followed by solubility in 2% citric acid.

The availability of copper in fractions of sheep digesta was assayed by Price and Chesters (1985) using rats. Samples of digesta were fed to copper depleted rats and the recovery of cytochrome C oxidase in the mucosa of the small intestine was measured.

Several studies have examined the disappearance rates of minerals from nylon bags suspended within the rumen of sheep or cattle (Playne *et al.*, 1978; Rooke *et al.*, 1983; Stone–Wigg, 1983). The rumen is not however an important site of absorption for many minerals and solubilities may well be increased in the acid digesta of the abomasum or proximal small intestine.

In spite of apparent successes of some chemical and *in vitro* procedures, most solubility measurements are regarded as unsatisfactory in predicting the availability of minerals from different sources. De Groote (1983) reported that *in vitro* tests explains less than 60% of the variation in bioavailability of phosphorus in different foods and supplements.

FACTORS INFLUENCING AVAILABILITY

Mineral elements exist in three forms in digesta, as mineral ions in solution, metallo-organic complexes in solution and constituents of insoluble substances.

The first group is readily absorbed and the third group is not absorbed at all. The metallo-organic complexes, some of which are chelates may in some cases be absorbed.

For some minerals there are opportunities for conversion from one form to another so, broadly speaking, the availability of an element will depend on the form in which it occurs in food and the extent to which conditions in the gut favour conversion from one form to another. Thus sodium and potassium which occur in digesta almost entirely as ions have availabilities close to 100%. Copper occurs almost entirely as soluble or insoluble complexes and its availability is generally less than 10%. Phosphorus is present in many foods as phytic acid and its availablity depends on the presence of phytases – of microbial or animal origin – in the digestive tract. A potent factor controlling the interconversion of soluble and insoluble forms of mineral elements is the pH of the digesta. In addition there may be specific agents which bind mineral elements and thus prevent their absorption, for example calcium with oxalates, copper with sulphide and/or molybdenum.

The availability of minerals appears to be lower in older animals than in young animals. The decline may be in part an artifact, due to changes in the diet of animals as they age and also may reflect the relative over-supply of minerals in the diet as animal requirements decline. Because the availability of the minerals in a particular food depends so much on other constituents of the diet and on the animal to which it is given, average values for individual foods are of limited significance.

Selected availability values in mineral sources and feedstuffs

PHOSPHORUS SOURCES

Mineral feed phosphates are derived from rock phosphates by various procedures and the major product is dicalcium phosphate which accounts for 70% of the Western European market. In general the orthophosphates are well absorbed and utilized by all species. Differences in availability between different forms of calcium orthophosphate can be largely explained by the heat treatments applied during the drying stages. High drying temperatures lead to an increase of the pyrophosphate and metaphosphate radicals which are of lower availability (Houseman, 1984).

Phosphorus availability in pigs seems to be similar to that in chicks but the values for ruminants are markedly different, especially from plant sources. The evidence indicates that the soluble orthophosphates are highly available to ruminants and many of the phosphate salts and defluorinated rock phosphates, with the exception of soft phosphate with colloidal clay, have been found to be satisafactory sources (Cohen, 1975). The availability of phosphorus from plant sources is relatively unaffected by phytate in ruminant diets (discussed below) and in forages. Although there are few direct measurements, there is evidence that a very large proportion of the total phosphorus becomes soluble in the digesta (Stone–Wigg, 1983). Values in straw and tropical grasses are low.

The availability values shown in Table 13.4 are maximum values relative to monosodium or dicalcium phosphate. The efficiency of absorption can be affected by drying and processing procedures and can be influenced by other nutrients and by a variety of animal factors such as age, pregnancy and lactation.

Phosphorus is included in many mineral-vitamin supplements for farm livestock and is usually the most expensive component of these supplements. Chandler

Table 13.4   AVAILABILITY OF PHOSPHORUS RELATIVE TO MONOSODIUM
PHOSPHATE OR DICALCIUM PHOSPHATE IN DIFFERENT SPECIES

	Chicks and broilers		*Pigs*		*Ruminants*	
Orthophosphates	87–112	(1)	88–100	(4)	100	(5)
Metaphosphates	0–28	(1)			70–97	(5)
Rock Phosphates	25–87	(1)				
Soft Phosphates	0–49	(2)			28	(2)
Phytate phosphorus			25–40	(2)	50–66	(2)
Maize	16	(3)	18	(4)		
Wheat	50	(3)	48	(4)	88	(6)
Barley	40	(3)	32	(4)		
Oats			30	(4)		
Bran	23	(3)	44	(4)		
Soyabean meal	28	(3)	26	(4)		
Cottonseed meal			2–21	(9,4)		
Sunflower			3	(9)		
Fishmeal	98	(1)			100	(7)
Meat & bone meal	90	(1)				
Temperate grasses					72	(6)
Alfalfa			100	(4)	69	(8)
Tropical grasses					44–66	(8)

1. De Groote (1983). Slope-ratio of mineralization of skeleton or parts of it (tibia, toes).
2. Peeler, H.T. (1972). Relative percentage of tibia ash in chicken; adjusted true digestibility in ruminants.
3. Cromwell, G.L. (1980). Slope-ratio studies (chick data only).
4. Jongbloed, A.W. (1987). Slope-ratio technique on bone ash or breaking strength for most data.
5. Chicco *et al.* (1965). Oral isotope technique.
6. TCORN (1989). True digestibility adjusted by × 1.25 assuming monosodium phosphate is absorbed with 80% efficiency by ruminants.
7. Suttle (1983). True digestibility adjusted by × 1.25.
8. Playne, M.J. (1977). True digestibility adjusted by × 1.25.
9. Burnell *et al.* (1988). Relative breaking strengths of leg bones.

(1989) estimated that in a mineral-vitamin supplement for dairy cows, five nutrients accounted for 90% of the total premix cost. These were as percentages of total costs, Phosphorus (54), Vitamin E (13), Calcium (10), Magnesium (8) and Vitamin A (6).

The availability of the element in different sources is clearly important in the economical use of phosphorus.

CALCIUM SOURCES

Calcium from inorganic sources appears to be utilized more efficiently than that from plant origin and the data for cattle can be classified into three groups with the calcium from bone meal, mono- or dicalcium phosphate being highly available and those from limestone or defluorinated phosphate being moderately available and those from hays and forages being of lowest availability. The calcium in cereals is in such small amounts that its availability is only of academic interest.

Investigations on the comparative values of different calcium sources for poultry have tended to show rather small differences between many of the minerals in common use (Table 13.5). Calcium in the soft phosphate sources has been reported

Table 13.5 AVAILABILITY OF CALCIUM FROM VARIOUS SOURCES RELATIVE TO CALCIUM CARBONATE

	Chicks		*Ruminants*	
Calcium carbonate	100	(1)	100	(2)
Ground limestone	102	(1)	85–90	(3)
Monocalcium phosphate			125–135	(3)
Dicalcium phosphate	100–113	(1)	95–125	(3)
Bone meal	109	(1)	138	(2)
Soft phosphate	69	(1)		
Alfalfa hay			78	(2)
Lespedeza			90	(2)
Orchard grass			98	(2)
Temperate forages			75	(4)

1. Peeler, H.T. (1972) Slope-ratio technique on bone ash or weight gain.
2. Hansard, Crowder and Lyke (1957). Cited by Peeler, H.T. (1972).
3. Chase, L.E. and Sniffen, C.J. (1982).
4. TCORN (1990). Based on an average true digestibility of 68% and assuming 90% of calcium carbonate is absorbed.

to have a low availability and in gypsum its availability is about 90% of that in calcium carbonate; in dolomitic limestone the average availability is 66% (Peeler, 1972).

Roland (1980) reported there was no difference between good quality limestone and oystershell in promoting egg shell quality in laying hens. Both should contain 38–39% calcium.

The absorption of calcium is regulated to a large extent by animal requirements and is inversely related to intake. It is particularly important with this element that true digestibilities or availabilities are measured where the diet is deficient and in many experiments low availability values reflect a high intake of the element and do not adequately indicate the potential of the food.

Calcium absorption is a closely regulated process involving the action of several hormones (parathyroid, calcitonin and 1.25 dihydroxy cholecalciferol). Interactions of other nutrients in the digesta can reduce availability generally as a result of the formation of insoluble complexes such as the phytates and oxalates. Insoluble phosphates are formed at wide ratios of calcium and phosphorus and can interfere with the absorption of phosphorus or of calcium. Age and lactation can profoundly affect the efficiencies of the absorption process and, in growing pigs for example, the apparent availability of a readily extractable source of calcium falls from 85% at 5 kg liveweight to 45% at 90 kg liveweight (Mudd and Stranks, 1981). The lower availability may also indicate a poor assessment of pig requirements for calcium as it increases in weight and is given an excessive supply of the element relative to need.

MAGNESIUM SOURCES

A variety of different techniques have been used to measure magnesium availability. Some methods have been based on blood plasma concentrations and some have been based on regression of urinary magnesium excretion against magnesium intake. Reservations have been made about these procedures and the balance trial with isotope dilution to measure endogenous losses appears to be the most widely accepted procedure.

The availability of dietary magnesium is considered to be about twice as high for simple stomached animals as for ruminants with values varying from 60% to 95% for young calves, rats or guinea pigs (Peeler, 1972). Dietary requirements for poultry are met with diets containing 0.4–0.6 g/kg (Simons, 1986) compared to dietary concentrations of 1.5–2 g/kg required by ruminants. Excess magnesium is more of a problem with practical diets for poultry rather than deficiency and the maximum tolerable level appears to be about 3 g/kg of feed (Simons, 1986).

Approximate availabilities of magnesium in ruminant diets from different sources are indicated in Table 13.6. These are true digestibilities measured by the balance trial technique and other methods and the same compounds or foods fed to poultry would have much higher values.

Table 13.6 APPROXIMATE AVAILABILITY OF MAGNESIUM FROM VARIOUS SOURCES FOR RUMINANTS

Source	Approximate availability (%)
Magnesium oxide	50–60
Magnesium chloride	40–60
Magnesium sulphate	40–70
Magnesium carbonate	45–75
Dolomitic limestone	10–20
Forages	10–30
Concentrates	30–40

(Courtesy of Chase and Sniffen, 1982)

Magnesium availabilities show much variation between feeds and, in cereal diets, it appears to be absorbed much more efficiently than from forages. The availability of magnesium from springs herbage is very low, frequently below 10%, and a high nitrogen content or a high potassium level can reduce magnesium availability further. Dutch studies (Committee on Animal Nutrition, 1973) have defined the relationships between the magnesium, potassium and nitrogen contents of spring grass and the likely incidence of hypomagnesaemic tetany. A variety of factors can interfere with magnesium absorption by ruminants (Reid and Horvath, 1980) including the energy supply, the protein to energy ratio, a variety of interfering ions and natural complexing compounds in diets and conservation and drying treatments.

Most of these effects are not well quantified and this has led to overfeeding of magnesium. High rates of excretion of magnesium in urine are frequently responsible for urinary calculi formation which can be particularly dangerous with young male lambs.

Parker *et al.* (1989) have recently reported on the availability of magnesium from calcined magnesite from different sources. Both temperature of calcination and particle size have important effects and may explain differences between different commercial products. In recent years there have been advances in understanding the mechanism of magnesium absorption in ruminants. Martens (1983) showed that magnesium absorption from the rumen is a saturable process and the transport system is believed to be a sodium-linked process with an energy supply derived from volatile acids (Martens and Rayssiguier, 1979).

SULPHUR

Sulphur deficiencies are not often evident in the diets of farm animals except in ruminant diets where a major part of the dietary nitrogen is supplied by urea. Sulphur deficiency has been found in urea-fed sheep where it has led to poor wool growth. The ratio of nitrogen to sulphur in animal tissue is 15:1 but Moir (1970) observed improved performance in sheep when the dietary ratio was narrowed to 9.5:1. Wool has a higher sulphur content and a nitrogen to sulphur ratio of 5:1. Sulphur deficiencies in ruminant diets can lead to deficiencies of the sulphur amino acids, cystine or methionine, whereas in non-ruminants such deficiencies may arise from dietary proteins of low biological value. Sulphur from sodium, potassium, calcium or magnesium sulphates has availabilities of 50–90% compared with elemental sulphur at 30–50% and methionine analogue at 65–80% (Chase and Sniffen, 1982). Sulphates are reduced in the rumen to sulphides and absorbed through the rumen wall. In non-ruminants sulphate is rapidly absorbed from the intestinal tract without altering its form.

Ruminants have a narrow threshold of tolerance for sulphur (see Table 13.2) compared with non-ruminants and, at higher levels of intake, sulphur will markedly affect copper absorption.

AVAILABILITIES OF TRACE ELEMENTS

The availabilities of trace minerals for chicks have been reviewed by Dewar (1986), and are summarized in Table 13.7. Availabilities have been determined for copper, manganese and zinc by measuring mineral retention over a short balance period and estimating endogenous losses with a purified diet (Nwokolo *et al.*, 1976; Aw–Yung *et al.*, 1983). The availability of zinc has also been determined from the growth response to test feeds added to low zinc diets, compared with zinc carbonate additions (O'Dell *et al.*, 1972). The availability of selenium is relative to its ability to prevent exudative diathesis compared to a similar weight of sodium selenite. Sodium selenite is assigned an availability of 100% in this comparison (Cantor *et al.*, 1975). These authors also investigated the availability of selenium in different inorganic and organic sources and concluded that selenium in sodium selenite, sodium selenate and selenocystine was more available than that in sele-nomethionine, sodium selenide, seleno-ethionine or selenopurine for the prevention of exudative diathesis.

The availabilities of the sulphate, chloride or nitrate salts of most of the trace elements are high, reflecting their solubility in the digestive tract but the availability of iron salts cannot be predicted from solubility alone and Fritz *et al.* (1970) has reported that iron in ferric carbonate is of very low availability in chick diets (2%) compared with ferrous sulphate (100%).

The availabilities of copper in ruminant diets are very much lower than the corresponding values for poultry (Table 13.8). The absorption coefficients or availabilities were measured by the plasma response in copper depleted sheep and different dietary sources were compared with the plasma response to infused

Table 13.7 AVAILABILITIES (AS PERCENTAGE OF TOTAL) OF COPPER, MANGANESE ZINC AND SELENIUM IN SOME FEEDSTUFFS FOR CHICKS

	Cu^*	Mn^*	Zn^*	$Zn\dagger$	$Se\ddagger$
Barley	78	55	49	–	–
Maize	87	60	58	–	86
Rice	–	–	–	62	–
Triticale	90	54	57	–	–
Wheat	77	53	49	59	71
Maize germ	–	–	–	55	–
Cottonseed meal	42	76	38	–	86
Palm kernel meal	45	46	14	–	–
Rapeseed meal	62	57	58	–	–
Sesame meal	–	–	–	59	–
Soyabean meal	51	76	67	67	60
Egg yolk	–	–	–	79	–
Fish meal	–	–	–	75	16–25
Dried skimmed milk	–	–	–	82	–
Oysters	–	–	–	95	–
Meat and bone meal	–	–	–	–	15

*Nwokolo, Bragg and Kitts (1976) and Aw–Yong, Sim and Bragg 1983);
†O'Dell, Burpo and Savage (1972);
‡Cantor, Scott and Noguchi (1975).
(After Dewar, 1986.)

Table 13.8 COPPER AVAILABILITY IN DIFFERENT FOODS FOR SHEEP

	Total Cu (mg/kg)	Absorption coefficient	Absorbable Cu (mg/kg)
Summer pasture	6.5	0.025	0.16
Autumn pasture	8.5	0.012	0.10
Silage	8.2	0.049	0.40
Hay	5.5	0.073	0.40
Roots	2.5	0.068	0.17
Kale/rape	3.9	0.132	0.51
Cereals	3.0	0.100	0.30

(Courtesy of Suttle, 1983.)

copper (Suttle, 1974). Copper absorption or utilization can be reduced by a variety of dietary factors. Suttle (1983) has discussed two of the more important ones – molybdenum and sulphur which can result in the formation of insoluble complexes in the rumen such as copper thiomolydbates. Iron and zinc have also been shown to reduce significantly the absorption of copper (Campbell *et al.*, 1974; Towers *et al.*, 1981). The interactions of these metals can be of particular use in the control of copper toxicity in sheep (Humpries *et al.*, 1983; Kincaid and White, 1989).

Bremner and Dalgarno (1973) found that the availability of iron in the soluble sulphate or citrate form was 72% in the diets of calves and Ammerman (1967) has reported that the oxide was of low availability to iron depleted calves (less than 20%).

Zinc availability in ruminant diets ranges from 16 to 51%, with an average of about 30% (ARC, 1980).

Cobalt is converted in the digestive tract of ruminants to vitamin B12 and the vitamin may be absorbed with an efficiency of up to 33% (Gardner 1977). The efficiency of conversion of cobalt to vitamin B12 depends on the cobalt content of the diet and may be as high as 13% in cobalt deficient sheep (ARC, 1980).

Phytate and mineral availability

Phytate is a common constituent of seeds and tubers and it has been known for 50 years to affect the availability of calcium, phosphorus and magnesium and in more recent years a number of trace elements.

Phytic acid (inositol hexaphosphate) readily forms complexes and mixed salts with calcium, magnesium, potassium, and trace elements such as zinc, iron and manganese. It will also complex with protein (Nelson, 1967). The primary role of phytates in plants is probably storage of phosphorus. Phytates are formed in grains as they mature and can be markedly reduced when seeds are soaked or allowed to germinate.

Certain phytates such as the calcium and iron salts are insoluble, unreactive and unavailable to animals (Bremner and Dalgarno, 1973; Morris and Ellis, 1976). The minerals of various phytates can be well absorbed such as the iron in monoferric phytate (Morris and Ellis, 1976) and the phosphorus in wheat bran is well utilized by pigs.

The availability of phytate minerals depends on the nature of these complexes and the activity of the enzyme phytase which hydrolyses phytates to yield inorganic phosphorus. Phytase is found in the seeds of most plants and in the intestines of some animals but its activity is very variable.

PHYTASES

It is generally accepted that wheat has a powerful phytase activity, barley less so and maize has little phytase activity. Williams and Taylor (1985) showed that, with rats fed on a diet low in phytase, the hydrolyses of phytate was low (5%) but when the diet contained wheat nearly half the phytate was hydrolysed. Plant phytases have optimum activity at pH 5.0 (Hill and Tyler, 1954) and are believed to be irreversibly inactivated at pH 2.5 or lower so they probably do not survive acid conditions of the stomach.

Intestinal phytases are produced by the microbial flora of the intestine and can be contributed to by phosphatases of animal origins including intestinal alkaline phosphatase (Davies and Flett, 1978). Intestinal phytase activity does not seem to be of great significance in pigs (Pointillart *et al.*, 1984; Williams and Taylor, 1985).

It is established that phytate is less of a problem in ruminants than in non-ruminants and that it is hydrolysed to a large extent in the reticulo-rumen so that the phosphorus of grains and seeds is readily utilized by ruminants (Reid *et al.*, 1947).

Bacteria present in the hind gut can hydrolyse phytates but whether the pig or chick can benefit from phosphorus released here is still uncertain.

Several metal ions affect phytase activity. Since insoluble complexes can be formed during digestion, diets high in calcium have depressed phytase activity, partly through the low solubility of the substrate and partly due to a direct unexplained effect of calcium.

Cooper and Gowing (1983) demonstrated responses in phytase activity to $ZnCl_2$ or $MgCl_2$ and concluded that the intestinal phytases were affected by the same factors as alkaline phosphatase but to a greater extent. This is a zinc containing enzyme which decreases in activity in many tissues during zinc deficiency (Kirchgessner and Roth, 1980); so zinc deficient animals may digest phytate less efficiently than normal animals.

Vitamin D has been shown to stimulate the activity of intestinal phytase in pigs and to have resulted in improved calcium absorption and increased intestinal calcium-binding protein (Pointillart and Fontaine, 1986). This suggests that in diets high in phytate, calcium availability may be depressed and calcium-binding protein may play an important role in adaptation.

The phytate and non-phytate content of various feedstuffs are shown in Table 13.9. It can be seen that the phytate phosphorus fraction is frequently in excess of 60% of the total phosphorus but this fraction does not correspond with unavailable phosphorus although many nutritionists and statutory bodies make this assumption (Houseman, 1984).

Table 13.9   TOTAL PHOSPHORUS PHYTATE PHOSPHORUS AND PHOSPHORUS AVAILABLE TO PIGS AND POULTRY

	Total P (%)	Phytate P as % total P	Available P as % total P
Grains:			
Maize	0.27	69	14
Wheat	0.37	71	48
Barley	0.38	65	37
Oats	0.37	60	23
Protein sources:			
Soyabean meal (50%)	0.61	61	21
Cotton seed meal (41%)	1.07	70	2–21
Sunflower meal			3
Other sources:			
Wheat bran	1.41	80	29
Rice bran	1.50	–	18

(Based on data from Houseman, 1984; Cromwell, 1980 and Burnell *et al.*, 1988.)

Phytases have given encouraging results when they have been added to the diet of chicks (Nelson *et al.*, 1971). Phytase from *Asperillus ficuum* was incubated with the feed and quantities and times could be regulated to hydrolyse all of the phytate phosphorus. Conclusive results have not been found with pigs when a yeast culture was added to a maize-soyabean meal diet (Cromwell and Stahly, 1987; Shurson *et al.*, 1984).

Chelates, sequestering agents and ligands

Metals react with many organic molecules to form soluble complexes and some of these are called chelates. A chelate is a heterocyclic ring generally containing 5–6 atoms including a metal ion and, according to Kincaid (1989), chelating agents must possess at least two groups each capable of donating an electron pair and located so that a ring structure is formed with the metal atom.

Naturally occurring chelates have been identified in various reactions in the utilization of minerals; for example, certain bacteria secrete chelating agents into the environment to bind iron; microbes also make cobalamins, including vitamin B12. Animals synthesize a porphyrin ring which forms complexes with iron and other metals, and ionophores are compounds which react with small cations to transport them across membranes. Synthetic ionophores may affect mineral metabolism in animals. Greene *et al.* (1988) have reported on a response in magnesium absorption to dietary monensin. The most interesting of the chelates appears to be the metal amino-acid chelates which results from the reaction of an amino acid with a soluble metal salt (Atherton, 1989).

There is no longer any doubt that various metal chelates and complexes play important roles in the absorption and utilization of minerals by animals. It is also evident that some synthetic metal chelates are effective in enhancing absorption and animal performance. There are, however, doubts about the usefulness of some of the chelates including the strongly bonded complexes with EDTA and DTPA.

Suttle (1983) found that a number of soluble chelated forms of copper (CuEDTA, CuDTPA, CuTETA and CuDOS) were no more available to sheep than the inorganic salts, basic copper carbonate or cuprous chloride.

There are also uncertainties on the mechanism of absorption of metal chelates and Kincaid (1989) was of the opinion that information on mechanisms and animal performance is badly needed. Generally an increase in bioavailability of an element will not justify the cost of a chelated element; there must also be an improvement in performance or safety.

When a metal is bonded by two atoms in the same ligand molecule a double heterocyclic ring structure may result. An example is the dipetide chelate of copper formed with glycine (Figure 13.4).

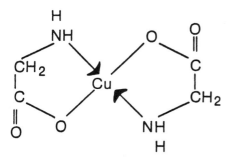

Figure 13.4   The structure of a dipeptide-like chelate of copper (After Ashmead *et al.*, 1985)

Ashmead *et al.*, (1985) have demonstrated that the amino-acid chelates and peptide chelates are absorbed by rat intestinal tissue and the intestinal uptake is markedly greater with the chelates than with the mineral salts (Table 13.10).

It was argued that the dipeptide-like amino acid chelates are absorbed into the mucosa by a different pathway than the carrier-dependent active mechanism

Table 13.10   INTESTINAL UPTAKE OF DIFFERENT MINERAL FORMS

Control		Animal protein chelate	Plant protein chelate	Whey chelate	Carbonate	Sulphate	Oxide
Copper	Trace	33	35	17	12	8	11
Magnesium	7	94	57	52	77	36	23
Iron	23	298	130	80	82	78	61
Zinc	14	191	191	126	87	84	66

Data are expressed as ppm of metal in dry mucosal tissue.
Rats adequately nourished with minerals were killed and 2 cm jejunal segments were prepared. Samples were incubated with 50 micrograms of metal as a salt or chelate for 2 minutes. Chelates were made using ligands from fishmeal, casein or soy protein.
(Courtesy of Ashmead *et al.*, 1985.)

Table 3.11   SERUM FERRITIN CONCENTRATIONS IN SOWS (ng/ml)

	Number of sows	4 weeks prior farrowing	At farrowing	At weaning
Control	4	7.3 ± 7.1	2.0 ± 2.6	3.0 ± 2.6
Dipeptide like iron amino acid chelate	4	13.0 ± 9.6	16.5 ± 10.4	11.5 ± 8.3
Ferrous fumerate	4	7.3 ± 6.7	4.5 ± 3.7	5.3 ± 4.7

(mean ± SD)
(Courtesy of Ashmead *et al.*, 1985)

identified for most minerals, and it was believed that intact chelates were absorbed by the dipepetide absorption pathway.

Dipeptide-like metal chelates have been shown to result in responses in appropriate tissues indicating that they are functional within the body.

Yamamota *et al.* (1982) (cited by Ashmead *et al.*, 1985) fed groups of gestating sows 600 mg of iron daily in addition to their regular diet, as either ferrous fumarate or a dipeptide-like amino acid chelate. Serum ferratin concentrations were measured and showed marked increases with the chelate treated sows from four weeks prior to farrowing right through to weaning (Table 13.11).

Zinc methionine supplementation has resulted in some unexplained benefits in animal performance which Kincaid (1989) suggests may be due to in part to greater uptake of methionine and also a zinc-β carotene and vitamin A interaction. Cows fed zinc methionine during the dry period and early lactation had higher plasma vitamin A and β carotene concentrations than those without supplements (Kincaid, 1989; Moore *et al.*, 1988). No benefits were found in similar trials with beef cows fed wheat straw/haylage based diets (Gibson and Males, 1988).

The topic of chelated minerals in diets is somewhat controversial but there is increasing evidence that they can make a useful contribution to animal nutrition. The evidence of benefit following a treatment needs to be considered with care since some products are of limited value. Chelates will not compete economically with inorganic sources of minerals but may have special applications with minerals which are toxic at low doses and in particular disorders of animals.

Conclusion

It is disturbing to have to concede in a paper on mineral nutrition that of all the constraints used in diet formulation, those concerning the minerals are the least exact. The precision of modern methods of chemical analysis for the elements themselves is obscured by the uncertainties about the exact numbers for availability to the animal. Even then, there is an even greater degree of uncertainity about the so-called net requirement of the animal. There is a tremendous tolerance in animals to fluctuations in mineral supply, and they are capable of strong minerals at well above the concentration which is the minimum for acceptable performance. In practice this uncertainty is met by ensuring that the supply of minerals is well in surplus to the requirement. This technique which has served so well in the past may not be regarded so favourably in the future. There is now considerable concern about the indiscriminate spread of minerals in the agrarian environment particularly in relation to phosphorus, copper and selenium. In addition costs, which in earlier times were regarded as trivial, have now become significant because of the much greater precision with which diets are formulated for other nutrients. It is clearly becoming more important to the diet manufacturer and to the consumer that our understanding of mineral nutrition should be refined and that it should no longer be regarded as a minor detail in the giant industry of diet formulation.

References

Agricultural Research Council (1980) *The Nutrient Requirement of Ruminant Livestock* Slough, UK, Commonwealth Agricultural Bureau

Ammerman, C.B., Wing, J.M., Dumavant, B.G., Robertson, W.K., Feaster, J.P. and Arrington, L.R. (1967) Utilisation of inorganic iron by ruminants as influenced by form of iron and iron status of the animal. *Journal of Animal Science*, **26**, 404–410

Ashmead, H.D., Graff, D.J. and Ashmead, H.H. (1985) *Intestinal Absorption of Metal Ions and Chelates*, Springfield, Illinois, Charles C. Thomas

Atherton, D. (1989) Personal Communication

Aw–Young, L.M., Sim, J.S. and Bragg, D.B. (1983) Mineral availability of corn, barley, wheat and triticale for the chick. *Poultry Science,* **62**, 659–664

Braithwaite, G.D. (1982) Endogenous faecal loss of calcium by ruminants. *Journal of Agricultural Science (Cambridge)*, **99**, 355–358

Bremner, I. and Dalgarno, A.C. (1973) Iron metabolism in veal calves. The availability of different iron compounds. *British Journal of Nutrition*, **29**, 229–243

Buckley, W.T. (1988) The use of stable isotopes in studies of mineral metabolism. *Proceedings of the Nutrition Society*, **47**, 407–416

Burnell, T.W., Cromwell, G.L. and Stahly, T.S. (1988) Bioavailability of phosphorus in high-protein feedstuffs for pigs. *Journal of Animal Science*, **66**, Suppl 1: 317, (Abstract).

Campbell, A.G., Coup, M.R., Bishop, W.H. and Wright, D.E. (1974) Effect of elevated iron intake on copper status of grazing cattle. *New Zealand Journal of Agricultural Research*, **17**, 393–399

Cantor, A.H., Scott, M.L. and Noguchi, T. (1975) Biological availability of selenium in feedstuffs and selenium compounds for prevention of exudative diathesis in chicks. *Journal of Nutrition*, **105**, 96–105

Chandler, P. (1989) Actual cost factors in mineral, vitamin nutrition defined. *Feedstuffs*, **61**, 16 and 45.

Chase, L.E. and Sniffen, C.J. (1982) Minerals in dairy cattle nutrition. *Proceedings 1982 Feed Dealer Seminars*, No. 66, pp. 45–57

Chicco, C.F., Ammerman, C.B., Moore, J.E., Van Walleghem, P.A., Arrington, L.R. and Shirley, R.L. (1965) Utilisation of inorganic ortho-, meta- and pyro-phosphates by lambs and by cellulolytic rumen micro-organisms *in vitro*. *Journal of Animal Science*, **24**, 355–363

Cohen, R.H.D. (1975) Phosphorus and the grazing ruminant. *World Review of Animal Production*, **11**, 27–43

Committee on Mineral Nutrition of the Hague (1973) *Tracing and Treating Mineral Disorders in Dairy Cattle*. Wageningen, Centre for Agricultural Publishing and Documentation

Cooper, J.R. and Gowing, H.S. (1983) Mammalian small intestinal phytase (EC 31.3.8.). *British Journal of Nutrition*, **50**, 673–678

Cromwell, G.L. (1980) Biological availability of phosphorus for pigs. *Feedstuffs*, **52**, 38–42

Cromwell, G.L. and Stahley, T.S. (1978) Study finds live yeast ineffective for swine use. *Feedstuffs*, **50**, 12

Davis, P.N., Norris, L.C. and Kratzer, F.H. (1968) Iron utilisation and metabolism in the chick. *Journal of Nutrition*, **94**, 407–417

Davies, N.T. and Flett, A.A. (1978) The similarity between alkaline phosphatase and phytase activities in rat intestine and their importance in phytate-induced zinc deficiency. *British Journal of Nutrition*, **39**, 307–316

De Groote, G. (1983) Biological availability of phosphorus in feed phosphates for broilers. *Proceedings of the Fourth European Symposium on Poultry Nutrition*, pp. 91–100, Tours

Dewar, W.A. (1986) Requirements for trace minerals. In *Nutrient Requirements of Poultry and Nutritional Research*. Poultry Science Symposium no. 19, (eds C. Fisher and K.N. Boorman), Butterworths, London, pp. 155–171

Duncan, D.L. (1958) The interpretations of studies of calcium and phosphorus balance in ruminants. *Nutrition Abstracts and Reviews*, **28**, 695–715

East, B.W., Preston, T. and Robertson, I. (1984) The potential of *in vivo* neutron activation analysis for body composition measurements in the agricultural sciences. In *In vivo Measurement of Body Composition in Meat Animals* (ed D. Lister) Amsterdam, Elsevier Applied Science Publishers, pp. 134–138

Fritz, J.C., Pla, G.W., Roberts, T., Boehne, J.W. and Hove, E.L. (1970) Biological availability in animals of iron from common dietary sources. *Journal of Agricultural and Food Chemistry*, **18**, 647–651

Gardner, M.R. (1977) *Cobalt in Ruminant Nutrition. A Review.* Technical Bulletin No. 36. Department of Agriculture, Western Australia

Gibson, M.L. and Males, J.R. (1988) Effect of crude protein level with or without zinc methionine on performance of first calf beef heifers. *Journal of Animal Science*, **66**, Suppl; 195 (Abstract)

Greene, L.W., May, B.J., Schelling, G.T. and Byers, F.M. (1988) Site and extent of apparent magnesium and calcium absorption in steers fed monensin. *Journal* of Animal Science, **66**, 2987–2991

Gueguen, L. (1977) A propos du controle de la qualité du phosphore des composes mineraux. *Elèvage porcin*, **65**, 33–35 (cited by Jongbloed 1987)

Hansard, S.L., Crowder, H.M. and Lyke, W.A. (1957) The biological availability of calcium in feeds for cattle. *Journal of Animal Science*, **16**, 437–443

Hegstead, D.M. (1976) Balance studies. *Journal of Nutrition*, **106**, 307–311

Hill, R. and Tyler, G. (1954) The influence of time, temperature, pH and calcium carbonate on the activity of the phytase of certain cereals. *Journal of Agricultural Science (Cambridge)*, **44**, 306–310

Houseman, R.A. (1984) Phosphorus, some aspects of phosphorus supply to farm livestock. *The Feed Compounder*, **4**, 15–18

Humphries, W.R., Phillippo, M., Young, B.W. and Bremner, I. (1983) The influence of dietary iron and molybdenum on copper metabolism in calves. *British Journal of Nutrition*, **49**, 77–86

Jongbloed, A.W. (1987) *Phosphorus in the Feeding of Pigs*. Drukkerij De Boer Lelysted

Kincaid, R.L. and White C.L. (1988) The effects of ammonium tetrathiomolybdate intake on tissue copper and molybdenum in pregnant ewes and lambs. *Journal of Animal Science*, **66**, 3252–3258

Kincaid, R.L. (1989) Availability biology of chelated, sequestered minerals explored. *Feedstuffs*, **61**, 22 and 58

Kirchgessner, M. and Roth, H.P. (1980) Biochemical changes of hormones and metallonenzymes in zinc deficiency. In *Zinc in the Environment. Part II. Health Effects* (ed J.O. Nhagu), Chichester, John Wiley and Sons, pp. 71–103

Little, D.A. (1972) Bone biopsy in cattle and sheep for studies of phosphorus status. *Australian Veterinary Journal*, **48**, 668–670

Martens, H. and Rayssiguier, Y. (1979) Magnesium metabolism and hypomagnesaemia. In *Digestive Physiology and Metabolism in Ruminants* (eds Y. Ruckebusch and P. Thivend), Lancaster, MPT Press, pp. 447–466

Martens, H. (1983) Saturation kinetics of magnesium efflux across the rumen wall in heifers. *British Journal of Nutrition*, **49**, 153–158

Mertz, W. (1985) Limitations to current estimates of trace element requirements and tolerance. In *Trace Elements in Man and Animals* (eds C.F. Mills, I. Bremner and J.K. Chesters), Slough, CAB, p. 895

Mertz, W. (1987) Use and misuse of balance studies. *Journal of Nutrition*, **117**, 1811–1813

Mills, C.F. (1985) Trace elements in the context of feedingstuffs evaluation. In *Feedingstuffs Evaluation – Modern Aspects – Problems – Future Trends* (ed R.M. Livingstone), Aberdeen, Rowett Research Institute, pp. 108–114

Moir, R.J. (1970) In *Sulphur in Nutrition* (ed O.H. Muth), West Port, CT. AVI

Moore, C.L., Walker, P.M., Jones, M.A. and Webb, J.W. (1988) Zinc methionine supplementation for dairy cows. *Journal of Dairy Science*, **71**, Suppl. 1, 152, (Abstract)

Morris, E.R. and Ellis, R. (1976) Isolation of monoferric phytate from wheat bran and its biological value as an iron source to the rat. *Journal of Nutrition*, **106**, 733–760

Mudd, A.J. and Stranks, M.H. (1981) Mineral and trace element requirements of pigs. In *Recent Advances in Animal Nutrition 1981* (ed W. Haresign), London, Butterworths, pp. 93–107

Nielsen, A.J. (1972) Deposition of calcium and phosphorus in growing pigs determined by balance experiments and slaughter investigation. *Acta Agriculturae Scandinavica*, **22**, 223–237

Nelson, T.S. (1967) The utilisation of phytate phosphorus by poultry – A review. *Poultry Science*, **46**, 862–871

Nelson, T.S., Shich, T.R., Wodzinski, R.J. and Ware, J.H. (1971) Effect of supplemented phytase on the utilisation of phytate phosphorus by chicks. *Journal of Nutrition*, **101**, 1289–1293

Nwokolo, E.N., Bragg, D.B. and Kitts, W.D. (1976) A method for estimating the mineral availability in feedstuffs. *Poultry Science*, **55**, 2217–2221

O'Dell, B.L., Burpo, C.E. and Savage, J.E. (1972) Evaluation of zinc bioavailability of feedstuffs of plant and animal origin. *Journal of Nutrition*, **102**, 653–660

Parker, E.R., Ritchie, N.S. and Hemingway, R.G. (1989) Dietary availability and rumen solubility of calcined magnesite source. *Proceedings of the Nutrition Society*, **48**, 7A

Playne, M.J. (1977) Availability of phosphorus in feedstuffs for utilisation by ruminants. *Reviews in Rural Science*, **III**, 155–164

Playne, M.J., Echevarria, M.G. and Megarrity, R.G. (1978) Release of nitrogen, sulphur, phosphorus, calcium, magnesium, potassium and sodium from four tropical hays during their digestion in nylon bags in the rumen. *Journal of the Science of Food and Agriculture*, **29**, 520–526

Peeler, H.T. (1972) Biological availability of nutrients in feeds: availability of major mineral ions. *Journal of Animal Science*, **35**, 695–712

Pointillart, A. and Fontaine, N. (1986) Effects of vitamin D in calcium regulation in vitamin D-deficient pigs given a phytate-phosphorus diet. *British Journal of Nutrition*, **56**, 661–669

Pointillart, A., Fontaine, N. and Thomasset, M. (1984) Phytate phosphorus utilisation and intestinal phosphatases in pigs fed low phosphorus wheat or corn diets. *Nutrition Reports International*, **29**, 473–483

Price, J. and Chesters, J.K. (1985) A bioassay for Cu availability and its application in studies on the distribution of available Cu in ruminant digesta. In *Trace Elements in Man and Animals* (eds C.F. Mills, I. Bremner and J.K. Chesters) Slough, CAB, pp. 682–685

Reid, R.L., Franklin, M.C. and Hallsworth, E.G. (1947) The utilisation of phytate phosphorus by sheep. *Australian Veterinary Journal*, **23**, 136–140

Reid, R.L. and Horvath, D.J. (1980) Soil chemistry and mineral problems in farm livestock. A review. *Animal Feed Science and Technology*, **5**, 95–167

Rogers, P.A.M. and Poole, D.B.R. (1977) The effects of Cu-EDTA injection on weight gain and copper status of calves and fattening cattle. TEMA-3. *Proceedings of the 3rd International Symposium* (ed M. Kirchgessner), Freising-Weihenstephen; Inst. Ernahrung Sphysiol, pp. 481–485

Roland, D.A. (1980) Limestone, oystershell particles help eggshell quality equally well. *Feedstuffs*, **52**, 31

Rooke, J.A., Akinsoyinu, A.O. and Armstrong, D.G. (1983) The release of mineral elements from grass silages incubated *in sacco* in the rumens of Jersey cattle. *Grass and Forage Science*, **38**, 311–316

Scott, D. (1989) Personal Communication.

Shurston, G.C., Ku, P.K. and Miller, E.R. (1984) Evaluations of a yeast phytase product for improving phytate phosphorus bioavailability in swine diets. *Journal of Animal Science*, **59**, Suppl. 1, 106 (Abstract).

Simons, P.C.M. (1986) Major minerals in the nutrition of poultry. In *Nutrient Requirements of Poultry and Nutritional Research* Poultry Science Symposium No. 19, (eds C. Fisher and K.N. Boorman), London, Butterworths, pp. 141–154

Southon, S., Fairweather–Tait, S.J. and Hazell, T. (1988) Trace element availability from the human diet. *Proceedings of the Nutrition Society*, **47**, 27–35

Stone–Wigg, R.A. (1983) Effect of grass species and conservation method on calcium phosphorus and magnesium solubilisation in the rumen of sheep. BSc Honours Thesis. University of Aberdeen.

Suttle, N.F. (1974) A technique for measuring the biological availability of copper to sheep using initially hypocupraemic ewes. *British Journal of Nutrition*, **32**, 395–405

Suttle, (1983) Assessment of the mineral and trace element status of feeds. In *Feed Information and Animal Production* (eds G.E. Robards and R.G. Packham), Slough, CAB

TCORN (1990) A reappraisal of the calcium and phosphorus requirements of sheep and cattle. *Nutrition Abstracts and Reviews* (In Press)

Thompson, A. (1964) Availability of minerals in food of plant origin. *Proceedings of the Nutrition Society*, **24**, 81–88

Thompson, J.K. and Scott, D. (1990) Recent developments in calcium and phosphorus nutrition in ruminants. *Journal of the Society of Food and Agriculture* (In Press)

Towers, N.R., Young, P.W. and Wright, D.E. (1981) Effect of zinc supplementation on bovine plasma copper. *New Zealand Veterinary Journal*, **29**, 113–114

Wan Zahari Mohamed (1988) Factors affecting body composition and mineral retention in growing lambs. PhD Thesis, University of Aberdeen.

Whineray, S., Thomas, B.J., Ternouth, J.H. and Davies, H.M.S. (1980) A transportable system for the determination of phosphorus in sheep bone by *in vivo* neutron activation analysis. *International Journal of Applied Radiation and Isotopes*, **31**, 443–445

Williams, P.J. and Tayler, T.D. (1985) A comparative study of phytate hydrolysis in the gastrointestinal tract of the golden hampster (*Mesocricetus auratus*) and the laboratory rat. *British Journal of Nutrition*, **54**, 429–435

Yoshida, M., Ishikawa, M., Nakajima, H. and Hotta, S. (1979) Solubility of phosphorus in citric acid solution as an index of biological availability. *Japanese Poultry Science*, **16**, 290–292 (Cited by Jongbloed, 1987).

14

EVALUATION OF VITAMIN CONTENT IN INGREDIENTS AND COMPOUND DIETS

M.E. TENNESON and K.R. ANDERSON
Peter Hand Animal Health Limited, Stanmore, Middlesex, UK

Introduction

This chapter attempts to *evaluate* the vitamin content of ingredients and compound diets. The key question is 'What does evaluation mean?' A dictionary definition of evaluate is 'to appraise, or find value of'. Therefore in considering vitamins all their various forms and the combined activity within the animal must be considered. As far as the animal is concerned this is the only true evaluation and methods of analysis of vitamins become critical in their evaluation. By selecting four vitamins and illustrating the various methods of analysis used an attempt will be made to show how difficult vitamin evaluation is.

What vitamin content do we generally attribute to feed ingredients?

Table 14.1 lists values for various vitamins in some common feed ingredients as given by the National Research Council (1985). The figures generally reflect very low concentrations and yet they appear precise. How accurate are they? What level of analytical variation can be expected? What variation in vitamin contribution is seen in feed ingredients? Do these values give an evaluation of vitamins? If the determination of data such as these is examined these questions can be answered.

There are two main areas of analytical expertise, namely chemical and biological. The important chemical techniques used in vitamin analysis are spectroscopic and chromatographic.

The spectroscopic methods used are mainly ultraviolet and visible (colorimetric).

Ultra violet based methods are mostly used for relatively pure compounds. Visible (colorimetric) based methods generally involve the reaction of the compound of interest with a reagent which gives rise to a colour, the intensity of which is measured. Limitation of this technique is mainly the interference with other compounds of similar formulae. The chromatographic methods used are high performance liquid chromatography (HPLC) and gas chromatography (GC). Both techniques are very specific for the quantitative determination of particular compounds. However, GC will not determine thermally unstable compounds.

The important biological techniques used in vitamin analysis are:

Table 14.1 VITAMIN CONTENT OF FEED INGREDIENTS

Raw material	A (i.u/%)	Carotene (mg/kg)	E (mg/kg)	B1 (mg/kg)	B2 (mg/kg)	B6 (mg/kg)	B12 (µg/kg)	Pantothenic additive (mg/kg)	Nicotinic additive (mg/kg)	Folic (mg/kg)	Cholesterol (mg/kg)	Biotin (mg/kg)
Cereals:												
Barley	–	2	25	5	1.8	7.3	–	9.1	94	0.6	1177	0.17
Wheat	–	–	17	4.8	1.6	5.6	1	11.4	64	0.5	1085	0.11
Maize	–	3	25	3.8	1.4	5.3	–	6.6	28	0.3	567	0.08
Oats	–	–	15	7.1	1.7	2.8	–	8.8	16	0.4	1116	0.31
Rye	–	–	17	4.2	1.9	2.9	–	9.1	21	0.7	479	0.06
Cereal by-products:												
Brewers grains	–	1	29	0.7	1.6	0.8	–	8.9	47	7.7	1757	0.68
Distillers grains	–	8	31	2.6	7.1	6.0	–	12.2	49	0.2	2584	–
D. grains and solution	–	3	43	3.1	10.0	5.4	–	15.3	79	1.0	2803	0.85
Wheat middlings	–	–	26	17.1	2.3	10.3	–	19.4	103	1.1	1387	0.27
Wheat bran	–	3	21	7.9	4.6	9.6	–	33.5	268	1.6	1797	0.32
Oatfeed	–	–	–	0.7	1.9	2.4	–	3.4	10.0	1.0	281	–
Pulps:												
Sugar beet	–	–	–	0.4	0.8	–	–	1.5	18	–	902	–
Citrus	–	–	–	1.6	2.5	–	–	15.4	24	–	867	–
Animal protein:												
Bloodmeal	–	–	–	0.4	2.2	4.8	49	2.6	34	0.1	854	0.09
Meat and bone	–	–	1	0.2	4.9	9.4	116	4.4	53	0.4	2196	0.11
Fishmeal (H)	–	–	24	0.4	11.0	5.2	467	18.2	93	0.4	5752	0.52
Whole milk	11.6	–	–	3.9	20.6	4.9	–	23.8	9	–	–	0.4
Skimmed milk	–	–	10	3.9	20.5	4.5	54	38.6	12	0.7	1480	0.35
Whey	0.5	–	–	4.3	29.4	3.6	20	49.6	11	0.9	1921	0.38
Vegetable protein:												
Cotton	–	–	17	7.3	5.2	6.2	–	15.0	45	1.5	3056	1.06
Lucerne	–	131	121	3.7	14.1	7.7	–	32.4	40	4.8	1494	0.36
Soya bean	–	–	3	6.2	3.2	6.7	–	18.2	31	0.7	2915	0.36
Sunflower	–	–	12	3.4	4.2	14.8	–	43.9	288	–	4430	–
Molasses	–	–	7	1.2	3.8	5.7	–	50.3	49.0	0.1	1012	0.92

1. Microbiological.
2. Animal.

The microbiological methods are Agar Diffusion or Turbidimetric analysis. In the case of vitamins the promotional effect of the growth to micro-organisms is a measurement of the amount of vitamin present. Unfortunately these methods are non-specific and are affected by the presence of other vitamins, leading to inaccuracies.

The use of animals to measure vitamin content is not advisable because of its inherent variability. This is unfortunate because animal response is the ultimate evaluation.

How do these methods help us to evaluate feed ingredients and compound diets? The four selected vitamins can be examined in more detail and they are vitamin A, vitamin E, vitamin B6 and vitamin K.

Vitamin A (retinol)

Vitamin A usually occurs in nature as the fatty acid ester. Six known isomers of vitamin A have been isolated of which only two are of practical importance, namely 'all-trans vitamin A' and the 13-cis isomer or 'neovitamin A'.

Natural sources contain about one-third neovitamin A while commercial products are mostly all-trans. Neovitamin A has a relative biological activity of 75% all-trans vitamin A. Commercial forms are added as retinol acetate or palmitate.

However, there are a series of compounds called provitamins A which are transformed in the body of the animal to vitamin A. There are in fact many of these provitamins A, the most common being beta-carotene and its derivatives.

Vitamin A is determined by three main methods: ultraviolet light absorption (British Pharmacopoeia, 1980), the Carr–Price method (AOAC, 1984), and HPLC (Peter Hand Animal Health, 1982).

Determination of pure vitamin A or simple concentrate mixtures can be carried out using the ultraviolet light absorption method. However, this is unsuitable for complex matrices such as feed and feed ingredients, and in these cases either the Carr–Price method or HPLC are used. These methods do not differentiate between the isomers and measure the total retinol content after saponification of any esters. HPLC is preferable since it does not suffer from any interference and is more robust than the Carr–Price method.

None of these methods measure provitamin A and therefore do not evaluate total vitamin A activity.

Vitamin E

Natural vitamin E comprises a series of compounds known as 'tocopherols' of which at least eight are known. The most important is α-tocopherol. Commercial forms are added as α-tocopherol acetate.

Vitamin E is determined by ultraviolet light absorption (British Pharmacopoeia, 1980), colorimetry (AOAC, 1984) and HPLC (Peter Hand Animal Health, 1982). Determination of pure vitamin E or simple concentrate mixtures can be carried out using ultra violet absorption. However, this is unsuitable for complex matrices such as feed and feed ingredients, in which case either the colorimetric or HPLC methods are used. These methods do not differentiate between the various isomers and measure the total alpha-tocopherol content after saponification of any esters. HPLC as in vitamin A, does not suffer from interference in the same way as the colorimetric method.

Although analysis of vitamin E appears a closer estimation of true vitamin E activity than was the case for vitamin A, the different activities of the various tocopherols means that again *analysis does not equate to evaluation.*

Vitamin B6

Vitamin B6 can be present as pyridoxine (alcohol), pyridoxal (aldehyde) and pyridoxamine (amine) and they all form salts with B6 activity.

The commercially used preparation is almost exclusively pyridoxine hydrochloride.

Vitamin B6 is determined by ultraviolet light absorption (Association of Vitamin Chemists, 1966), microbiological (AOAC, 1984) and HPLC (Peter Hand Animal Health, 1982). In pure solutions and high potency products of simple composition pyridoxine can be determined by ultraviolet absorption. However, this is unsuitable for complex matrices such as feed and feed ingedients or varying combinations of the B6 compounds mentioned above when present together. Generally this is applicable for the added pyridoxine hydrochloride only. However, if the other two compounds are also pure they can be measured independently by this method. HPLC has the benefit of separating the three B6 compounds and measurement is possible of all three in a mixture by this method. Unfortunately use of HPLC in complex matrices is limited because of the large interference problems that result. Therefore, with respect to feed or feed ingredients the method of choice is generally microbiological. While the choice of organisms used can measure one or other of the B6 compounds, generally a method is used that measures all three compounds together. Microbiological methods are notoriously inaccurate, and in the case of vitamins have the added complication of the effects of the presence of other vitamins, which can lead to inaccurate estimations.

Unlike vitamins A and E the analysis of vitamin B6 is the biggest limitation to an accurate evaluation and this is obviously further complicated by the presence of three active compounds.

Vitamin K

Vitamin K can be present as vitamin K1, vitamin K2, vitamin K3 and various other derivatives of menadione. Commercial preparations are also derivatives of menadione, generally vitamin K3 derivatives or salts.

Vitamin K is determined by colorimetry (Association of Vitamin Chemists, 1966), GLC (AOAC, 1984) and biological test. In pure solutions of menadione or a menadione derivative, the colorimetric methods can be used. For complex matrices such as feed and feed ingredients, the GLC procedure can be used. As well as this, menadione can be assayed by a biological method involving the curative test in chicks. However, this procedure is extremely variable and no vitamin activity standard has been defined. There is a fundamental difference in activity between menadione and its derivatives, therefore accurate evaluation of vitamin K is probably more difficult than any of the previous examples already mentioned.

Declarations

In practice, the levels of vitamins declared by law are those added (Tables 14.2 and 14.3), and any naturally occurring vitamins in addition to the declaration are considered an overage. The tables show that only three vitamins are required by law (A, D and E). It may not be surprising therefore that the other two examples (B6 and K) are far more difficult to analyse and evaluate since they have received less attention due to their legal status.

Table 14.2 LEGAL DECLARATION: AN EXAMPLE OF SUPPLEMENT/PREMIX DECLARATION (A DOG PRE−MIX)

Each 2.5 kg contains:		
Ash	68.89% calcium	17.74%
Vitamin A	16.00 MIU vitamin D3	2.00 MIU
Vitamin E	65.00 KIU copper	7.00 g

Table 14.3 AN EXAMPLE OF A FEED DECLARATION

Oil 6%
Protein 40%
Fibre 10.0%
Ash 16%

Vitamins and minerals guaranteed present until end of October 1988. Contains:
  Vit A (i.u/kg) 60 000
  Vit D3 (i.u/kg) 12 000
  Vit E (i.u/kg) 30
  Sodium selenite (mg/kg) 0.6
  Copper sulphate (mg/kg) 100

Evaluation of vitamin status of supplements and feed

The vitamins within supplements are of all commercially available forms. Generally speaking, these forms are relatively pure and of a single compound. Therefore, analysis as mentioned previously can be generally carried out by simplified methods that are both reasonably accurate and precise.

Evaluation of supplementary vitamins is the most useful. However, in the case of feeds as have been shown earlier, this can be much further complicated by the type of matrix and the presence of the other forms of vitamins and interfering compounds. Therefore analysis of feeds tends to be for the added compounds rather than the total vitamin contents. This means that the natural vitamins are essentially ignored albeit that they will provide additional activity to the animal.

Conclusions

The ultimate evaluation of a vitamin must be animal performance. However, in this chapter it has been shown that no single vitamin activity can be truly evaluated. At best an estimation of individual vitamin compounds can only be undertaken when a greater understanding of how and what one is measuring is available. Until we have this better understanding it is felt that requirements of the animal must be met by the vitamin supplement.

References

Association of Official Analytical Chemists (1984)*AOAC Official Methods of Analysis* 14th edn. Vitamin A, pp. 832–834; Vitamin E, pp. 855–858; Vitamin B6, pp. 870–873. Arlington, VA, AOAC

Association of Vitamin Chemists (1966) *Methods of Vitamin Assay*.

HMSO (1980) *British Pharmocopoeia* Vitamin A, p A95, Vitamin E, p 21. London, HMSO

National Research Council (1985). *Nutritional Requirements of Dogs*, Washington DC, National Academy Press, pp. 48–57

Peter Hand Animal Health (1982)*Peter Hand Internal Methods of Analysis*; Vitamin A: PH100, Vitamin E: PH102, Vitamin B6: Ph105. Leyland, Peter Hand Animal Health

15

PREDICTION OF THE DIETARY ENERGY VALUE OF DIETS AND RAW MATERIALS FOR PIGS

E.S. BATTERHAM
North Coast Agricultural Institute, Wollongbar, New South Wales, Australia

Introduction

Information on dietary energy content is needed in the formulation of diets of specified energy concentration for a particular class of stock, to allow the formation of amino acids (and other nutrients) relative to energy concentration and to provide adequate energy intake to achieve production targets. It is also desirable to be able to assess the energy content of diets for stockfeed regulation purposes. It follows that there is a need initially to (a) have information on the energy concentration of raw materials to enable accurate diet formulation and (b) a means of assessing the energy concentration of formulated diets to ensure that the desired specifications have been supplied.

Determining the energy content of diets and raw materials

In the past, a number of systems have been used to assess the feeding value of raw materials and diets. These include Total Digestible Nutrients (TDN), Starch Equivalents and Scandinavian Feed Units. These systems have gradually been replaced by the direct measurement of energy in feeds and its expression in terms of Gross, Digestible, Metabolizable or Net energy.

GROSS ENERGY (GE)

This is the energy released on combustion of a feed. It is determined by assessing the heat of combustion of the sample, in ballistic or adiabatic bomb calorimeters. Gross energy is the potential amount of energy in a feed. However, it may bear little relationship to the amount of usable energy.

DIGESTIBLE ENERGY (DE)

This is the amount of energy in a feed after subtracting the energy in the faeces. It is normally determined with pigs housed in metabolism crates and using 5–7 day collection periods. Energy losses in the faeces may account for 10–30% of the gross energy. Most of the loss in digestible energy is due to the fibre component of the raw material or diet. Values are considered to be additive, but this is an over-simplification, as the inclusion level may influence the digestibility of fats (Wiseman and Cole, 1987), and proteins and fibre are over-estimated.

METABOLIZABLE ENERGY (ME)

In addition to losses in digestion, metabolizable energy takes into account the energy lost in urine and gases. Losses from gases in pigs are only small (0.5–1%) and are usually ignored. Energy lost in urine is not constant, but will depend on the surplus amino acid concentration of the raw material or diet. As such, ME values are not additive. In order to overcome this problem, the values are normally corrected to a given nitrogen balance (normally zero, 30% or 50%). The energy lost in urine is relatively small for cereals (2%), higher in protein concentrates (10%) and intermediate in balanced diets (4–5%). As ME values are 95–96% of DE values for balanced diets, there is little difference between the two systems, and both are widely used.

NET ENERGY (NE)

NE accounts for the losses in ME plus the energy lost during the digestion of nutrients (the heat increment of a feed). The amount of heat energy generated during digestion depends on the site of absorption and the metabolic pathway the nutrient follows during digestion. Nutrients such as fats are absorbed in the small intestines and deposited with little modification, generating less heat than fibre, which is fermented in the hind gut and absorbed as volatile fatty acids. NE is thus the energy available for maintenance and production. It is not a characteristic of a particular raw material as such, but more a characteristic of the compound diet. It is measured by feeding a particular diet and determining the energy lost in the heat increment, either by calorimetry or by comparative slaughter technique. Of all the systems, net energy more closely reflects the energy available for production, and is the information most desired by nutritionists when formulating diets or allocating rations. However, it is also the most elusive value to determine.

It is evident from the above that, due to the time and resources involved, it is not normally practical to determine the energy content of raw materials or diets for commercial pig production. Accordingly, considerable research has been directed towards methods of predicting the energy contents of diets and raw materials.

Prediction of the energy content of diets

PREDICTION OF DE AND ME

The prediction of DE and ME in diets has been approached by attempting to define the factors that influence these values. The approach has been twofold:

1. To investigate the relationship between the digestibility of nutrients, and DE and ME content.

2. To determine whether they can be predicted using regression equations involving the various chemical fractions of the diet.

Digestibility studies

Studies have shown that the digestible crude protein, digestible fat, digestible fibre and digestible nitrogen-free extract all influence the DE and ME contents of diets (Hoffman and Schiemann, 1980; Just, 1982). It has been shown that the influence of these four factors of the Weende proximate analysis system of describing feeds account for the majority of variation in DE and ME content. As a consequence, it is possible to use these relationships to predict energy contents of mixed diets (Table 15.1).

Table 15.1 PREDICTION EQUATIONS OF ENERGY VALUE OF SINGLE OR MIXED FEEDS (MJ/kg FEED) FROM DIGESTIBLE NUTRIENTS (g/kg FEED) FOR PIGS (AFTER HENRY *ET AL.*, 1988)

Energy system	Regression coefficients				CV %	R^2	References
	Digestible crude protein	Digestible fat	Digestible crude fibre	Digestible nitrogen-free extract			
DE	0.0242	0.0394	0.0184	0.0170			Hoffmann and Schiemann (1980)
	0.0239	0.0363	0.0211	0.0167	0.8	1.0	Just (1982)
ME	0.021	0.0374	0.0144	0.0171			Hoffmann and Schiemann (1980)
	0.0215	0.0377	0.0197	0.0173	0.4	0.98	Just (1982)

The limitation to the use of these equations is that they are based on a knowledge of the digestibility of these four fractions. It is easier to determine actual DE or ME content of an unknown diet than it is to determine the digestibility of these fractions. Thus to use these relationships as predictors of energy content it is necessary to use reference values for the various fractions. As these have wide variations in diets of unknown composition, this limits the accuracy of such equations.

Alternatively, if the ingredient composition of the diet is known, then it is possible to calculate energy content using reference values for the digestibility of the individual components from the literature. This should be associated with less error than using digestibility coefficients for diets of unknown composition.

Chemical components of the diets

The alternative approach is to predict the energy content of diets based on their crude chemical analyses. A number of investigations have been conducted to define the relationship between chemical composition and energy content. As a result, a considerable number of prediction equations have been developed (for example

Morgan *et al.*, 1975; Wiseman and Cole, 1980; Just *et al.*, 1984; Morgan *et al.*, 1987). These relationships have shown:

1. Crude protein, fat, and nitrogen-free extract (or starch and sugars) contribute positively to the DE and ME content of feeds.
2. Ash tends to act as an energy dilutent and thus has a negative influence.
3. Fibre contributes in a negative manner.

In general there are two main types of prediction equations: (1) those that account for all the chemical fractions that contribute towards the energy content of a diet; and (2) those that have a constant term and include one or more modifiers of this term.

The former are normally based on equations involving crude protein, fat, crude fibre and nitrogen-free extract (Table 15.2). These equations are relatively easy to apply as they are based on the principal components of the proximate analysis system of diet or feeds, which is routinely conducted in many laboratories.

The latter may be simple equations with a constant term, gross energy, to represent the energy components and a chemical constituent (normally an estimator of fibre content) to act as a modifier to gross energy (Table 15.2). Alternatively, more comprehensive equations with a number of chemical constituents may be used (Table 15.2). Obviously, as the number of constituents of the equations increases, so does the likelihood that the equations will have application to diets having a wider range of raw material components. However, the costs associated with the use of these analyses also increases, especially for the less routine types of analyses.

The dilemma facing nutritionists confronted with the wide array of prediction equations that have been developed is to choose the equation that is most useful for predicting the DE and ME of diets. One approach has been to try to ensure that the equations are biologically meaningful and therefore more likely to apply over a wider range of conditions (Batterham *et al.*, 1980a; Fisher, 1982; Wiseman and Cole, 1983; Whittemore, 1987). In this regard Whittemore (1987) proposed a theoretical equation based on the assumption that (a) all the energy in a diet can be accounted for by a knowledge of the crude protein, fat, fibre and starch fractions of a feed and (b) from a knowledge of the GE of these components and their likely digestibility, a theoretical equation could be constructed which describes the relationship betwen these nutrients and overall dietary DE content. For example, if the GE of oil was $39 \, MJ/kg^{-1}$, and the digestibility of energy in the oil 90%, then the coefficient for oil in an equation to estimate DE would be $39 \times 0.9 = 35$. Using this approach Whittemore (1987) estimated the contribution made by each of the four components and constructed the following theoretical equation:

$$DE \ (MJ/kg \ DM) = \ 0.016 \ starch \ (g/kg) + 0.035 \ oil \ (g/kg)$$
$$+ \ 0.019 \ protein \ (g/kg) + 0.001 \ fibre \ (g/kg).$$

Furthermore, Whittemore (1987) proposed that in effective models it was probable that:

1. The constant term would be close to zero if either gross energy or carbohydrate was included in the model.
2. The constant term would be close to the average DE of a diet if either gross energy or carbohydrate was not included in the model.

Table 15.2 PREDICTION EQUATIONS OF ENERGY VALUE FOR DIETS (MJ/kg DM) FROM CRUDE CHEMICAL COMPOSITION (g/kg DM) FOR PIGS (MOST EQUATIONS ARE FROM HENRY et al., 1988)

Energy system	Regression coefficients								rsd (CV*, %)	R^2	References
	Constant	Crude protein	Fat	Crude fibre	Nitrogen free extract	Gross energy	Neutral-detergent fibre	Ash			
DE	−4.54	0.0033				1.177	−0.0168	−0.0185	0.44	0.94	King and Taverner (1975)
	0.662	0.0066				0.986		−0.035	0.33	0.85	Wiseman and Cole (1980)
	5.75	0.0078				0.68	−0.016	−0.0325	0.27	0.887	Perez et al. (1984)
	17.50		0.0157	−0.0442			−0.0149		0.32		Morgan et al. (1987)
ME	5.62	0.0040				0.69	−0.016	−0.0223	0.34	0.956	Morgan et al. (1987)
	5.53	0.0028				0.70	−0.016	−0.034	0.25	0.893	Perez et al. (1984)
	5.74	0.0021				0.69	−0.017	−0.024	0.33	0.960	Morgan et al. (1987)
		0.0203	0.0252	−0.0178	0.0162				0.54 (5.0)	0.76	Just et al. (1984)
		0.0180	0.0315	−0.0149	0.0163				(3.7)	0.8	E.A.A.P. Working Group: provisional equation

*CV = coefficient of variation in %

3. Fat and fibre would have large effects within the equation and ash and protein would have less.
4. The coefficients would bear some relationship to the magnitude of those established in the theoretical equation.

It is interesting that the equations reported in Table 15.1 which account for the four chemical parameters as outlined by Whittemore (1987), have fairly similar coefficients except for fibre, which are higher than postulated.

It is also important to ensure in the development of prediction equations that the range of values used for each chemical constituent is similar to, and covers the same range of values for those constituents in the diets or feedstuffs to be evaluated (Thorne *et al.*, 1989).

Normally it is assumed that individual raw materials are additive in their contribution to DE and most equations only include linear terms. However, this is not strictly correct, as the DE content of fats may be under-estimated and proteins and fibre over-estimated. However, inclusion of quadratic effects for crude fibre and fat had only very small effects in equations developed by Wiseman and Cole (1983). Recently, Morgan *et al.* (1987) also reported small beneficial effects of both interaction terms for a number of nutrients and quadratic terms for ether extract in equations for DE. However, Morgan *et al.* (1987) indicated the adoption of such equations would be a departure from the normally assumed linearity in DE, which is one of the factors in favour of this system. As with the results of Wiseman and Cole (1983), the overall effects were small and appeared to make no substantial improvement over linear equations.

A number of investigations have been undertaken to improve the accuracy of prediction equations by attempting to improve the chemical techniques used to assess the energy contribution of the various components. These investigations have mainly focused on the various techniques for measuring fibre but have also included the oil and soluble carbohydrate components of the diets.

In the Weende system, carbohydrates are divided into crude fibre and nitrogen-free extract substances (Fernandez and Jorgensen, 1986). Attempts have been made to replace these two terms with techniques that measure more nutritionally defined entities. Most effort has been directed towards more appropriate techniques for determining the fibre content. This is not surprising as fibre has a substantial negative effect on digestibility and therefore DE and ME content. For example, Fernandez and Jorgensen (1986) reported that a single fraction (crude fibre or neutral-detergent fibre) accounted for about 70% of the variation in digestibility of diets and about 85% of the variation in raw materials. However, fibre is a broad term for a number of insoluble carbohydrate fractions including cellulose, hemicellulose, lignin, pectin and glucans (Fernandez and Jorgensen, 1986) and is determined by various chemical techniques.

Crude fibre is the indigestible fraction following acid and alkaline hydrolysis. It consists largely of cellulose and lignin. The main criticism of the technique in animal feeding is that it does not represent a defined nutritional entity. Despite this limitation, it is still widely used and many studies have been in its favour.

Neutral-detergent fibre is the residue remaining after extraction with neutral detergent and contains mainly cellulose, lignin and hemicellulose. It is thought to reflect the components of the cell wall of a plant which is considered the indigestible fraction for pigs. For this reason the assay is preferred by many nutritionists as it represents a more nutritionally defined component than crude fibre. However, the assay was originally developed to measure cell walls in forages (Van Soest and Wine, 1967) and has been adapted for use with cereals, grain, legumes, etc. Unfortunately, starches interfere with the filtration process and this has led to difficulties with the precision in the assay for many feeds. To overcome this, pretreatment of feeds rich in starches with enzymes is often undertaken. Nevertheless, because of variability associated with the technique, the application of prediction equations involving this assay, outside the original data set, remains a problem.

Other fibre techniques that have been investigated in prediction equations include acid-detergent fibre (mainly cellulose and lignin which remains after digestion with acid detergent), and hemicellulose. The latter may be calculated as the difference between neutral-detergent fibre and acid-detergent fibre. Neutral-detergent fibre has also been estimated as (nitrogen-free extract + crude fibre) − (starch and sugar) (Fernandez and Jorgensen, 1986).

The majority of studies have shown that all fibre components are negatively related to DE and ME content but there is no clear agreement as to the relative advantages of the different techniques. In some studies neutral-detergent fibre has been superior to crude fibre (e.g. Morgan *et al.*, 1987) but in others, crude fibre has been more efficient (e.g. Just *et al.*, 1984). It is possible that these differences may be due to between-laboratory differences in the estimation of the various fibre fractions used in these studies.

Nitrogen-free extract in the Weende analyses is calculated as 100 − (crude protein + ether extract + crude fibre + ash). It represents largely starch, sugars and hemicellulose. There is some concern with using nitrogen-free extract as an independent variable in equations involving the other components of the Weende system of analyses as it is not independent. Attempts have been made to use more descriptive analyses for the components of nitrogen-free extract in equations by determining starch or starch plus sugar (soluble sugars) in the analyses.

There have also been different techniques examined to estimate the oil and soluble carbohydrate components of diets and raw materials. Oil has been determined by techniques such as ether extraction, or acid hydrolysis prior to ether extraction. The starch and sugar components of the nitrogen-free extract have also been determined by different methods.

Despite these undertakings, there does not seem to be any consistent, clear-cut advantage of any particular assay procedure in prediction equations. Nor does there appear to be any consistent advantage of prediction equations based on a range of analyses compared with those based on the traditional system of proximate analysis (Table 15.2).

PREDICTION OF THE NE CONTENTS OF DIETS

There have been two types of prediction equations developed for assessing the NE content of diets. They are based on the prediction of NE for fattening or for production.

A prediction equation of the NE of fattening (NEF) was developed by Schiemann *et al.* (1971) and Hoffmann and Schiemann (1980). The concept is based on the measurement of NE for fattening in heavy castrated pigs depositing predominantly fat. The NEF was related to the amounts of digestible nutrients according to the Weende system, after making allowances for maintenance energy requirements.

NEF (kJ/kg) = 10.70 digestible crude protein + 35.70 digestible fat +
           12.37 digestible crude fibre +
           12.37 digestible nitrogen-free extract

Prediction of the NE for growth in pigs was developed by Just *et al.* (1983a). It was based on the observation that energy concentration accounted for the largest part of the difference in net energy for maintenance and production in diets fed to growing/finishing pigs.

NE (MJ/kg DM) = 0.8 ME (MJ/kg DM) − 2.0

The equation developed by Just *et al.* (1983a) appears more applicable for estimating the net energy content of diets for growing/finisher pigs, as it was determined with pigs depositing protein. However, neither equation fully accounts for all of the factors effecting NE.

A comprehensive review of the various prediction equations used by the different European countries to predict the DE, ME or NE content of diets was undertaken by Henry *et al.* (1988). Their review has been the basis of the majority of equations that have been included in Tables 15.1 and 15.2.

Prediction of the energy content of raw materials

Whilst considerable work has been undertaken to assess the energy contents of diets, fewer studies have been undertaken to assess the energy content of raw materials. This is surprising considering that it is necessary to know the ingredient composition of raw materials before diets can be formulated to a specific energy concentration. It most probably reflects the interest in developing equations for predicting the energy content of diets for regulation purposes. Of the work that has been undertaken with raw materials there have been three main approaches adopted: (1) the use of reference tables of feed composition, (2) prediction equations based on digestibility coefficients, and (3) prediction equations based on chemical components.

REFERENCE TABLES OF ENERGY VALUES

Reference values for most individual raw materials are available in the literature. In a number of countries these values have been collated into tables of feed composition. The tables have most application if the energy contents are based on determined energy values, the range of raw materials examined is extensive, and adequate descriptions of individual raw materials are included. Such tables have more application when the variation in energy concentration for a particular class of raw material is narrow. This has been the case with cereals in the UK (Wiseman and Cole, 1980) and Australia (Batterham *et al.*, 1980a) where there was little overall variation in the DE content of different samples or varieties within the one type of cereal. Where the variation in energy concentration within a class of raw material is wide (as is the case with weather-damaged wheats, wheat by-products (Batterham *et al.*, 1980a) and meat and bone meals (Batterham *et al.*, 1980b)), reference Tables of Feed Composition have less application. However, the tables are more useful for these meals if it is possible to subdivide the one class of raw material into smaller units according to a particular attribute, i.e. ash content of meat and bone meals, as reported by Just *et al.* (1982).

USE OF PREDICTION EQUATIONS

As with diets, prediction equations have potential for use with raw materials. The factors affecting energy content in diets apply to raw materials, so similar types of prediction equations are used.

Prediction equations based on digestibility coefficients

The use of digestibility coefficients for the different components of the proximate analysis of a feed – for crude protein, crude fat, ash, crude fibre and nitrogen-free extract, has been widespread in the past with the TDN system to describe feeds. TDNs can be converted to DE by the relationship:

1 kg TDN $= 18.43$ MJ DE (Henry *et al.*, 1988).

However, the relationship between TDN and DE is affected by the crude protein content of the feed and separate equations may need to be used for feeds of varying crude protein content (Wiseman and Cole, 1983). Nowadays the TDN system has been replaced by the direct determination of DE, ME or NE. However, a number of values for DE in the literature are based on conversion from TDN values. Care needs to be taken when using these values, as they may not be as applicable as those determined directly.

The equations presented in Table 15.1 for diets also have application for single raw materials. For the latter, it is possible that the digestibility coefficients in the literature for the different nutritional components may be less variable. This is especially so if the factors affecting digestibility within a class of raw material has been defined. For example, Just *et al.* (1982) defined the effects of chemical composition of meat and bone meals on digestibility coefficients.

Prediction equations based on chemical analyses

There have been two approaches to the development of prediction equations based on chemical equations: (1) equations for an individual class of raw materials and (2) equations based on a range of raw materials.

Attempts to develop prediction equations for a number of individual types of feed ingredients have been made by Wiseman and Cole (1980), Batterham *et al.* (1980a, b), Just *et al.* (1983b, 1984) and Lukule *et al.* (1989). With a number of classes of feeds, the variation in the individual type of raw material was not very great and the prediction equations developed were not of sufficient accuracy to have any real advantage over the use of tabulated values (Wiseman and Cole, 1980; Batterham *et al.*, 1980a). In other cases, there has been little relationship established between chemical techniques and DE or ME content (Batterham *et al.*, 1980a; Just *et al.*, 1983b, 1984; Lukule *et al.*, 1989). It is only in cases where the variation in DE content was wide, such as in weather-damaged wheats, wheat by-products and meat and bone meals, that the equations may have application

(Batterham *et al.*, 1980a, b). Examples of some of the prediction equations developed for cereals or cereal by-products are given in Table 15.3. For meat and bone meals, the following relationship between DE (MJ/kg) and chemical constituents (g/kg) may be used (Batterham *et al.*, 1980b):

$$DE = -2.97 + 0.77 \text{ GE(MJ/kg)} + 0.020 \text{ ether extract} + 0.080 \text{ Ca} - 0.159P$$
$$\text{rsd} = 0.53. \ R^2 = 0.89$$

An alternative approach to the use of individual equations for particular classes of raw materials is the use of equations developed over a wide range of feed ingredients (Table 15.3). Similarly, Just *et al.* (1984) developed the following equation between ME (MJ/kg DM) and the proximate analysis (g/kg DM) of 300 samples of vegetable feedstuffs.

$$ME = 0.0197 \text{ crude protein} + 0.0194 \text{ fat} - 0.0125 \text{ crude fibre} + 0.0158$$
$$\text{nitrogen-free extract} \qquad CV = 7.1\%, \ R^2 = 0.83$$

Table 15.3 REGRESSION EQUATIONS FOR DE CONTENT IN CEREALS AND CEREAL BY-PRODUCTS (AIR-DRY BASIS) FOR GROWING PIGS (FROM BATTERHAM *et al.*, 1980a)

	Regression coefficients					*rsd*	R^2
	Constant	*Gross energy (MJ/kg)*	*Density (kg/m³)*	*Crude fibre (g/kg)*	*Acid-detergent fibre (g/kg)*	*Neutral-detergent fibre (g/kg)*		
Barley	5.04	0.32	0.004				0.12	0.73
Wheat	−4.35	1.17		−0.052			0.13	0.94
Weather-damaged wheat	9.67		0.007				0.37	0.79
	−3.92	1.10		−0.024			0.09	0.99
	−1.24	0.95			−0.019		0.08	0.99
Wheat by-products	27.76		−0.016		−0.085		0.45	0.89
All samples	−2.93	1.07		−0.039			0.38	0.88
	−7.52	1.36				−0.012	0.36	0.89

Again, whether these equations have advantages over tabulated values is doubtful.
Predicting the ME content of raw materials using chemical analyses had a substantially greater error term than that based on digestible nutrients (Just *et al.*, 1984). Again, this difference would be expected to be considerably less with the use of tabulated digestibility coefficients for the latter equations.

Prediction of net energy in raw materials

Although net energy is more a characteristic of a diet than a single raw material, it is possible to estimate the potential net energy in a raw material using equations such as those developed by Just *et al.* (1983a) and reported earlier.

Accuracy of equations for predicting energy concentration in diets or raw materials outside the original data set

It is surprising that despite the number of studies that have been conducted to determine the relationship between digestibility of nutrients and/or chemical composition and energy content of diets, there have been relatively few reports of studies to assess the applicability of the prediction equations developed. It is therefore difficult to assess the degree of confidence which may be placed in the use of the equations currently available. This is especially so as the rsd and R^2 values for an individual prediction equation give an estimation of the goodness of fit to the original data, but this does not necessarily indicate the accuracy of the equation for predicting values in diets or feedstuffs outside the original data set.

Morgan *et al.* (1975) developed regression equations involving 37 individual feeds and whole diets with various chemical constituents. The two most promising equations were then tested on 16 whole diets and the results compared with values computed by summation of the determined DE or ME values in the individual feedstuffs. Overall, the values determined by summation were closer to the determined DE values of the diets than those predicted from the chemical analyses.

In a comprehensive study, Morgan *et al.* (1987) formulated 36 diets from a range of individual raw materials to a broad range of nutrient compositions. The specification for the nutrient composition of the raw materials was taken from reference tables. The actual DE of the diets was then determined for growing/finishing pigs and the relationship between chemical composition of the diets and DE content assessed. In addition, the relative accuracy of some earlier reported equations developed by Morgan *et al.* (1975), and from their own centre were also tested. The diets were analysed for chemical composition in a number of collaborative laboratories and the effect of between laboratory variation on DE content for the different chemical tests assessed. The results indicated:

1. The diets were formulated to approximately 0.22 MJ/kg of DE of actual determined values. This again confirms that reference values of DE contents for raw materials have application in diet formulation.
2. A previously determined equation of Morgan *et al.* (1975) underpredicted the DE of the diets whilst two equations previously developed at their own centre gave better fits.
3. Based on the prediction equations developed in their current study the authors suggested that it should be possible to estimate the DE content of diets to within 0.5 MJ/kg with about an 80% probability of success or to 0.75 MJ/kg with a 95% probability of success.

This is somewhat similar to the possible accuracy of the other prediction equations presented in Table 15.2. However, in the application of these equations to diets of unknown composition by other laboratories, it is possible that the accuracy of the equations may be less. This could arise due to, (a) greater between laboratory variation in techniques than that experienced by Morgan *et al.* (1987), and (b) the inclusion of ingredients in the test diets which were not included in the original diets used to develop the prediction equations. For example, the inclusion of lupin seed meal (King and Taverner, 1975) and feedstuffs of animal origin (Just *et al.*, 1984) both decreased the accuracy of prediction equations developed by these workers.

It is doubtful whether the prediction equations for individual raw materials would be any more accurate than those for compounded diets and may even be less. This level of accuracy is far less than the error associated with direct determination of DE or ME. Thus where a more accurate assessment of the energy content of a raw material is required, then direct determination is, unfortunately, still the only reliable method.

Future studies

IMPROVING THE ACCURACY OF PREDICTION EQUATIONS

There is a need to improve the accuracy of prediction equations, not only for diets but especially for raw materials. This will only arise from a more accurate description of the nutrient composition of raw materials and diets. In particular, there is a need for more definitive studies on the energy yielding components in feeds. The current techniques do not give a sufficiently accurate resolution of the various fibre fractions in raw materials and diets and their effect on energy utilization. There is also a need for a more comprehensive understanding of the factors affecting energy utilization in diets of animal origin. Once this is achieved, it should be possible to develop relationships that apply to a more heterogeneous range of feed ingredients than is currently possible. However, for the prediction equations to be economically feasible to use, the number of components and complexity of analyses will have to be kept to a reasonable level.

There also seems considerable potential for the use of near infra-red reflectance analyses in the prediction of the energy content of feeds. This technology has the potential to quickly and inexpensively predict the chemical composition of a diet or feed ingredient, provided accurate equations can be established. It appears worthwhile to include this technique in future studies, not only for predicting the chemical composition of raw materials and perhaps diets but also to predict actual DE, ME and NE contents. If this technique could be shown to have application, it would considerably reduce the costs associated with, and time involved, in predicting the energy contents of raw materials and diets.

ENERGY SYSTEMS

More work is needed to determine the most suitable system for describing the energy content of raw ingredients and diets for pigs. It is evident that the DE system has limitations regarding the effect of inclusion level and under and over-estimation of some ingredients. Furthermore, these limitations are not all overcome by the ME system. There is a need for the more widespread adoption of a NE system for production. However, as net energy is more a function of the prediction system than an individual raw ingredient or compound diet, then this will be difficult.

In this regard, the use of computer models such as the 'Auspig' model, (developed by CSIRO, Australia) which simulates pig response, appears to be a logical way to assess the NE of production. Such models are capable of accounting for the various factors which determine NE – digestion of nutrients, efficiency of nutrient utilization, maintenance costs, lean and fat deposition, environmental factors and genotype. Thus it is possible to simulate the NE content of a compound diet from the raw ingredient composition and to estimate the potential NE for production of an individual feed ingredient. The 'Auspig' model is integrated with a least-cost feed matrix and a profit maximizing model, and this enables an assessment of the most profitable diet to feed from a set of raw materials varying in nutrient density and costs.

RAW MATERIALS

There is a need for continuous update on the DE, ME and NE content of new sources of raw materials, as they are developed. It would be best if the range in values for the above factors could be determined for each class of ingredient. Such an approach was taken by Thorne *et al.* (1989) in the development of prediction equations for copra meal. As discussed earlier, for many ingredients such as cereals, the energy concentration has been shown to be relatively uniform for each particular type. A similar situation would most probably apply to ingredients such as grain legumes etc. Where the raw material is a by-product, such as meat and bone meals, then greater variation would be expected. For such materials, prediction equations relating easily determined chemical constituents with energy concentration, need to be developed. With raw materials, individual prediction equations for each class of raw material, rather than general equations, need to be developed. This will not be easy, as studies to-date have not been very successful. A similar situation also applies with poultry (Fisher, 1982). It is probable that little progress will be possible until more quantitative techniques for fibre determinations are developed for meals of vegetable origin, and there is a better prediction of the energy yielding components in meals of animal origin.

Conclusions

The energy content of diets may be predicted from the chemical composition using previously developed prediction equations. DE, ME, and NE can be predicted. It has been estimated that it should be possible to predict the DE or ME content to within about 0.5 MJ/kg in mixed diets in about 80% of cases or to 0.75 MJ/kg in 95% of cases. Further refinements are unlikely to be achieved until more quantitive techniques for determining the various fibre fractions of raw materials and diets are developed. The potential of near infra-red reflectance spectrophotometry for assessing not only chemical composition but also DE and ME contents of raw materials and perhaps diets should be explored.

More emphasis should be given to providing information on the energy content of raw materials rather than mixed diets as the former is needed as the basis of formulating diets. Tabulated values for DE and ME appear more applicable than the prediction equations developed so far for most types of raw materials. It is doubtful whether much progress will be made in this area until (a) more quantitive techniques for fibre determinations are developed for materials of vegetable origin, and (b) there is a greater understanding of the factors that effect energy utilization in materials of animal origin.

Whilst NE can be predicted from relationships with ME it appears that there is more potential in the future to use computer simulation models to predict the NE of diets or raw materials. This, combined with near infra-red analyses to predict the chemical components of the raw materials, appears to offer more scope in the future to rapidly and inexpensively predict the NE contents of both raw materials and diets.

Acknowledgements

The author is grateful to Dr A. Just, National Institute of Animal Science, Foulum; Dr J. Wiseman, Faculty of Agriculture and Food Science, University of Nottingham; Professor C.T. Whittemore, Edinburgh School of Agriculture and Dr Y. Henry, Station de Recherches Porcines, Saint–Gilles, for information and assistance in the preparation of this review.

References

Batterham, E.S., Lewis, C.E., Lowe, R.F. and McMillan, C.J. (1980a) Digestible energy content of cereals and wheat by-products for growing pigs. *Animal Production*, **31**, 259–271

Batterham, E.S., Lewis, C.E., Lowe, R.F. and McMillan, C.J. (1980b) Digestible energy content of meat meals and meat and bone meals for growing pigs. *Animal Production*, **31**, 273–277

Fernandez, J.A. and Jorgensen, J.N. (1986) Digestibility and absorption of nutrients as affected by fibre content in the diet of the pig. Quantitative aspects. *Livestock Production Science*, **15**, 53–71

Fisher, C. (1982) Energy evaluation of poultry rations. In *Recent Advances in Animal Nutrition – 1982*. (ed W. Haresign) London, Butterworth, pp. 113–139

Henry, Y., Vogt, H. and Zoiopoulos (1988) Feed evaluation and nutritional requirements III. 4. Pigs and poultry. *Livestock Production Science*, **19**, 299–354

Hoffmann, L. and Schiemann, R. (1980) Von der Kalorie zum joule: Neue Groszenbeziehungen bei Messungen des Energieeumsatzes und bei der Berechnung von Kenzahlen der energetischen Futterbewertung. *Archiv für Tierenährung*, **30**, 733–742

Just, A. (1982) The net energy value of balanced diets for growing pigs. *Livestock Production Science*, **8**, 541–555

Just, A., Fernandez, J.A. and Jorgensen, H. (1982) *The Value of Meat and Bone Meal for Pigs*. Report 525, Copenhagen, National Institute of Animal Science

Just, A., Jorgensen, H. and Fernandez, J.A. (1983a) Maintenance requirement and the net energy value of different diets for growth in pigs. *Livestock Production Science*, **10**, 487–506

Just, A., Jorgensen, H., Fernandez, J.A., Bech–Andersen, S. and Hansen, N.E. (1983b) *The Chemical Composition, Digestibility, Energy and Protein Value of Different Feedstuffs for Pigs*. Report 556, Copenhagen, National Institute of Animal Science

Just, A., Jorgensen, H. and Fernandez, J.A. (1984) Prediction of metabolizable energy for pigs on the basis of crude nutrients in the feeds. *Livestock Production Science*, **11**, 105–128

King, R.H. and Taverner, M.R. (1975) Prediction of the digestible energy in pig diets from analyses of fibre contents. *Animal Production*, **21**, 275–284

Lekule, F.P., Jorgensen, H., Fernandez, J.A. and Just, A. (1989) Nutritive value of some tropical feedstuffs for pigs. Chemical composition, digestibility and metabolizable energy content. *Animal Feed Science and Technology*, (in press).

Morgan, D.J., Cole, D.J.A. and Lewis, D. (1975) Energy values in pig nutrition II. The prediction of energy values from dietary chemical analysis. *Journal of Agricultural Science, Cambridge*, **84**, 19–27

Morgan, C.A., Whittemore, C.T., Phillips, P. and Crooks, P. (1987) The prediction of the energy value compounded of pig foods from chemical analysis. *Animal Feed Science and Technology*, **17**, 81–107

Schiemann, R., Nehring, K., Hoffmann, L., Jentsch, W. and Chudy, A. (1971) *Energetische Futterbewertung und Energienormen. VEB*. Landwirtschaftsverlag, Berlin, Deutscher

Thorne, P.J., Wiseman, J., Cole, D.J.A. and Machin, D.H. (1989) The digestible and metabolizable energy values of copra meals and their prediction from chemical composition. *Animal Production*, **49**, 459–466

Van Soest, P.J. and Wine, R.H. (1967) Use of detergents in the analysis of fibrous feeds. IV. Determination of plant cell-wall constituents. *Journal of the Association of Official Analytical Chemists*, **50**, 50–55

Whittemore, C.T. (1987) *Elements of Pig Science*, 1st edn. Harlow, Longman, pp. 75–104

Wiseman, J. and Cole, D.J.A. (1980) Energy evaluation of cereals for pig diets. In *Recent Advances in Animal Nutrition – 1980* (ed W. Haresign), London, Butterworths, pp. 51–67

Wiseman, J. and Cole, D.J.A. (1983) Predicting the energy content of pig feeds. In *Recent Advances in Animal Nutrition – 1983*. (ed W. Haresign), London, Butterworths, pp. 59–70

Wiseman, J. and Cole, D.J.A. (1987) The digestible and metabolizable energy of two fat blends for growing pigs as influenced by level of inclusion. *Animal Production*, **45**, 117–122

16

PREDICTING THE DIETARY ENERGY VALUE OF POULTRY FEEDS

B. CARRÉ
Institut National de la Recherche Agronomique, Station de Recherches Avicoles, Nouzilly 37380, France

Introduction

The equations predicting the apparent metabolizable energy (AME) value of poultry feeds have shown increasing improvements in their accuracy for half a century. Mitchell (1942) reported an r^2 value of 0.83 for an equation predicting the metabolizability of diets, based on crude fibre (CF) determination. The equation proposed by Carpenter and Clegg in 1956, based on crude protein (CP), lipids (L), starch (St) and sugar (Su), gave an r^2 value of 0.90. R^2 values of 0.77 and 0.97 respectively were obtained by Sibbald *et al.* (1963) and Härtel *et al.* (1977), using these same four parameters. Using a combination of five parameters including the unsaturated : saturated fatty acid ratio, Fisher (1982) proposed an equation with an r^2 value of 0.985. With three parameters, including water-insoluble cell-wall content, an r^2 value of 0.968 was obtained by Carré and Brillouet (1989). This constant progress has resulted largely from progress in the design of experiments. In the first experiments (Carpenter and Clegg, 1956; Sibbald *et al.*, 1963), samples of ingredients and compound feeds were taken together for calculation of linear regressions. Subsequent studies tended to separate samples of ingredients and those of compound feeds (Härtel *et al.*, 1977). In addition, progress probably came from improvements in the method of *in vivo* determination of AME values with, for instance, the correction to zero nitrogen balance (AMEn) as described by Hill and Anderson (1958). Proposals for rapid *in vivo* assays as described by Farrell (1978), Kussaïbati and Leclercq (1985) or Sibbald (1986) have allowed the collection of a large amount of data (Sibbald, Price and Barrette, 1980). However, it must be appreciated that such rapid methods, as the dry tube feeding procedure developed by Sibbald, can lead to more variable (Härtel, 1986) and overestimated AMEn values (Fisher and McNab, 1987).

Replacement of simple linear regression (Carpenter and Clegg, 1956; Sibbald *et al.*, 1963) by multiple linear regression allowed greater degrees of significance in the values of coefficients assigned to variables (Härtel, 1979). Computers now make these calculations very easy and even allow adjustment of the value of the exponent applied to variables (Carré, Prévotel and Leclercq, 1984). Other progress relates to improvement in chemical determinations, especially those employed for carbohydrates. The nitrogen free extract (NFE) has been replaced by starch and sugar measurements (Carpenter and Clegg, 1956; Härtel *et al.*, 1977). Water-insoluble cell-wall (Carré and Brillouet, 1989) is now proposed to replace crude fibre determination. Changes applied to analytical methods have been adopted with the aim of obtaining a better applicability to poultry nutrition (Carpenter and Clegg, 1956; Carré, Prévotel and Leclercq, 1984).

Reproducibility of analytical methods is an important topic which has been developed for energy declaration purpose (Alderman, 1985; Bailey and Henderson, 1990), but which probably requires further investigation.

Target of prediction equations

Four interdependent factors need to be accounted for in formulating diets for poultry:

1. The animals to be fed (young chicks, rooster, laying hens, pullets, etc.).
2. The nutritional characteristics of ingredients (energy value, amino acid contents, mineral contents, vitamin contents, etc.).
3. Palatability of diets.
4. Price of ingredients.

Accordingly, from the first factor, the type(s) of animal to which an energy predicting equation can be assigned should be specified. This specification is required in poultry production principally because of the effect of age on the energy value of diets (Sibbald, Summers and Slinger, 1960; Petersen, Meyer and Sauter, 1976; Fisher and McNab, 1987).

It is probable that two or three equations would be necessary dependent upon age of birds. However, at the present time, it seems that consistent equations exist only for adult birds (Fisher and McNab, 1987). For AMEn values determined with young birds, a limited number of equations are available. According to one of the two studies described by Fisher and McNab (1987), most of the differences between adult and young AMEn values were explained in terms of utilization of lipids. However, from the other study, it would appear that lipids are probably not the only nutrient involved in the differences in AMEn values between adult and young birds. In addition, Moir, Yule and Connor (1980) provided equations for chicks. Unfortunately, the AMEn value of diets did not appear in this paper; moreover, crude fibre determinations were not performed following a standardized procedure. Accordingly, no comparison with other laboratories are possible.

This emphasizes the need for standardization of procedures both *in vitro* and *in vivo*, to allow comparison between laboratories. In this respect, the 2nd AMEn European ring-test conducted in 1988 has provided clear results, in part because of standardization of procedures. From this study, it appears that lipids are probably not the only nutrients involved in the difference between adult and young birds, since the two low fat diets of this study (<27.7 g lipid/kg dry matter) displayed, on average, a difference of 0.56 MJ/kg dry matter between adult and young. Significant differences between adult and young AMEn values have also been reported before with low fat feedstuffs such as wheat (Lessire, 1985; Mollah *et al.*, 1983) and barley (Petersen *et al.*, 1976). For both of these raw materials, carbohydrates are the main nutrients. Härtel (1986) detected significant differences in the digestibility of sugars between four week old chickens and adult cockerels.

More studies are required to investigate carbohydrate digestibility in both adult and young birds using a wide range of raw materials to investigate the possibility of interactions. Without this comprehensive approach, which is in progress for lipid digestion (Kussaïbati, Guillaume and Leclercq, 1982; de Groote, Ketels and Huyghebaert, 1987; Wiseman and Lessire, 1987a,b) but not for carbohydrate digestion, it will remain very hazardous to propose prediction equations for young birds. One of the features of studies comparing adult and young AMEn values is the inconsistency between experiments. For instance, no difference in the AMEn value of maize was detected between adult and young birds by Hochstetler and Scott (1975), while Lessire (1985) found a difference of 0.55 MJ/kg dry matter. These discrepancies between experiments probably depend on some uncontrolled unknown factors. It is important to note that the status of the digestive tract changes rapidly from 0 to 40 days of age in terms of capacity for lipid digestion (Polin and Hussein, 1982) and also in terms of microbial fermentation (Annison, Hill and Kenworthy, 1968). Thus, it would not be surprising to note that the physiological status of birds varies according to experiments, even if age of animals is similar. Conditions acting on physiological status of poultry have to be identified in order to incorporate them in a standardized method of AMEn determination in young chickens. Then, with the advent of reproducible AMEn determinations in young chickens, it will be possible to propose accurate equations to predict AMEn appropriate for them.

Improvement in the palatability of diets is obtained by some processing techniques, one of which, pelleting, is widespread. Even if pelleting is often considered as only promoting a small effect on the AMEn value of diet (Calet, 1965), its effect is probably not zero (Carré, Prévotel and Leclercq, 1984; Leclercq and Escartin, 1987). Clear beneficial effects of pelleting have been observed on the AMEn value of some ingredients such as wheat bran (Janssen, 1976) and legume seeds (Carré *et al.*, 1987; Lacassagne *et al.*, 1988). Thus specifying the process applied to feeds (grinding, pelleting, etc.) is needed for a full assessment of prediction equations, even those proposed for compound feeds.

Most of the equations predicting AMEn are based on chemical measurements of crude nutrients. So, theoretically, these equations cannot take account of variations in digestibility of nutrients. In spite of this, equations predicting 97–98% of the variation in AMEn values (adult) of compound feeds have been reported in the literature (Härtel *et al.*, 1977; Fisher, 1982; Carré, Prévotel and Leclercq, 1984). Such an accuracy observed for compound feeds is not surprising, because formulating diets is based on many conditions (1, 2, 3, 4; see above) which limits the inclusion rate of numerous raw materials. Raw materials with low nutrient digestibility such as rapeseed meal, lucerne meal, etc. are in most cases incorporated at low rates. Accordingly, it may be expected that the digestibilities of nutrients occurring in balanced compound feeds would vary within a limited range. This is the reason why equations predicting AMEn have to be divided into those assigned to compound feeds and those assigned to raw materials because, in contrast to compound feeds, digestibility of nutrients in raw materials can be extremely variable.

Predicting energy values of compound feeds

As stated above, equations predicting AMEn proposed for compound feeds are potentially very efficient. These equations can eventually be improved slightly by adding factors able to predict variations in the digestibility of some nutrients such as lipids. In this field, adding a measurement of the ratio unsaturated: saturated fatty acids (USR) can improve the prediction (Fisher, 1982). Measurements of some crude nutrients can also lead to the possibility of predicting variations in digestibility, due to their correlation with factors acting on digestibility. For instance, many raw materials with low nutrient digestibility also have a high content of plant cell walls. Thus, cell wall related parameters provide an indirect approach to estimate a part of the variation in digestibility. It is probable that measurement of ash also provides an indirect method of predicting part of the variations (see below). However, when examining the different types of equations, it is evident that they are very similar in terms of efficiency: very different combinations of parameters such as (L, CP, St, Su), (L, Ash(As), Water-insoluble cell wall (WICW)) or (Gross energy (GE), CP, WICW) provide very similar residual standard deviations (Carré, Prévotel and Leclercq, 1984). The choice of which equation to employ is dependent upon the use to which it is put. If equations are used for internal purposes, the choice will be made according to feasibility and price of analyses. In this regard, the feasibility of the WICW measurement has been considerably improved (Carré and Brillouet, 1989) and is now quite similar to that of the neutral detergent fibre (NDF) determination.

Thus the combination (L, As, WICW) is probably the most convenient in terms of ease of practice. The combination (GE, CP, WICW) is very convenient for laboratories measuring *in vivo* AMEn values, because only one measurement (WICW) has to be added to obtain the predicted AMEn value, since the determination of AMEn *in vivo* requires GE and CP to be measured. It could be concluded in fact that a predicted AMEn value is not necessary when the *in vivo* value has already been measured. However, it will be seen below that predicted values can provide very useful information.

If a choice had to be made for energy declaration purposes, it is not definite that the EEC equation would be the best solution. The problem of the EEC equation does not relate to its potential accuracy. The second European ring-test of *in vivo*

Table 16.1 EQUATIONS TO PREDICT AMEn VALUE (MJ/kg DRY MATTER; IN ADULT COCKERELS) OF COMPOUND FEEDS USING CHEMICAL PARAMETERS EXPRESSED AS PERCENTAGE OF DRY MATTER

Equations	References	
EEC	Fisher and McNab (1987)	AMEn = 0.1551 CP + 0.3431 L + 0.1669 St + 0.1301 Su
CW1	Carré and Brillouet (1989)	AMEn = 0.9362 GE − 0.0643 CP − 0.105 $WICW^{1.2}$ R.S.D. = 0.22 MJ/kg D.M.
CW2	Carré and Brillouet (1989)	AMEn = 16.66 + 0.197 L − 0.222 Ash − 0.187 WICW R.S.D. = 0.26 MJ/kg D.M.
CW2′		AMEn = 16.06 + 0.197 L − 0.122 $Ash^{1.20}$ − 0.113 $WICW^{1.14}$ R.S.D. = 0.26 MJ/kg D.M.

Recommended ranges for CW equations:
10.90 MJ/kg D.M. < AMEn < 16.70 MJ/kg D.M.
10% D.M. < CP < 30% D.M.
 9% D.M. < WICW < 25% D.M.

CP: crude protein; L: lipid; ether extract without hydrolysis for CW equations; ether extract after hydrolysis for EEC equation; St: starch (polarimetric method); Su: sugars; GE: gross energy; WICW: water-insoluble cell walls (Carré and Brillouet, 1989).

Table 16.2 SECOND EUROPEAN RING-TEST ON AMEn DETERMINATION: PREDICTION OF AMEn VALUES (ADULT COCKERELS) OF COMPOUND FEEDS. DATA EXPRESSED AS MJ/kg DRY MATTER

Diet number (pelleted)	*in-vivo* AMEn values (y)		Predicted AMEn values (xi)					
	Mean	Number of laboratory	EEC (x_0)	Number of laboratory	CW1 (x_1)	Number of laboratory	CW2 (x_2)	Number of laboratory
1	13.77	8	13.70	5–9	13.51	5–9	13.36	2–5
2	12.16	8	12.26	5–9	11.97	4–9	11.98	2–4
3	14.69	8	14.55	5–9	14.47	5–9	14.41	2–5
4	14.73	8	14.56	5–9	14.50	5–9	14.45	2–5
5	13.13	6	13.11	5–7	12.95	4–7	12.96	1–4
6	13.18	6	13.00	5–7	12.87	4–7	12.82	1–4
7	13.39	7	13.47	5–8	13.30	4–8	13.33	2–4
Means	13.58		13.52		13.37		13.33	
$y = f_{(xi)}$		$y=1.086 x_0$ -1.105			$y=1.008 x_1$ $+0.103$		$y=1.03 x_2$ -0.150	
R^2		0.988			0.993		0.980	

Data from Bourdillon *et al.*, 1990

ME determination showed that the EEC equation was able to give predicted values very close to the *in vivo* values (Tables 16.1 and 16.2). Equations based on WICW determination gave slightly lower values compared to *in vivo* values with a mean difference of 0.23 MJ/kg DM (Table 16.2). Bearing in mind that the WICW based equations were calculated mainly from unpelleted diets (Carré, Prévotel and Leclercq, 1984), it is not impossible that the slight underestimation obtained with these equations was partly due to the pelleted form of the experimental diets, since a slight beneficial effect of pelleting on AMEn value of diets has often been observed (Carré, Prévotel and Leclercq, 1984; Leclercq and Escartin, 1987).

The main problem of the EEC equation proposed for energy declaration relates to its variability induced by chemical measurements. An analytical tolerance of ±1 MJ/kg is often proposed for the EEC equation (Cooke, 1987). It is quite difficult to have a good estimate of the analytical tolerance because this depends largely on samples and laboratories involved in the ring-tests. On the basis of analytical variabilities reported before (Cooke, 1987; Fisher, 1982; Carré, Perez and Lebas, 1988; Fournier, 1988) the analytical tolerance for the EEC equation would be, respectively, ±0.45, ±0.60, ±0.60 or ±0.85 MJ/kg. The ring-tests reported by Cooke (1987) involved a high number of laboratories (10–20) but only one or two samples. In contrast, the data reported by others involved a limited number of laboratories (3–6) and numerous samples (11–56). The analytical tolerance of ±0.60 MJ/kg between laboratories seems quite reasonable. It is probable that a high tolerance such as ±0.85 MJ/kg (Fournier, 1988) is due either to some analytical procedures which differ slightly between some laboratories or to errors in sampling.

According to previous studies (Carré, Perez and Lebas, 1988; Fournier, 1988), the analytical tolerances of the WICW based equations are lower compared with those of the EEC equation. In the first study (Carré, Perez and Lebas, 1988), the analytical tolerances of EEC, CW1 and CW2 equations were, respectively, ±0.60, ±0.50 and ±0.50 MJ/kg. In the second study (Fournier, 1988), the tolerances of

EEC and CW2 equations were respectively ±0.85 and ±0.60 MJ/kg. Accordingly, the WICW based equations would probably be more adapted to energy declaration than the EEC equation.

The other aspect to take into consideration for the choice of prediction equations is their biological basis. Thus empirical equations which do not illustrate logical coefficients and variables are not entirely acceptable. With such equations, the range of recommendation remains unclear. Moreover, empirical equations are difficult to understand when compared with other equations. In this respect, the approach of the EEC equation is very straightforward. Thus the variables of the EEC equation represent the amounts of available nutrients; their coefficients represent their gross energy corrected for their mean digestibility and metabolizability. According to the EEC equation, the mean digestibility of fat, crude protein, starch and sugars would be, respectively, 0.87, 0.86, 0.96 and 0.79. These data do not seem unrealistic. Adding cell wall related parameters to the combination (L, CP, St, Su) does not improve the accuracy of the prediction (Härtel *et al.*, 1977), suggesting that cell wall components are not available to poultry (Carpenter and Clegg, 1956). Very low digestibility of cell wall components is usually found in poultry either using conventional methods such as crude fibre and nitrogen-free-extract (Härtel *et al.*, 1977) or using specific methods such as gas-chromatography of derivatized cell wall sugars (Carré and Leclercq, 1985). Organic components other than lipids, protein, starch, sugars and cell walls represent minute amounts in poultry compound feeds. So, the combination (L, CP, St, Su) is able, in most cases, to account for practically all the available nutrients; hence the risk of bias by omitting some nutrients is low with this combination.

The fact that cell wall components are not available to poultry explain why cell wall parameters are very often among the variables which display the highest correlations with metabolizable energy value (Sibbald, Price and Barrette, 1980; Fisher, 1982; Carré, Prévotel and Leclercq, 1984; Campbell, Salmon and Classen, 1986). Equations using classical cell wall parameters such as crude fibre (CF) or neutral detergent fibre (NDF) can be almost as efficient as some equations based on available nutrients (Fisher, 1982; Campbell, Salmon and Classen, 1986). However, the CF or NDF based equations fail to reach the accuracy of the (L, CP, St, Su) combination (Carré, Prévotel and Leclercq, 1984) unless starch or sugar is associated with CF or NDF (Fisher, 1982). Under such conditions, the interest in using CF or NDF is reduced considerably, because their main interest would be in the possibility of removing the need to determine available carbohydrates (Sibbald, Price and Barrette, 1980). Interest in using CF or NDF is also reduced because of difficulties in explaining the significance of coefficients appearing in the CF or NDF based equations.

The introduction of the WICW parameter instead of CF or NDF has brought progress on three points:

1. The analytical variability of the WICW based equations is lower than that of the (L, CP, St, Su) combination (see above).
2. The introduction of WICW allows St and Su to be removed from the equations without decrease of the accuracy (Carré, Prévotel and Leclercq, 1984).
3. A logic in understanding approach of the WICW based equations can be undertaken.

This approach is based on the proposal that WICW represents the whole undigested fraction of plant cell walls. Adult cockerels display very low ability for the digestion of lignin, cellulose, water-insoluble hemicelluloses and water-insoluble pectic substances (Carré and Leclercq, 1985); WICW represents the sum of these components (Carré and Brillouet, 1989) and, so, it represents the whole undigested fraction of plant cell walls in poultry feeds. This is not the case for CF and NDF since CF does not contain hemicelluloses and pectic substances, and NDF determination does not take account of pectic substances (Van Soest and Wine, 1967). Water-insoluble pectic substances are major components in dicotyledonous plants such as legume seeds (Brillouet and Carré, 1983; Carré, Brillouet and Thibault, 1985). Thus, great differences can appear between NDF and WICW contents in these plant materials (Carré and Brillouet, 1986, 1989). Accordingly, variation in the amount of legume meal in diets is the origin of the lower efficiency of NDF as a predictor of dietary energy compared to WICW (Carré, Prévotel and Leclercq, 1984).

In general the WICW parameter appears within two combinations: (GE, CP, WICW) and (L, As, WICW). Both can be interpreted on the basis of the principle that WICW is undigestible and acts as diluter of available nutrients, namely lipids, proteins, starch and sugars.

A logical analysis of the first combination can be described as follows:

$$[\text{Thx}]: \text{AMEn} = A'(\text{GE} - \text{GE}_{CP} \times \text{CP} - \text{GE}_{CW} \times \text{WICW}) + C \times \text{CP}$$

Where:
A' = mean digestibility of gross energy assigned to lipids and non-cell wall carbohydrates
GE = gross energy content of diet
GE_{CP} = gross energy of crude protein (23.41 MJ/kg)
CP = crude protein content
GE_{CW} = gross energy of cell walls
WICW = WICW content
C = metabolizable energy value of crude proteins.

The A' coefficient can be calculated using the mean digestibilities of lipids, starch and sugars given by the EEC equation (see above) and the mean nutrient contents of compound feeds (Carré, Prévotel and Leclercq, 1984):

$$A' = \frac{39.50 \times 0.054 \times 0.87 + 17.43 \times 0.457 \times 0.96 + 16.51 \times 0.035 \times 0.79}{39.50 \times 0.054 + 17.43 \times 0.457 + 16.51 \times 0.035} = 0.933$$

where 39.50, 17.43 and 16.51 are the respective gross energy values (MJ/kg) of lipids, starch and sugars.

Assuming a GE_{CW} value equivalent to that of wheat cell wall (19.23 MJ/kg; Carré and Rozo, 1988; unpublished data), and a C coefficient equal to that given by the EEC equation (15.51 MJ/kg), a theoretical equation can be proposed. The use of the wheat cell wall gross energy for the GE_{CW} value is more realistic than the polysaccharide gross energy value (17.16 MJ/kg) previously used (Carré, Prévotel and Leclercq, 1984), since cell walls also contain lignin whose gross energy value is much higher than that of polysaccharides. The theoretical equation for compound feeds is as follows:

$$\text{AMEn (MJ/kg)} = 0.933 \times \text{GE (MJ/kg)} - 0.063 \times \text{CP (\%)} - 0.179 \times \text{WICW (\%)}$$

The theoretical coefficients of GE (0.933) and CP (0.063) do not differ significantly from the experimental ones (0.936 and 0.064; see Equation CW1). The mean value of the term $[0.105 \times \text{WICW}^{0.2}]$ appearing from the experimental equation CW1 can be used as a comparison with the theoretical coefficient of WICW. It appears that this mean value (0.174) is very close to the theoretical coefficient of WICW (0.179). Thus a very good agreement exists between experimental and theoretical equations based on GE, CP and WICW showing that a precise and logical analysis can be performed on WICW based equations. Other cell wall parameters such as CF or NDF have never allowed such an approach.

A logical analysis of the (L, As, WICW) combination can also be undertaken. Using the theoretical equation previously reported (Carré, Prévotel and Leclercq, 1984) and the following expression of GE (dry matter basis):

$$\text{GE (MJ/kg)} = 0.3950 \times \text{L}(\%) + 0.2341 \times \text{CP}(\%) + 0.1743 \times \text{NFE}_{CW}(\%) + \text{GE}_{CW} \times \text{WICW}$$

where: $\text{NFE}_{CW}(\%) = 100 - \text{L}(\%) - \text{CP}(\%) - \text{A}(\%) - \text{WICW}(\%)$
    17.43 = gross energy (MJ/kg) of NFE_{CW} taken from that of starch.

$$\text{AME (MJ/kg)} = \text{A}(17.43 + 0.221 \times \text{L} + 0.060 \text{XCP} - 0.1743 \times \text{As} - 0.1743 \times \text{WICW}) - \text{BxCP}$$

where A = 0.914 when calculated in the same way as that described for A′
    B = 0.048 MJ/% CP (see Carré, Prévotel and Leclercq, 1984)

$$\text{AME (MJ/kg)} = 15.93 + 0.202 \times \text{L} - 0.159 \times \text{A} - 0.159 \times \text{WICW} + (0.055 - 0.048) \times \text{CP}$$

The variation of the term $(0.007 \times \text{CP})$ is considered negligible and, accordingly, this is taken as a constant (0.126) with CP equal to 18% D.M. Thus, the theoretical equation for compound feeds is as follows:

$$\text{AME (MJ/kg D.M.)} = 16.06 + 0.202 \times \text{L}(\%) - 0.159 \times \text{As}(\%) - 0.159 \times \text{WICW}(\%)$$

No significant difference exists between the experimental (0.197) and theoretical (0.202) coefficients assigned to the L variable (in CW2). Some differences appear between experimental (CW2) and theoretical factors for other parameters with a more pronounced difference for the As variable. These discrepancies are probably a result of curvilinear response of As and WICW variables. After adjusting the intercept and the coefficient of WICW on the theoretical values and optimizing the exponents of the variables WICW and As, the experimental equation CW2′ (Table 16.1) is obtained without change of the residual standard deviation compared to the equation CW2. According to the mean value of the term $(0.122 \text{As}^{0.20})$ (0.176) (see Equation CW2′, Table 16.1) which is quite near to the theoretical coefficient of As (0.159), it can be concluded that the theoretical scheme is consistent with the experimental equation.

In both CW equations, curvilinear responses of variables are detected. The significance of these responses can be interpreted in two ways:

1. Both As and WICW variables are inversely correlated with starch. Thus the higher the contents of ash or cell walls, the lower the starch content; accordingly, for a high content of ash or cell walls, lower values of A and A′ factors (see theoretical equations) can be expected, since these values largely depend on starch content. In this way, the curvilinear responses of As and WICW variables could correspond to changes in the values of A and A′ factors induced by variation in starch content.

2. As and WICW variables could be inversely correlated with the digestibility of
   some nutrients, either because of their direct negative effect or because of their
   correlation with other negative factors.

It would not be surprising for ash to have a direct negative effect, since calcium
was demonstrated to decrease lipid digestibility (Kussaïbati, Leclercq and
Guillaume, 1983) and ME values (Sibbald, Slinger and Ashton, 1961). A direct
negative effect of cell walls can also be expected with some ingredients, such as
wheat bran, because they surround nutrients and reduce accessibility there
(Saunders, Walker and Kohler, 1968). Negative correlations between cell wall
content and protein digestibility can also be expected since numerous protein
sources with high cell wall contents such as rapeseed or lucerne meals have low
protein digestibility.

The curvilinear response of variables raises the question of the recommended
range of variables for prediction equations. Extrapolation of a curvilinear response
outside the range of variables observed in the experiment may be inappropriate,
because the mathematical fitting of a curvilinear response depends largely on the
specific conditions of the experiment. Thus the range of variables employed in the
experiment as the basis for the regression calculation has to be adopted for practical
use if a curvilinear response is observed in the regression calculation: this is the case
for the CW equations. Moreover, it has been seen above that CW equations
depend on the digestibility value (A and A') of gross energy assigned to available
nutrients, and that this value depends on the nutrient composition. Thus if the
nutrient composition of a feed is very different from the experimental conditions
under which equations were derived (Carré, Prévotel and Leclercq, 1984), there is
some risk that a large difference will appear in the A and A' values between the
feed values and the mean values observed in the experimental conditions of
equations. It is the case, for instance, with protein meals such as soyabean,
sunflower or rapeseed meals which are impossible to be correctly predicted by both
CW equations (Table 16.1) because the A and A' values of these feeds are much
lower than the values observed in CW equations (Table 16.1).

It is probable that curvilinear response of variables is much more likely than had
been imagined. In most cases, investigations of curvilinear responses are not
undertaken and only linear responses are fitted. Comparison with theoretical
equations as described above allows detection of curvilinear responses. The
existence of a significant intercept where no such intercept is given in the
corresponding theoretical equation, is probably a strong indication. Additionally
large differences in the value of coefficients between experimental and theoretical
equations is also an indication of curvilinear responses. In this way, the EEC
equation does not seem to provide evidence of curvilinear responses and it is
probable that its recommended range of variables is wider than the range of variables
in the CW equations. However CW equations are probably more efficient, within their
range of variables, for detection of interactions due to cell walls or ashes.

Empirical equations: there are no theoretical equations with which empirically
derived can be compared, and accordingly it is difficult to assess the recommended
range of variables for the latter. Ranges for empirical equations must be limited
strictly to the specific conditions of the experiment used for their calculation. These
specific conditions include not only ranges of parameter values but also correlations
observed between 'independent' variables. Moreover, coefficients of empiric
equations cannot be adapted to situations outside the recommended range of
variables.

Equations predicting feed energy value can be useful not only for practical purposes but also for research studies. For instance, observations of discrepancies between measured and predicted AMEn values may indicate some abnormalities which require further investigations. Such a discrepancy observed with a pea-based diet (Carré, Prévotel and Leclercq, 1984) has led to several studies conducted with legume seeds (Carré *et al.*, 1987; Lacassagne *et al.*, 1988; Conan and Carré, 1989), showing abnormally low digestion of legume starch. Another approach for research is the calculation of the A and A' coefficients (see theoretical equations) with measurements of AMEn value and WICW content. Subsequently an attempt to relate A and A' values to some negative factors can be undertaken. Such an approach can be particularly useful for investigations on raw materials.

Predicting energy values of raw materials

Predicting energy values of raw materials requires initially the derivation of regression equations from accurate *in vivo* AMEn values. Obtaining these values for some raw materials may prove difficult either because of their low palatability or because of non-linear responses based on their rate of inclusion in diets. Very often these two drawbacks are encountered together. The question of the non-linear response has to be investigated and taken into account when designing experiments for the derivation of equations predicting AMEn values. Curvilinear responses have been clearly identified for meat meals (Lessire *et al.*, 1985) and fats (Shannon, 1971; Sibbald and Kramer, 1978; Mateos and Sell, 1980; Wiseman *et al.*, 1986; De Groote, Ketels and Huyghebaert, 1987; Wiseman and Lessire, 1987a). Accordingly, equations have been proposed relating AMEn values to inclusion rate, for fats (Wiseman and Lessire, 1987a,b) and meat meals (Lessire *et al.*, 1985). Fats are probably one of the most difficult raw materials for derivation of prediction equations for AMEn values, because these depend on many factors such as age of birds, inclusion rate, interactions with other nutrients, fatty acid composition of triglycerides, free fatty acid content (Wiseman, 1984; De Groote, Ketels and Huyghebaert, 1987), hydroperoxide content (Andrews *et al.*, 1960). Accordingly, attempts to predict AMEn value of fats by encompassing all situations are practically impossible. Huyghebaert, De Munter and De Groote (1988), and Janssen (1985) reported equations based on chemical variables such as iodine value, non-polar fraction content and fatty acid composition for fat included in diets for young chickens at a rate of about 100 g/kg. Additionally Janssen (1985) reported equations based on iodine value and fatty acid composition for animal fats measured in adult cockerels.

With respect to meat meals, Lessire and Leclercq (1983) proposed an equation based on ash and fat determinations, with rates of inclusion in the range of 100–200 g/kg. Farrell (1980) obtained another equation based on the same parameters, but providing lower AME values probably because of the high rate of inclusion level (500 g/kg) employed. In both cases energy values were determined on adult cockerels.

For many raw materials containing anti-nutritional factors such as glucosinolates, medicagenic acid or water-soluble arabino-xylans, which are encountered respectively in 'single zero' rapeseed meal (Fenwick and Curtis, 1980; Sukhija *et al.*, 1985), lucerne meal (Cheeke, 1976) and rye (Ward and Marquardt, 1987), a decrease in ME values with an increase in rate of inclusion has to be expected. This has already been observed by McAuliffe and McGinnis (1971) with rye given to young chicks. Thus, for such raw materials, reliable ME values will be obtained by utilizing rates of inclusion as near as possible to those found in practice. For low rates of inclusion, higher numbers of replicates will be required for determining the *in vivo* ME value of raw materials, since standard deviations increase with decreasing rate of inclusion. Thus, for these raw materials, establishing equations is not very easy and often disappointing with respect to their accuracy: this was the case for rapeseed meals (Lessire and Conan, personal communication; Table 16.3).

Table 16.3 EQUATIONS TO PREDICT THE AME VALUE OF RAW MATERIALS GIVEN AS MASH TO ADULT COCKERELS (DATA AS MJ/kg AND %DM BASIS)

Raw material	Range of compositions	Rate of inclusion (g/kg)	AME	R.S.D.	r^2	Number of samples
Barley	4.1<C.F.<6.2	900	15.80−0.476×CF	0.21	0.41	25
Sorghum	0.27<Tan.<1.61	940	16.18−1.66×Tan.	0.26	0.91	13
Rapeseed meal 'single zero' type	1.3<Lip<5.7 37.1<CP<41.0	300	−20.18+1.73×GE +0.153×L−0.903×As	0.56	0.60	18
Sunflower meal	34.9<CP<41.5 0.6<Lip<2.8	300	2.52+0.114×CP +0.127×L	0.23	0.60	12
Corn-gluten feeds and wheat products	2.7<As<9.6 21.5<WICW<49.3	400	16.25−0.217As −0.157×WICW	0.23	0.97	6+6

RSD: residual standard deviation; CF: crude fibre; Tan.: tannins according to Daiber (1975) with tannic acid as standard; L: lipid as ether extract without acid hydrolysis; GE: gross energy; As: Ashes; CP: crude protein; WICW: water-insoluble cell walls according to Carré and Brillouet (1989).
(Data relating to barley, sorghum, rapeseed and sunflower meals are from Lessire and Conan.)

This low accuracy probably arose in part from high variability in the *in vivo* AME values assigned to rapeseed meal fractions. However, this was probably not the only reason. Variation in protein digestibility between samples could also be expected to be responsible. This hypothesis may be used to explain the low correlations between analytical parameters and AME values of pea seeds (Lessire *et al.*, 1980), since large variations in nutrient digestibilities have been observed between samples of this crop (Conan and Carré, 1989). For legume seeds, it is likely that derivation of prediction equations with pelleted feeds would be much more effective because this treatment tends to decrease the variability in legume starch digestibility (Lacassagne *et al.*, 1988). Development of methods able to predict nutrient digestibilities could help in improving the accuracy of prediction equations applied to raw materials. Unfortunately, attempts in this direction have concentrated mainly on compound feeds (Fisher, 1982; Clunies, Leeson and Summers, 1984) while the main interest of such methods is in the field of raw materials. Some methods have been proposed to predict digestibilities of amino acids but it seems that the application for these methods remains limited (Papadopoulos, 1985).

The final difficulty of predicting energy value of raw materials relates to the limited range of analytical variables, which is often encountered within the experimental group of samples. Attempts to extend the range of analytical variables often require the collection of data over several years and then necessitate the *in vivo* AME determination to be carefully calibrated to avoid any slight shift of AME data due to time. Frequently, however, even extending the time of collection is not sufficient to improve greatly the range of analytical variables. Thus, owing to the low variation in analytical parameter values, the coefficients of multiple linear equations often display high standard deviations and low reliability. Moreover, analytical parameters for a specific type of raw material often display high correlations between themselves. As a consequence, only a few parameters can be introduced into the equations and the use of these equations must be strictly limited to the range of analytical values observed in the experiment which was the basis for deriving the equation.

The use of digestibility values of nutrients for predicting the energy values of raw materials (Titus, 1955; Janssen, 1976; Härtel, 1979) is probably an efficient approach to solve the problem of the low variability in analytical parameter values. With methods based on nutrient digestibilities, the equations take account of all nutrients, the coefficients are logical and the ranges of variables are not strictly limited. Examples of this type of approach are given in the European table of energy values for poultry feedstuffs (Janssen, 1986). In this field, Härtel (1979) recommended the use of starch and sugar digestibilities instead of nitrogen-free-extract digestibility, since the former parameters display much less variation between samples. The drawback of this type of approach is in the need to measure the *in vivo* utilization of six parameters, namely lipid, protein, starch, sugars, crude fibre and energy (Härtel, 1979) for establishing the linear equations. Accordingly, an approach based on the determination of nutrient digestibilities appears efficient but time consuming for establishing equations.

An alternative solution would be to derive regression equations from combinations of several types of raw materials in order to obtain wider range of analytical values and lower correlations between independent variables. Theoretically, this approach is possible with cereal products because it would be expected that the digestibilities of their nutrients do not vary greatly according to type of raw material, since their main nutrient, namely starch, is, in most cases, fully digested (Moran, 1982). However, reports combining several types of raw material for establishing AME prediction equations are rarely found in the literature. This is not surprising because it would be expected that the independent variables which are commonly used in practice cannot be effective for all nutrients. For example, when using the combination (CP, EE, St, Su) (Sibbald and Price, 1976a,b; 1977), the water-soluble non-starch polysaccharides (WS-NSP) and organic acids are not taken into account. NFE or CF variables (Coates *et al.*, 1977; Janssen and Carré, 1985) could directly or indirectly take account of these latter components, but these variables have the disadvantage of not relating to precise nutritional entities: digestibility of NFE will vary according to the proportions of starch and hemicelluloses present in NFE fractions, and these proportions can vary greatly according to type of raw material; crude fibre and even neutral detergent fibre relate only to a fraction of undigested plant cell walls (Carré and Leclercq, 1985) and their ratios to cell walls will vary between raw materials (Carré and Brillouet, 1986). WS-NSPs are assumed to be digested in adult cockerels (Carré and Leclercq, 1985) and then have to be accounted for together with other nutrients, especially when their amounts vary according to type of raw material, as has been shown for whole grains and by-products from cereals (Nyman *et al.*, 1984). For example, the latter authors found levels of WS-NSP of nearly zero, 0.5%, 0.5%, 1.4%, 3.8% and 3.9% D.M. respectively in whole grains of maize, sorghum, rice, wheat, rye and barley. Saini and Henry (1989) obtained similar data of 1.8% and 3.6% D.M. of water-soluble arabinoxylans respectively in wheat and rye.

Organic acids are also components which can occur in some raw materials and not in others: wet processes employed for corn-gluten feeds can promote organic acid formation in this raw material, in contrast with wheat bran which is expected to contain only traces of organic acids because of the absence of wet processing employed for this latter raw material. As a result of the high digestion rate observed for organic acids in adult birds (Bolton and Dewar, 1962; 1965), organic acids have to be considered as energy yielding nutrients.

The WICW variable is indirectly able to take account of these latter minor components (WS-NSP and organic acids) and it does not have the drawback of low biological significance. In other words, WICW is a universal variable, from the point of view of adult bird digestion, and this can then be applied effectively for combinations of raw materials provided that the digestibilities of main nutrients vary inside a limited range. An example of this is given in Table 16.3 where a multiple regression line based on ash and WICW variables is applied to a combination of six wheat brans and six corn-gluten feeds: a wide analytical range is observed, the correlation ($r^2 = 0.11$) between ash and WICW variables is not significant and the accuracy of prediction of AMt obtained is high ($r^2 = 0.97$; R.S.D. $= 0.23$ MJ/kg D.M.). No bias appeared in the deviations of prediction according to type of raw material. In contrast, the best equation based on classical variables (AME $= 15.21 - 0.70 \times$ CF) led to a much lower accuracy ($r^2 = 0.90$; R.S.D. $= 0.42$ MJ/kg D.M.) for this combination of raw materials.

Prediction equations applied to combinations of raw materials which display differences in the digestibility of their nutrients are more complicated: this would be the case for combination of protein meals such as soya-bean and rapeseed meals since these raw materials display large differences in the digestibility of protein (Guillaume and Gomez, 1980). For these cases, the use of the theoretical equations (see above) will be of help to calculate the mean digestibility of nutrients. These digestibility values will be applied to correct the value of analytical variables before calculation of regression lines. This approach will be investigated in the future for developing equations adapted to protein meals.

Summary

The target of the equations predicting the apparent metabolizable energy (AME) value of poultry feeds has to be defined precisely. The more this is precise, the more the equations will be efficient. Targets of equations relate to age of birds, choice between compound feeds and raw materials, and type of processing applied to feeds. The equations for compound feeds given to adult birds, even using different combination of independent variables, often display similar high efficiencies. The main differences between equations for compound feeds concern the easiness of analyses to be performed and their reproducibility between laboratories. In this regard, the equations using the simplified method of water-insoluble cell wall (WICW) determination are probably the most efficient. The equations based on WICW parameter are not empiric and have comprehensible coefficients.

Efficient equations for raw materials are often difficult to be obtained because of many factors including difficulties in measuring precisely the AME values of raw materials, variation in digestibility values of nutrients, and limited range of analytical variables. Solutions to improve this situation are discussed. A new approach is proposed consisting in combining several types of raw materials and using WICW as predicting variable.

References

Alderman, G. (1985) Prediction of the energy value of compound feeds. In *Recent Advances in Animal Nutrition – 1985*, (eds W. Haresign and D.J.A. Cole) London, Butterworths, pp. 3–52

Andrews, J.S., Griffith, W.H., Mead, J.F. and Stein, R.A. (1960) Toxicity of air-oxidized soybean oil. *Journal of Nutrition*, **70**, 199–210

Annison, E.F., Hill, K.J. and Kenworthy, R. (1968) Volatile fatty acids in the digestive tract of the fowl. *British Journal of Nutrition*, **22**, 207–216

Bailey, S. and Henderson, K. (1990) Consequences of inter-laboratory variation in chemical analysis. In *Feedstuff Evaluation* (eds J. Wiseman and D.J.A. Cole). Butterworths, London, pp. 353–363

Bolton, W. and Dewar, W.A. (1962) The absorption of lactate by the fowl. In *Proceedings of the 12th World's Poultry Congress, 10–18 August, 1962, Sydney*. North-Sydney, Australian Branch of W.P.S.A., pp. 117–119

Bolton, W. and Dewar, W.A. (1965) The digestibility of acetic, propionic and butyric acids by the fowl. *British Poultry Science*, **6**, 103–105

Bourdillon, A., Carré, B., Conan, L., *et al.* (1990) European reference method of *in vivo* determination of metabolizable energy in poultry: reproducibility, effect of age, comparison with predicted values. *British Poultry Science* (in press)

Brillouet, J.M. and Carré, B. (1983) Composition of cell walls from cotyledons of *Pisum sativum*, *Vicia faba* and *Glycine max*. *Phytochemistry*, **22**, 841–847

Calet, C. (1965) The relative value of pellets versus mash and grain in poultry nutrition. *World's Poultry Science Journal*, **21**, 23–52

Campbell, G.L., Salmon, R.E. and Classen, H.L. (1986) Prediction of metabolizable energy of broiler diets from chemical analysis. *Poultry Science*, **65**, 2126–2134

Carpenter, K.J. and Clegg, K.M. (1956) The metabolizable energy of poultry feeding-stuffs in relation to their chemical composition. *Journal of the Science of Food and Agriculture*, **7**, 45–51

Carré, B. and Brillouet, J.M. (1986) Yield and composition of cell wall residues isolated from various feedstuffs used for non-ruminant farm animals. *Journal of the Science of Food and Agriculture*, **37**, 341–351

Carré, B. and Brillouet, J.M. (1989) Determination of water-insoluble cell walls in feeds: interlaboratory study. *Journal of the Association of Official Analytical Chemists*, **72**, 463–481

Carré, B., Brillouet, J.M. and Thibault, J.F. (1985) Characterization of polysaccharides from white lupin (*Lupinus albus L.*) cotyledons. *Journal of Agricultural and Food Chemistry*, **33**, 285–292.

Carré, B. and Leclercq, B. (1985) Digestion of polysaccharides, protein and lipids by adult cockerels fed on diets containing a pectic cell wall material from white lupin (*Lupinus albus L.*) cotyledon. *British Journal of Nutrition*, **54**, 669–680

Carré, B., Perez, J.M. and Lebas, F. (1988) Mesure des fibres végétales dans les aliments pour animaux. *Rapport Final de la Convention de Recherches DIAA/IRTAC* No 86/02. Institut de Recherches Technologiques Agro-alimentaires des céréales, Paris

Carré, B., Prévotel, B. and Leclercq, B. (1984) Cell wall content as a predictor of metabolizable energy value of poultry feedingstuffs. *British Poultry Science*, **25**, 561–572

Carré, B., Escartin, R., Melcion, J.P., Champ, M., Roux, G. and Leclercq, B. (1987) Effect of pelleting and associations with maize or wheat on the nutritive value of smooth pea (*Pisum sativum*) seeds in adult cockerels. *British Poultry Science*, **28**, 219–229

Cheeke, P.R. (1976) Nutritional and physiological properties of saponins. *Nutrition Reports International*, **13**, 315–324

Clunies, M., Leeson, S. and Summers, J.D. (1984) *In vitro* estimation of apparent metabolizable energy. *Poultry Science*, **63**, 1033–1039

Coates, B.J., Slinger, S.J., Ashton, G.C. and Bayley, H.S. (1977) The relation of metabolizable energy values to chemical composition of wheat and barley for chicks, turkeys and roosters. *Canadian Journal of Animal Science*, **57**, 209–219

Conan, L. and Carré, B. (1989) Effect of autoclaving on metabolizable energy value of smooth pea seed (*Pisum sativum*) in growing chicks. *Animal Feed Science and Technology,* **26**, 337–348

Cooke, B.C. (1987) The impact of declaration of the metabolizable energy (ME) value of poultry feeds. In *Recent Advances in Animal Nutrition – 1987* (eds W. Haresign and D.J.A. Cole) Butterworths, London, pp. 19–26

Daiber, K.H. (1975) Enzyme inhibition by polyphenols of sorghum grain and malt. *Journal of the Science of Food and Agriculture*, **26**, 1399–1411

De Groote, G., Ketels, E. and Huyghebaert, G. (1987) The energy evaluation of fats in poultry diets: new developments. In *Proceedings of the Sixth European Symposium on Poultry Nutrition, 11–15 October 1987, Königslutter, F.R. Germany*, Celle, German branch of the W.P.S.A., pp. C21–C30

Farrell, D.J. (1978) Rapid determination of metabolizable energy of foods using cockerels. *British Poultry Science*, **19**, 303–308

Farrell, D.J. (1980) The rapid method of measuring the metabolizable energy of feedstuffs. *Feedstuffs*, **November 3**, 24–25

Fenwick, G.R. and Curtis, R.F. (1980) Rapeseed meal in rations for laying hens: a review of the effect on egg quality. *Journal of the Science of Food and Agriculture*, **31**, 515–525

Fisher, C. (1982) *Energy Values of Compound Poultry Feeds*, Occasional Publication no 2, Roslin, Midlothian, Institute for Grassland and Animal Production, Poultry Division

Fisher, C. and McNab, J.M. (1987) Techniques for determining the metabolizable energy (ME) content of poultry feeds. In *Recent Advances in Animal Nutrition – 1987* (eds W. Haresign and D.J.A. Cole) Butterworths, London, pp. 3–18

Fournier, L. (1988) Energie métabolisable des aliments composés pour poulet de chair et pondeuses. *Rapport de Maîtrise de Sciences et Techniques en Production Animale*. Tours, France, Université F. Rabelais

Guillaume, J. and Gomez, J. (1980) Digestibilité de la protéine brute de quelques matières premières chez le coq. II. Résultats sur 14 produits. Moyenne et variabilité. *Archiv für Geflügelkunde*, **44**, 55–56

Härtel, H. (1979) Methods to calculate the energy of mixed feeds. In *Proceedings of the Second European Symposium on Poultry Nutrition, 8–11 October 1979, Beekbergen, The Netherlands* (eds C.A. Kan and P.C.M. Simons) Beekbergen, The Netherlands, Dutch branch of the W.P.S.A., Spelderholt Institute for Poultry Research

Härtel, H. (1986) Influence of food input and procedure of determination on metabolisable energy and digestibility of a diet measured with young and adult birds. *British Poultry Science*, **27**, 11–39

Härtel, H., Schneider, W., Seibold, R. and Lantzsch, H.J. (1977) Beziehungen zwischen der N-korrigierten Umsetzbaren Energie und den Nährstoffgehalten des Futters beim Huhn. *Archiv für Geflügelkunde*, **41**, 152–181

Hill, F.W. and Anderson, D.L. (1958) Comparison of metabolizable energy and productive energy determinations with growing chicks. *Journal of Nutrition*, **64**, 587–603

Hochstetler, H.W. and Scott, M.L. (1975) Metabolizable energy determinations with adult chickens. In *Proceedings of Cornell Nutrition Conference (1975), 28–30 October 1975*, Cornell University, Ithaca, pp. 81–86

Huyghebaert, G., de Munter, G. and de Groote, G. (1988) The metabolizable energy (AMEn) of fats for broilers in relation to their chemical composition. *Animal Feed Science and Technology*, **20**, 45–58

Janssen, W.M.M.A. (1976) *Research on the Metabolizable Energy of Feedstuffs for Poultry and on the Relation Between the Metabolizable Energy and the Chemical Composition*, Beekbergen, The Netherlands, Rapport 121.76 Afdeling Voeding. Spelderholt Centre for Poultry Research and Extension

Janssen, W.M.M.A. (1985) Prediction of the energy content of fats from chemical analysis. In *Proceedings of the Fifth European Symposium on Poultry Nutrition, 27–31 October, 1985* (ed. S. Bornstein), Ma'ale Hachamisha. Israel Branch of the W.P.S.A., Jerusalem, pp. 191 A–191D

Janssen, W.M.M.A. (ed.) (1986) *European Table of Energy Values for Poultry Feedstuffs*. The Netherlands, W.P.S.A., Beekbergen

Janssen, W.M.M.A. and Carré, B. (1985) Influence of fibre on digestibility of poultry feeds. In *Recent Advances in Animal Nutrition – 1985* (eds W. Haresign and D.J.A. Cole), Butterworths, London, pp. 71–86

Kussaibati, R., Guillaume, J. and Leclercq, B. (1982) The effects of age, dietary fat and bile salts, and feeding rate on apparent and true metabolizable energy values in chickens. *British Poultry Science*, **23**, 393–403

Kussaibati, R. and Leclercq, B. (1985) A simplified rapid method for the determination of apparent and true metabolizable energy values of poultry feed. *Archiv für Geflügelkunde*, **49**, 54–62

Kussaibati, R., Leclercq, B. and Guillaume, J. (1983) Effets du calcium, du magnésium et des sels biliaires sur l'énergie métabolisable apparente et la digestibilité des lipides, de l'amidon et des protéines chez le poulet en croissance. *Annales de Zootechnie*, **32**, 7–20

Lacassagne, L., Francesch, M., Carré, B. and Melcion, J.P. (1988) Utilization of tannin-containing and tannin-free Faba beans (*Vicia faba*) by young chicks: effects of pelleting feeds on energy, protein and starch digestibility. *Animal Feed Science and Technology*, **20**, 59–68

Leclercq, B. and Escartin, R. (1987) Further investigations on the effects of metabolizable energy content of diet on broiler performances. *Archiv für Geflügelkunde*, **51**, 93–96

Lessire, M. (1985) Faut-il remettre en cause la valeur énergétique du maïs? In *Comptes rendus de la Conférence Avicole W.P.S.A.-S.I.M.A.V.I.P. du 18 octobre 1985, Paris. Cahier No 1: valeur énergétique et qualité des aliments*, Nouzilly – Monnaie, France, Groupe français de la W.P.S.A., pp. 26–36

Lessire, M. and Leclercq, B. (1983) Metabolizable energy content of meat meal for chicken. *Archiv für Geflügelkunde*, **47**, 1–3

Lessire, M., Sauveur, B., Guillaume, J., Conan, L. and Mallet, S. (1980) Premiers résultats sur la valeur énergétique du pois protéagineux chez le coq et le rat. *Industries de l'Alimentation Animale*, **337**, 47–52

Lessire, M., Leclercq, B., Conan, L. and Hallouis, J.M. (1985) A methodological study of the relationship between the metabolizable energy values of two meat meals and their level of inclusion in the diet. *Poultry Science*, **64**, 1721–1728

MacAuliffe, T. and McGinnis, J. (1971) Effect of antibiotic supplements to diets containing rye on chick growth. *Poultry Science*, **50**, 1130–1134

Mateos, G.G. and Sell, J.L. (1980) True and apparent metabolizable energy value of fat for laying hens: influence of level of use. *Poultry Science*, **59**, 369–373

Mitchell, H.H. (1942) The evaluation of feeds on the basis of digestible and metabolizable nutrients. *Journal of Animal Science*, **1**, 159–173

Moir, K.W., Yule, W.J. and Connor, J.K. (1980) Energy losses in the excreta of poultry: a model for predicting dietary metabolizable energy. *Australian Journal of Experimental Agriculture and Animal Husbandry*, **20**, 151–155

Mollah, Y., Bryden, W.L., Wallis, I.R., Balnave, D. and Annison, E.F. (1983) Studies on low metabolizable energy wheats for poultry using conventional and rapid assay procedures and the effect of processing. *British Poultry Science*, **24**, 81–89

Moran, E.T. (1982) Starch digestion in fowl. *Poultry Science*, **61**, 1257–1267

Nyman, M., Siljeström, M., Pedersen, B., Bach Knudsen, K.E., Asp, N.G., Johansson, C.G. and Eggum, O. (1984) Dietary fiber content and composition in six cereals at different extraction rates. *Cereal Chemistry*, **61**, 14–19

Papadopoulos, M.C. (1985) Estimations of amino acid digestibility and availability in feedstuffs for poultry. *World's Poultry Science Journal*, **41**, 64–71

Petersen, C.F., Meyer, G.B. and Sauter, E.A. (1976) Comparison of metabolizable energy values of feed ingredients for chicks and hens. *Poultry Science*, **55**, 1163–1165

Polin, D. and Hussein, T.H. (1982) The effect of bile acid on lipid and nitrogen retention, carcass composition, and dietary metabolizable energy in very young chicks. *Poultry Science*, **61**, 1697–1707

Saini, H.S. and Henry, R.J. (1989) Fractionation and evaluation of triticale pentosans: comparison with wheat and rye. *Cereal Chemistry*, **66**, 11–14

Saunders, R.M., Walker, H.G. and Kohler, G.O. (1968) The digestibility of steam-pelleted wheat bran. *Poultry Science*, **47**, 1636–1637

Shannon, D.W.F. (1971) The effect of level of intake and free fatty acid content on the metabolizable energy value and net absorption of tallow by the laying hen. *Journal of Agricultural Science, Cambridge*, **76**, 217–221

Sibbald, I.R. (1986) *The T.M.E. System of Feed Evaluation: Methodology, Feed Composition Data and Bibliography*. Technical Bulletin 1986-4 E. Animal Research Centre, Research Branch – Direction générale de la recherche, Agriculture Canada, Ottawa

Sibbald, I.R. and Kramer, J.K.G. (1978) The effect of the basal diet on the true metabolizable energy value of fat. *Poultry Science*, **57**, 685–691

Sibbald, I.R. and Price, K. (1976a) Relationship between metabolizable energy values for poultry and some physical and chemical data describing Canadian wheats, oats and barleys. *Canadian Journal of Animal Science*, **56**, 255–268

Sibbald, I.R. and Price, K. (1976b) True metabolizable energy values for poultry of Canadian barleys measured by bioassay and predicted from physical and chemical data. *Canadian Journal of Animal Science*, **56**, 775–782

Sibbald, I.R. and Price, K. (1977) True and apparent metabolizable energy values for poultry of Canadian wheats and oats, measured by bioassay and predicted from physical and chemical data. *Canadian Journal of Animal Science*, **57**, 365–374

Sibbald, I.R., Price, K. and Barrette, J.P. (1980) True metabolizable energy values for poultry of commercial diets measured by bioassay and predicted from chemical data. *Poultry Science*, **59**, 808–811

Sibbald, I.R., Slinger, S.J. and Ashton, G.C. (1961) Factors affecting the metabolizable energy content of poultry feeds. 4. – The influence of calcium, phosphorus, antibiotic and pantothenic acid. *Poultry Science*, **40**, 945–951

Sibbald, I.R., Summers, J.D. and Slinger, S.J. (1960) Factors affecting the metabolizable energy content of poultry feeds. *Poultry Science*, **39**, 544–556

Sibbald, I.R., Czarnocki, J., Slinger, S.J. and Ashton, G.C. (1963) The prediction of the metabolizable energy content of poultry feedingstuffs from a knowledge of their chemical composition. *Poultry Science*, **42**, 486–492

Sukhija, P.S., Loomba, A., Ahuja, K.L. and Munshi, S.K. (1985) Glucosinolates and lipid content in developing and germinating cruciferous seeds. *Plant Science*, **40**, 1–6

Titus, H.W. (1955) *The Scientific Feeding of Chickens*, 3rd edn. Danville, USA, The Interstate

Van Soest, P.J. and Wine, R.H. (1967) Use of detergents in the analysis of fibrous feeds. IV. Determination of cell-wall constituents. *Journal of the Association of Official Analytical Chemists*, **50**, 50–55

Ward, W.T. and Marquardt, R.R. (1987) Antinutritional activity of a water-soluble pentosan-rich fraction from rye grain. *Poultry Science*, **66**, 1665–1674

Wiseman, J. (1984) Assessment of the digestible and metabolizable energy of fats for non-ruminants. *Fats in Animal Nutrition* (ed. J. Wiseman), London, Butterworths, pp. 277–297

Wiseman, J., Cole, D.J.A., Perry, F.G., Vernon, B.G. and Cooke, B.C. (1986) Apparent metabolizable energy values of fats for broiler chicks. *British Poultry Science*, **27**, 561–576

Wiseman, J. and Lessire, M. (1987a) Interactions between fats of differing chemical content: apparent metabolizable energy values and apparent fat availability. *British Poultry Science*, **28**, 663–676

Wiseman, J. and Lessire, M. (1987b) Interactions between fats of differing chemical content: apparent availability of fatty acids. *British Poultry Science*, **28**, 677–691

17

PREDICTING THE NUTRITIVE VALUE OF COMPOUND FEEDS FOR RUMINANTS

P.C. THOMAS
West of Scotland College, Auchincruive, Ayr, UK

The purchase of compound concentrate feeds represents a substantial part of the variable cost of ruminant animal production on most UK farms, and has a major influence on their economic profitability. In many instances, a change in the amount or composition of the concentrate used also provides an important management control over product yield and quality. Thus, assessment of the value of compound feeds in nutritional terms is of considerable practical and economic significance.

Nutritive value and feeding value

It may be useful at the outset to draw a distinction between the terms *nutritive value* and *feeding value* since these terms are sometimes regarded as synonymous. In this paper the term nutritive value is used specifically to describe the nutrient content of a feed per unit of mass. By comparison, the term feeding value is used to embrace the wider concept of the capacity of a feed to contribute directly or indirectly to the nutrient supply of the animal and to influence nutrient utilization. This distinction is important. In practice, where animals may be offered forage *ad libitum* with a combination of supplementary feeds, the influence of a given feed on forage intake and the interaction of one feed with others may be crucial in determining animal performance. Under these circumstances two concentrate feeds may be of similar nutritive value but of differing feeding value, because of their effects on forage intake or their synergistic or antagonistic interaction with other feeds in the total diet.

Nutritive value

Leaving aside the assessment of the vitamin and mineral components of feeds, which have been dealt with earlier in this book (see Chapters 13 and 14), the value of concentrate feeds is measured in terms that are compatible with the feeding systems that are used as a guide for rationing. The systems presently used in Europe and North America have been reviewed by Van der Honing and Steg

(1990) and Alderman and Jarrige (1987). In detail there are differences between the systems adopted in each country, but all have the following features:

1. They use 'unifying currencies' of energy and protein to relate minimal animal requirements and dietary supply.
2. They adopt a factorial approach to the estimation of animal requirements.
3. They consider feeds to be digested to supply energy and protein, which is absorbed by the animal and used with variable efficiency for maintenance, tissue synthesis, lactation, etc.

They also conceptually regard dietary energy supply as the primary 'driving force' for animal production and dietary protein supply as permissive in allowing the influences of energy supply to be expressed. Thus within the framework of these systems prediction of nutritive value of feeds focuses on prediction of the energy and protein value of the feed.

These topics are considered in the following sections where, for simplicity, reference has been made mainly to the metabolizable energy (ME) system which is the principal energy rationing system used in Europe.

Prediction of feed energy value

For the 'straight' concentrate feeds that are used as ingredients in compounded feeds there are extensive published databases of 'typical' feed analyses, including metabolizable energy values (e.g. MAFF, 1986). In recent years the reliability of these databases in the UK has improved considerably as more and more feeds have been subjected to *in vivo* study at the Feed Evaluation Units at the Rowett Research Institute (RRI) and the ADAS station at Drayton (MAFF, 1986). However, it should be noted that many published feed tables still contain representative feed energy values which have been estimated by calculation rather than by direct determination. Furthermore, even for feeds generally considered to have relatively constant composition there may be significant sample by sample variation in energy value (Table 17.1). Thus, if precise energy values for compound feeds are the objective, it is important that ingredient feed samples are tested to ensure that they are correctly valued. Additionally, there is a need for laboratory evaluation to provide quality control on the compounded product.

Table 17.1 RANGE OF OBSERVED METABOLIZABLE ENERGY (ME) VALUES BETWEEN SAMPLES OF SOME COMMON FEEDS

Feed	Number of samples evaluated	Range of ME values reported (MJ/kgDM)
Barley	16	11.70–14.60
Wheat	16	11.90–14.70
Maize	16	11.69–15.40
Sugar beet pulp	12	10.80–13.00
Soya bean meal	6	12.55–13.95

Results are from DAFS, 1975; Wainman, Dewey and Boyne, 1978; Wainman, Dewey and Brewer, 1984

METHODS FOR THE ESTIMATION OF FEED ENERGY VALUES

The methods that may be used for the estimation of the energy value of feeds can be categorized under three headings:

1. Older methods based on chemical composition and digestibility.
2. Newer methods based on regression relationships with feed composition.
3. Prospective methods based on models of digestion and nutrient metabolism.

Each type of method has its advantages and limitations and each may find application in specific circumstances. So far as the author is aware, there has been no comprehensive comparison of the relative accuracies of the different approaches across the full range of concentrate feedstuffs.

Methods based on chemical composition and digestibility

A number of approximations can be used to provide an estimate of the ME content of test feeds from limited information about their digestibility and chemical composition (see MAFF, 1975, 1984; Morgan, 1980). The starting point for the calculations is either the gross energy (GE) content of the feed determined by bomb calorimetry or a gross energy content estimated from the chemical composition of the feed. The latter is most commonly achieved through application of the following equation, which is based on the work of the Oskar Kellner Institute (Nehring, 1969) and has a reported residual standard deviation of $\pm0.2\,$MJ/kg DM.

$$GE = 0.0226\,CP + 0.0407\,EE + 0.0192\,CF + 0.0177\,NFE \qquad (17.1)$$

Where:

GE is gross energy (MJ/kgDM),
CP is crude protein (N $\times$ 6.35 g/kgDM),
EE is ether extract (g/kgDM),
CF is crude fibre (g/kgDM) and
NFE is nitrogen free extract (g/kgDM).

Estimates of the digestibility of organic matter (OMD) in the feed can be derived from *in vitro* determination using a modified Tilley and Terry (1963) procedure (see Wainman, Dewey and Brewer, 1984). Since this value is very close to the corresponding digestibility of energy, the ME content of a feed can be estimated from:

$$ME = 0.84\,(GE \times OMD) \qquad (17.2)$$

Where:

ME is metabolizable energy (MJ/kgDM),
GE is gross energy (MJ/kgDM) and
OMD is *in vitro* organic matter digestibility.

It should be noted that the factor 0.84 which is used to convert digestible energy (DE) to ME is greater than the 0.81 normally applied to forage feeds (MAFF, 1984), since the ME/DE ratio for concentrates is higher than that for forages.

A simplified version of this approach is to calculate the ME (MJ/kgDM) content of the feed from the content of digestible organic matter in the dry matter (DOMD, g/kgDM) using the equation ME = 0.16 × DOMD. This makes the assumption that the average ME content of digested organic matter is 19 MJ/kgDM, and again allows for an ME/DE ratio of 0.84. The errors of this approach will increase with diets containing significant amounts of added fat because the ME value of DOMD is raised.

Based on an extensive series of calorimetric studies in sheep and cattle, Nehring (1969) and Hoffman (1969) published a series of equations to estimate the DE, ME and NE contents of feeds from information on their content and digestibility of proximate constituents. These equations have been widely adopted and used for the estimation of ME content. For example, MAFF (1984) advises the generalized equation:

$$ME = 0.0152\,DCP + 0.0342\,DEE + 0.128\,DCF + 0.0159\,DNFE \qquad (17.3)$$

Where:

ME   is metabolizable energy (MJ/kgDM),
DCP   is digestible crude protein (g/kgDM),
DEE   is digestible ether extract (g/kgDM),
DCF   is digestible crude fibre (g/kgDM),
DNFE is digestible nitrogen free extract (g/kgDM).

Under circumstances where information on the digestibility of the proximate constituents is not available the equation can be applied through use of typical digestibility values: 0.8 for crude protein; 0.9 for ether extract; 0.4 for crude fibre; and 0.9 for nitrogen free extract (MAFF, 1975). Alternatively, MAFF (1984) recommends an approach using the feed DOMD value:

$$ME = (0.0152\,CP + 0.0342\,EE + 0.128\,CF + 0.0159\,NFE)$$
$$\times\ \frac{DOMD}{1000 - TA} \qquad (17.4)$$

Where the terms are as defined in equation 17.3, TA is the total ash content (g/kgDM) and DOMD is the digestible organic matter in the dry matter (g/kg).

Methods based on regression relationships with feed composition

Regression relationships to estimate the ME content of compound feeds from laboratory determinations of their chemical composition were initially developed on the basis of the studies carried out at the RRI in the late 1970s (Wainman, Dewey and Boyne, 1981). In these experiments a total of 24 compound feeds, formulated by selection of ingredients from a catalogue of 35 common raw materials, were subjected to systematic laboratory analysis and were also evaluated calorimetrically to determine their ME content in sheep at the maintenance level of feeding. The 24 compounds were formulated to cover a range of composition as specified by the following criteria:

1. Two ranges of ether extract, 20–39 g/kgDM and 50–70 g/kgDM.
2. Two ranges of crude fibre, 40–60 g/kgDM and 80–120 g/kgDM.
3. Three ranges of crude protein, 120–149 g/kgDM, 150–179 g/kgDM and 180–209 g/kgDM.

The ME content of each of the feeds was determined using a 'by difference' method with each feed given together with hay or silage and at levels of 25%, 50% and 75% of the total diet. Observations on each compound feed and forage combination were made in duplicate using two sheep and thus the experimental design provided twelve independent observations for each compound. This approach allowed initial examination of the data for interactions between the concentrate and type of forage, and for linearity with respect to the level of inclusion of concentrate in the diet. However, neither of these effects was found to be significant and it was therefore possible to fit the data to a simple linear model that allowed the ME content of the concentrate to be derived by regression analysis. For the feeds examined, the range of ME content varied widely, from 9.8 MJ/kgDM to 13.7 MJ/kgDM.

On the basis of these results Wainman, Dewey and Boyne (1981) developed a total of 73 regression equations which would allow the ME content of compound feeds to be predicted from laboratory analysis with an error of less than 0.5 MJ/kgDM. However, it was clear that in practice some equations would be more suitable for routine use than others. This question was examined by a UKASTA/ADAS/COSAC Working Party (1985) who concluded that three equations should be recommended for adoption by the feed industry and advisory services. These equations most closely met the criteria established by the Working Party of low residual standard deviations taking account of regression and inter-laboratory analytical variances, of few analytical determinations, and of high speed and low cost of analysis. The details of the equations, designated U1, U2 and U3, are given in Table 17.2.

These equations offered a significant advance in the routine prediction of feed ME content, and U1 in particular was widely used since its constituent terms were coincident with the analytical declaration required under the UK Feedingstuffs Regulations. However, as time passed several reservations about the equations began to emerge.

Table 17.2 RECOMMENDED EQUATIONS FOR THE PREDICTION OF METABOLIZABLE ENERGY CONTENT FROM THE CHEMICAL ANALYSIS OF COMPOUNDED FEEDS

Equation designation	Recommended use	Equation*	Standard deviation (MJ/kgDM)†
U1	Legal and voluntary	$ME = 11.78 + 0.0654\,CP + 0.0665\,EE^2 \\ -0.0414\,EE \times CF - 0.118\,TA$	0.36
U2	Reference purposes	$ME = 11.56 - 2.37\,EE + 0.030\,EE^2 \\ + 0.030\,EE \times NCD - 0.034\,TA$	0.32
U3	Voluntary declaration	$ME = 13.83 - 0.488\,EE + 0.0394\,EE^2 \\ \times CP - 0.0085\,MADF \times CP - 0.138\,TA$	0.35

*ME is MJ/kgDM, all other units are g/100 g DM. CP = crude protein; EE = ether extract; CF = crude fibre; TA = total ash; NCD = cellulase digestible organic matter in the dry matter following neutral detergent extraction; MADF = modified acid detergent fibre (for details see Wainman, Dewey and Brewer, 1984)

†Residual standard deviation taking account of between laboratory variances in chemical analysis of feed composition (see UKASTA/ADAS/COSAC, 1985)

First amongst these was the fact that the ME values predicted by the equations for analysed compounded feeds tended to diverge systematically from values calculated from ingredient proportions using the ME contents of the ingredients as given in the contemporary UKASTA and ADAS feed databases (UKASTA/ADAS/COSAC, 1985).

Second, although the range of feeds examined by Wainman, Dewey and Boyne (1981) was representative of that being manufactured in the late 1970s, developments in technology in the 1980s had led to a greater use and wider variety of added fat and fibre sources than previously. It was thus questioned whether the ME content of these new types of feed would be satisfactorily estimated by the existing prediction equations. The changes in feed design were recognized by the analytical chemists in 1982, when to accommodate the increased use of fatty acid oils, the Feedingstuffs Regulations were modified to include an acid extraction procedure for fats and oils (Method B) in addition to the existing ether extract method for neutral fats (Method A).

Finally, there was an unanswered question as to whether predicted values for feed ME based on measurements made in mature sheep at a maintenance level of feeding were accurate for compounded dairy feeds, which are used for lactating cows at much higher levels of feeding.

To address these issues a new series of calorimetric studies with sheep and dairy cows was initiated at the RRI and Hannah Research Institute (HRI) in 1986. Details of these experiments are given elsewhere (Thomas *et al.*, 1988) but briefly the studies consisted of a coordinated series of calorimetric evaluations of compound feeds together with associated laboratory analyses similar to, but more extended than, those undertaken by Wainman, Dewey and Boyne (1981). During the experiments, a total of 100 compound feeds was investigated in sheep at the maintenance level of feeding and 9 of the feeds were selected for additional calorimetric studies in lactating dairy cows.

When compared with the earlier studies there were several differences in experimental approach. Firstly, the compound feeds were given with silage as the only forage and at one level of inclusion (50%). As before, the ME content of the compound was determined 'by difference' but this was achieved through measurements of the ME content of the silage when given as a sole feed. Secondly, although the compound feeds used were formulated from a raw material catalogue of twenty ingredients, the ingredients and formulations were carefully selected to provide a 'matrix' of treatments varying in the type and level of added fat and fibre sources (Table 17.3). Thirdly, in contrast to the balanced design used in the earlier sheep work the experimental treatments were unequally replicated (Table 17.3) to strengthen the statistical model in areas of particular responsiveness.

The dairy cow experiments were designed to test predictive models developed from the sheep work, and thus the 9 compound feeds examined represented treatment 'extremes' within the matrix (Table 17.3). With sheep studies, the compounds were fed in equal proportions with silage and the 'by difference' calculations of concentrate ME were achieved through determinations of the ME content of the basal silage when given as a sole feed.

In comparison with the diets used in the initial RRI study those in the later work covered a greater range of composition and included diets containing higher concentrations of fat and fibre (Table 17.4). However, with some adjustments in statistical approach to take account of the differences in experimental design, it proved possible to combine the earlier and later experimental results to provide a

Table 17.3 A SUMMARY OF THE FAT AND FIBRE SOURCES AND THE LEVELS OF INCLUSION FOR THE COMPOUNDED FEEDS USED AND THE NUMBER OF EXPERIMENTAL OBSERVATIONS ON EACH FEED

Fat sources	(g/kg) in feed	Fibre source (g/kg) in feed						
		None	Straw		NIS†		Sugar beet and citrus pulp	
			200	400	200	400	200	400
None	0	(10)*	3	(3)	3	5†	3	(3)
	15	–	–	–	3	–	–	–
	30	2	2	2	2	2	2	2
Palm acid oil	60	(3)	2	(3)	4	3	2	(3)
	90	3	–	–	3	3	–	–
Maize/soya oil	30	2	2	2	2	2	2	2
	60	3	2	3	2	3	2	3
	90	3	–	–	3	3	–	–
FP1†	30	2	4	2	4	2	2	2
	60	(3)	2	(3)	2	3	4	(3)
Megalac†:	30	2	2	3	2	2	2	2
	60	3	2	4	2	3	2	3
Fat prills†	15	–	–	–	–	–	–	–
	30	2	2	2	2	2	2	2
	60	3	2	3	2	3	2	3
	90	–	–	–	–	–	3	–

*Values in parentheses indicate treatments which were common to the sheep study and the dairy cow study
†NIS is nutritionally improved straw, prepared through NaOH treatment. FPI is a protein coated, 'rumen-protected' fat supplement. Megalac and fat prills are two commercially prepared 'free flow' fat products

Table 17.4 THE MEAN CHEMICAL COMPOSITION AND THE RANGE IN CHEMICAL COMPOSITION FOR THE COMPOUNDED FEEDS USED IN THE STUDY OF WAINMAN, DEWEY AND BOYNE (1981) AND IN THE ROWETT RESEARCH INSTITUTE/HANNAH RESEARCH INSTITUTE (RRI/HRI) COLLABORATIVE STUDY

	Wainman et al. (1981)		RRI/HRI study	
	Mean (n=24)	Range	Mean (n=100)	Range
Gross energy (MJ/kgDM)	18.12	16.30–19.15	18.66	17.50–20.30
Crude protein (g/kgDM)	167	127–209	200	186–232
Oil (g/kgDM)*	44	27–73	67	19–127
Crude fibre (g/kgDM)	88	37–167	116	48–218
Neutral detergent fibre (g/kgDM)	224	116–363	252	116–428
Acid detergent fibre (g/kgDM)	134	57–239	160	50–288
Sugar (g/kgDM)	69	50–100	96	58–179
Starch (g/kgDM)	320	162–437	194	42–397
NCD (g/kgDM)†	787	666–881	777	620–849

*Oil by method B.
†Cellulase digestible organic matter after neutral detergent extraction.

comprehensive data set for regression analyses based on a total of 124 compound feeds ranging in ME content from 9.80 MJ/kgDM to 14.98 MJ/kgDM. Since in both studies between laboratory comparisons had been undertaken to assess analytical variances on the estimation of chemical constituents, errors arising from this source could also be taken into account in identifying preferred prediction equations.

The results of the data analysis are described at length by Thomas *et al.* (1988) but the main conclusion can be simply summarized here. The analysis showed that the ME content of the compound feeds could be predicted with a very low residual error (0.24 MJ/kgDM) using the following two-term equation, designated Equation E3:

$$ME = 0.25\,Oil + 0.14\,NCD \qquad\qquad (17.5)$$

Where:
ME is metabolizable energy (MJ/kgDM),
Oil is oil as determined by Method B of the Feedingstuffs Regulations, (g/kgDM)
NCD is cellulase digestible organic matter in the dry matter following neutral detergent extraction (g/kgDM).

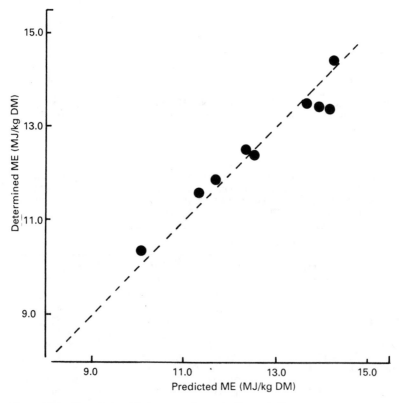

Figure 17.1   The relationship between determined metabolizable energy values and predicted metabolizable energy values using equation E3 for nine compound feeds evaluated in lactating dairy cows (Thomas *et al.*, 1988). The broken line is the line of equivalence

Moreover, prediction of ME by equation E3 provided values close to those calculated from contemporary UKASTA and ADAS database values for ingredient feeds. For example, for a group of 77 compound feeds used in the RRI/HRI study the mean ME predicted by E3 was 12.41 ± 0.13 MJ/kgDM; the corresponding figure calculated from summation of ingredient ME values was 12.35 ± 0.11 MJ/kgDM on the basis of the ADAS database and 12.40 ± 0.11 MJ/kgDM on the basis of the UKASTA database. Also, ME values predicted from E3 were in good agreement with those determined calorimetrically in lactating dairy cows (Figure 17.1).

Equation E3 has now been formally adopted in the UK as the preferred equation for the prediction of ME from laboratory analysis of compound feed composition (MAFF, 1989). The equation is robust and widely applicable although it has been recognized that there may be special problems in the analysis of NCD in feeds containing high levels of palm kernel meal, and this may lead to errors. Further research is being undertaken to resolve this problem.

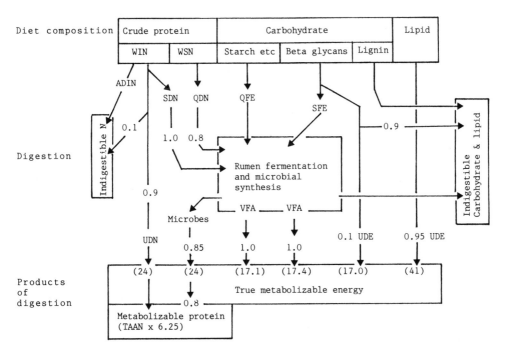

Figure 17.2 A model (MENTOR) to describe the digestion and utilization of feed components in the ruminant (after Webster, Dewhurst and Waters, 1988). WIN, water insoluble nitrogen; WSN, water soluble nitrogen; ADIN, acid detergent insoluble nitrogen; SDN, slowly digested nitrogen; QDN, quickly digested nitrogen; UDN, undegraded digestible nitrogen; QFE, quickly fermented energy; SFE, slowly fermented energy; UDE, unfermented digestible energy; VFA, volatile fatty acids; TAAN, total amino acid nitrogen. Numerical values for transfer coefficients are shown together with the enthalpy values (MJ/kgDM) for absorbed products, which are in parentheses

Predictive methods based on models of digestion and metabolism

An interesting recent approach to the prediction of feed energy value is outlined by Webster, Dewhurst and Waters (1988). This is based on the use of a computer model ('MENTOR') to predict the main products of digestion in the rumen and intestines from analytical details of feed composition, and to describe the energetics of utilization of the products in animal tissues by reference to their enthalpy. The principles of the approach are outlined in Figure 17.2, the main 'energy' components in the diet being described by reference to ruminal digestion as quickly fermented energy (QFE), slowly fermented energy (SFE) and unfermented digestible energy (UDE). The model can be used to calculate true metabolizable energy (ME_t), which is defined as gross energy minus the dietary energy losses in faeces, methane and heat of rumen fermentation. Agreement between ME_t values derived from the model and those estimated directly from calorimetric studies was found to be very close for a series of 121 forage samples evaluated calorimetrically at the RRI (Dewhurst *et al.*, 1986). However, when the model predictions were compared with the directly estimated ME_t values for the 24 compound feeds evaluated by Wainman, Dewey and Boyne (1981) the results showed good correlation but a systematic overprediction such that:

$$ME_{tp} = 3.03 + 0.846\,ME_{to}\ (r = 0.89, SD = 0.76) \tag{17.6}$$

Where
  ME_{tp} is predicted ME_t (MJ/kgDM), and
  ME_{to} is observed ME_t (MJ/kgDM)

The implication of these results is that the model at present fails adequately to describe the fermentation of one or more of the compound feed components. Taking account of the characteristics of the model Webster, Dewhurst and Waters (1988) suggested that the shortcomings were likely to be in the treatment of fibre fermentation or in the failure to account for associative effects of starch on fibre digestion.

Protein

Because of the changes in protein rationing systems that have taken place during the past decade the evaluation of dietary protein sources in terms of crude protein content and digestibility of crude protein has now been superseded by methods which provide an assessment of the rumen-degradability of the dietary protein source (see Alderman and Jarrige, 1987). Thus the pepsin/HCl test (AOAC, 1980) traditionally used for digestibility evaluation has progressively been supplemented by other tests designed to determine rumen degradability. These include laboratory methods based on the solubility of feed proteins in water, mineral buffers or proteolytic enzyme solutions (see NRC, 1985), but the 'standard' method remains the intraruminal incubation test using the Dacron bag technique (Orskov and MacDonald, 1979).

   The protein degradation of ruminant diets has been reviewed in detail in Chapter 4 and will therefore not be considered further here. However, it is relevant to point out that when compared with the methods now available for the prediction of dietary energy values the methods for assessing the characteristics of dietary protein sources remain less than satisfactory. For example, Oldham (1987) in a report of two 'ring tests' using the Dacron bag technique has shown disturbingly large between laboratory variations in estimated protein degradability (Table 17.5). Against this background it is unrealistic to do other than accept broad degradability categories of 0.71–0.90, 0.51–0.70, 0.31–0.50 and ≤0.31, as proposed by AFRC (1984) and adopted by MAFF (1986). However, such categorization by its very nature tends to obscure variations between different consignments of a given type of feed, although these variations may be of considerable nutritional and economic significance.

Table 17.5 MEAN VALUES AND RANGES OF VALUES BETWEEN LABORATORIES FOR THE ESTIMATION OF THE RUMEN DEGRADABILITY OF FEED PROTEIN USING THE DACRON BAG TECHNIQUE (RESULTS ARE TAKEN FROM OLDHAM (1987) AND REFER TO TWO BETWEEN LABORATORY RING TESTS)

Ring test	Feed	Number of laboratories	Mean degradability %	Range of degradability values (%) between laboratories
1*	Hay	13	43	18–57
	Soya meal	12	85	53–99
2†	Hay	5	44	39–51
	Barley	5	75	60–89
	Soya meal	5	61	56–69
	Rape seed meal	5	64	56–68

*Values are mean loss of nitrogen after 24 hours of incubation.
†Values are minimal nitrogen degradabilities calculated according to Orskov and McDonald (1979) with an assumed rumen outflow rate (k) of 0.08.

Feed composition and feeding value

It is self-evident that the chemical composition of compound feeds has a determining influence on their nutritive value as defined and discussed above. However additionally, the chemistry of the feeds may have distinctive influences on their feeding value. These influences arise through effects of the compound feed on forage intake and digestion or on the balance of digestion products absorbed from the gut and utilized in animal tissues. These effects are not explicitly recognized within present rationing systems. However, they are often a consideration in diet design, particularly for dairy cows, and they merit some comment here.

PROTEIN COMPOSITION

There is ample evidence that DM intake, dry matter digestion and milk production are enhanced through increasing the protein content of the diet. Moreover, the effects are observed at dietary protein levels well above those needed to meet the animals theoretical requirements for rumen degradable protein (RDP) and undegraded dietary protein (UDP) (see Oldham, 1984; Thomas and Rae, 1988; Chamberlain, Martin and Robertson, 1989). A variety of studies has shown that milk production responses to animal protein sources such as fishmeal are greater than those to vegetable sources such as soyabean meal. For example, in a summary of work at the HRI, Chamberlain, Martin and Robertson (1989) calculated milk yield responses (g/g change in crude protein intake) were 3.1 ± 0.4 for soya bean meal supplements and 4.1 ± 0.6 for supplements containing a mixture of fishmeal, bloodmeal, and meat and bone meal.

It can be argued that these differential responses to specific protein sources simply reflect differences in rumen degradability. However, when compared with soya, fishmeal supplements were associated with a more efficient transfer of duodenal crude protein to milk (Oldham *et al.*, 1985) and this suggests an effect linked to improvements in the balance of amino acid supply. Methionine and lysine are particularly implicated since these amino acids are present in fishmeal in comparatively high concentration and have been identified as potentially limiting amino acids for milk production.

Against this background, there has been some movement amongst feed manufacturers to select protein sources on the basis of amino acid composition or amino acid composition of the UDP fraction determined on Dacron bag residues, and in some instances specifications for a desired 'duodenal amino acid profile' have been adopted (see Dennison and Phillips, 1983a, 1983b). Whilst this is theoretically sound, the practical benefits have yet to be satisfactorily assessed and at present it would be difficult to give unequivocal support for the approach. Direct supplementation of the diet with lysine and/or methionine in rumen protected form has not been associated with any significant increase in milk yield (Figure 17.3).

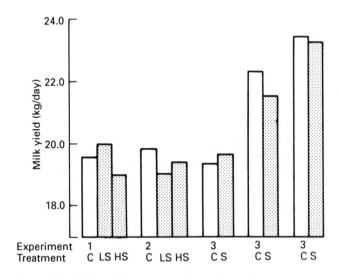

Figure 17.3   Milk yield in cows given diets with or without dietary supplements of methionine (Experiment 1) or methionine and lysine (Experiment 2 and 3) in rumen-protected form. C – control unsupplemented diets; S – diets corresponding to control but with supplements of methionine or methionine and lysine at low (L) or high (H) rates of inclusion (Results are from Shamoon (1983) and Girdler, Thomas and Chamberlain (1987)

CARBOHYDRATE COMPOSITION

There is an increasing trend for compound feed designers to specify criteria for the chemical make-up of the carbohydrate fraction of feeds, either directly or indirectly through the use of constraints on the inclusion of specific raw materials. The specifications are variously influenced by concerns that the diet should have: sufficient digestible fibre to maintain optimum conditions of pH, cellulolysis and volatile fatty acid production in the rumen; sufficient rapidly fermented starch or sugar to optimize rumen microbial protein synthesis; and a proportion of slowly fermented starch to provide a supply of glucose to the small intestine. The emphasis placed on each of these specifications reflects the market niche for which the feed is intended, but it also depends heavily on the design technologist. There is a good deal of private opinion about the optimum make-up of the carbohydrate fraction of compound feeds but relatively little that can be regarded as definitive and unquestionable; in practice most commercial compound formulations reflect conservatism and caution, extremes of composition being avoided.

Experimental evidence indicates that there are benefits in milk fat yield from limiting the starch and increasing the digestible fibre content of feeds for dairy cows, although it should be noted that the most pronounced effects are observed where the proportion of the concentrate in the diet is high (see Thomas and Rae, 1988; Chamberlain, Martin and Robertson, 1989). Similarly, there is clear evidence that where the forage is high-protein silage, replacement of starch in the supplementary feed by sugar can increase microbial protein synthesis in the rumen (Chamberlain *et al.*, 1985), although the effects may not be fully expressed unless the diet is also suitably buffered and supplemented with protein (Newbold, Thomas and Chamberlain, 1988). In experimental studies, sugar inclusions in the diet have given a variable effect on milk production, benefits being obtained in some instances but not in others (Chamberlain *et al.*, 1984; Thomas and Rae, 1988). In view of the known differences between sugars in fermentation rate and pattern of volatile fatty acid production (Sutton, 1968, 1969) all sugars may not be equivalent in their effects. Finally, whilst there is evidence that an increase in the slowly fermented starch content of the concentrate achieved by replacement of barley with ground maize can have beneficial effects on milk yield and/or fat content, the effects are not consistent and appear to vary with the proportion of maize in the concentrate and the proportion of concentrate in the diet (Sutton, Oldham and Hart, 1980; Martin, 1986).

The foregoing examples may lead to the conclusion that if the diet contains a 'balanced mixture' of carbohydrates, small changes in make-up are unlikely to be of significance. However, this is probably too complacent a conclusion. Many nutritionists will relate anecdotally instances where a small change in diet led to unexpected effects on animal performance, and this suggests that subtle changes in formulation may be important under certain circumstances.

As an illustration, reference can be made to a study by Martin (1986). In an experiment conducted over the first ten weeks of lactation two carefully balanced groups of ten cows received a diet of hay (5.8 kgDM/d) and concentrate (9.0 kgDM/d) which was given either with or without 0.73 kgDM/d of a mixture of highly fermentable sugar derivatives, which were included in the concentrate in replacement for barley. In the group of cows given the control concentrate milk production was 'normal', the cows had low blood 3-hydroxybutyrate levels and there were no incidences of ketosis. However, in the cows receiving the sugar derivatives blood 3-hydroxybutyrate levels were high and milk production was impaired because 60% of the animals suffered from clinical bovine ketosis. Subsequently it was demonstrated that the sugar derivatives increased the proportion of butyrate in the rumen, and this was regarded as a factor contributing to their effect on the metabolism of cows.

LIPID COMPOSITION

The inclusion of free fats and oils in dairy compound feeds is now a general practice and the source and composition of the fats and oils is an important consideration. Utilization of lipid supplements for milk production in the dairy cow has been reviewed fully elsewhere (Storry, 1980; Clapperton and Steele, 1983; Palmquist, 1984). In brief, the lipids have beneficial effects on animal performance through increasing the energy supply in the diet and through the provision of preformed fatty acids for milk fat synthesis. However, these benefits may be offset by the adverse effects of the lipids on forage intake, on fibre digestion and lipid synthesis in the rumen, on the ruminal ratio of acetate: propionate and on the synthesis of short- and medium-chain fatty acids *de novo* in the mammary gland. In broad terms, the adverse effects are least with saturated fats fed at low rates and greatest with unsaturated fats fed at high rates, although the adverse effects may be reduced if fats are given in frequent meals throughout the day.

There is no recommended nutritional standard which can be used to specify the fatty acid profile of compound feeds. In practice a considerable range in fatty acid composition may occur and this can make a significant contribution to variations in feeding value between feeds of similar ME content. The adverse nutritional influences associated with supplementary lipids are best avoided by specifying feed formulations with a rather saturated fatty acid profile, and the fatty acid composition of red palm oil is often seen as a useful guide because of the beneficial

Table 17.6 THE INTAKE OF SILAGE AND CONCENTRATE DRY MATTER AND THE YIELD AND COMPOSITION OF MILK IN DAIRY COWS GIVEN SILAGE *AD LIBITUM* AND CONCENTRATES CONTAINING NO ADDED FIBRE SOURCE, GROUND STRAW OR A MIXTURE OF SUGAR BEET PULP AND CITRUS PULP IN COMBINATION WITH NO ADDED FAT, PALM ACID OIL (F) AND A RUMEN-PROTECTED FORM OF PALM ACID OIL (PF) (RESULTS ARE FROM THOMAS AND ROBERTSON (1987). FIBRE SOURCES WERE ADDED AT A RATE OF 400 g/kg AND FAT SOURCES AT 60 g/kg)

Fibre inclusion:	*None*			*Straw*			*Pulp*			*SE of difference*
Fat inclusion:	O	F	PF	O	F	PF	O	F	PF	
Concentrate intake (kgDM/d)	6.75	6.84	6.83	6.88	6.82	6.77	6.74	6.85	6.86	
Silage intake (kgDM/d)	9.86	8.69	8.67	9.90	9.32	9.01	9.40	9.33	9.21	0.27***
Total DM intake (kgDM/d)	16.61	15.53	15.50	16.78	16.14	15.78	16.14	16.18	16.07	0.27*
Milk yield (kg/d)	19.9	21.0	24.4	19.3	20.9	21.7	21.7	21.1	22.4	1.1**
Milk fat (g/kg)	40.1	42.3	35.2	41.9	41.3	37.0	43.4	40.0	34.7	1.5***
Milk protein (g/kg)	33.7	30.9	29.3	32.3	30.6	29.4	32.5	31.2	30.0	0.5***
Milk lactose (g/kg)	46.9	46.2	48.4	46.2	46.3	45.4	47.6	45.8	46.4	0.8*

* P$\leqslant$0.05; ** P$\leqslant$0.01; *** P$\leqslant$0.01. Statistically significant differences by F test

influences of this oil on milk yield and milk fat yield (Storry, Hall and Johnson, 1968). Alternatively, fats may be incorporated in the diet in a protected form (Storry and Brumby, 1980), thus avoiding both microbial biohydrogenation of the fats in the rumen and, more importantly, the adverse effects of the fats on rumen microbial metabolism.

However, irrespective of the form of inclusion of fat, recent evidence suggests that there are important nutritional interactions between fat and fibre sources in compound feeds and that these may modulate the influences of the fats on milk yield and composition (Table 17.6).

Conclusions

The past two decades have seen major advances in the development of energy and protein rationing systems for ruminant farm livestock and a considerable commitment of effort and resources to the evaluation of common feedstuffs. As a result, there is now a sound framework of scientific understanding underpinning the nutritional management of livestock in practice.

As has been indicated in the foregoing sections of this paper, traditional methods for the estimation of the ME content of compound feeds from their proximate composition and digestibility have now been largely superseded by regression methods based on established links between the chemical composition of the feed and its ME content determined *in vivo*. The recent development of equation E3, in particular, offers a rapid and reliable laboratory method that will predict ME with good accuracy across a wide range of feed formulations.

Progress has also been made in the evaluation of dietary protein sources, although concerns must remain about the accuracy of estimation of rumen degradability whilst between laboratory reproducibility of the Dacron bag technique remains poor.

Finally, whilst methods to predict the nutritive value of compound feeds have progressively advanced, many questions relating to the feeding value of the feeds remain to be addressed. As the accuracy with which nutritive value can be predicted improves and precise recording of animal performance on farms becomes more common, questions related to feeding value will gain prominence, perhaps heralding the research challenge of the next decade.

References

Alderman, G. and Jarrige, R. (1987) *Protein Evaluation of Ruminant Feeds*, Luxembourg, Commission of European Communities

Agricultural and Food Research Council (1984) *The Nutrient Requirements of Ruminant Livestock, Supplement 1.* Farnham Royal, Slough, Commonwealth Agricultural Bureaux

AOAC (1980) *Association of Official Agricultural Chemists*, 13th edn.   Washington, DC, AOAC, pp. 130–132

Chamberlain, D.G., Martin, P.A. and Robertson, S. (1989) Optimising compound feed use in dairy cows with high intakes of silage. In *Recent Advances in Animal Nutrition – 1989* (eds W. Haresign and D.J.A. Cole). London, Butterworths, pp. 175–194

Chamberlain, D.G., Thomas, P.C., Wilson, W.D., Kassem, M.E. and Robertson, S. (1984) The influence of the type of carbohydrate in the supplementary concentrate on the utilization of silage diets for milk production. In *Proceedings of the Seventh Silage Conference* (eds F.J. Gordon and E.F. Unsworth). Belfast, The Queens University, pp. 37–38

Chamberlain, D.G., Thomas, P.C., Wilson, W., Newbold, C.J. and Macdonald, J.C. (1985) The effect of carbohydrate supplements on ruminal concentrations of ammonia in animals given diets of grass silage. *Journal of Agricultural Science, Cambridge*, **104**, 331–340

Clapperton, J.L. and Steele, W. (1983) Fat supplementation in animal production – ruminants. *Proceedings of the Nutrition Society*, **42**, 343–350

DAFS (1975) *First Report of the Feedingstuffs Evaluation Unit*, Edinburgh, Department of Agriculture and Fisheries for Scotland, pp. 1–55

Dennison, C. and Phillips, A.M. (1983a) Estimation of the duodenal amino acid supply in ruminants by amino acid analysis of the products of fermentation *in vitro*. *South African Journal of Animal Science*, **13**, 120–126

Dennison, C. and Phillips, A.M. (1983b) Balancing the duodenal amino acid supply in ruminants with practical feed ingredients. *South African Journal of Animal Science*, **12**, 229–235

Dewhurst, R.J., Webster, A.J.F., Wainman, F.W. and Dewey, P.J.S. (1986) Prediction of the true metabolizable energy concentration in forages for ruminants. *Animal Production*, **43**, 183–194

Girdler, C.P., Thomas, P.C. and Chamberlain, D.G. (1987) Responses to dietary rumen-protected amino acids and abomasally infused protein and amino acids in lactating cows given silage diets. In *The Eighth Silage Conference* (ed. C. Thomas). Hurley, Institute of Grassland and Animal Production, pp. 73–74

Hoffman, L. (1969) The suitability of energy index numbers for performance prediction. In *Energy Metabolism of Farm Animals* (eds K.L. Blaxter, J. Kielanowski and G. Thorbek). Newcastle Upon Tyne, Oriel Press, pp. 51–58

MAFF (1975) *Energy Allowances and Feeding Systems for Ruminants*. Technical Bulletin No. 33. London, HMSO

MAFF (1984) *Energy Allowances and Feeding Systems for Ruminants*, London, HMSO

MAFF (1986) *Feed Composition*, Marlow, Chalcombe Publications

MAFF (1989) *The Prediction of the Metabolizable Energy Values of Compound Feedingstuffs for Ruminant Animals*. Summary of the Recommendations of a Working Party Sponsored by the Ministry of Agriculture, Fisheries and Food. London, Ministry of Agriculture, Fisheries and Food

Martin, P.A. (1986) *The Effects of Polyols and Selected Starch Sources on the Metabolism and Milk Production of Dairy Cows*, PhD Thesis, University of Glasgow, UK

Morgan, D.E. (1980) The advisers approach to predicting the metabolizable energy value of feeds for ruminants. In *Recent Advances in Animal Nutrition – 1979* (eds W. Haresign and D. Lewis). London, Butterworths, pp. 93–106

National Research Council (1985) *Ruminant Nitrogen Usage*, National Research Council Sub-Committee on Nitrogen Usage in Ruminants. Washington, DC, National Academy Press, pp. 23–27

Nehring, K. (1969) Investigations on the scientific basis for the use of net energy for fattening as a measure of feed value. In *Energy Metabolism of Farm Animals* (eds K.L. Blaxter, J. Kielanowski and G. Thorbek). Newcastle upon Tyne, Oriel Press, pp. 5–20

Newbold, C.J., Thomas, P.C. and Chamberlain, D.G. (1988) Effect of dietary supplements of sodium bicarbonate on the utilization of nitrogen in the rumen of sheep receiving a silage based diet. *Journal of Agricultural Science, Cambridge,* **110**, 383–386

Oldham, J.D. (1984) Protein–energy interrelationships in dairy cows. *Journal of Dairy Science,* **67**, 1090–1114

Oldham, J.D. (1987) Testing and Implementing the Modern Systems: UK. In *Feed Evaluation and Protein Requirements Systems for Ruminants* (eds R. Jarrige and G. Alderman). Luxembourg, Commission of the European Communities Office of Official Publications, pp. 269–281

Oldham, J.D., Phipps, R.H., Fulford, R.J., Napper, D.J., Thomas, J. and Weller, R.F. (1985) Response of dairy cows to rations varying in fishmeal on soyabean meal in early lactation. *Animal Production,* **40**, 519 (Abstract).

Orskov, E.R. and Macdonald, I. (1979) The estimation of protein degradability in the rumen from incubation measurements weighted according to rate of passage. *Journal of Agricultural Science, Cambridge,* **92**, 499–523

Palmquist, D. (1984) Use of fats in diets for lactating dairy cows. In *Fats in Animal Nutrition* (ed. J. Wiseman). London, Butterworths, pp. 357–281

Shamoon, S.A. (1983) *Amino Acid Supplements for Ruminant Farm Livestock with Special Reference to Methionine.* PhD Thesis, University of Glasgow

Storry, J.E. (1980) Influence of nutritional factors on the yield and content of milk fat: non-protected fat in the diet. In *Factors Affecting the Yields and Contents of Milk Constituents of Commercial Importance,* Brussels, International Dairy Federation, Document No. 125, pp. 88–95

Storry, J.W. and Brumby, P.E. (1980) Influence of nutritional factors on the yield and content of milk fat: protected non-polyunsaturated fat in the diet. In *Factors Affecting the Yields and Contents of Milk Constituents of Commercial Importance,* Brussels, International Dairy Federation Document No. 125, pp. 105–125

Storry, J.E., Hall, A.J. and Johnson, V.W. (1968) The effect of increasing amounts of dietary red palm oil on milk fat secretion in the cow. *British Journal of Nutrition,* **22**, 609–614

Sutton, J.D. (1968) The fermentation of soluble carbohydrates in rumen contents of cows fed diets containing a large proportion of hay. *British Journal of Nutrition,* **22**, 89–712

Sutton, J.D. (1969) The fermentation of soluble carbohydrates in rumen contents of cows given diets containing a large proportion of flaked maize. *British Journal of Nutrition,* **23**, 567–583

Sutton, J.D., Bines, J.A. and Napper, D.J. (1985) Composition of starchy and fibrous concentrates for lactating dairy cows. *Animal Production,* **40**, 533 (Abstract)

Sutton, J.D., Oldham, J.D. and Hart, I.C. (1980) Products of digestion, hormones and energy utilization in milking cows given concentrates containing varying proportions of barley or maize. In *Energy Metabolism* (ed. L.E. Mount), London, Butterworths, pp. 303–306

Thomas, C. and Rae, R.C. (1988) Concentrate supplementation of silage for dairy cows. In *Nutrition and Lactation in the Dairy Cow* (ed. P.C. Garnsworthy). London, Butterworths, pp. 327–354

Thomas, P.C. and Robertson, S. (1987) The effect of lipid and fibre source and content on silage intake milk production and energy utilization. In *The Eighth Silage Conference* (ed. C. Thomas). Hurley, AFRC Institute for Grassland and Animal Production, pp. 173–174

Thomas, P.C., Robertson, S., Chamberlain, D.G., Livingstone, R.M., Garth-waite, P.H., Dewey, P.J.S., Smart, R. and Whyte, C. (1988) Predicting the metabolizable energy (ME) content of compound feeds for ruminants. In *Recent Advances in Animal Nutrition – 1988* (eds W. Haresign and D.J.A. Cole). London, Butterworths, pp. 127–146

Tilley, J.M.A. and Terry, R.A. (1963) A two stage technique for the *in vitro* digestion of forage crops. *Journal of the British Grassland Society*, **18**, 104–111

UKASTA/ADAS/COSAC (1985) *Prediction of the Energy Value of Compound Feeds*. Report of the UKASTA/ADAS/COSAC Working Party. London, United Kingdom Agricultural Suppliers and Traders Association

Wainman, F.W., Dewey, P.J.S. and Boyne, A.W. (1978) *Second Report of the Feedingstuffs Evaluation Unit*, Edinburgh, Department of Agriculture and Fisheries for Scotland, pp. 1–35

Wainman, F.W., Dewey, P.J.S. and Boyne, W. (1981) *Compound Feedingstuffs for Ruminants. Third Report of the Feedingstuffs Evaluation Unit*, Edinburgh, Department of Agriculture and Fisheries for Scotland, pp. 1–49

Wainman, F.W., Dewey, P.J.S. and Brewer, A.C. (1984) *Fourth Report of the Feedingstuffs Evaluation Unit*, Edinburgh, Department of Agriculture and Fisheries for Scotland, pp. 1–85

Webster, A.J.F., Dewhurst, R.J. and Waters, C.J. (1988) Alternative approaches to the characterisation of feedstuffs for ruminants. In *Recent Advances in Animal Nutrition – 1988* (eds W. Haresign and D.J.A. Cole), London, Butterworths, pp. 167–191

18

EVALUATION AND PREDICTION OF THE NUTRITIVE VALUE OF PASTURES AND FORAGES

C. THOMAS
West of Scotland College, Auchincruive, Ayr, UK

D. G. CHAMBERLAIN
Hannah Research Institute, Ayr, UK

Introduction

The chemical composition of forages and pastures varies widely as a result not only of intrinsic differences between species and varieties but also of modifications caused by the climatic environment, the management of the crop and the subsequent treatment of the herbage prior to feeding to ruminants.

Legumes are characterized by their lower cell wall content compared with grasses of similar digestibility. However, within the cell wall there is a lower ratio of hemicellulose to cellulose and a higher content of lignin. Legumes also have higher protein contents and, although levels of non-structural carbohydrates are equivalent between legumes and grasses, a much lower proportion of this is in the form of water soaluble carbohydrate in the legume (Osbourn, 1980; Smith, 1973). Marked differences also occur between grass species, varieties and genotypes (Munro and Walters, 1986). Nevertheless, within species, it is stage of maturity that has the most marked effect on composition. As grass plants mature the proportion of cell wall increases whilst that of cell contents is reduced and, within both the cell wall and cell contents, there are changes in chemical composition and structure (Osbourn, 1980). If the material is cut for conservation rather than grazed, then further changes occur which modify the composition of the cell wall and cell contents. For example, if the crop is left to dry in the field then continuing respiration in the plant leads to oxidation of carbohydrate (Carpintero, Henderson and McDonald, 1979). Plant proteolysis also occurs and there are increases in the proportions of non-protein N, amino-N, amide-N and ammonia-N (Brady, 1960).

A subsequent period of ensilage leads to further changes which result in the partial degradation of both the carbohydrate and the remaining protein. These changes are dealt with in detail by McDonald (1981) and Woolford (1984). Attempts to classify the range of fermentation activity (e.g. Wilkinson *et al.*, 1981) serve to illustrate the wide range that can be encountered and it is important to recognize that the products of fermentation overlay the differences in crop composition ensuing from the standing crop.

The objective of this chapter is to provide a basis for the evaluation and prediction of the nutritive value of herbage given the wide variation in composition. It is not the intention to review in detail the mechanisms controlling the intake of herbage nutrients and their utilization for growth and lactation. Rather, the aim is to outline problem areas with a view to providing improved methods of prediction relevant to the needs of the producer. Two areas will be examined: voluntary intake of herbage and forage and the utilization of energy and nitrogen yielding substrates.

Voluntary intake

MECHANISMS OF INTAKE CONTROL

Dried forage

Much of the research work over the past 40 years has concentrated on the role of physical factors. In this respect Balch and Campling (1962) outlined the concept of forage intake being mediated by physical distension of the reticulo-rumen. Thus intake of hay increased with increasing digestibility as a consequence of reducing contribution to fill (Blaxter, Wainman and Wilson, 1961). However, the relationship between intake and digestibility over a wide range of forages has been poor (ARC, 1980) and the accuracy and precision of digestibility as an index of intake is insufficient for predictive purposes.

The first problem area, therefore, is the definition of a suitable parameter to describe the influence of forage on rumen fill. There is ample evidence that legumes are consumed in greater quantities than grasses of similar digestibility and further that there are, even within grasses, species and variety differences in the relationship between digestibility and intake (Osbourn, 1980). In the tropics differences between species and varieties are even more marked and it appears that rumen retention time provides a better definition of the ingestibility of a forage than digestibility *per se* (Thornton and Minson, 1973). Rumen retention time is one of the components determining *in vivo* digestibility, the other being the degradation characteristics of the forage. This latter aspect has been examined by Hovell, Ngambi, Barber and Kyle (1986) using the nylon bag technique. The authors found that the rate of degradation was relatively unimportant but, on the other hand, potential degradability was closely related to intake (Table 18.1). A subsequent study by Miller and Oddoye (1989) provided similar evidence again demonstrating that the fractional rate of disappearance of the insoluble but potentially degradable DM or NDF was not related to intake. In contrast, using forages with a wider range of intake characteristics, Lopez, Carro, Gonzalez and Ovejero (1989) found little evidence of a relationship with potential degradability but a close agreement with the fractional rate.

From the results examined thus far, there is clear evidence that digestibility does not adequately describe the 'fill' characteristics of a forage. Rather, there is a need to consider estimates of rumen retention time coupled with a knowledge of the degradation characteristics of the insoluble but potentially degradable fraction. It is not clear at present which are the most relevant characteristics nor indeed whether the form of the Orskov and McDonald (1979) equation provides meaningful estimates of degradation rates that are applicable to forage cell walls. Nevertheless, it is vital that the relevant degradation characteristics are linked to defined chemical entities to provide an understanding of the mechanisms.

In addition it must be recognized that factors associated with physical fill are not the sole regulators of forage intake. Food is eaten in discrete meals and intake is the

Table 18.1 VOLUNTARY INTAKE AND DIGESTIBILITY OF HAYS

	Hay			
	A	*B*	*C*	*D*
Organic matter intake (g/kg $M^{0.75}$ per day)	66	57	48	40
Organic matter digestibility	0.64	0.59	0.47	0.47
Potential degradability (a + b)* of DM	0.76	0.66	0.54	0.46

*p = a + b (1 − e^{-ct}) where p = DM loss after t h incubation and a, b and c are constants.

product of the number of meals and the size of each meal. Further it must be assumed that a number of factors can be simultaneously transmitting signals and it is the integration of these which determines the size of each meal and the stimulation of the next (Forbes, 1983). This concept was demonstrated experimentally by Thiago and Gill (see Gill, Rook and Thiago, 1988). In young cattle given grass hay *ad libitum* the weight of rumen contents after the first (main) meal was not the maximum, higher values being obtained at 8 and 14 h after feeding. Further, rumen fill declined markedly between 14 and 24 h after feeding with no stimulation of eating. These results suggest that the weight of rumen contents was not the sole factor influencing the intake of hay. Even so it remains a valid argument that fill can be a dominant, though not necessarily the only, factor mediating the intake of fresh and dried herbage.

The foregoing discussion has concentrated on the evaluation of dried herbage and forage. However, dried herbage no longer represents a major dietary input to productive ruminants. Rather, forage is more commonly conserved as silage and furthermore herbage itself is grazed. The utilization of forage and pasture in these forms results in major modifications in the control of voluntary intake such that evaluations of dried herbage using ruminants fed indoors may be of little relevance to the prediction of intake.

Ensilage

There is considerable evidence that DM intake is less with silage than with corresponding fresh or dried herbage (Moore, Thomas and Sykes, 1960; Harris and Raymond, 1963). However, the extent of the depression varies greatly. Demarquilly (1973) noted a range from 1 to 64% and there is evidence to indicate that factors other than weight of reticulo-rumen contents have a considerable influence on the intake of silage (e.g. Campling, 1966; Thiago and Gill, 1986).

Attempts to relate the low intakes of silages to their pH or contents of specific fermentation end-products have given variable and equivocal results. McLeod, Wilkins and Raymond (1970) and Thomas and Wilkinson (1975) observed that the ingestion of low pH silages leads to disturbances in acid–base balance and reductions in blood and urine pH, with little effect on pH within the rumen.

Intraruminal infusions or dietary additions of silage acids have also been investigated without a clear picture emerging. Ulyatt (1965) noted a reduction in the intake of herbage when acetic acid was given, whereas Hutchinson and Wilkins (1971) could detect no effect independent of the effect of pH *per se*. Similarly,

McLeod, Wilkins and Raymond (1970) found silage intake to be reduced by dietary additions of lactic acid but Morgan and L'Estrange (1977) did not detect a significant reduction in intake when lactic acid was added to dried grass diets. There is evidence that the depressant effect of lactic acid on silage intake is overcome by use of fishmeal to increase dietary protein supply (Thomas, Gill and Austin, 1980), although the protein has no associated effects on digestive efficiency or acid-base balance. The mechanism involved in this interaction and that reported by Cottrill *et al.* (1976) has not been elucidated and indeed the mechanism through which lactic acid itself reduces silage intake remains to be established.

Recent studies of lactic acid metabolism (Chamberlain, Thomas and Anderson, 1983; Gill *et al.*, 1984) have shown that in silage-fed animals both L (+) and D (−) isomers are rapidly fermented in the rumen to acetate, propionate and butyrate, the proportions of the acids formed varying with the relative numbers of bacteria and protozoa in the rumen. Studies on the effects of nitrogenous silage-fermentation products have indicated that intake is not reduced by dietary or intraruminal administration of tyramine, tryptamine (Neumark, Bondi and Volcani, 1964; Thomas *et al.*, 1961), histamine (McDonald, MacPherson and Watt, 1963) or γ-amino butyric acid (Clapperton and Smith, 1983). On the other hand, intraruminal infusions of ammonium salts have reduced silage intake (Thomas *et al.*, 1961) and in animals given high-concentrate diets results in shorter and less frequent meals (Conrad, Baile and Mayer, 1977). These intake effects appear to relate to an excessive uptake of ammonia from the rumen and could be important when silages with a high N content and high proportion of non-protein N are consumed. In experiments both with lactating and non-lactating dairy cows given grass silage diets and intraruminal infusions of urea, marked depressions of silage intake occurred when the CP concentration of the total diet exceeded 220 g CP/kg DM (Choung, Chamberlain and Thomas, 1989a).

The data reviewed suggest that there is no simple post ingestive mechanism and ruminal effects may interact with other factors. The mechanism involved in responses to VFA may not simply be due to chemoreceptors since osmotic receptors have also been implicated (Buchanan-Smith and Phillip, 1986). Oropharyngeal effects too have been suggested although the evidence for this is not strong (Buchanan-Smith, 1989). In addition, silage intake may be influenced by the balance between energy and amino acid supply to the tissues (cf. Egan, 1977). Intake has been increased by dietary supplements of fishmeal (Gill and England, 1983) but whether this reflects an interaction with lactic acid (see above) or an influence on amino acid supply is uncertain. However, recent experiments with dairy cows given silage diets have shown clearly that increases in silage intake can be mediated via an increase of amino acid supply to the abomasum (Girdler, Thomas and Chamberlain, unpublished observations).

Although there is little evidence to support direct physical constraints, they must not be discounted in the control of silage intake. In work with a highly digestible silage of high fermentation quality, Farhan and Thomas (1978) found that intake was reduced by the insertion of water-filled bladders into the rumen and concluded that a 'rumen-fill' mechanism was in operation. Consistent with this, intake has been shown to be increased when silage is minced before feeding (Thomas, Kelly and Wait, 1976; Deswysen, Vanbelle and Focart, 1978) and a similar response has been obtained with grass chopped short during silage-making (Castle, Retter and Watson, 1979; Castle, Gill and Watson, 1981) although it is important here to separate the influence of chop-length *per se* from the indirect effect through silage fermentation quality. Animals receiving silage have a low rumination activity and a long 'latency period' after feeding (Deswysen, Vanbelle and Focart, 1978) and this may impose special limitations on the physical breakdown of silages in the rumen. Clancy, Wangsness and Baumgardt (1977) showed that intraruminal infusions of silage juice reduced rumen motility and rate of eating but the results have not been reproduced consistently (Phillip, Buchanan-Smith and Grovum, 1981a, 1981b; Smith and Clapperton, 1981).

Grazed herbage

There is clear evidence to demonstrate that herbage intake increases with increasing digestibility even up to high levels of digestibility (Hodgson, Rodriques Capriles and Fenlon, 1977). Superficially, such data would suggest that physical factors were dominant in the control of herbage intake. However, other changes in the characteristics of swards occur in parallel with those associated with herbage digestibility. These changes in sward characteristics have implications for the grazing behaviour of the ruminant and these behavioural effects combine with metabolic and physical factors to influence intake. In the context of the reaction of the ruminant to changes in sward state, intake is best considered as the product of the weight of herbage harvested per bite, the rate of biting and total grazing time (Spedding, Large and Kydd, 1966). The ranges in these components of intake are shown in Table 18.2.

Table 18.2   COMPONENTS OF INGESTIVE BEHAVIOUR

	Sheep	*Cattle*
Grazing time (h/day)	6.5–13.5	5.8–10.8
Biting rate (bites/min)	22–94	20–66
Intake/bite (mg OM/kg LW)	0.4–2.6	0.3–4.1
Rate of intake (mg OM/kg LW per min)	22–80	13–204

The mechanisms involved have recently been examined by Hodgson (1986) and by Mayne and Wright (1988) and it is intended here only to provide a brief outline of behavioural response to changing sward conditions. Intake per bite, biting rate and grazing are interrelated and react to changes in sward structure to varying extents (Penning, 1985). As sward surface height or herbage mass decreases then intake per bite decreases and biting rate increases, although this latter effect may not be a direct compensatory mechanism. Nevertheless, on tall swards a combination of these two effects tends to maintain intake but as sward height falls to low levels any increase in biting rate will not compensate for the reduction in intake per bite. Also there may be a limit on biting rate (Phillips and Leaver, 1985) and further, the ability to enhance grazing time to compensate for reduced intake rate may be limited (Stobbs, 1973). Intake per bite is therefore the most variable and critical component in the determination of intake and this in turn is markedly affected by the volume of the canopy removed by a bite and its bulk density. The effects of the bulk density of a sward on intake per bite are most apparent in tropical species (Stobbs, 1973).

Nevertheless, even with temperate species sward structure can be influenced to a considerable extent by grazing management. Continuous grazing results in the development of dense swards with high tiller populations (up to 40 000 tillers/m^2) whereas rotational grazing results in a more open sward containing approximately 10–15 000 tillers/m^2. Thus although sward height may be the best single description of the relationship between behaviour and sward structure, different relationships must exist between height and intake in continuous and rotationally grazed swards (Table 18.2). Furthermore, there are difficulties in the measurement and interpretation of sward height (Gibb and Ridout, 1986) and also its usefulness in defining behavioural responses in rotationally grazed swards is less clear.

It must be recognized in the evaluation of grass for grazing that sward structure, through its influence on grazing behaviour, influences intake. A framework exists to describe these effects, at least as a first approximation. What is less clear is how these sward characteristics interact with physical and metabolic factors associated with the herbage ingested. These relationships are further complicated by the ability of the ruminant to select material for ingestion which is of different composition to the average of the herbage mass. This may not always be a positive selection since it may merely be a reflection of the differences in the distribution of leaf and stem and of dead material through a vertical section of the sward (Hodgson, 1985).

PREDICTION OF VOLUNTARY INTAKE

The importance and sensitivity of accurately predicting intake has recently been pointed out by Barber, Offer and Givens (1989). In practical terms it is the prediction of intake of silage and grazed herbage which is of greatest significance.

The data presented so far indicate that both attributes of the crop and the conservation process influence voluntary intake of silage and the rate at which silage will be substituted for by supplements. However the majority of equations to

Table 18.3   RANKING OF EQUATIONS TO PREDICT THE VOLUNTARY INTAKE OF SILAGE BASED DIETS BY DAIRY COWS

Source	Prediction error (kg DM^2)	Independent variables
Vadiveloo and Holmes (1979) (equation 1)	2.1	Milk yield, LW, lactation week, concentrate intake
Lewis (1981)	2.5	Silage characteristics, milk yield, LW, concentrate intake
Vadiveloo and Holmes (1979) (equation 3)	2.8	As in equation 1
MAFF (1975)	3.3	Milk yield, LW
ARC (1980)	4.0	Milk yield, LW, lactation month
INRA (1978)*	4.9	Milk yield, LW, lactation week, silage characteristics, concentrate intake, substitution rate

*From Report of Joint Working Party on the Prediction of the Voluntary Intake of Ruminants.

predict voluntary intake of forage-based diets by ruminants do not take into account differences in the composition of silages. These equations were examined by Neal, Thomas and Cobby (1984) in relatioan to observed intakes by a dairy herd (Table 18.3). Only the equations derived by Lewis (1981) and INRA (1979) include aspects of fermentation quality and physical characteristics of the forage. However, the equation of Lewis (1981) was not markedly different in its predictive ability from others which relied primarily on attributes of the cow rather than the forage. The most extensive description of silage is encompassed in the INRA (1979) equation and yet this equation gave the greatest errors of prediction, although much of this was in the form of bias as a result of an inadequate description of the influence of stage of lactation.

Given the inadequate description of silage characteristics it is not surprising that the best equations outlined in Table 18.3 result in an error of prediction of ± 4.0 kg milk at zero energy balance. It is clear from an examination of the mechanisms involved that a single index such as ammonia-N is unlikely to provide an accurate prediction of silage intake. Indeed, even in studies where ammonia-N has proved to be a significant factor it has only accounted for 38% of the variation in intake when considered alone. The commonly assessed additional indices of DM content and pH are unlikely to contribute to any increase in predictive ability.

Further, the qualitative rather than quantitative descriptions of silage as used in the INRA system do not appear to increase precision. The problems of using regression analysis to produce predictive equations have been discussed by Gill, Rook and Thiago (1988). Difficulties arise due to colinearity among independent variables. To overcome this, Rook, Gill and Dhanoa (1989) have used ridge regression techniques. Errors of prediction were less using the technique than with conventional methods and the coefficients were biologically more apt. Nevertheless, errors of prediction remain high although some improvement was observed over existing models.

Regression analysis has also been used to develop equations to predict the intake of grazed herbage. The most comprehensive study is that of Caird and Holmes (1986). Equations that included variables of herbage allowance (Meijs and Hoekstra, 1984; Stockdale, 1985) were compared with those derived from indoor studies (Table 18.3). The authors, however, concluded that the equations that included sward characteristics resulted in higher errors of prediction. As a result, Caird and Holmes (1986) derived improved equations based on herbage mass, herbage allowance and sward height in addition to animal characteristics and concentrate input. Interestingly, the influence of height differed when the sward surface was below 5 cm. Nevertheless, when these equations were then compared with simple models that relied entirely on animal characteristics, the accuracy of prediction was not consistently improved.

Prospects for improving predictability and for adopting a dynamic approach will be discussed below.

Utilization of energy and protein

UTILIZATION OF ENERGY

The wide variation in composition of forages and pasture has been emphasized. In terms of metabolizable energy (ME), the UK Tables of feed composition (MAFF,

1986) lists a range of 7.0–13.5 for pasture, 6.7–15.6 for silages and 5.9–12.9 MJ ME kg DM for hays. The techniques and methods used to estimate the ME content of forages have recently been reviewed by Barber, Offer and Givens (1989). The objective of this section is to examine the problems associated with prediction of the utilization of ME for growth and lactation.

There is clear evidence that the ME from diets of forage alone is used less efficiently for maintenance and growth compared with mixed diets (ARC, 1980), although for growth the differences in efficiency diminish with increasing metabolizability (qm). The reasons for these differences in efficiency have yet to be fully explained. Physical factors associated with movement of fibrous digesta along the tract have been largely discounted (Webster, 1980). Rather, the lower efficiency of utilization of ME from forage diets has been associated with the end products of digestion, principally acetate. However the mediation of the effects may be more complex than originally envisaged. For example, the efficiency of acetate utilization differs according to basal diet (Tyrrell, Reynolds and Moe, 1979). Further, with ensiled rather than fresh or dried forages there is evidence that efficiency of utilization for growth can be further reduced. The size of the discrepancy in the k_f values varies, however, and while determined and calculated values are in agreement for some silages, for others they are widely divergent. Where low determined k_f values are recorded, they reflect an abnormally high dietary-induced thermogenesis, the cause of which remains to be explained (Thomas and Thomas, 1985). The effect may arise because of a greater nutritional imbalance in silage diets, resulting perhaps from a poor utilization of silage nitrogen in the rumen.

Consistent with this, the agreement between determined and predicted k_f values for silage-barley diets appears to be better than for silage alone diets. Thomas *et al.* (1988) compared the response of beef cattle to increased digestibility of silage with that achieved by enhancing the proportion of barley in the supplement (Table 18.4). Higher gains were achieved by the steers given silage of high rather than low digestibility. However, despite the fact that the steers given the early-cut silage consumed the most ME, their gains were lower than those achieved by the animals receiving the mixture of late-cut silage and the high level of barley. It can be calculated that 20% more ME above maintenance was required by steers given the

Table 18.4   INTAKE AND GAIN BY BEEF CATTLE GIVEN EARLY-CUT SILAGE ALONE OR LATE-CUT SILAGE ALONE OR WITH ROLLED BARLEY

	Treatments			
	Late-cut silage			Early-cut silage
Barley (g/kg DM)	0	280	560	0
Digestibility of gross energy	0.619	0.668	0.705	0.735
ME intake (kJ/kg $W^{0.75}$ per day)	758	834	871	932
Gains				
Empty body (g/day)	292	552	800	696
Energy (MJ/day)	5.48	9.23	14.58	12.24

(From Thomas *et al.*, 1988.)

early-cut silage in order to achieve the same energy retention as those given the mixed diet of late-cut silage and barley. Similarly it can be shown that supplementation with fishmeal increases efficiency (see Thomas and Thomas, 1985). More importantly perhaps, increasing the dietary supply of protein changed the composition of gain. It is also possible that the higher efficiencies noticed with legumes rather than grass may in part be associated with greater protein supply (Glenn *et al.*, 1987).

The low efficiencies in growing cattle associated with high forage diets are also apparent in lactating dairy cows. In this respect calorimetric studies suggest lower than expected efficiencies (k_1) when silages comprise a relatively high proportion of the diet (Table 18.5). Analysis of feeding experiments where silages were given as a sole feed (Rae *et al.*, 1987) indicates k_1 values ranging from 0.54 to 0.58 compared with predictions of 0.61 to 0.63. Thomas and Castle (1978), on the basis of their feeding trials, have calculated k_1 values of between 0.39 and 0.64 for diets containing a high proportion of silage. Although such estimates are subject to considerable error, Thomas and Castle (1978) pointed out that estimated k_1 values were consistently higher in diets containing supplementary protein.

Table 18.5 DETERMINED AND PREDICTED VALUES FOR THE UTILIZATION OF METABOLIZABLE ENERGY FOR MILK PRODUCTION

Silage		*Efficiency* (k_1)	
		Determined	*Predicted**
(a)	Unwilted	0.56	0.65
	Wilted	0.58	0.64
(b)	Unwilted	0.50	0.64
	Wilted	0.53	0.65

*From $k_1 = 0.35\,ME/GE + 0.42$ (ARC, 1980)
(From Unsworth and Gordon, 1985.)

The foregoing has emphasized the low and variable efficiency of energy use for growth from forage alone diets. This effect is present in fresh and dried forages but is even more marked with individual silages. To some extent the depression in efficiency is associated with imbalances in nutrient supply and presumably in the ratio of the supplies of energy and protein.

UTILIZATION OF PROTEIN

Of overriding importance in determining the utilization of protein from forage is the net outcome of nitrogen transactions in the rumen; whether there are net 'losses' or 'gains' of nitrogen between the mouth and the duodenum. In recent years, information from digestion studies with sheep and cattle has become available and some tentative general statements can be made.

Forages in the fresh state and as silage are more susceptible to net losses of nitrogen across the rumen wall than are dried forages (Table 18.6). The efficiency

Table 18.6   THE 'CRITICAL' CONCENTRATION OF NITROGEN (gN/kg OM) ABOVE WHICH LOSSES OF NITROGEN BETWEEN MOUTH AND DUODENUM OCCUR. (RESULTS ARE FROM EXPERIMENTS WITH SHEEP AND YOUNG CATTLE GIVEN GRASSES AND CLOVERS)

Form of forage	*Critical N concentration* g/kg OM	*Reference*
Artificially dried	35	Thomas (1982)
Hays	33	Thomas (1982)
Silage	26	Thomas and Chamberlain (1982)
Fresh pasture	26	Beever and Gill (1987)

Table 18.7   THE EXTENT OF DEGRADATION OF FORAGE PROTEIN IN THE RUMEN. VALUES ARE FROM *IN VIVO* MEASUREMENTS IN SHEEP AND CATTLE GIVEN GRASSES AND CLOVERS

Form of forage	*Proportion of dietary protein degraded*	*Reference*
Hay	0.55–0.65	Thomas, 1982
Silage	0.75–0.85*	Thomas and Thomas, 1985
	0.3–0.5†	Thomas and Thomas, 1985
Fresh	0.7–0.85	Beever *et al.*, 1986

*No additive or formic acid.
†Additives containing formaldehyde.

of transfer of dietary nitrogen from mouth to duodenum depends on the extent of ruminal degradation of forage protein and the efficiency of incorporation of the released nitrogen into microbial protein. Dehydration reduces the extent of degradation of forage protein in the rumen (Table 18.7), this effect being especially marked with high-temperature drying (Thomson and Beever, 1980). However, this may not be the complete explanation of the higher flows of protein to the duodenum since, in some cases, there have been indications of increased rates of microbial synthesis with dried grass compared with the same material given either fresh or frozen (Thomson and Beever, 1980).

On the surface, the data in Tables 18.6 and 18.7 suggest a general similarity in the pattern of digestion of fresh and ensiled forages (silages treated with formaldehyde are, of course, a special case and they are not included in the following discussion). However, the data for silages in Table 18.6 should be viewed cautiously. These data are derived mostly from experiments with sheep and there is an increasing amount of evidence to indicate that the pattern of digestion in cattle may differ in that substantial losses of nitrogen between the mouth and duodenum occur at nitrogen concentrations much lower than that indicated in Table 18.6 (see Chamberlain, Martin and Robertson, 1989). That there are differences in the ruminal digestion of protein from fresh forage and ensiled forage is further illustrated by the marked differences in the rates of microbial synthesis: average values for silage are around 23 g N/kg DOMR (ARC, 1984) whereas recent studies have observed values for fresh herbage of around 50 g N/kgDOMR (Beever, Dhanoa, Losada, Evans, Cammell and France, 1986).

On the evidence available, the ruminal digestion of fresh forage and ensiled forage is similar in that the protein of both is very degradable and imbalances of nitrogen and energy supply in the rumen can result in substantial losses of nitrogen before the small intestine. However, with silages the inefficiency of transfer of dietary nitrogen to the duodenum is exacerbated by limitations on microbial synthesis (see Thomas and Chamberlain, 1982). This is understandable since ensilage, unless carried out with formaldehyde-containing additives, entails a substantial breakdown of protein to NPN constituents which normally contribute 500–750 g N/kg total N. These NPN constituents are very rapidly degraded to ammonia in the rumen. Moreover, most of the sugar present in the original crop has been converted to fermentation products which are of low ATP yield to the rumen microbes (see Chamberlain, 1987) which magnifies the imbalance between the supplies of available energy and nitrogen in the rumen.

Whether protein degradability of forages estimated from the rate of disappearance of nitrogen from nylon bags incubated in the rumen equates with degradation *in vivo* remains to be established (Thomas and Thomas, 1985). However, using this technique, the degradability of crude protein (CP) of silages has been correlated with the concentration of modified acid detergent fibre (MADF) in the forage (Webster, Simmons and Kitcherside, 1982):

Degradability = 0.9 (CP − 0.1 MADF)/CP

Such a relationship suggests a link between fibre content and the availability of nitrogen and may derive from the association of part of the nitrogen with structural carbohydrate (Acid Detergent Insoluble Nitrogen, ADIN) a fraction which, it has been suggested (Van Soest, 1982), is indigestible. The implication is that stage of maturity of the crop will influence the rumen degradability of its protein.

As for the composition of protein reaching the duodenum, Thomson (1982) concluded that there was little evidence to indicate that amino acid composition of duodenal digesta differed in animals given fresh grass and dried forages. However, for grass silages, the duodenal concentrations of methionine and lysine are lower than for diets of hay (Thomas and Chamberlain, 1982).

There is no clear evidence of consistent differences between fresh, dried, and ensiled forages in the absorbability of amino acids entering the small intestine but there is a fairly wide range in reported values (ARC, 1980). There are insufficient data from *in vivo* experiments to allow comment on the effect of stage of maturity of forage on the absorbability of amino acids. Although there is no direct *in vivo* evidence to support the claim (Van Soest, 1982) that ADIN is completely indigestible, there is circumstantial evidence in its favour (Webster, Dewhurst and Waters, 1988). Again, as for rumen degradability of protein, the implication is that the increasing concentration of fibre as the plant matures could influence the absorbability of amino acids entering the small intestine.

The results from digestion experiments would suggest that with fresh and with ensiled forages, production responses to protein supplementation of these diets might be expected in growing and in lactating ruminants and evidence is accruing to show that this is the case. Abomasal infusion of casein and methionine in growing lambs, zero-grazed on a ryegrass : clover mixture, markedly increased liveweight gain and protein deposition (Table 18.8). In lactating cows consuming fresh forage, supplements of formaldehyde-treated casein have increased milk production (Rogers, Porter, Clarke and Stewart, 1980; Minson, 1981). For diets of grass silage, supplementation with fishmeal produced marked increases in growth rate of beef cattle (Table 18.9) and a number of studies have shown big increases in milk production in response to protein supplementation (see Chamberlain *et al.*, 1989). Furthermore, studies with dairy cows given diets containing a high proportion of grass silage have shown clear differences in the magnitude of milk production responses to proteins from different sources, the milk production increase from soya supplements being about 0.7 of that from an equivalent amount of fishmeal protein (Chamberlain *et al.*, 1989). Whether these differences between protein

Table 18.8   THE EFFECT OF ABOMASAL INFUSIONS OF CASEIN AND METHIONINE IN GROWING LAMBS ZERO-GRAZED ON RYEGRASS : WHITE CLOVER

	Control	*Casein + methionine*
ME intake (MJ/kg $^{0.75}$/d)	0.92	0.94
Liveweight gain (g/d)	79	99
Protein deposition (g/d):		
Body	8.7	14.3
Wool	3.9	6.7
Fat deposition	21.2	18.5

(From Barry *et al.*, 1982.)

Table 18.9   THE EFFECT OF SUPPLEMENTS OF FISHMEAL (150 g/d) ON GROWTH RATE OF BEEF CATTLE RECEIVING A DIET OF GRASS SILAGE

	Control	*+ fishmeal*
Empty bodyweight gain (g/d)	569	787
Carcass-weight gain (g/d)	359	505
Composition of body gain:		
Protein (g/d)	95	145
Fat (g/d)	92	98
Energy (MJ/d)	5.8	7.2

(From Beever and Gill, 1987.)

Table 18.10   MILK PRODUCTION OF COWS GIVEN *AD LIBITUM* ACCESS TO A COMPLETE MIX DIET OF GRASS SILAGE, BARLEY AND SOYA BEAN MEAL (680 : 290 : 30 g/kg DM) AND SUPPLEMENTS OF 1.1 kg DM/d OF BARLEY (B) OR A MIXTURE OF FISHMEAL, BLOODMEAL AND MEAT AND BONEMEAL (500 : 300 : 200 g/kg DM) (P)

	Supplement	
	B	*P*
DM intake (kg/d)	17.0	17.2
Milk (kg/d)	19.8	23.0
Fat (g/d)	853	949
Protein (g/d)	653	758

(From Girdler, Thomas and Chamberlain, 1988b.)

sources relate to differences of amino acid composition or to differences of ruminal degradability is unclear but supplementation of soya with rumen-protected methionine and lysine did not increase milk production (Girdler, Thomas and Chamberlain, 1988a). Despite the low concentrations of methionine and lysine in duodenal digesta of silage-fed animals, it has yet to be established that the supplies of these acids limit milk production. In this respect, results of recent experiments with abomasal infusion of proteins and amino acids in dairy cows given grass silage diets have focused attention on the role of tryptophan, methionine and phenylalanine (Choung, Chamberlain and Thomas, 1989b).

The interpretation of the responses to protein supplementation of silage diets is not straightforward. Increases of milk production of dairy cows are usually, but not always, accompanied by increases of silage intake (Chamberlain *et al.*, 1989) and a similar comment would apply to responses of beef cattle (Thomas and Thomas, 1985). Although big increases of milk production have been observed in the absence of increases of silage intake (Table 18.10), in most reports much of the milk production response can be explained on the basis of the extra ME supply from the increased voluntary intake of silage and the probable increases of diet digestibility (Oldham, 1984).

Conclusions

In terms of intake prediction, grazing presents special problems arising from the complex interactions of sward characteristics and grazing behaviour. However the prediction of silage intake is also beset with major problems, not least of which is the lack of any detailed understanding of the influence of the fermentation process on intake characteristics.

For both fresh and conserved forages, progress continues to be made in identifying the nutritional limitations arising from their pattern of digestion and the subsequent supply of nutrients for absorption. However, much more experimentation is needed if these features of digestion are to be related quantitatively to the chemical composition of the forage in a way that will allow accurate prediction of its nutritive value.

What should, perhaps, be a cause for even greater concern is that, at present, there is no evidence that a more detailed chemical analysis of silage will allow prediction of its nutritional value with any acceptable degree of precision.

References

ARC (1980) *The Nutrient Requirements of Ruminant Livestock*, Slough, Commonwealth Agricultural Bureaux

ARC (1984) *The Nutrient Requirements of Farm Livestock*, Suppl. No 1, Slough, Commonwealth Agricultural Bureaux

Balch, C.C. and Campling, R.C. (1962) Regulation of voluntary food intake in ruminants. *Nutrition Abstracts and Reviews*, **32**, 669–686

Barber, G.D., Offer, N.W. and Givens, D.I. (1989) Predicting the nutritive value of silage. In *Recent Advances in Animal Nutrition, 1989* (eds W. Haresign and D.J.A. Cole). London, Butterworths, pp. 141–158

Barry, T.N., *et al.* (1982) Protein metabolism and responses to abomasal infusion of casein plus methionine in growing lambs fed fresh primary growth ryegrass/clover pasture *ad libitum*, In *Forage Protein in Ruminant Animal Production* (eds D.J. Thomson, D.E. Beever and R.G. Gunn). Occasional Publication of the British Society of Animal Production No. 6, pp. 146–148

Beever, D.E., *et al.* (1986) The effect of forage species and stage of harvest on the processes of digestion occurring in the rumen of cattle. *British Journal of Nutrition*, **56**, 439–454

Blaxter, K.L., Wainman, F.W. and Wilson, R.S. (1961) The regulation of food intake by sheep. *Animal Production*, **3**, 51–61

Brady, C.J. (1960) Redistribution of nitrogen in grass and leguminous fodder plants during wilting and ensilage. *Journal of the Science of Food and Agriculture*, **11**, 276–284

Buchanan-Smith, J.G. (1989) Effect of soluble constituents of silage on oro-pharyngeal intake by sheep. In *Winter Meeting of the British Society of Animal Production* (in press)

Buchanan-Smith, J.G. and Phillip, L.E. (1986) Food intake in sheep following intraruminal infusion of extracts from lucerne silage with particular reference to organic acids and products of protein degradation. *Journal of Agricultural Science, Cambridge*, **106**, 611–617

Caird, L. and Holmes, W. (1986) The prediction of voluntary intake of grazing dairy cows. *Journal of Agricultural Science, Cambridge*, **107**, 25–39

Campling, R.C. (1966) The intake of hay and silage by cows, *Journal of the British Grassland Society*, **21**, 41–49

Carpintero, C.M., Henderson, A.R. and McDonald, P. (1979) The effect of some pretreatments on proteolysis during the ensiling of herbage. *Grass and Forage Science*, **34**, 311–315

Castle, M.E., Gill, M.S. and Watson, J.N. (1981) Silage and milk production: a comparison between barley and dried sugar beet pulp as silage supplements. *Grass and Forage Science*, **36**, 319–324

Castle, M.E., Retter, W.C. and Watson, J.N. (1979) Silage and milk production: comparisons between grass silages of three different chop lengths. *Grass and Forage Science*, **34**, 293–301

Chamberlain, D.G. (1987) Silage fermentation in relation to the utilization of nutrients in the rumen. *Process Biochemistry*, **22**, 60–63

Chamberlain, D.G., Martin, P.A. and Robertson, S. (1989) Optimizing compound feed use in dairy cows with high intakes of silage. In *Recent Advances in Animal Nutrition, 1989* (eds W. Haresign and D.J.A. Cole). London, Butterworths, pp. 175–193

Chamberlain, D.G., Thomas, P.C. and Anderson (1983) Volatile fatty acid proportions and lactic acid metabolism in the rumen in sheep and cattle receiving silage diets. *Journal of Agricultural Science, Cambridge*, **101**, 47–58

Choung, J.J., Chamberlain, D.G. and Thomas, P.C. (1989a) Effect of rumen ammonia concentration on silage intake in cows. In *Proceedings of the 14th International Congress of Nutrition* (in press)

Choung, J.J., Chamberlain, D.G. and Thomas, P.C. (1989b) Effect of intra-abomasal infusion of different protein sources on milk production and silage intake in dairy cows. In *Proceedings of the 14th International Congress of Nutrition* (in press)

Clancy, M., Wangsness, P.J. and Baumgardt, B.R. (1977) Effect of silage extract on voluntary intake, rumen fluid constituents and rumen motility. *Journal of Dairy Science*, **60**, 580–590

Clapperton, J.L. and Smith, E.J. (1983) Unpublished observations

Conrad, H.R., Baile, C.A. and Mayer, J. (1977) Changing meal patterns and suppression of feed intake with increasing amounts of dietary non-protein nitrogen in ruminants. *Journal of Dairy Science*, **60**, 1725–1733

Cottrill, *et al.* (1976) The effect of dietary pH and nitrogen supplementation on the intake and utilization of maize silage by young calves. *Animal Production*, **22**, 154–155 (Abstract)

Demarquilly, C. (1973) Composition chemique characteristiques fermentaires, digestibilite et quantitee ingress des ensilage de fourages: modifications par rapport au fourrage initial. *Annales Zootechnie*, **22**, 1–35

Deswysen, A., Vanbelle, M. and Focant, M. (1978) The effect of silage chop length on the voluntary intake and rumination behaviour of sheep. *Journal of the British Grassland Society*, **33**, 107–115

Egan, A.R. (1977) Nutritional status and intake regulation in sheep, VIII. Relationship between the voluntary intake of herbage by sheep and the protein/energy ratio in the digestion products. *Australian Journal of Agricultural Research*, **28**, 907–915

Farhan, S.M.A. and Thomas, P.C. (1978) The effects of partial neutralization of formic acid silages with sodium bicarbonate on their voluntary intake by cattle and sheep. *Journal of the British Grassland Society*, **33**, 151–158

Forbes, J.M. (1983) Physiology of regulation of food intake. In *Nutritional Physiology of Farm Animals* (eds J.A.F. Rook and P.C. Thomas). Harlow, Longman Group, pp. 177–202

Gibb, M.J. and Ridout, M.S. (1986) The fitting of frequency distribution to height measurements on grazed swards. *Grass and Forage Science*, **41**, 247–250

Gill, *et al.* (1984) Lactate metabolism in the rumen of silage-fed sheep. *Canadian Journal of Animal Science*, **64** (Supplement), 169–170

Gill, M. and England, P. (1983) The effect of level of fishmeal and sucrose supplementation on the voluntary intake of silage and live-weight gain in young cattle. *Animal Production*, **36**, 513 (Abstract)

Gill, M., Rook, A.J. and Thiago, R.S. (1988) Factors affecting the voluntary intake of roughages by the dairy cow. In *Nutrition and Lactation in the Dairy Cow* (ed. P.C. Garnsworthy). London, Butterworths, pp. 262–279

Girdler, C.P., Thomas, P.C. and Chamberlain, D.G. (1988a) Effect of abomasal infusions of amino acids or of a mixed animal protein source on milk production in the dairy cow. *Proceedings of the Nutrition Society*, **47**, 50A

Girdler, C.P., Thomas, P.C. and Chamberlain, D.G. (1988b) Effect of rumen-protected methionine and lysine on milk production from cows given grass silage diets. *Proceedings of the Nutrition Society*, **47**, 82a

Girdler, C.P., Thomas, P.C. and Chamberlain, D.G. (unpublished observations)

Glenn, *et al.* (1987) Duodenal nutrient flow in growing steers fed alfalfa and orchard grass silages at two intakes. In *Proceedings of the 8th Silage Conference*, pp. 67–68

Harris, C.F. and Raymond, W.F. (1963) The effect of ensiling on crop digestibility. *Journal of the British Grassland Society*, **18**, 204–212

Hodgson, J. (1985) The control of herbage intake in the grazing ruminant. *Proceedings of the Nutrition Society*, **44**, 339–346

Hodgson, J. (1986) Grazing behaviour and herbage intake. In *Grazing* (ed. J. Frame). British Grassland Society Occasional Symposium No. 19, pp. 51–54

Hodgson, *et al.* (1977) The influence of sward characteristics on the herbage intake of grazing calves. *Journal of Agricultural Science, Cambridge*, **89**, 743–750

Hovell, *et al.* (1986) The voluntary intake of hay by sheep in relation to its degradability in the rumen as measured in nylon bags. *Animal Production*, **42**, 111–118

Hutchinson, K. and Wilkins, R.J. (1971) The voluntary intake of silage by sheep, II. The effects of acetate on silage intake. *Journal of Agricultural Science, Cambridge*, **77**, 539–543

INRA (1978) *Alimentation des Ruminants*, Versailles, Institut National de la Recherche Agronomique

Lewis, M. (1981) Equations for predicting silage intake by beef and dairy cattle. In *Proceedings of the 6th Silage Conference* (ed. A.R Henderson). University of Edinburgh, pp. 35–36

Lopez, *et al.* (1989) The possibility of predicting the sheep voluntary dry matter intake of hays from their rumen degradation characteristics. *Animal Production*, **48**, 638

McDonald, P. (1981) *The Biochemistry of Silage*, Chichester, John Wiley and Sons

McDonald, P., MacPherson, H.T. and Watt, J.A. (1963) The effect of histamine on silage dry matter intake. *Journal of the British Grassland Society*, **18**, 230–242

McLeod, D.S.L., Wilkins, R.J. and Raymond, W.F. (1970) The voluntary intake by sheep and cattle of silages differing in their free acid content. *Journal of Agricultural Science, Cambridge*, **75**, 311–319

Mayne, C.S. and Wright, I.A. (1988) Herbage intake and utilization by the grazing dairy cow. In *Nutrition and Lactation in the Dairy Cow* (ed. P.C. Garnsworthy). London, Butterworths, pp. 280–293

Meijs, J.A.C. and Hoekstra, J.A. (1984) Concentrate supplementation of grazing dairy cows, I. Effect of concentrate intake and herbage allowance on herbage intake. *Grass and Forage Science*, **39**, 59–66

Miller, E.L. and Oddoye, E.O.K. (1989) Prediction of voluntary intake of conserved forages by cattle and sheep from degradability characteristics determined using synthetic fibre bags in sheep. *Animal Production*, **48**, 637–638

Ministry of Agriculture, Fisheries and Food (1986) *Feed Composition*, Marlow, Bucks, Chalcombe Publications

Minson, D.J. (1981) The effects of feeding protected and unprotected casein on the milk production of cows grazing ryegrass. *Journal of Agricultural Science, Cambridge*, **96**, 239–241

Moore, L.A., Thomas, J.W. and Sykes, J.F. (1960) The acceptability of grass/legume silage by dairy cattle. In *Proceedings of the 8th International Grassland Congress*, Oxford, Alden Press, pp. 701–704

Morgan, D.J. and L'Estrange, J.L. (1977) Voluntary feed intake and metabolism of sheep when lactic acid is administered in the feed or intraruminally. *Journal of the British Grassland Society*, **32**, 217–224

Munro, J.M.M. and Walters, R.J.K. (1986) The feeding value of grass. In *Grazing* (ed. J. Frame), Occasional Publication, British Grassland Society, pp. 65–78

Neal, H.D.S., Thomas, C. and Cobby, J.M. (1984) Comparison of equations for predicting voluntary intake by dairy cows. *Journal of Agricultural Science, Cambridge*, **103**, 1–10

Neumark, H., Bondi, A. and Volcani, R. (1964) Amines, aldehydes and keto-acids in silages and their effect on food intake by ruminants. *Journal of the Science of Food and Agriculture*, **15**, 487–492

Oldham, J.D. (1984) Protein-energy interrelationships in dairy cows. *Journal of Dairy Science*, **67**, 1090–1114

Osbourn, D.F. (1980) The feeding value of grass and grass products. In *Grass, its Production and Utilization* (ed. W. Holmes). London, Blackwells, pp. 70–124

Penning, P.D. (1983) A technique to record automatically some aspects of grazing and ruminating behaviour in sheep. *Grass and Forage Science*, **38**, 89–96

Phillip, L.E., Buchanan-Smith, J.G. and Grovum, W.L. (1981a) Effects of infusing the rumen with acetic acid and nitrogenous constituents in maize silage extracts on food intake, ruminal osmolality and blood acid–base balance in sheep. *Journal of Agricultural Science, Cambridge*, **96**, 429–438

Phillip, L.E., Buchanan-Smith, J.G. and Grovum, W.L. (1981b) Food intake and ruminal osmolality in sheep: differentiation of the effect of osmolality from that of the products of maize silage fermentation. *Journal of Agricultural Science, Cambridge*, **96**, 439–445

Phillips, C.J.C. and Leaver, J.D. (1986) Seasonal and diurnal variation in the grazing behaviour of dairy cows. In *Grazing* (ed. J. Frame), Occasional Publication of the British Grassland Society, pp. 98–104

Rae, *et al.* (1987) The potential of an all-grass diet for the late-winter calving dairy cow. *Grass and Forage Science*, **42**, 249–257

Rogers, G.L., Bryant, A.M. and McLeay, L.M. (1979) Silage and dairy cow production, III. Abomasal infusions of casein, methionine and glucose, and milk yield and composition. *New Zealand Journal of Agricultural Research*, **22**, 533–541

Rook, A.J., Gill, M. and Dhanoa, M.S. (1989) Prediction of voluntary intake of grass silage by growing cattle using ridge regression. *British Society of Animal Production, Winter Meeting* (in press)

Smith, D. (1973) The non-structural carbohydrates. In *Chemistry and Biochemistry of Herbage*, Vol. 1, (eds G.W. Butler and R.W. Bailey). New York, Academic Press, pp. 106–155

Smith, E.J. and Clapperton, J.L. (1981) The voluntary food intake of sheep when silage juice is infused into rumen. *Proceedings of the Nutrition Society*, **40**, 22A

Spedding, C.R.W., Large, R.V. and Kydd, D.D. (1966) Evaluation of herbage species by grazing animals. In *Proceedings of the 10th International Grassland Congress*, Helsinki, Finnish Grassland Association, pp. 479–483

Stobbs, T.H. (1973) The effect of plant structure on the intake of tropical pastures. 1. Variation in the bite size of grazing cattle. *Australian Journal of Agricultural Research*, **24**, 809–819

Stockdale, C.R. (1985) The influence of some sward characteristics on the consumption of irrigated pastures grazed by lactating dairy cattle. *Grass and Forage Science*, **40**, 31–39

Thiago, L.R.S. and Gill, M. (1986) The effect of conservation method and frequency of feeding on the removal of digesta from the rumen. *Proceedings of the Nutrition Society*, **45**, 97A

Thomas, *et al.* (1961) A study of factors affecting rate of intake of heifers fed silage. *Journal of Dairy Science*, **44**, 1471–1483

Thomas, C., Gill, M. and Austin, A.R. (1980) The effect of supplements of fishmeal and lactic acid on voluntary intake of silage by calves. *Grass and Forage Science*, **35**, 275–279

Thomas, C. and Wilkinson, J.M. (1975) The utilization of maize silage for intensive beef production, 3. Nitrogen and acidity as factors affecting the nutritive value of ensiled maize. *Journal of Agricultural Science, Cambridge*, **85**, 255–261

Thomas, P.C. (1982) Utilization of conserved forages. In *Forage Protein in*

Ruminant Animal Production (eds D.J. Thomson, D.E. Beever and R.G. Gunn), Occasional Publication No. 6, Thames Ditton, British Society of Animal Production

Thomas, P.C. and Castle, M.E. (1978) *Annual Report of the Hannah Research Institute*, pp. 108–117

Thomas, P.C. and Chamberlain, D.G. (1982) The utilization of silage nitrogen. In *Forage Protein Conservation and Utilization* (eds T.W. Griffiths and M.F. Maguire). Brussels, Commission of the European Communities

Thomas, P.C., Kelly, N.C. and Wait, M.K. (1976) The effect of physical form of a silage on its voluntary consumption and digestibility in sheep. *Journal of the British Grassland Society*, **31**, 19–21

Thomson, D.J. (1982) The nitrogen supplied by and the supplementation of fresh or grazed forage. In *Forage Protein in Ruminant Animal Production* (eds D.J. Thomson, D.E. Beever and R.G. Gunn). Occasional Publication No. 6, Thames Ditton, British Society for Animal Production, pp. 53–66

Thomson, D.J. and Beever, D.E. (1980) The effect of conservation and processing on the digestion of forages by ruminants. In *Digestive Physiology and Metabolism in Ruminants* (eds Y. Ruckebasch and P. Thivend). Lancaster, MTP Press, pp. 291–318

Thornton, R.F. and Minson, D.J. (1973) The relationship between apparent retention time in the rumen, voluntary intake and apparent digestibility of legume and grass diets in sheep. *Australian Journal of Agricultural Research*, **24**, 889–898

Ulyatt, M.J. (1965) The effects of intraruminal infusions of VFA on food intake by sheep. *New Zealand Journal of Agricultural Research*, **8**, 397–408

Van Soest, P.J. (1982) *Nutritional Ecology of the Ruminant*, Cornvallis, Oregon, O.B. Books

Webster, A.J.F., Dewhurst, R.J. and Waters, C.J. (1988) Alternative approaches to the characterization of feedstuffs for ruminants. In *Recent Advances in Animal Nutrition – 1988* (eds W. Haresign and D.J.A. Cole). London, Butterworths, pp. 167–191

Webster, A.J.F., Simmons, I.P. and Kitcherside, M.A. (1982) Forage protein and the performance and health of the dairy cow. In *Forage Protein in Ruminant Animal Production* (eds D.J. Thomson, D.E. Beever and R.G. Gunn), Occasional Publication No. 6, Thames Ditton: British Society of Animal Production, pp. 89–98

Wilkinson, *et al.* (1983) In *Proceedings of the 14th International Grassland Congress*. Boulder, Col., Westerview Press, pp. 631–634

Woolford, M.K. (1984) *The Silage Fermentation*, New York, Marcel Dekker

19

THE APPLICATION OF NEAR INFRA-RED SPECTROMETRY TO FORAGE EVALUATION IN THE AGRICULTURAL DEVELOPMENT AND ADVISORY SERVICE

C.W. BAKER
Agricultural Development and Advisory Service, Starcross, Devon, UK

R. BARNES
Perstorp Analytical Ltd., Bristol, UK

Introduction

Traditionally, Nutrition Chemists have used data derived from wet chemical analysis of feeding stuffs, for the characterization of feeds and for the prediction of nutritive value; for example, the prediction of Metabolizable Energy (ME) from Modified Acid Detergent Fibre (MADF). The most commonly requested determinands for forages are Crude Protein (CP), MADF, Neutral Detergent Fibre (NDF) and Water Soluble Carbohydrates (WSC). The determination of these by traditional wet chemical methods is relatively slow and costly in terms of both labour and chemicals. Modern automatic techniques have to a large extent overcome these objections for some determinands, but the analysis, for fibre fractions particularly, still has to be done by essentially manual methods.

In 1983 the Agricultural Development and Advisory Service (ADAS) purchased a model 6350 Near Infra Red Spectrometer (NIRS). The initial object of this acquisition was to reduce the amount of labour and costs involved in the daily routine and to provide a much faster turnaround of samples. In addition, it would be used in conjunction with *in vivo* data, to derive equations enabling the direct prediction of the digestibility of forages, rather than via a chemical predictor.

Since 1984 the NIRS has been used to predict CP, MADF and WSC in fresh grass, and CP and MADF in silages, as a routine service to farmers. In addition it has also been used for Ministry Agronomy and Soil Science research and development projects. During 1988, with the cooperation of the North and West of Scotland Agricultural Colleges, the Rowett Research Institute, and the Department of Agriculture for Northern Ireland (DANI), predictive NIRS equations were successfully derived for the *in vivo* digestibility of the organic matter (OMD) in silages. This has enabled ADAS to move away from the only previously practical but unsatisfactory derivation of ME from MADF. Since July 1988, over 10 000 silage samples have been analysed.

A predictive equation has also been derived for *in vivo* DOMD of straw, with quite acceptable accuracy and precision. One of the many variables that can have an effect on the validity of NIRS predicted data, is that of the seasons from year to year. The equations developed have data from more than one season, and are proving to be valid in subsequent seasons. Each year more data has been added to the database and this will continue. For CP in particular, an experimental predictive equation has been derived that will function, for both grass and silage, over more than one season.

Recent developments in data transformation techniques (Barnes *et al.*, 1989), have been investigated for routine use, and show great promise in overcoming the effects of particle size on the spectra, and the often quoted criticism of multi-collinearity within the spectra. One result of this is the improved robustness of the derived equations. The NIRS prediction of ash and soil contamination in silages, which has never been satisfactorily resolved, is being investigated using this technique. This report is not an exhaustive description of the development of NIRS predictive equations, but one of the progressive moves towards improving the services that are available to farmers.

Near infra-red diffuse reflectance spectrometry

Near infra-red diffuse reflectance spectroscopy covers the spectral range from 1100 to 2500 nm. Energy in this spectral range is directed via a monochromator grating onto a cell containing dried and milled sample. No other sample preparation is required. The energy reflected from the sample cell is collected by lead sulphide detectors. This reflected energy consists of two major components, being (a) specular reflection and (b) diffuse reflection.

The specularly reflected energy carries information about the physical nature of the sample particle surfaces and is the result of random scatter of light at those surfaces. The diffuse reflectance carries information about the chemical nature of

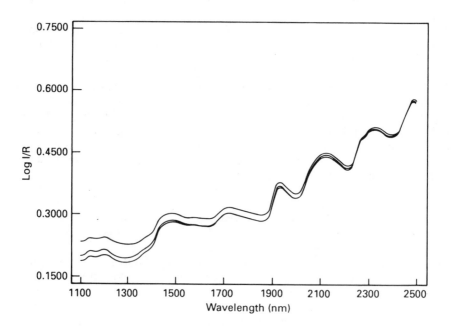

Figure 19.1 Log 1/R spectral – the effect of particle size

the sample. The chemical bonds within the sample (–CH, –OH, –NH, –SH) absorb energy at specific wavelengths and modify the intensity of the radiation emerging from the sample. The reflected energy is collected at 2 nm intervals across the spectral range and stored as the reciprocal logarithm (log 1/R) of the reflected energy collected. It is important to maximize the collection of the diffuse reflected energy. The geometry of the monochromator and detector system used is designed to meet this requirement. The system is analogous in some ways to transmission spectroscopy except that the radiation emerges from the same side of the sample cell as the incident radiation. Some light energy will pass between sample particles striking only their surfaces, whilst some of the light energy penetrates the sample particles where chemical bonds absorb and modify the energy. The result of this is that the path length through the sample cell is not constant and is dependent on the sample particle size and packing density. This gives rise to spectra of the same material being of the same shape but displaced from each other (Figure 19.1). First or second order derivative spectroscopy (Stark, 1988) overcomes this effect of variable path length. Figure 19.2 illustrates second order derivative spectra of the same three spectra in Figure 19.1. Derivative spectroscopy also results in some resolution of overlapping absorption bands. The absorption bands due to water at 1450 and 1940 nm, carbohydrate 2100 nm, and peptide linkages 2140 to 2180 nm can be seen. Calibration is achieved by relating reference wet laboratory chemical analysis values, or *in vivo* data, to the log 1/R absorbance for a selected set of samples using multi-linear regression and testing the regression equations produced against other analysed samples. Proper selection of calibration sample sets and good quality reference laboratory data are essential to the success of this technique.

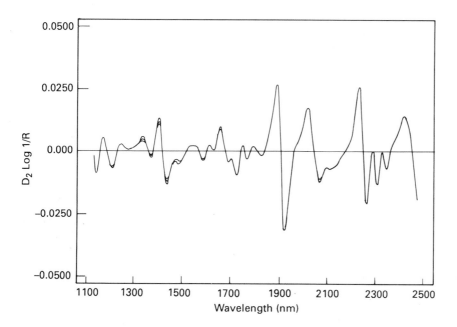

Figure 19.2 Second derivative spectra – the effect of particle size

Calibration development

All the calculations for equation derivation were performed using a specific software program from Infrasoft International (Shenk *et al.*, 1981). All the equations have been derived by multiple linear regression on first or second derivative segments of the spectral data.

GRASS SILAGE

The present equations for crude protein and MADF were developed from a set of 198 samples, of which 151 were used for calibration and 47 for validation. Samples from both 1987 and 1988 are included. During 1988, the validation was confirmed by analysing seven different subsets, totalling 177 for CP and 168 for MADF, at regular intervals throughout the year.

The ashed NDF equation, introduced in 1988, was derived from 195 randomly selected samples, and initially validated with a further 47 samples. All the wet chemistry was performed by the standard ADAS analytical methods (Bailey, 1985).

Table 19.1   NIRS EQUATION PERFORMANCE PARAMETERS FOR GRASS SILAGE

	N	Mean	SEC	RSQ	Principal wavelengths (nm)		
Calibration variable							
OMD	101	70.91	2.77	0.808	1666	2266	1350
%CP	151	14.15	0.72	0.935	2138	1752	1316
%MADF	151	34.32	1.15	0.897	2322	2296	1810
%NDF	105	57.07	1.69	0.893	2302	2266	1926
Validation variable							
Set1	N	SEV(C)	Bias	RSQ	Slope		
OMD	26	2.14	1.10	0.833	1.038		
%CP	46	0.59	0.14	0.936	0.953		
%MADF	47	1.19	−0.32	0.804	0.948		
%NDF	47	1.17	0.24	0.931	1.067		
Set2							
OMD	38	2.72	1.56	0.832	0.926		
%CP	177	0.80	0.03	0.923	0.927		
%MADF	168	1.37	0.38	0.827	1.050		

SEC = Standard error of calibration.
SEV(C) = Standard error of validation corrected for bias.

An *in vivo* OMD equation was derived from 101 samples and validated with a further 26. These samples and the *in vivo* data were provided by the ADAS Feed Evaluation unit, the North of Scotland Agricultural College and the Rowett Research Institute. A further validation set of 40 samples were provided by the Department of Agriculture for Northern Ireland. The samples covered the years 1978 to 1988, and include different cuts of clamp and big bale silages. The equation performance parameters are given in Table 19.1.

The OMD equation is based on a second derivative segment and has 9 terms. The remainder are first derivative and each has 7 terms in the equation.

FRESH GRASS

The Analytical Chemistry department provides three different grass analyses services to ADAS Nutrition Chemists, throughout the year:

1. The Herbage Quality Monitoring service, for primary growths, which provides a general regional picture of the probability of achieving good fermentation on ensiling.
2. The Grass for Ensiling service, which is a farm specific service continuing throughout the year, mainly for primary and second regrowths.
3. Analysis of fresh grass from experimental field sites, which covers the year until the end of October.

No single equation has been derived which is satisfactory to meet all three requirements. Each of these services has required the derivation of a new equation each year, as it has not proved possible to account for the seasonal variations. This has meant relying on wet chemistry data at the very beginning of each season, for the calibration data, and a high percentage of validation throughout the year. The target of 48 hour response to request for information for advisory purposes, is met in this manner. There is generally less urgency for R and D data, which can be treated as a closed set each year. In a typical year about 2000 R and D samples are received. By using a technique of spectral dissimilarity selection of the whole set, it is possible to select a subset for wet chemical analysis, for the calibration and validation exercise. This way the amount of wet chemistry required to produce the calibration and validation sets, is minimized.

Water soluble carbohydrate data are required for the HQM and Grass for ensiling services, for which the samples are dried for 2 hours at 100°C. Other than

Table 19.2 NIRS EQUATION PERFORMANCE PARAMETERS FOR FRESH GRASS

	N	*Mean*	*SEC*	*RSQ*	*Principal wavelengths* (nm)		
Herbage quality monitoring and grass for ensiling service							
Calibration variable							
%CP	146	20.23	0.89	0.985	2184	1644	1980
%MADF	146	22.64	0.67	0.956	1824	1872	1662
%WSC	146	19.00	1.25	0.973	2276	2368	2290
Validation variable	*N*	*SEV(C)*	*Bias*	*RSQ*	*Slope*		
%CP	184	0.88	0.01	0.989	0.993		
%MADF	184	0.78	−0.06	0.973	1.015		
%WSC	184	1.22	−0.13	0.981	0.993		
R and D grasses							
Calibration variable							
	N	*Mean*	*SEC*	*RSQ*	*Principal wavelengths* (nm)		
%CP	88	18.31	0.94	0.971	2166	2032	2020
%MADF	89	26.23	1.01	0.908	1564	2066	1808
Validation variable							
	N	*SEV(C)*	*Bias*	*RSQ*	*Slope*		
%CP	74	1.12	−0.02	0.948	0.978		
%MADF	67	0.99	−0.41	0.835	0.921		

this wet chemical methods are the same standard ADAS (Bailey, 1985). The equation performance parameters for each service are given in Table 19.2.

CEREAL STRAW

Attempts to predict digestibility *in vivo* in cereal straws, on a practical routine basis from cell-wall fractions and *in vitro* measurements have met with only limited success. During 1988 an equation was derived for the direct prediction of *in vivo* DOMD, using NIRS (Givens *et al.*, 1989). Seventy untreated and 14 ammonia-treated straws were used for calibration and 31 untreated and eight ammonia-treated straws used for validation. The equation derived is of first derivative and contains eight terms. The equation performance parameters are given in Table 19.3. Comparison of the NIRS predictive method with those of rumen fluid, neutral detergent fibre and pepsin cellulase, showed that the NIRS method has similar accuracy to cellulase based methods. NIRS can adequately describe the *in vivo* DOMD of both untreated and ammonia-treated straws, and has the advantage of speed and cost over the other methods.

Table 19.3   NIRS EQUATION PERFORMANCE PARAMETERS FOR CEREAL STRAW

Calibration variable	*N*	*Mean*	*SEC*	*RSQ*	*Principal wavelengths* (nm)		
DOMD	84	44.83	2.94	0.784	2240	1694	2340
Validation variable	*N*	*SEP(C)*	*Bias*	*RSQ*	*Slope*		
DOMD	39	3.33	1.53	0.623	0.83		

Recent developments

Control of sample preparation, especially grinding, has been a long standing problem in NIR. It is not possible to collect diffuse reflectance only, without including some interfering specular reflectance. In some cases it is actually desirable to collect specular reflectance to give information on physical factors other than particle size, such as soil contamination which has a significant effect on both wet chemistry and the NIR spectrum. The variation in particle size can account for as much as 90 per cent of the variance in a set of NIR spectra (Cowe, 1985) and is largely responsible for the high level of multi-collinearity between data points in NIR spectra. Derivative spectroscopy removes multi-collinear effects to some extent, but introduces complexity and is sensitive to the background signal to noise ratio of the instrument used (Figures 19.3 and 19.4). As well as derivative spectroscopy, various normalization mathematical transforms all based on the mean spectrum of the calibration set, have been proposed to overcome the particle size variation. Recently the use of standard normal variate (SNV) and de-trending (DT) have been introduced (Barnes *et al.*, 1989), to correct for path length and to some extent scatter effects in diffuse reflectance spectra. This mathematical transformation standardizes the variance of the spectrum to unity with a mean of

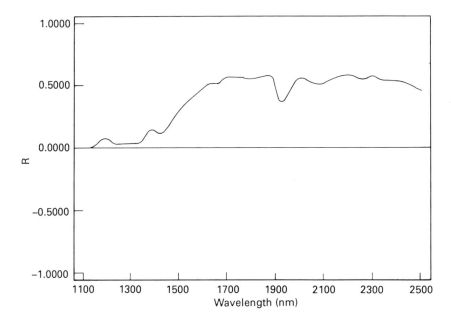

Figure 19.3 Correlellogram of Log 1/R spectral data and *in vivo* OMD in silages

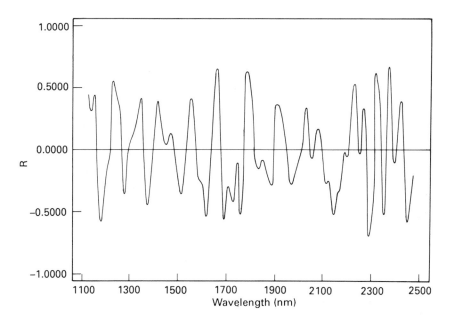

Figure 19.4 Correlellogram of second derivative of spectral data and *in vivo* OMD in silages

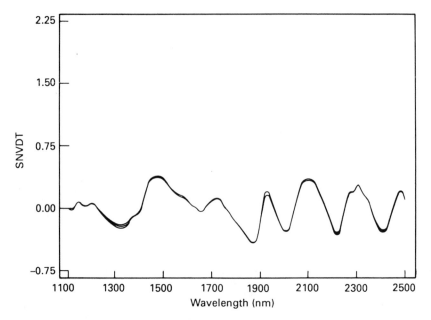

Figure 19.5 SNV and DT of Log 1/R spectral – the effect of particle size

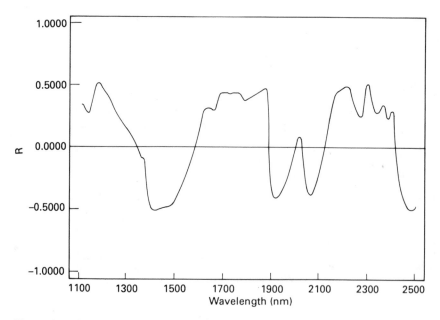

Figure 19.6 Correlellogram of SNV and DT of Log 1/R spectral data and *in vivo* OMD in silages

zero, and removes curvilinearity of the spectruam by use of a second order polynomial correction. It is independent of the sample set mean spectra. The spectra of the silage shown in Figure 19.1, are shown in Figure 19.5 after application of SNV and DT. A correlogram of *in vivo* OMD and NIRS spectra following SNV and DT is shown in Figure 19.6, and demonstrates the removal of the multi-collinearity shown in Figure 19.3. SNV and DT transformation is carried out using the program NIRTOOLS, on an IBM PC.

To investigate the effect of soil contamination of silages on the NIRS predicted data, soil was added to a silage sample to produce a sample set with a range of added soil from 5% to 75%. The samples were then subjected to the normal routine NIRS analysis. The data are given in Table 19.4. It is evident that the

Table 19.4 NIRS PREDICTED DATA FROM SOIL–SILAGE MIXTURES

% Added soil	OMD	%CP	%MADF	%Ashed NDF
0	69.34	14.30	34.88	60.42
5	70.31	13.77	35.18	60.89
10	66.00	13.90	35.56	61.04
20	67.27	13.65	36.39	60.66
30	66.63	13.37	37.19	59.58
50	66.6	12.47	38.26	61.96
75	62.14	11.53	43.05	59.10

addition of soil has little effect upon the predicted values for all the determinands, until it exceeds about 50%. The log 1/R data were transformed using SNV and DT. Figure 7–9 show the log 1/R, the SNV and the SNV and DT spectra of these samples. In the log 1/R figure, the increase in curvature and baseline shift with decrease in overall absorbance can be seen as the amount of soil added increases. For the SNV spectra the scaling effect of the mathematical transform can be seen and also the increasing curvature of the spectra with increasing soil level. In Figure 19.9, where both SNV and DT transformations have been applied, the removal of the curvature from the spectra is apparent. When, for each level of added soil the SNV spectra is subtracted from the SNV and DT spectra, a series of curves is obtained (Figure 19.10). The curvature increases with increasing level of soil contamination. Second order polynomial curve coefficients calculated during the DT process were found to be exponentially related to the amount of soil added (Dhanoa, 1989). Figure 19.11 is a plot of added soil versus NIRS predicted added soil. The same analysis was applied to a set of silage samples known to have a wide range of soil contamination, and the ash data for these samples regressed against the predicted soil contamination data (Figure 19.12). The exponential curve fit was not as good as that found with the previously demonstrated known soil additions, but the data will contain sub-sampling error due to the soil contamination. Curvilinear regression of ash data, with interference data in NIR spectra may be the route to follow for the determination of soil contamination by NIR.

During a calibration exercise, some decision has to be made as to the number of terms that are selected for the equation, in order not to 'overfit' the equation. There is a danger of tailoring the equation so closely to the calibration set that it

346

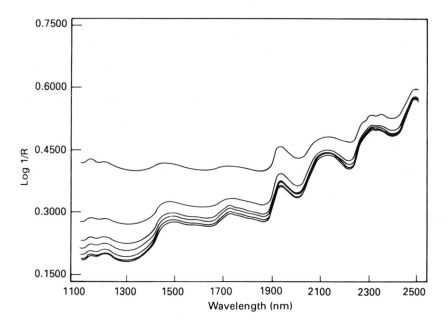

Figure 19.7 Log 1/R spectra of silage/soil mixtures

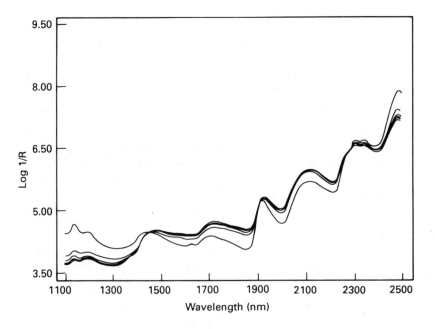

Figure 19.8 SNV of Log 1/R spectra of silage/soil mixtures

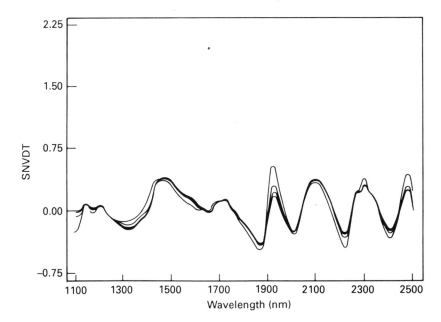

Figure 19.9   SNV and DT of Log 1/R spectra of silage/soil mixtures

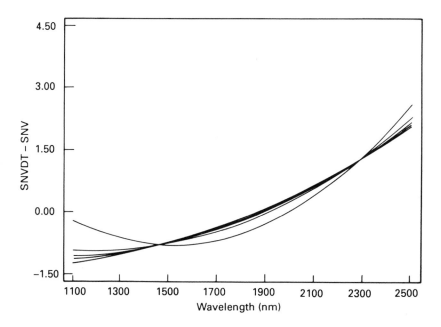

Figure 19.10   [SNV and DT]-[SNV] of Log 1/R spectra of silage/soil mixtures

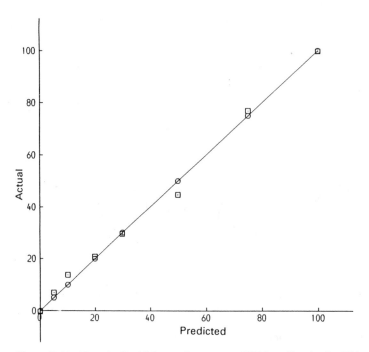

Figure 19.11   Plot of soil addition to silage, versus NIRS predicted soil addition

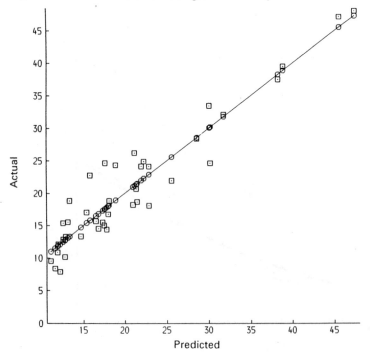

Figure 19.12   Plot of ash data on routine silage samples, versus NIRS predicted soil contamination

effectively becomes a closed set, and may not function as well as predicted on future samples. For this reason it is vital that separate prediction and validation sets are available to check the 'robustness' of the equation. As the number of terms in the equation increases, the correlation coefficient will increase and the standard error of calibration will decrease. A decision has to be made at which point the equation adequately describes the samples without overfitting. A general rule is that the F number (variance ratio) describing the degree of contribution of a particular term, should not be less than 10. Using SNV and DT transforms of the data, the large contribution to the spectral variance caused by particle size variations is very much reduced, and the number of terms required in the equation is also reduced.

An experimental, second derivative of SNV and DT, equation, having six terms, was derived for predicting the *in vivo* OMD of silages. Compared with the present equation (see Table 19.1) an improvement was observed in both the SEC and R^2, being 2.39 and 0.840 respectively. To test the ability of the equation it was used to predict the *in vivo* OMD of the set of silages from DANI, which had not been used in any of the previous calibrations. A comparison of the resulting data from equations derived using SNV and DT, (six term) and those not using this technique (nine term) is summarized in Table 19.5. Both the SEP and R^2 again show improvement, however the bias of the SNV and DT equation has almost doubled. The difference in the bias may be due to the differences in this population of silage which were not represented in the calibration, and they are a different population, or the variance due to particle size may have been obscuring the real bias in the log 1/R data.

The present situation of having three different predictive equations for grass is unsatisfactory because of the large amount of wet chemistry validation that is required. Previous attempts to produce a comprehensive grass equation had failed.

Table 19.5 COMPARISON OF MATHEMATICAL TRANSFORMS ON NIRS EQUATIONS

	N	T	SEP	RSQ	Bias	Slope
Second derivative log 1/R	37	9	2.91	0.826	1.42	0.955
Second derivative SNV/DT	37	6	2.18	0.895	2.22	1.020

Variable – *in vivo* OMD of silages

Table 19.6 NIRS COMPREHENSIVE GRASS EQUATION PERFORMANCE PARAMETERS

	N	Mean	SEC	RSQ	Principal wavelengths		
Calibration variable							
%CP	196	19.07	1.00	0.957	2224	2044	1984
%MADF	194	24.51	1.07	0.912	2284	2258	1702
Validation variable	N	SEV(C)	Bias	RSQ	Slope		
%CP	95	1.16	−0.10	0.972	1.013		
%MADF	95	1.04	0.18	0.933	1.005		

Using the SNV and DT transformation, experimental equations were derived for CP and MADF from over 2000 samples covering the years 1987–8. The equation performance parameters are shown in Table 19.6. These equations have so far been tested on four subsets, and the results are very encouraging. During 1989 further validation of the equations will be tested.

Calibration equation selection based on SEC, SEP, R^2, bias, validation slope, and F number, may appear to provide satisfactory data. In addition to these parameters, the equations should also be tested for repeatability and sensitivity to moisture content and/or temperature. Repeatability is sensitive to the packing density of the sample in the scanning cell, and therefore it might be expected that using the SNV and DT treatment would improve it. A random sample of silage was packed and scanned and repacked and scanned 19 times by the same operator. The 95% repeatability figures were 3.36 and 2.18, with mean values of 73.8 and 73.0 for the normal, and SNV and DT treatments respectively. By applying experimental equations to a spectral file of a sample which has been allowed to pick up moisture slowly over a period of months, and has been NIRS scanned at regular intervals, any moisture and/or temperature sensitivity can be easily detected.

The prediction of *in vivo* DOMD in silages by MADF and NIRS

Until recently MADF, determined by wet chemistry, and NIRS were used most widely to predict *in vivo* DOMD, of silages. Other methods such as acetyl bromide lignin, pepsin cellulase, and *in vitro* rumen fluid digestibility measurements have been used routinely (Barber, 1989), but when determined by normal laboratory procedures have severe limitations in providing a fast response and a high sample throughput service.

MADF prediction of *in vivo* DOMD to calculate ME, has serious shortcomings in its ability to provide accurate data. In correlating MADF with *in vivo* data, the standard error of calibration and the account of variability are poor (Barber *et al.*, 1989).

Wet chemistry MADF data are affected by the amount of ash and soil contamination of the samples. For silages that are not soil contaminated, the error is negligible and ADAS uses unashed MADF data. For silages that have significant amounts of soil contamination, correcting for the ash content will only partially solve the problem, and the residual error will vary with soil type. For NIRS predicted MADF data the situation is similar because the amount of ash or soil contamination has no significant effect on the predicted MADF value below about 50% contaminant present (see Table 19.4). Correcting the NIRS predicted data will present the same problems as that for the wet chemistry data.

Since July 1988, ADAS has routinely used direct NIRS prediction of *in vivo* OMD, and wet chemistry determination of the ash content, as a route to the means of determining the ME of silages. Over 10000 samples have been analysed in that time, reaching a peak of 300 samples a day. The present OMD equation has some minor water sensitivity that is corrected by predicting a 'water index' during the analysis and normalizing the predicted OMD data to the mean water content of the calibration set. The adherence to a strict regimen of drying, the use of the same type of cylone mill at each preparation site to ensure adequate repeatability of particle size distribution and the 'water index' correction, have resulted in satisfactory results for greater than 98% of the samples. Repeatability has been a problem for a very small number of samples, and can be attributed to sample homogeneity, suspected occasional random inconsistency in the instrument and to the odd sample that is probably an outlier and not represented in the calibration population.

The investment in NIRS technology has proved a great success in terms of more efficient use of resources, a better service due to more rapid response, and more precise prediction of the nutritive values by direct prediction of *in vivo* data.

NIRS is not only a tool for replacing wet chemistry techniques, it is potentially a powerful one for quickly showing organic chemical structures, changes in those structures with time, and for following chemical and biological processes. The complete unravelling of the spectral data enabling identification of specific molecular groups and compounds is still in its early days. NIRS has the potential to describe biological materials such as feeds in more detail than hitherto, and could be very valuable in its application to research studies of the mechanics of digestion.

Acknowledgements

We thank I. Givens, G. Barber, I. Murray, E. Unsworth, A. Adamson and B. Cottrill, for provision of samples, *in vivo* data, and technical discussions and information during the NIRS equation derivation, and for assistance in the preparation of this paper. M.S. Dhanoa for discussions and the analysis of data concerning the determination of soil contamination and ash.

References

Bailey, S. (1985) *The Analysis of Agricultural Materials. A Manual of the Analytical Methods used by the Agricultural Development and Advisory Service*, 3rd edn. Reference Book 427. London, HMSO

Barber, G.D., Givens, D.I., Kridis, M.S., Offer, N.W. and Murray, I. (1989) Prediction of the organic matter digestibility of grass silage. *Animal Science and Technology* (in press)

Barnes, R.J., Dhanoa, M.S. and Lister, S. (1989) *Applied Spectroscopy*, **43**, 7

Cowe, I.A. (1985) *Applied Spectroscopy*, **39**, 257

Dhanoa, M.S. (1989) Private communication

Givens, D.I., Everington, J.M. and Baker, C.W. (1989) *British Society of Animal Production. Winter Meeting* (A comparison of NIRS with three *in vitro* digestibility methods to predict digestibility *in vivo* in cereal straws)

Shenk, J.S., Landa, I., Hoover, M.R. and Westerhaus, M.O. (1981) Description and Evaluation of a near infra red reflectance spectro-computer for forage and grain analysis. *Crop Science*, **21**, 355–358

Stark, E.W. (1988) In *Analytical Applications of Spectroscopy* (eds Creaser and Davies). London, Royal Society of Chemistry, p. 21

20

CONSEQUENCES OF INTER-LABORATORY VARIATION IN CHEMICAL ANALYSIS

SIDNEY BAILEY
Agricultural Development and Advisory Service, Nobel House, London

KEITH HENDERSON
Agricultural Development and Advisory Service, Block C, Government Buildings, Cambridge, UK

Introduction

Animal feedstuffs are subjected to chemical analysis for a number of reasons. These include statutory requirements in respect of compositional statements on labels, livestock advisers' requirements for data on which to base advice related to various aspects of performance, quality assurance testing by manufacturers, feed surveillance and research workers requirements for data on materials used in experimental programmes. In all these cases the requirement is generally for quantitative data and since all quantitative measurements are subject to variability it is important that the extent of this variability is appreciated. There are limitations in all measurements and consequently no absolutes in analytical results. What is therefore needed is some indication of the reliability of the result, i.e. what is the precision and accuracy of the reported value and whether these are acceptable for the end use of the data. This situation applies both to established methods and newly developed procedures, although the latter may require a more rigorous assessment. Banes (1969) considered that authors of reports of new and improved methods are rarely unbiased reporters. They frequently exaggerate the virtues of their newly developed methods or modifications and over-emphasize the deficiencies of other methods as justification for their work. Similarly they tend to understate or omit the faults of their own method and ignore the strong points of previous methods. Since Banes (1969) considered the situation there has been considerable movement to the inclusion of precision data in published methods, but many papers on biological performance still report analytical data without reference to the method of their generation or indication of the achieved precision of the laboratory.

Collaborative studies

PURPOSE AND PROCEDURES

The best method of assessing the performance of a method of chemical analysis is by means of a collaborative study. During such a study a number of laboratories each analyse a number of samples using the method under study. The results from the study are collated and submitted to statistical treatment to measure the within-laboratory and between-laboratories variability. Many organizations from international bodies to individual organizations have conducted studies and it is not the purpose of this chapter to present an in-depth review or criticism of the procedures employed in them. Much of this has already been undertaken in the Workshops organized by the International Union of Pure and Applied Chemists (IUPAC) on the Harmonization of Collaborative Analytical Studies. The recommendations from the Workshop held in Geneva in 1987 have been incorporated into the Guidelines published by the Association of Official Analytical Chemists (1988). The statistical treatment of results for such a study utilizes ISO 5725 (International Standards Organisation, 1986) to determine values for the within-laboratory precision (repeatability, r) and between-laboratory precision (reproducibility, R).

Although methods are available for the assessment of precision there is a paucity of reported figures for methods utilized in feedstuffs analysis. Even in the United Kingdom Feeding Stuffs Regulations (1982, 1985) there is reference in Section 1b, Schedule 2 to satisfactory repeatability but a figure is given only for four methods. Reproducibility is quoted only for two methods concerning aflatoxins. There is also no reference to the study which generated these data.

ESSENTIAL CONSIDERATIONS

Before considering the use of the results from any study of a method it is necessary to consider certain aspects of collaborative studies in more detail. It is most important to appreciate that collaborative studies assess the performance of the analytical procedure. They depend on experience with the type of determination within a laboratory. These studies do not take account of any sampling errors or variation due to heterogeneity because the samples involved are carefully prepared and sub-sampled. Laboratories involved in collaborative studies frequently utilize their best analyst who probably works very carefully. Therefore the results reflect in all probability the best that can be expected of a method, although some evidence does exist (Horwitz, Kamps and Boyer, 1980) that precision can improve with practice until it reaches a constant value.

One criterion for collaborative studies which frequently receives less attention than necessary is the need for strict adherence to the method. Whilst failure to carry out the method exactly may be due on occasions to poor drafting of the instructions leading to misinterpretation, experience shows that analysts frequently change procedures to suit themselves, either through a lack of some reagent or equipment or perhaps through the belief that they 'can do better'. The procedure must be properly written and have sufficient detail both in the description of equipment and procedure. It is important that participants follow the method exactly in order to identify the true precision. Modifications frequently lead to bias in results and this increases the variability between laboratories beyond that normally attainable through correct use of the method. It is therefore important that analysts undertaking this type of determination continue to follow the method exactly for routine analysis particularly when the method itself defines the property or component, e.g. crude fibre.

INTERPRETATION OF RESULTS

When collaborative studies are undertaken within the discipline of Guidelines such as those of AOAC (1988), statistical procedures are applied to results in order to identify and remove outliers. These atypical results can occur and may or may not be explainable by the participants concerned. The case for consideration of outliers has been put by Horwitz (1977) and frequently discussed before incorporation into the Guidelines. The 'acceptable' results are utilized in the calculation of repeatability and reproducibility figures. These are calculated by the procedure of the International Standards Organisation (ISO-1986) with a 95% confidence limit. Thus in 19 out of 20 cases of comparison between either two results within a laboratory or two results obtained by different laboratories these will fall within the variability of the method calculated as repeatability or reproducibility. It should be noted that r and R may be dependent on concentration or may be consistent across a range of concentrations of analyte.

For a method where results from a published collaborative study are available the relevant figures can be compared with the performance of a laboratory initiating or continuing the use of the method. There remains however the problem of assessing performance where no study has been published for the method, or the method itself is new, possibly having recently been developed in that laboratory. It may not be feasible to establish a collaborative study to determine the precision of the method. In these cases results from other studies may be used as a guide, as demonstrated in a series of papers. In the first of these papers Horwitz (1977) compared the precision of the results from collaborative studies within the pharmaceutical sector, published by the AOAC. From the available results it was found that automated methods did not appear to be any more precise than manual methods. The comparison of methods was extended by Horwitz, Kamps and Boyer (1980) who examined the results of over 50 studies conducted by AOAC on various commodities for numerous analytes at a wide range of concentrations. The paper reported that in general the inter-laboratory precisions as a function of concentration appeared to be independent of the nature of the analyte or of the analytical technique that was used.

Use of coefficient of variation (relative standard deviation)

Horwitz (1982) showed a relationship between a mean coefficient of variation expressed as powers of 2, with the mean concentration measures expressed as powers of 10. The same studies also revealed a relationship between the within-laboratory and between-laboratory errors. The within-laboratory was found to be approximately one-half to two-thirds of the between-laboratory error.

The mathematical relationship was further developed by Horwitz (1982) into the equation:

$$CV(\%) = 2^{(1-0.5 \log C)} \qquad (20.1)$$

Where C is the concentration expressed as powers of 10 with 100% = 1. Boyer, Horwitz and Albert (1985) changed this equation by the substitution of relative standard deviation (RSD) for CV.

Horwitz and Albert (1984, 1985) investigated results from published studies of methods of analysis used for regulatory purposes as Drug Dosage Forms. The papers covered chromatographic/spectrophotometric measurement (1984a), gas chromatographic methods (1984b) automated methods (1985a), High Pressure Liquid Chromatographic methods (1985b) and miscellaneous methods (1985c). Boyer, Horwitz and Albert (1985) also investigated the results from 18 methods for trace elements. Each paper utilized the same computer program for handling the statistical appraisal, including identification of outliers. The investigations confirmed the remarkable agreement between the different methods of analysis involving different sample matrices and supported the exponential equation. In addition the ratio between the within-laboratory and between-laboratory errors was confirmed.

Boyer, Horwitz and Albert (1985) commented that there is currently no scientific basis or theory to support the validity of the equation. Nevertheless they have found that it serves as a useful 'bench mark' to judge the precision characteristics of previously unevaluated methods. In addition they considered that when the Relative Standard Deviation produced by a method under collaborative study conditions exceeds more than twice the value predicted by the equation the method may not be applicable for the determination of that analyte in the matrix being analysed. Furthermore when the method is applied below the lower concentration limit of a measurement technique resulting RSD values can be expected to depart radically from the value predicted by the equation.

Application to feedstuffs analysis

Accepting the Horwitz curve as an indication of acceptability of performance, the Relative Standard Deviations for between-laboratory variability appropriate to

Table 20.1 PRECISION DATA CALCULATED FROM HORWITZ EQUATION (HORWITZ, 1982)

Percentage analyte concentration	CV_R	R	r
100	2.00	5.66	3.74
60	2.16	3.67	2.42
50	2.22	3.14	2.07
40	2.30	2.60	1.72
30	2.40	2.04	1.34
20	2.55	1.44	0.95
15	2.66	1.13	0.75
10	2.83	0.80	0.53
5	3.14	0.44	0.29
1	4.00	0.11	0.07
0.1	5.67	0.016	0.011
0.01	8.00	0.0023	0.0015
0.001	11.31	0.00032	0.00021
0.0001	16.00	0.000045	0.000030
0.00001	22.63	0.0000064	0.0000042
0.000001	32.00	0.0000009	0.0000006

concentrations is given in Table 20.1. Reproducibility at the 95% confidence level is related to the RSD % by the equation

$$R = 2\sqrt{2}\, m \frac{RSD}{100} \tag{20.2}$$

Where m is the mean % concentration.

The calculated figures are quoted in Table 20.1 together with the corresponding repeatability values calculated as two-thirds of the reproducibility figures. These values are average estimates and one cannot say that a particular analyst will achieve these degrees of precision. They can however be interpreted as targets but the final achievement of analysts depends on their skill and experience as well as factors beyond their control. Achievement of these levels of performance should however be treated as acceptable, although figures of twice this value are still considered by Boyer, Horwitz and Albert (1985) to indicate a satisfactory analytical procedure.

Performance during routine analysis

Having proposed the means by which performance can be quantified approximately, it is necessary to consider the methods presently in use in the United Kingdom for analysis of animal feedstuffs. Wainman, Dewey and Boyne (1981) reported coefficients of variation for the methods used for generating data for the study of ruminant feeds which was designed to obtain regression equations for use in determining Metabolizable Energy values. Fisher (1982) quoted similar figures from a corresponding study on poultry feeds whilst Morgan *et al.* (1984) quoted the figures appropriate to a study of pig feeds. Alderman (1985) collated the results from all these studies which were coordinated by a joint working party of the United Kingdom Agricultural Supply Trade Association, the Agricultural Development and Advisory Service and the Council of Scottish Agricultural Colleges, whose report (UKASTA/ADAS/COSAC-1985) had been published. Although the figures reported for the estimated reproducibility standard deviations vary between the studies on pig feeds and the ruminant/poultry feed studies they are of the same order. Thomas *et al.* (1988) reported figures relative to a second study of ruminant feeds.

Annually a comparison of performance in ADAS and UKASTA laboratories is made with currently six ADAS and nine UKASTA laboratories participating. Every year four samples each of hay, compound feed and fishmeal are analysed by routine procedures. The average coefficients of variation over the past 10 years for the methods under comparison are reported in Table 20.2, together with the corresponding average analyte concentration. Additional figures are given from other ADAS/UKASTA collaborative studies. It will be noted that many of these results are below the figures predicted by the Horwitz equation. There are however three exceptions which qualify as acceptable methods only under the expanded criteria. These determinations are the subject of legal requirements in the United Kingdom and therefore consideration must be given to improving their performance. Crude Fibre determinations are frequently undertaken with modifications made to the filtration conditions, which may account for some of the variability. In addition the ashing process is subject to variability which may also influence the result. Whilst improvement in the technique might be desirable, the

Table 20.2 PRECISION DATA ACHIEVED (UKASTA/ADAS LABORATORIES)

Study	Analyte	Matrix	Average concentration (%)	Average $CV_R\%$
A	NCD*	Compound	76.4	2.0
A	Protein	Fishmeal	72.5	1.3
A	MADF†	Hay	37.0	2.4
A	Protein	Compound	21.0	1.6
A	Protein	Hay	12.0	2.4
A	Ash	Compound	9.4	2.6
A	Crude fibre	Compound	7.9	4.9
A	Ash	Hay	7.7	3.0
A	Oil (Acid Hydrolysis)	Compound	6.8	5.3
A	Oil (Petroleum Ether Extract)	Compound	5.0	4.8
B	Starch	Compound (Poultry)	31.2	2.9
B	NDF	Compound (Pig)	20.6	2.3

*NCD – Cellulase digestibility following neutral detergent extraction.
†MADF – Modified acid detergent fibre.
A = Routine monitoring of performance.
B = Collaborative studies.

Agricultural and Food Research Council Technical Committee on Responses to Nutrients (1987) have recommended that this determination should be discontinued and replaced by the use of neutral detergent fibre (NDF).

The lack of precision in the determination of oil, particularly the acid hydrolysis oil method which is now applicable to most compound feeds, is cause for greater concern. Experience with the procedure reveals that it is possible to lose oil during filtration following acid hydrolysis without this being apparent to the operator. Although the method permits the use of solvent extraction before the hydrolysis step, the procedure is not precisely written and there is urgent need for improvement in drafting of the method and the organization of a full collaborative study to determine the true precision of this method, which is the subject of European Community Regulations. The tolerance attached to this determination in the Regulations is 4% relative at concentrations between 5 and 10% and 0.4% absolute at higher concentrations. Whilst these figures reflect the lack of precision in this determination a need for improvement is indicated before assurance can be given that this tolerance will not be frequently exceeded.

The hydrolysis stage in the method for starch determination is extremely sensitive to temperature and care must be taken to ensure that the reaction mixture is fully raised to temperature throughout this stage.

Consequence of analytical variability on prediction of energy

The UKASTA/ADAS/COSAC Working Party Report (1985) considered the impact of permitted tolerances on the predicted energy values using various regression equations derived from studies on ruminant, poultry and pig feeds. It also reported on the use of residual standard deviations as an acceptable estimate of the precision of an estimated DE or ME value. The expanded residual standard deviation, designated S″ took into account the between-laboratory variability in analytical data generated in the various studies in addition to other technical variations associated with manufacturing and sampling.

A simpler method of calculating the sole effect of between-laboratory variability on the predicted energy using recommended regression equations for poultry feeds was adopted by Cooke (1987). This method can be applied utilizing the Relative Standard Deviations calculated from the Horwitz equation to indicate what variability would be considered acceptable. Applying this philosophy to the three current Energy equations for poultry, pig and ruminant feeds results in the following.

POULTRY

$$AME = 0.343 \text{ oil} + 0.155 \text{ CP} + 0.167 \text{ starch} + 0.130 \text{ sugar}$$

For a feed with a mean oil content of 6.3%, protein 18.4%, starch 46% and sugar 4% all on a dry matter basis, the Horwitz RSD values would be 3.03, 2.57, 2.25 and 3.25. The standard deviations for the concentrations are therefore 0.19, 0.47, 1.04 and 0.13 respectively with 95% confidence limits of twice these figures. By use of the energy equation coefficients, the analytical variability might result in the following contributions to overall energy:

Oil	$= \pm 0.130$
Crude protein	$= \pm 0.146$
Starch	$= \pm 0.347$
Sugar	$= \pm 0.034$

Calculation of the variance of this combination of analyses at the 95% confidence limit is given by the square root of the sum of the squares of the individual variabilities. Thus the total variation in energy $= \pm 0.4 \text{ MJ/kg}$. The mean energy content of this feed would therefore be 13.22 MJ/kg with an analytical range of 12.82–13.63 MJ/kg on a dry matter basis.

PIGS

Consider the equation:

$$DE = 17.47 + 0.16 \text{ oil} + 0.078 \text{ CP} - 0.33 \text{ ash} - 0.14 \text{ NDF}.$$

For a feed of mean composition 5.75% oil, 18.5% protein, 9.2% ash and 18.4% NDF the variability in energy calculates as:

Oil	$= \pm 0.056$
Protein	$= \pm 0.074$
Ash	$= \pm 0.174$
NDF	$= \pm 0.132$

The total variability in energy $= \pm 0.24$ for this feed. The mean energy content is calculated as 14.22 MJ/kg with an analytical range of 13.98–14.46 on a dry matter basis.

RUMINANT FEEDS

Consider the equation:

$$ME = 0.14\ NCD + 0.25\ oil$$

For a feed of composition 80% NCD and 6% oil, the variability in energy calculates as:

$$NCD = \pm0.464$$
$$Oil\ \ = \pm0.092$$

The total variation in energy is calculated as ±0.47. Thus the energy content of this feed calculates as 12.70 MJ/kg with an analytical range 12.23–13.17 MJ/kg on a dry matter basis.

These ranges refer to feeds on a dry matter basis, if predictions of ME are on an as-fed basis there would be a corresponding reduction in the factor, e.g. 0.87 would be appropriate for a DM content of 87%. The figures quoted are appropriate to between-laboratory comparisons and also relate purely to analytical variability. No estimate is made of errors due to sampling. These errors are likely to exceed analytical variability even when due precautions are taken in sampling. Where manufacturers have formulated feeds utilizing database figures there will be an additional variability between their calculated value and that predicted from analysis. Figures relating to repeat analysis within a laboratory would be between one-half and two-thirds of these ranges.

The analytical ranges resulting from the use of precision data, which is considered normal for acceptable methods, may be considered wide by some, particularly legislators or customers. It is therefore important to ensure that these levels are met, or preferably bettered by day-to-day performance. Precision may be improved to some extent by replication of analysis but this greatly increases cost. The more cost effective method is by measurement of the precision of each method and investigation of ways to reduce variability by identification of critical stages in the analytical methods. After these stages have been optimized, methods must be drafted so that all analysts may follow the correct procedure to ensure best precision. Studies are being undertaken particularly through the International Standards Organisation, British Standards Institute, the AOAC, the Analytical Methods Committee of the Royal Society of Chemistry and the European Commission. These studies tend to be prolonged due to current lack of resources in laboratories and publication of results adds further delay. It is hoped that all studies will in future report data to a common format so that the precision of alternative methods can be compared. It should however be noted that Zaalberg (1989) has recently challenged the method of calculation of the repeatability standard deviation used in ISO 5725. This might delay a uniform approach whilst statisticians debate the relative merits of the alternatives.

When studies, particularly international studies, result in standard methods it is most important that analysts adhere strictly to the method. This paper has addressed the subject of precision throughout since in many cases it is not possible to comment on accuracy of methods. Accuracy may be measured where alternative methods for measuring the same property are available. Thus total element composition may be determined through different digestion and determination techniques and the results compared. Protein is measured by determination of nitrogen and standard substances, which are difficult to chemically digest, are available in pure form for checking performance. In other cases, particularly fibre fractionation, the method defines the property. Any modification to apparatus, time of digestion, temperature or reagent strength can have a profound effect on the result. Although such modifications might lead to the same result as the standard method for a number of test samples there is no guarantee that this will apply to all samples in the future. Even within laboratories where strict adherence to the method is observed there is a need for quality assurance procedures to ensure that any bias can be identified, investigated and eliminated quickly. There is a need for the establishment of a collection of certified reference materials related to animal feedstuffs. These should cover the nutritional parameters and could be used for the comparison of methods or measurement of laboratory performance. There are however technical problems to this proposal in respect of changes to matrix structure during procedures necessary to prevent deterioration in storage.

Conclusions

Wherever possible precision data should be obtained on methods to be used for feedstuffs analysis. The methods should be investigated in detail to determine procedures which optimize performance and reduce variability to a minimum. There is an urgent need to undertake this work with the present United Kingdom Regulatory methods for oil and sugar determination since these methods have relatively poor precision but must be used when making a labelling declaration in addition to being used for energy content estimation. Laboratories should ensure that methods are strictly adhered to and quality control procedures are followed so that performance meets the Horwitz criteria as a minimum. Samples submitted to laboratories must be taken with full regard to homogeneity of sample and representation of the total lot.

Variability will always exist and even with careful practice the difference between laboratories will from time to time exceed the figures calculated in this chapter.

References

Agricultural and Food Research Council Technical Committee on Responses to Nutrients (1987) Report No. 1 – Characterisation of feedstuffs: energy. *Nutrition Abstracts and Reviews, Series B. Livestocks Feeds and Feeding*, **57**, 507–523

Alderman, G. (1985) Prediction of the energy value of compound feeds. In *Recent Advances in Animal Nutrition* (eds W. Haresign and D.J.A. Cole). London, Butterworths, pp. 3–52

Association of Official Analytical Chemists (1988) Guidelines for collaborative study procedure to validate characteristics of a method of analysis. *Journal of the Association of Official Analytical Chemists*, **71**, 161–171

Banes, D. (1969) The collaborative study as a scientific concept. *Journal of the Association of Official Analytical Chemists*, **52**, 203–206

Boyer, K.W., Horwitz, W. and Albert, R. (1985) Inter-laboratory variability in trace element analysis. *Analytical Chemistry*, **57**, 454–459

Cooke, B.C. (1987) The impact of declaration of the metabolizable energy (ME) value of poultry feeds. In *Recent Advances in Animal Nutrition 1987* (eds W. Haresign and D.J.A. Cole). London, Butterworths, pp. 19–26

Fisher, C. (1982) Energy Values of Poultry Compound Feeds. *Occasional Publication No. 2*, Roslin, Midlothian, Poultry Research Centre

Horwitz, W. (1977) The variability of AOAC methods of analysis as used in analytical pharmaceutical chemistry. *Journal of the Association of Official Analytical Chemists*, **60**, 1355–1363

Horwitz, W., Kamps, L.R. and Boyer, K.W. (1980) Quality assurance in the analysis of foods for trace constituents. *Journal of the Association of Official Analytical Chemists*, **63**, 1344–1354

Horwitz, W. (1982) Evaluation of analytical methods used for regulation of foods and drugs. *Analytical Chemistry*, **54**, 67A–76A

Horwitz, W. and Albert, R. (1984a) Performance of methods of analysis used for regulatory purposes. 1. Drug dosage forms. A. Chromatographic separation/spectrophotometric measurement. *Journal of the Association of Official Analytical Chemists*, **67**, 81–90

Horwitz, W. and Albert, R. (1984b) Performance characteristics of methods of analysis used for regulatory purposes. 1. Drug dosage forms. B. Gas chromatographic methods. *Journal of the Association of Official Analytical Chemists*, **67**, 648–652

Horwitz, W. and Albert, R. (1985a) Performance characteristics of methods of analysis used for regulatory purposes. 1. Drug dosage forms. C. Automated methods. *Journal of the Association of Official Analytical Chemists*, **68**, 112–121

Horwitz, W. and Albert, R. (1985b) Performance characteristics of methods of analysis used for regulatory purposes. 1. Drug dosage forms. D. High pressure liquid chromatographic methods. *Journal of the Association of Official Analytical Chemists*, **68**, 191–198

Horwitz, W. and Albert, R. (1985c) Performance characteristics of methods of analysis used for regulatory purposes. 1. Drug dosages forms. E. Miscellaneous methods. *Journal of the Association of the Official Analytical Chemists*, **68**, 830–838

International Standards Organisation (1986) *Precision of test methods – Determination of Repeatability and Reproducibility for a Standard Test Method by Inter-Laboratory Tests.* ISO 5725. Geneva, Switzerland. Available from British Standards, Linford Wood, Milton Keynes MK14 6LE, UK and other national standards organizations

International Union of Pure and Applied Chemistry (1988) Protocol for the design, conduct and interpretation of collaborative studies. *Pure and Applied Chemistry*, **60**, 856–864

Morgan, C.A., Whittemore, C.T., Phillips, P. and Crooks, P. (1984) *The Energy Value of Compound Foods for Pigs.* Edinburgh, Edinburgh School of Agriculture

The Feeding Stuffs (Sampling and Analysis) Regulations (1982) Statutory Instruments No. 1144. London, HMSO

The Feeding Stuffs (Sampling and Analysis) (Amendment) Regulations (1985) Statutory Instruments No. 1119. London, HMSO

Thomas, P.C., Robertson, S., Chamberlain, D.G., Livingstone, R.M., Garthwaite, P.H., Dewey, P.J.S., Smart, R. and Whyte, C. (1988) Predicting the metabolizable energy (ME) content of compounded feeds for ruminants. In

Recent Advances in Animal Nutrition 1988 (eds W. Haresign and D.J.A. Cole). London, Butterworths, pp. 127–146

UKASTA/ADAS/COSAC (1985) *Prediction of Energy Values of Compound Feed.* The Report of an UKASTA/ADAS/COSAC Working Party. London, MAFF

Wainman, F.W., Dewey, P.J.S. and Boyne, A.W. (1981) *Feedingstuffs Evaluation Unit, Rowett Research Institute, Third Report 1981.* Edinburgh, Department of Agriculture and Fisheries for Scotland

Zaalberg, J. (1989) Experimental design for inter-laboratory precision experiments: comparison of ISO 5725 with Draft NEN 6303. *Journal of the Association of Official Analytical Chemists*, **72**, 34–37

21

DEVELOPMENT AND APPLICATION OF A FEED DATABASE

JEANNIE M. EVERINGTON
Agricultural Development and Advisory Service, Feed Evaluation Unit, Stratford-upon-Avon, UK
S. SCHAPER
Centraal Veevoederbureau, Postbus 1076, 8200 BB, Lelystad, The Netherlands
D.I. GIVENS
Agricultural Development and Advisory Service, Feed Evaluation Unit, Stratford-upon-Avon, UK

Tables of feed composition and nutritive value of animal feeds have been in use for some 180 years. In recent years, however, the volume of data has increased manyfold and computer databases have been developed. This chapter describes the need for databases, important features of their construction, the data to be stored and how databases are most effectively utilized.

Introduction

An assessment of the nutritional and economic value of feedstuffs is an essential prerequisite for their efficient utilization by animals. The degree of development achieved by a country largely depends upon the extent and utilization of its resources. Although there are limitations to the ability to increase supplies of natural feed resources, many ways are available to improve their utilization and consequently reduce dependence upon imported feedstuffs.

In developed countries, considerable progress has been made in feed evaluation and storage of the resulting data and to this end increasing reliance has been placed on the use of computerized databases. These have the ability to store large quantities of data, give rapid access, and allow for more accurate and economic diet formulation as well as giving an overview of feed resource utilization within the country. Alderman *et al.* (1983) have given a comprehensive account of databases of animal feeds and their present application.

In many developing countries, there is a lack of basic information on the compositional and nutritive value of many of their feed resources. Although many feeds are unique to a particular region, much of the information held in the feed databases of developed countries would be helpful to them and indeed the utilization of developed countries databases in this way was one of the initial aims of the International Network of Feed Information Centres (INFIC, 1977).

Computerized systems of storing data on the composition and nutritional value of feedstuffs have, for some of the purposes outlined, developed rapidly and are in regular use in many countries and some examples have been given by Leche (1983).

This chapter describes the most important aspects of the development and maintenance of a feed database. The term database in the present context refers essentially to the definition cited by Hain (1983) as 'a collection of interrelated data stored together with controlled redundancy to serve one or more applications in an optimal fashion; the data are stored so that they are independent of the programs which use the data; a common and controlled approach is used in adding new data and modifying and retrieving existing data within the database'. A most important aspect in this definition is the relational nature of the database. This excludes from our consideration the use of spreadsheet type software or storage in sequential files which may store data in a non-relational way. In this paper the development of databases is therefore concerned solely with the use of Database Management Systems (DBMS) and Fourth Generation Languages. Aspects of both of these have been described by Hain (1983).

The need for databases

Tables of chemical composition and nutritive values of animal feeds have been in use for some 180 years, the first recorded publication being that of Thaer (1809). The development of these tables over the years has spread around the world and some of the history has been recorded by Alderman *et al.* (1983) and the Ministry of Agriculture, Fisheries and Food (MAFF, 1986).

The widespread development of tables of feed composition and nutritive values has highlighted the demand for this sort of information. However, as described by Topps (1989), the past 30 years has seen the amount and type of data available increase enormously. This has resulted from more sophisticated and rapid analytical procedures, the introduction of new feeds and the expansion in many countries of a more intensive animal production industry. Whilst the use of feed tables will continue for some time, they are no longer suitable for the storage of large quantities of data. The availability of computers and software to take over and extend this role has been fortuitous.

The major users of feed databases tend to fall into five categories, namely feed manufacturers, extension workers, government policy makers, those involved with nutritional research and farmers. Feed manufacturers require rapid access to information on feedstuffs to enable them to make informed decisions regarding commodity purchase and sale, together with information which will allow them to formulate nutritionally adequate diets or compound feeds within the prevailing economic restrictions. Extension workers require reliable and meaningful information on the nutritive value of feedstuffs which will assist them to formulate diets to give economic and efficient animal production. In both of the foregoing examples any inadequacies in the relevance or quality of data will manifest itself fairly rapidly.

It is clearly necessary for government policy makers to be able to plan the utilization of national feed resources and to understand the interrelationships in the feed-livestock economy. One way in which agricultural policy makers can assess changes in the feed-livestock economy is through the use of a feed-utilization matrix (FUM) which indicates the utilization of all feed types by different classes of livestock. A full description of the development of FUMs for member countries of the Organisation of Economic Cooperation and Development has been given by Parris and Tisserand (1988). The operation of FUMs is heavily dependent on the availability of databases for the provision of nutritional information on the feedstuffs relevant to each country.

Whilst research workers are often heavily involved in experimental work which results in information becoming available for inclusion in databases, they are also users of database contents. For example, the availability of up to date values is helpful in the formulation of experimental diets, in the teaching of students and very importantly, to enable identification of areas where knowledge is lacking. However, it should be noted that, in experimental work, full characterization of feedstuffs being used is preferable to the use of database values.

The information presented in traditional tables of feed composition has been shown to have two major drawbacks. Firstly it has concentrated heavily on the chemical composition of feedstuffs with little reference to biological or nutritionally meaningful values. As pointed out by Topps (1989), the usefulness of chemical composition data depends, amongst other things, on the 'availability' of the component to the animal. In the authors' opinion, much more effort should be made to include biological measurements in new feed databases and where appropriate in feedingstuff tables. In an attempt to begin this process the UK Tables of Feed Composition and Nutritive Value for Ruminants (MAFF, 1986) presented digestibility and energy values for feeds which were almost entirely based on measurements made *in vivo*.

The publication of MAFF (1986) also attempted to correct the second weakness in traditional tables by providing information for each feedstuff considered on the likely variability of both the biological and chemical determinations. To provide the users described above with nutritionally meaningful information, it is essential that feedstuff databases are also able to provide details on variability and on how this and the mean values may change with time.

Purpose of databases

The major purpose of a feed database must be to collect centrally as much information as possible about available feedstuffs with due regard for the quality of that data. The database can then provide information in terms of, for example, nutritional values, range of nutrient contents, variability and feed quality. At present this type of data is produced by government and private laboratories, government research organizations, private feed manufacturers and is often available from scientific publications.

Information to be gathered

In a perfect world all this information would be freely available and could contribute towards a national database. Indeed every measurement made on every feedstuff would be gathered and considered for entry on to the database. Data stored must be of a certain quality in terms of the samples chosen and the methods used and should relate to what the user needs or wants to know now or in the future. This falls broadly into three main areas, being background information, chemical composition and biological measurements.

BACKGROUND

A knowledge of the background of a raw material is very important to give a clear picture of that feedstuff and to enable it to be correctly named, classified, stored and hence retrieved. This will differ between feeds, for example with a silage one needs to know whether its parent material was grass, maize or other type, its variety, when was it cut, whether it was wilted, what was the chop length, and whether additives were used. However for a maize gluten feed all that it may be possible to find out is whether or not it was processed in the United Kingdom. For imported materials and by-products accurate feed description may be very difficult as information is often limited. Indeed the same by-product is often known under several slightly different common names which makes it very difficult when trying to decide whether or not feeds can be combined together under one description for retrieval. For example, grain distillers dark grains may be from wheat or maize precursors, and hence the final products will be significantly different, and should not be combined as one feed.

CHEMICAL COMPOSITION

There are a very great number of analyses that may be carried out on every raw material. In practice of course in most cases, particularly within the feed industry where routine monitoring of many batches of raw materials is necessary, only a few basic analyses are performed, for example protein, oil, fibre. Thus a database could soon become swamped with thousands of protein values for, say, wheat, but no starch, neutral detergent-cellulase digestibility or cell wall content values. Rigorous screening of the information before inserting and deletion of redundant data is necessary. For all analyses, a reference to the analytical method should be stored with the data and only results from recognized and approved methods should be allowed. Analyses should be collated according to method of analysis, and then compared before combining to give one value as they may not actually be measuring exactly the same component. For example lignin measured by the potassium permanganate method (Goering and Van Soest, 1970) has a very different value when measured using the acetyl bromide method (Morrison, 1972) and indeed is usually reported in different units. Thus in this instance, the two lignin results cannot be combined. It is equally important to ensure that the results of the same determination produced by different laboratories are not substantially different (see Chapter 20).

In essence what is required is a wide range of appropriate analyses on a set of materials that cover the range of that material which is available. As is described in a later section the quality of these data may be closely monitored to check, for example, for differences between years, or geographical location.

An additional procedure, which is carried out routinely for some of the materials used in the Dutch CVB database, is a microscopic analysis. This can often highlight the reason for anomalous analytical results and, in some cases of contamination, has been very revealing.

BIOLOGICAL MEASUREMENTS

These should be performed on all relevant species, for example ruminants, pigs and poultry, to give digestibility and dietary energy values and digestibility coefficients for relevant parameters. *In vivo* measurements are time consuming and expensive, and thus only a small number of samples can be tested in this way. They are, however, very desirable. It is important that the information chosen to go into the database is gathered from a reliable source in which standard referenced methodology is used. If possible the samples evaluated should cover a representative range of material in each class available to the compounder or farmer. For example, sunflower seed meals should be chosen to cover the range of crude fibre contents available since it is upon this basis that they are currently classified for sale. As an extra requirement it should be possible to relate the biological measurements to the chemical composition for a specific sample so that, for example, an unusually high metabolizable energy content could be related back to originating from a sample with a very high oil level.

All the above information may be gathered either directly from the source laboratory or research unit or, alternatively, abstracted from the published literature. It is the quality of the data that is important as upon this hangs the quality of the database, and hence a minimum of background information and references to methods must be available. This is often lacking or incomplete in scientific publications.

A fourth set of information may be added to wrap-around that discussed above, regarding broader more general factors about a feed that may be of use to the user. This would mostly be in the form of text, and could include information on anti-nutritive factors, maximum rates of inclusion, availability, storage and handling considerations.

Database construction

OVERALL PLAN

Despite claims by some computer and software companies, building a database is not an easy job, and requires much effort and time. The best way to begin is to first make a detailed study of the overall task, preferably with a systems analyst. This study has to result in a masterplan, essentially a detailed description of the system to be built.

The first step of the study is to identify clearly the user requirements. This information can be gathered by interviewing the potential users, and an idea of some of the questions to be answered are:

1. What is the general aim of the database?
2. What information has to be stored? (types of analysis, additional information).
3. What is the origin of the information? (laboratories, literature).
4. How many samples will be stored?
5. How many samples must be added each year/month/week?
6. Which values must be calculated? (mean, standard deviation, other statistical analyses).
7. What output must be processed? (tables and/or graphics).
8. How many users will there be? (altogether and at the same time).

Using this information it is possible to make a preliminary design of the system; a datamodel. A datamodel describes the information to be stored and the relationships between the various attributes. It is a general plan, which is not specific to any one computer system.

The final masterplan gives a review of the user requirements, a description of the datamodel, advises about hardware and software and provides a time schedule for the project. This plan must also contain information (number and qualifications) concerning the technical staff which are needed initially, for the period of system development and thereafter for the technical maintenance. Another point of importance is the staff required to load the data to get the database operational and for maintaining the stored data. This masterplan forms the basis for the construction of the database.

A widely used method in developing user applications is termed 'prototyping'. After the programmers have tested the programs, the users start testing the prototype-database with a special set of data, the testset. After a certain period, an evaluation takes place. Discussions are then held relating to specific problems in using the database including response time and results of calculations. If the prototype proves unsatisfactory, a new one is made. If the problems are only minor, the programs are adapted accordingly and, when the users are completely satisfied, the database is delivered.

At this point the users start inserting real data using the facilities. It is advisable to perform a further evaluation after the first few months of operation as, sometimes, improvements to the programs will be needed as a result of suggestions made whilst using the system in this initial period. Another point of importance is the response time. During the test period relatively small sets of data are used, so the response time will always be short but, after inserting a large number of samples (hundreds or thousands), the response time may be longer. When the response time becomes too lengthy it may become advantageous to run the large programs (checking data, large calculations) at night in batches.

SOFTWARE

Database Management Systems (DBMS) are special packages for building databases. These packages, which comprise several programs, have many features for data manipulation:

1. System-programs; manage the internal storage and retrieval of data.
2. Query language; a language to manipulate the stored data. A well known query language is SQL (pronounced 'sequel'). SQL was developed and defined by IBM Research, and has been proposed by the American National Standards Institute (ANSI) as the basis for a standard language for relational database systems.
3. Screen programs; for direct access to the data (for example, insert, delete, update).
4. Menu system; for choosing the different programs.
5. Back-up facilities; for making back-up copies of data to be used in case of calamities such as a computer crash.
6. Host language interface; an interface that makes it possible to access the database with programs written in the programming language of the host computer, for instance Fortran, Pascal or C.
7. A data-loader; for transferring ASCII-files into database format.
8. A report generator; for producing reports with a specified layout.
9. A worksheet; for making *ad hoc* calculations with data stored in the database.

One of the aims of a feed database is to apply statistical calculations to the stored data. Most database packages have only limited facilities for such calculations, thus there is a need for a software interface with a statistical package, such that data can then be retrieved from the database and loaded into the statistical program. The most frequently used calculations or options in a feed database are mean, standard deviation, median, minimum/maximum values, regression analyses, histograms and graphs. The numerical results from these calculations can then be stored in the database if required.

HARDWARE

In most cases databases are built on mini or mainframe computers. A DBMS requires a large amount of disc storage capacity (depending on the amount of data to be stored, 200–500 Mb including the operating programs) and at least 3 Mb of internal memory. Where the capacity of the CPU (Central Processor Unit) is insufficient, the response time will be too long, especially in case of multi-user systems.

When a multi-user system is not required and the amount of data to be stored is limited, it is possible to build the database on a micro-computer (e.g. an IBM-AT or compatible, with at least 1 Mb internal memory and, preferably, a 40 Mb hard disc). Unfortunately DBMSs on micro-computers often have fewer features than systems on mini and mainframe computers. Output is produced by a printer, and for printing graphical output of a statistical package, a plotter can also be useful.

CODING

In order to create a logical database, some systems of coding are required. Application of short numerical codes instead of complete feed names can save valuable disc space, and a further advantage is that these codes are better suited to queries than names or words. A fixed format numerical code can only be written in one way, whereas when words are used it is easier to make mistakes (thus for a computer 'Sugarbeet' is not equal to 'sugarbeet' or 'sugar-beet'). Use of a numerical code will avoid this problem. When designing a code it is essential to be aware of the range of possibilities that may occur, otherwise it is possible that the code will be exhausted within several years.

The International Network of Feed Information Centres (INFIC) has developed an international code for all kinds of feeds (Harris *et al.*, 1980). Other attributes that can be coded include methods of analysis, country names, source laboratory and soil types.

Data acquisition and quality

Data may be generated specifically for a database, it may be generated by laboratories for their own purposes, it may be abstracted from literature or transferred from other databases. Each of these four major methods of acquiring data is discussed below.

SPECIFIC GENERATION

Data produced specifically for a database has the advantage of meeting all the standards and requirements for that application. Background information is available, the feed can be accurately named and described, methods of analysis can be chosen in advance to ensure they are robust and adequately referenced and approved, quality can easily be controlled and the data produced is very relevant.

However, this can tend to become rather introverted and isolated unless the data and methodology undergoes regular comparison with laboratories undertaking similar projects. Ways in which this can be monitored include the use of ring tests for both *in vivo* and *in vitro* methods in which standard samples are circulated to a group of workers involved in similar work at different centres. The results may then be compared and outliers identified. The use of standard samples which allow centres to do internal batch to batch monitoring is also valuable. For example, all sheep before entering the experimental flock at the ADAS Feed Evaluation Unit are fed a standard diet of known digestibility. If the results for any individual fall outside the normal range for this diet, then that animal is discarded. In addition this diet has been used by several other units in the United Kingdom undertaking digestibility trials thus ensuring that the results from all centres are comparable.

FROM OTHER LABORATORIES

Much of these data will be chemical composition. The analytical methods used will vary considerably and may include local, undocumented 'improvements'. It is often difficult to obtain sufficient background information to describe the material adequately and the data available is often limited to, for example, proximate analysis, as it has been produced for a specific purpose. Obtaining information from other sources may be difficult as it may only be in laboratory notebooks or on several report forms and hence will need to be specially written out again for loading on to the database. This is very time consuming. In addition, parts of the information may be confidential and hence subject to special arrangements in terms of coding and storage.

LITERATURE SOURCES

Accurate data transcription from scientific papers can provide a significant volume of data. Decisions that need to be made prior to taking this approach are largely defined by the number of publications that will be screened routinely, hence the size of geographical area involved, the range of feeds of interest and types of animal production all need consideration (Leche, 1983). The advantages of this type of data acquisition are the large amount of data available, a relatively low staff input and the ability to combine searches for both data and bibliographic information on particular feeds. However, the cost is fairly high and continuous as continuity of literature surveillance is desirable but, perhaps most importantly, descriptions in papers of sampling and methodology may be vague and thus it is questionable as to when data may be pooled.

OTHER DATABASES

Data may be acquired from other databases, for example INFIC holds information from many different countries which may, if suitable, be combined. Assuming the complexities of electronic data transfer can be solved and the two systems are compatible in terms of electronics and coding, the two major problems are, firstly, relating feeds between the two databases such that feeds classified according to one convention in one database may be combined with their exact counterparts in the other and, secondly, the lack of control or knowledge regarding the quality and relevance of the data. INFIC can play an important role in this process of data exchange by defining codes and file formats for the data.

Maintenance of a database

MAINTENANCE OF THE PROGRAMS

The effort involved in the maintenance of a database is often underestimated. When a database is delivered it does not necessarily mean that it will work totally satisfactorily straight away and, during the first operational months, errors will be found that were not observed during the testing period.

An additional problem is that many computer and software companies periodically update their system programs. In most cases this means that the applications need some adaption, and this will require further input from a programmer. A new version of a DBMS can have more features than the old one, thus old sections of the database programs can be rewritten in improved ways, sometimes faster or with more facilities. A database is a constantly evolving entity and, therefore, the continued assistance of a good programmer (possibly part-time) is essential.

The database manager is also responsible for the safety and security procedures in the system. Not all users have the same privileges, some of them may only look up data, while others may also be permitted to insert and delete data and update mean values. The database manager has to give the correct privileges to new users and has to revoke them from those who are no longer allowed access to the database.

MAINTENANCE OF DATA

The stored data in a database also require maintenance. Back-up copies have to be made and stored securely and the integrity of the database has to be controlled (for example no false data, no double data, proper use of codes). Other tasks include defining codes for new feeds and new methods of an analysis. The database manager has also to decide how long samples are kept in the database. Old samples may not be used in the current calculations, and hence it may not be necessary to keep them in the database all the time. The old samples can be stored on tape, so that if required they can be restored on to the database.

Another area which must be continually addressed relates to the quality of the stored data. Before the data for a sample are inserted, the database manager has to provide the correct codes and must be aware of the units in which the results are presented. The most suitable way to store the data is on a dry matter (DM) basis. If the data for a new sample are not presented in that form they must be recalculated, which can be done by a specially written computer program.

After inserting the data, it must be checked. The best way to do this is to store for each feed the minimum and maximum values for each different variable. After the data have been inserted all values are compared with the stored criteria and when the result of an analysis does not fall between the minimum and maximum value, this is reported to the database manager who has to check whether a mistake was made during the coding or the insertion of the data or even in the source laboratory. It is also possible that the feedstuff has not been adequately defined, it could be a mixed product or even a falsification. Such samples must not be used in the calculations. The criteria used for data validation must be up to date, because populations can change, for example as a result of plant breeding programmes or new processing methods. Information for updating the minimum and maximum values can be displayed in the form of histograms and graphs for the different feeds. Outliers can easily be identified and one can also see if the population as a whole is changing. It is also possible to calculate the range automatically.

If there are a certain number of samples of a particular feed in the database, it will be appropriate to calculate the standard deviation (probably multiplied 2 or 3 times) as an indication of the range. Large changes in the range should be reported to the database manager. The calculation programs can also contain checks to trace incorrect data. It is possible to check, for example, whether toluene dry matter content is greater than oven dry matter content and, if this is not the case, it is likely that an analytical error has been made.

It is important to realize that the computer is not able to do all the checks. In the last resort a person with a sound knowledge of feeds and feeding has to examine the calculated results.

Output

Data may be output as tabulated values, in the form of a book or on a floppy disc. The period between updating these publications must be considered. Alternatively, information may be obtained by direct on-line search of the database either by the database manager, or an informed user who can either access regularly updated summary tables or access the whole database to give a wider range of information. Data are usually required either for immediate practical use such as for diet formulation, or for longer term research purposes. The kind of application will normally define the form of output as to how up to date and detailed it needs to be.

Retrieval of data has to be undertaken according to feed description and this is why it is important to be able to classify feeds accurately and as completely as possible. For example, a simple request for soyabean meal might include full fat, hi-pro, 44/9, 44/7, solvent and mechanically extracted products, the mean results of which would be meaningless, hence as detailed and accurate a classification as possible must be given.

The way the output is presented depends on the users of it. For practical use a very detailed output is not normally necessary, but when the results are used for research purposes, there may be a demand for more detailed information. The most frequently presented values are:

1. Mean value; the average value calculated for a certain feed or for a quality class of a feed (for example palm kernel expeller $<$ 220 g/kg crude fibre and palm kernel expeller $>$ 220 g/kg crude fibre).
2. Standard deviation; this is calculated when five or more samples are used. Using the standard deviation, a range can be calculated in which 95% of the whole population will fall. The use of a standard deviation only applies to normally distributed populations, and accordingly in some cases it will not give an adequate description of the distribution of the population.
3. Median; in populations which do not have a normal distribution, the median (in combination with the range) gives more information than the standard deviation.
4. Minimum and maximum values; give the full range in which the results fall.
5. Number of values; shows how many samples are used in the calculations.

Other forms of output that can be produced are:

1. Histograms, a histogram can be useful to see how the population is distributed and if there are any outliers in it. It is also possible to see whether a population consists of two different groups. In this case the product can be split up into two quality classes.
2. Graphs; a graph shows how two different analyses are related to each other, for example the protein and crude fibre content of palm kernel expellers. From the graph, outliers and different groups in a population are easily identified.
3. Regression analyses; when two or more different variables seem to be related to each other, the relationship can be tested by performing a regression analysis. A regression model can also be used to predict values from other analyses, for example in The Netherlands regression relationships are used to predict the content of phytate phosphorus in a feed from the analysed total phosphorus content (CVB, 1977).

Conclusions

A database is a highly valuable source of information, providing that simple requirements regarding data quality, background information and data checking are met and that the data held are relevant and up to date. There are benefits to be gained by all branches of the agricultural and animal feed industry as well as government policy makers by having access to such data, presented in the right way.

Building and maintaining a database is not an easy task and an outline of how to set about designing a system together with some of the problems involved have been highlighted in this paper. Workers in The Netherlands have been using a database of this type for several years now, and the United Kingdom National database is due to launch a new publication shortly.

References

Alderman, G., Barber, W.P. and Stranks, M.H. (1983) Databases for the composition and nutritive value of animal fats. In *Recent Advances in Animal Nutrition 1982* (ed. W. Haresign). London, Butterworths, pp. 91–110

CVB (1977) *Veevoedertabel, gegevens over voederwaarde, verteerbaarheid en samenstelling.* Lelystad, Centraal Veevoederbureau in Nederland

Goering, H.K. and Van Soest, P.J. (1970) *Forage Fiber Analysis.* Agriculture Handbook No. 379, USDA, Washington DC.

Hain, D.L. (1983) How data are acquired by feed information centres: problems and solutions. In *Feed Information and Animal Production* (eds G.E. Robards and R.G. Packham). Farnham Royal, Slough, Commonwealth Agricultural Bureaux, pp. 45–56

Harris, L.E., Haendler, H., Riviere, R. and Rechaussat, L. (1980) *International Feed Databank System; an Introduction into the System with Instructions for Describing Feeds and Recording Data.* International Network of Feed Information Centres. Publication 2. Prepared on behalf of INFIC by the International Feedstuffs Institute, Utah Agricultural Experiment Station, Utah State University, Logan, Utah, USA 84322

INFIC (1977) *International Network of Feed Information Centres*, Publication Number 1, Utah State University Press, Logan, Utah, USA

Leche, T. (1983) How data are acquired by feed information centres: problems and solutions. In *Feed Information and Animal Production* (eds G.E. Robards and R.G. Packham). Farnham Royal, Slough, Commonwealth Agricultural Bureaux, pp. 36–44

Ministry of Agriculture, Fisheries and Food (1986) *Feed Composition. UK Tables of Feed Composition and Nutritive Value for Ruminants.* Marlow Bottom, Bucks, Chalcombe Publications

Morrison, I.M. (1972) A semi-micro method for the determination of lignin and its use in predicting the digestibility of forage crops. *Journal of Science, Food and Agriculture*, **23**, 455–463

Parris, K.P. and Tisserand, J.L. (1988) A methodology to complete a national feed utilisation matrix using European data. *Livestock Production Science*, **19**, 375–388

Thaer, A. (1809) *Grundsatze der rationelle Landwirtschaft*, Vol. 1 Sec 275. Berlin, Die Realschulbuchhandlung

Topps, J.H. (1989) Databases of feed composition and nutritive value. In *Ruminant Feed Evaluation and Utilization* (eds B.A. Stark, J.M. Wilkinson and D.I. Givens). Marlow Bottom, Bucks, Chalcombe Publications, pp. 41–50

22

NATURALLY OCCURRING TOXIC FACTORS IN ANIMAL FEEDSTUFFS

IRVIN E. LIENER
Department of Biochemistry, College of Biological Sciences, University of Minnesota, Minnesota, USA

Introduction

Practical experience has shown that proteins of vegetable origin, particularly those derived from legumes and oilseeds, can provide a valuable source of protein in animal feeds. Experience has also revealed that the nutritional value of these proteins is sometimes limited by the presence of substances which can produce negative effects on growth, digestion, metabolism and the general health status of the animal. Some examples of these so-called antinutritional factors in the more commonly employed ingredients used in animal feedstuffs that are derived from plant materials are shown in Table 22.1. Each of these factors will be briefly discussed with respect to their distribution, chemistry, nutritional significance, mode of action, and the analytical techniques most commonly employed for their evaluation.

Table 22.1 EXAMPLES OF ANTINUTRITIONAL FACTORS THAT OCCUR IN PLANT MATERIALS COMMONLY USED AS INGREDIENTS IN ANIMAL FEEDSTUFFS

Antinutritional factor	Distribution	Physiological effect
Protease inhibitors	Most legumes	Impaired growth Pancreatic hypertrophy Pancreas carcinogen
Lectins	Most legumes	Impaired growth/death
Goitrogens	Rapeseed	Hyperthyroidism
Cyanogens	Sorghum Lima beans	Respiratory failure
Vicine/convicine	Faba bean	Adverse effect on egg production
Phytate	Most legumes	Interference with mineral availability
Tannins	Oilseeds Most legumes	Interference with protein digestibility
Alkaloids	Lupins	Depressed growth

377

Protease inhibitors

Substances that have the ability to inhibit mammalian digestive enzymes such as trypsin and chymotrypsin are ubiquitous in plants and animals but contents are particularly high in legumes. Historically, and because of their economic importance, the protease inhibitors in soybeans have been the object of considerable study, and have served as models for all other plant protease inhibitors. The soybean protease inhibitors fall into two classes – the so-called Kunitz inhibitor which has a molecular weight of about 20000 with only two disulphide bonds and a specificity directed mainly towards trypsin, and the so-called Bowman-Birk inhibitor with a molecular weight of about 8000, seven disulphide bonds and possessing the capability of inhibiting both trypsin and chymotrypsin at independent binding sites. Variants of these two types of inhibitors are found in most of the other legumes. More detailed information on the chemistry and mechanism of inhibition may be obtained in reviews by Liener and Kakade (1980) and Rackis, Wolf, and Baker (1986).

The inactivation of trypsin inhibitor activity by heat treatment is accompanied by a concomitant increase in the protein efficiency ratio (PER) as measured with rats (Figure 22.1). Similar results demonstrating the beneficial effect of heat treatment on most legumes have also been consistently observed for other non-ruminant animals including the chick, hen, turkey, and piglets (reviewed by Waldroup and Smith, 1989). In general the extent to which these inhibitors can be inactivated by heat is a function of temperature, duration of heating, moisture content, and particle size – variables that demand careful control by the feed processor in order to obtain a product possessing maximum nutritive value. Modes of heat treatment or processing which have proved effective include the following: (1) toasting, (2) boiling in water, (3) dry roasting, (4) microwave radiation, (5) gamma irradiation, (6) extrusion cooking, (7) infra-red radiation, and (8) micronization (Liener, 1983).

The simplest explanation for the growth inhibition produced by the protease inhibitors would be that they interfere with the digestibility of dietary protein. But

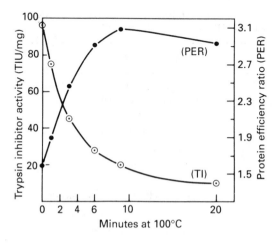

Figure 22.1   Effect of heat on the trypsin inhibitor activity and protein efficiency ratio (PER) of soybean meal (From Anderson, Rackis, and Tallent (1979); courtesy of Academic Press.)

this does not explain why purified preparations of these inhibitors are capable of inhibiting growth even when incorporated into diets containing predigested protein or free amino acids (Liener, Deuel and Fevold, 1949). Perhaps the most significant observation which led to a better understanding of the mode of action of trypsin inhibitors was the finding that raw soybeans, or the inhibitor itself, caused hypertrophy and hyperplasia of the pancreas (Yanatori and Fujita, 1976), accompanied by an increase in the secretion of enzymes by the pancreas (Nitsan and Liener, 1976). Thus, the growth depression induced by the trypsin inhibitor is most likely due to an endogenous loss of amino acids. Since trypsin and chymotrypsin are proteins which are rich in the S-containing amino acids, their loss by faecal excretion serves to further exacerbate the fact that soybean protein is already critically deficient in these amino acids.

It is now generally accepted that pancreatic secretion is controlled by a negative feedback mechanism whereby the secretory activity of the pancreas is regulated by the level of trypsin present in the small intestine (Green and Lyman, 1972). As the level of trypsin in the duodenum falls due to complexation with the trypsin inhibitor, the pancreas is induced to produce more enzyme in order to compensate for this loss. The hormone, cholecystokinin (CCK), which is secreted by the intestinal mucosa, is believed to act as a mediator between trypsin and the pancreas (Figure 22.2). An increase in the plasma of CCK has in fact been demonstrated in rats fed raw soy flour (Loser *et al.*, 1988).

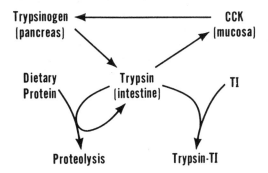

Figure 22.2 Negative feedback mechanism whereby trypsin inhibitor stimulates the secretory activity of the pancreas (From Anderson, Rackis, and Tallent (1979); courtesy of Academic Press.)

The mechanism whereby trypsin suppresses CCK release from the intestine is not clear. The answer to this question may come from the experiments by Iwai, Fushiki and Fukuoka (1988) who have isolated from pancreatic juice a trypsin sensitive peptide with a molecular weight of around 6000 which they believe controls the release of CCK from the intestinal mucosa. Since this peptide is inactivated by tryptic cleavage, a drop in the level of trypsin in the gut (as might be expected in the presence of the inhibitor) enables this peptide to escape inactivation so that it is free to stimulate the intestinal mucosa to produce more CCK.

The stress on the pancreas resulting from the long term feeding of raw soy flour eventually leads to the formation of pancreatic nodules and acinar adenomas (McGuinnis *et al.*, 1980). Moreover the incidence of tumours is positively correlated with the level of trypsin inhibitor in the diet (Liener *et al.*, 1985). This carcinogenic effect, however, may be species specific since the mouse and the hamster, unlike the rat, do not develop pancreatic lesions even when fed raw soybeans for an extended period of time (Liener and Hasdai, 1986).

The most commonly employed procedure for determining the trypsin inhibitor activity of plant materials is based on the inhibition of the activity of bovine trypsin on the synthetic substrate N-alpha-DL-arginine-p-nitroanilide as originally proposed by Kakade, Simons, and Liener (1969). This method has been subject to numerous modifications designed to increase its accuracy and reproducibility (Kakade *et al.*, 1974; Smith *et al.*, 1980; Hamerstrand, Black and Glover, 1981). Affinity chromatography using immobilized trypsin has also been employed as a means of avoiding interference from non-protein type of inhibitors (Roozen and de Groot, 1987). The recent introduction of monoclonal antibodies towards the Kunitz and Bowman-Birk inhibitors (Brandon, Bates and Friedman, 1988, 1989) should prove useful for the specific quantitation of these two inhibitors by an immunochemical approach.

It is significant to note that most of the trypsin inhibitor assays referred to here involve the measurement of the extent to which bovine trypsin is inhibited. This is frequently done despite the fact that the investigator may be interested in the nutritional effects that might be expected in a completely unrelated species of animal whether it be the chicken, pig, or even humans. *In vitro* studies on the inhibition of the proteases in the pancreatic juice of different animals have revealed marked differences in the degree to which these enzymes are inhibited by the protease inhibitors of various legumes (Krogdahl and Holm, 1983; Rascon *et al.*, 1985). Another point to be considered is the fact that not all protease inhibitors retain their full activity after exposure to gastric juice; for example, the Kunitz soybean inhibitor is readily inactivated by human gastric juice whereas the Bowman-Birk soybean inhibitor retains its activity under the same conditions (Krogdahl and Holm, 1981). For these reasons any attempt to extrapolate the results of *in vitro* assays for protease inhibitor activity to their true physiological effects must be viewed with caution.

Lectins

Paralleling the distribution of protease inhibitors in legumes is a class of proteins referred to as lectins. These substances exhibit the unique property of being able to agglutinate the erythrocytes of the blood from various species of animals. More importantly, however, they have the general property of being able to bind to glycoproteins located on the surface of cell membranes regardless of their origin. One of the most characteristic properties of these lectins is the specificity which they display towards specific sugars which comprise the structure of these glycoproteins. A detailed description of the lectins is beyond the scope of this presentation but may be found in a book devoted to this subject (Liener, Sharon and Goldstein, 1986).

Of immediate relevance to the subject at hand are the adverse nutritional effects evoked in animals by the consumption of plant materials in which the lectins are found. The toxicity of these lectins is quite variable and ranges from the extreme toxicity of the lectin in the castor bean (ricin) to the relatively innocuous lectin in the soybean. Those legumes which have been demonstrated to contain lectins which are toxic upon oral ingestion are listed in Table 22.2. Notably absent from

Table 22.2 LEGUMES KNOWN TO CONTAIN LECTINS WHICH ARE TOXIC UPON ORAL INGESTION

Scientific name	Common name
Canavalia ensiformis	Jack bean, sword bean
Dolichos biflorus	Horse gram
Dolichos lablab	Hyacinth bean
Phaseolus lunatus	Lima bean
Phaseolus vulgaris	Kidney bean, navy bean, pinto bean, black bean, etc.
Psophocarpus tetragonolobus	Winged bean
Ricinus communis	Castor bean

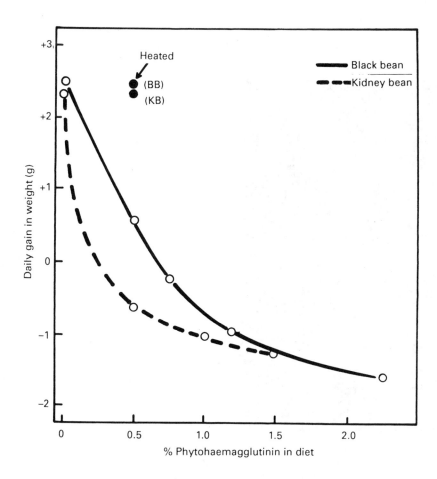

Figure 22.3 Effect of black bean and kidney bean lectins on the growth of rats. Curves have been constructed from data taken from Honavar, Shih, and Liener (1962)

this list are such legumes as lentils, garden peas, lupins, and cowpeas which are not particularly rich in lectins, or, if present, are non-toxic.

Illustrative of the toxicity of some lectins is the experiment shown in Figure 22.3, where the incorporation of the lectin from *Phaseolus vulgaris* led to a marked impairment in the growth of rats. Death of the animals usually ensued within 7–10 days after the animals were placed on diets containing the raw bean or the purified lectin at a level equivalent to that present in the raw bean. Also demonstrated in Figure 22.3 is the observation that the toxicity of the lectin could be effectively destroyed by autoclaving, which is consistent with the observation that the nutritive value of this bean for most other monogastric animals is considerably improved by heat treatment under the same conditions which have proved effective for inactivating the protease inhibitors (Waldroup and Smith, 1989).

The mechanism whereby the lectins exert their toxicity is related to their ability to bind to the epithelial cells lining the small intestine. For example, the *in vivo* binding of the kidney bean lectin to the intestinal mucosa can be demonstrated by using a fluorescent labelled antibody to the lectin. As shown in Figures 22.4 and 22.5 the lectin ingested by the rat in the form of the raw bean binds to the luminal surface of the microvilli in the proximal region of the small intestine resulting in

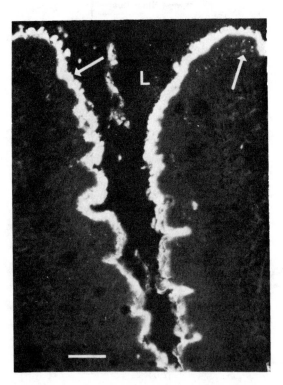

Figure 22.4 Immunofluorescence micrograph of a part of a transverse section through the duodenum of a rat fed on a diet containing raw kidney beans. Incubation with rabbit antilectin IgG showing fluorescence in brush border region and within apical cytoplasm of mature enterocytes (arrow). Scale bar: 50 mm (From King, Pusztai, and Clarke (1980); courtesy of Chapman and Hall, Ltd.)

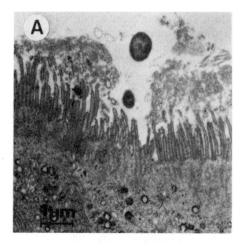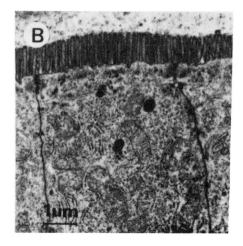

Figure 22.5   Electron micrographs of sections through apical regions from rats fed diets containing (A) 5% raw kidney beans and 5% casein and (B) 10% casein (From Pusztai *et al.* (1976b); courtesy of Society of Chemical Industry.)

disruption of the brush border. As a consequence of this damage, there is a serious impairment in the absorption of nutrients across the intestinal barrier (Donatucci, Liener and Gross, 1987). Other detrimental effects which have been attributed to the action of lectins are the colonization of the small intestine by coliform bacteria, systemic absorption of the lectin itself, an inhibition of brush border hydrolases, and an endogenous secretion of mucin from the intestinal lining (Liener, 1986).

The agglutinating activity of lectins is most frequently determined by serial dilution with visual observatioan of the highest dilution showing agglutination of red blood cells from one or more species of animals. Although this technique is simple and rapid, it is at best only semiquantitative and will detect only those lectins which have multiple binding sites. Attempts have been made to improve the precision of such assays by a spectrophotometric technique that measures the decrease in turbidity of a suspension of cells resulting from the settling of the agglutinated cells (Liener, 1955). An immunochemical assay has also been proposed to take advantage of the specificity of the lectin towards specific sugars (Howard and Shannon, 1977).

A serious limitation of the haemagglutination assay is the fact that what one is actually measuring is the ability of a particular lectin to agglutinate the red blood cells of a particular animal, rabbit erythrocytes being the one most commonly employed. This may have little or no relevance to the manner in which a lectin may or may not exert its *in vivo* toxic effect in a particular animal species. Assuming that the toxicity of lectins towards a given species of animal is associated with its ability to bind to the intestinal mucosa of that species, what is needed is an *in vitro* method that will reflect this binding. To this end Hendricks *et al.* (1987) have described an ELISA assay which permits the quantitative determination of the lectin-binding capacity of the small intestinal brush-border membrane. Although their model system involved the use of the small intestine of the cow and the soybean agglutinin in the form of a conjugate with peroxidase, this method could conceivably be adapted to the measurement of the binding of the lectins of various plant extracts to the brush-border membrane of any animal species. The predictive value of such an assay should, however, be verified by feeding the same plant material to an appropriate animal with special emphasis on the histopathological effects on the small intestine. As in the case of the protease inhibitors one should also take into account the possible vulnerability of the lectin to inactivation in the stomach; for example the soybean agglutinin is readily digested by pepsin (Liener, 1958), whereas the *Phaseolus vulgaris* lectin retains full activity even after transit through the stomach (Pusztai, Clarke and King, 1979a).

Goitrogens

Goitre producing agents in the form of thioglucosides (referred to as glucosinolates) are present in most cruciferous plants including rapeseed which is a commonly used feed ingredient in many parts of the world (Fenwick, Heaney and Mullin, 1983). This use as a feed ingredient has been limited, however, by the fact that the glucosinolates which, although innocuous in themselves, are enzymatically hydrolysed to yield products that are goitrogenic and act as growth depressants. Other undesirable consequences include cytotoxicity and the tainting of poultry eggs and dairy milk. These goitrogenic products, primarily isothiocyanates and oxazolidinethione (goitrin), are liberated from their parent glycosides by an enzyme (thioglucosidase) normally present in the plant tissue. These then act on their substrate (progoitrin) when the seed is crushed and masticated by the animal (Figure 22.6).

The goitrogens in rapeseed inhibit the uptake of iodine by the thyroid gland so that iodine supplementation is relatively ineffective. Heat treatment is only partially effective in eliminating the toxic effects of these goitrogens. Aside from the availability of strains of rapeseed which are very low in glucosinolates, the toxicity can be reduced by one or a combination of several different processing techniques including the removal of the glucosinolates or their end products by extraction with hot water, dilute alkali, or acetone, or by decomposition with iron salts (Liener, 1983).

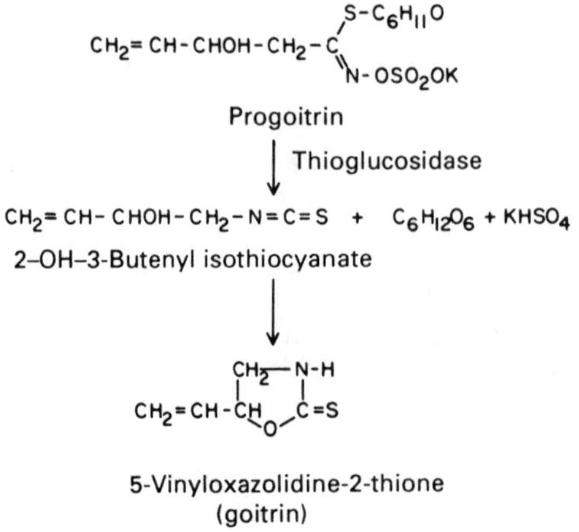

$$CH_2= CH-CHOH-CH_2-C \overset{\displaystyle S-C_6H_{11}O}{\underset{\displaystyle N-OSO_2OK}{}}$$

Progoitrin

$\downarrow$ Thioglucosidase

$$CH_2= CH- CHOH-CH_2-N=C=S \quad + \quad C_6H_{12}O_6 + KHSO_4$$

2-OH-3-Butenyl isothiocyanate

$\downarrow$

$$CH_2=CH-\underset{\displaystyle O}{CH} \underset{\displaystyle}{\overset{\displaystyle CH_2-N-H}{\diagdown\diagup}} C=S$$

5-Vinyloxazolidine-2-thione
(goitrin)

Figure 22.6   The goitrogen principle in rapeseed

Glucosinolates have traditionally been analysed indirectly by measurement of one or more of the products resulting from hydrolysis with a thioglucosidase (myrosinase), namely glucose or sulphate (McGregor, Mullin and Fenwick, 1983). This approach, although simple and rapid, does not provide any information regarding the identity of the aglycone moiety which can be quite diverse. The aglycone component of the glucosinolates can be measured directly by techniques involving gas chromatography of their trimethylsilyl derivatives (Mossaba *et al.*, 1989), infra-red spectrometry (Yang, 1988), or by enzyme-linked immunoassay (Hassan *et al.*, 1988).

Mention should also be made of the fact that soybeans and ground nuts have also been reported to exert a thyrotoxic effect in rats, although the identity of the causative factors in these plants is somewhat uncertain. In the case of soybeans the goitrogenic principle has been attributed to a low molecular weight oligopeptide (Konijn, Gershon and Guggenheim, 1973) or to some other unidentified component which is not destroyed by heat treatment (Fillisetticozzi and Lajola, 1988). In the case of ground nuts the thyrotoxic principle is believed to be due to a phenolic glycoside which resides in the skin of the nut (Srinivasan, Mougdal and Sarma, 1957).

Cyanogens

Although cyanogens are present to a greater or lesser extent throughout the plant kingdom, the only plant that may be of concern with respect to cyanide intoxication in farm animals is the young sorghum plant which is particularly rich in the cyanogenic glycoside, dhurrin (Conn, 1973). Most legumes contain low, non-toxic levels of cyanogenic glycosides with the possible exception of lima beans (see below). Maceration of the fresh plant tissue initiates the enzymatic breakdown of the glycoside, which itself is non-toxic, resulting in the release of HCN (Figure 22.7). It is common knowledge among farmers in the USA that their cattle must not be permitted to graze on young sorghum plants until 'the cane is belly high on the

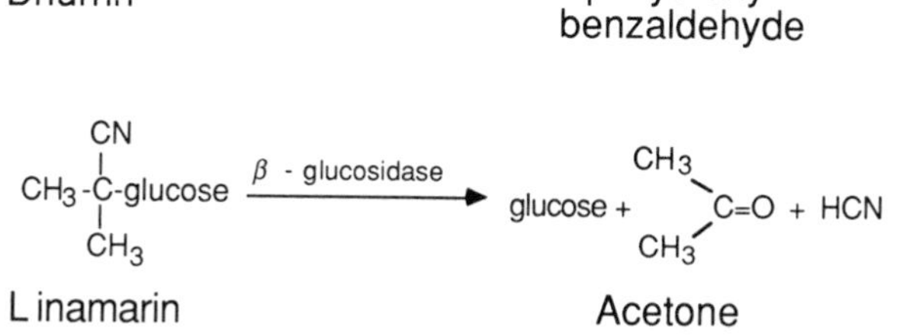

Figure 22.7 Cyanogenic factors in sorghum (dhurrin) and lima beans (linamarin)

cow'. It is only in the older plant that the concentration of dhurrin becomes low enough to permit grazing.

Less likely to be used as a feed ingredient are certain varieties of the lima bean and cassava, both of which may also contain high levels of the cyanogenic glycoside, linamarin (Figure 22.7). The cyanide released from this glycoside by an endogenous enzyme also present in these plants has been implicated primarily in cases of human poisoning (Montgomery, 1980; Oke, 1980).

The most common methods for determining the cyanide content of plant materials involve the colorimetric determination of cyanide following its release by linamarinase, a beta-glucosidase (Honig et al., 1983). Other more sophisticated techniques for the direct determination of the cyanogens have been proposed including molecular absorption spectrometry (Kupchella and Syty, 1984), high performance liquid chromatography (Brimer and Dalgaard, 1984), and a bioreactor incorporating immobilized linamarinase (Naresingh, Laipersad and Cheng-Yen, 1988).

Vicine and convicine

Although the faba bean or field bean (*Vicia faba*) is known to contain a number of antinutritional factors including trypsin inhibitors and lectins, the principal limitation to the use of faba beans as a feed ingredient has been the presence of vicine and convicine (Figure 22.8). Vicine and convicine are the causative agents of a haemolytic disease in humans known as favism, which is particularly prevalent in

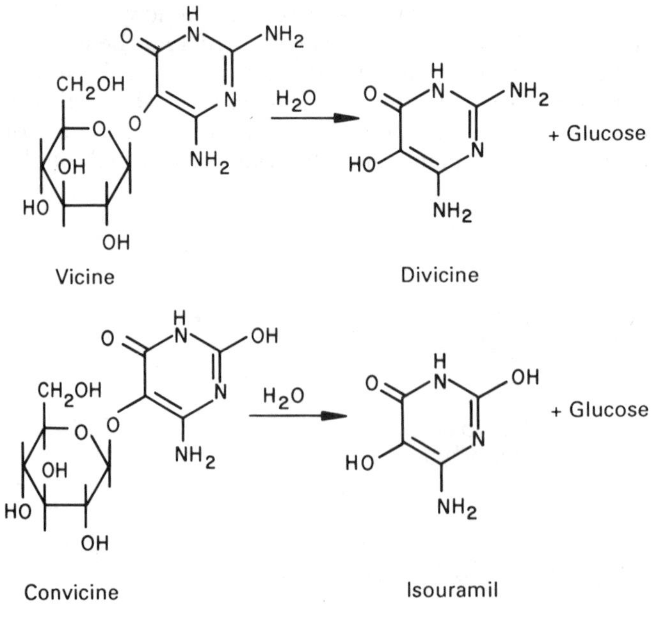

Figure 22.8 Structure of vicine and convicine in faba beans and their aglycones, divicine and isouramil, respectively, which are produced upon hydrolysis

the Middle East (Mager, Chevion and Glaser, 1980). The toxins which are directly responsible for this disease are the aglycones, divicine and isouramil, products that are formed when vicine and convicine, respectively, undergo hydrolysis by intestinal anaerobic microflora (Figure 22.8).

The only animal which appears to be sensitive to these two compounds is the laying hen. In this case these compounds decrease egg size with an increased incidence of blood spots, the latter presumably being due to erythrocyte haemolysis (Muduuli, Marquardt and Guenter, 1982). Unlike many of the other antinutritional factors which are proteinaceous in nature, vicine and convicine are thermostable and unaffected by heat treatment but may be reduced by 56% and 34% respectively if cooking is preceded by soaking (Hussein *et al.*, 1986). These compounds can also be extracted from faba beans with dilute acetic acid or water. The introduction of genetic variants of the faba bean with reduced levels of vicine and convicine remains a possibility, but so far efforts in this direction have met with only limited success.

Vicine and convicine may be determined by UV absorption after separation by thin layer chromatography (Jamalian, 1978) or by reverse-phase high pressure liquid chromatography (Marquardt and Frohlich, 1981).

Phytate

Phytate, a cyclic compound (inositol) containing six phosphate groups (Figure 22.9), occurs in most legumes and oilseeds to the extent of 1% to 5% of the dry weight. Phytate is generally regarded as an antinutritional factor because it interferes with the bioavailability of minerals (Reddy, Sathe and Salunkhe, 1982). Not only is much of the phosphorus in the plant unavailable because it is tied up in the phytate molecule (40–60%), but phytate also chelates with di- and tri-valent metals such as calcium, magnesium, zinc, and iron to form poorly soluble compounds that are not readily absorbed from the intestine. Phytate thus serves to increase the requirement of the animal for minerals which are essential for optimum growth.

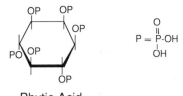

Phytic Acid

Figure 22.9 Structure of phytic acid

Although the ability of phytate to interfere with the availability of minerals no doubt accounts for its major antinutritional effect, phytate has also been shown to interact in a non-specific fashion with the basic residues of proteins. It is not surprising, therefore, that phytate will inhibit a number of digestive enzymes such as pepsin, pancreatin, and amylase, leading to a negative effect on the digestibility of proteins and carbohydrates. Inhibition of digestion may also result from the chelation of calcium ions which are essential for the activity of trypsin and amylase.

The complexing of phytate with proteins could also render these substrates more resistant to enzymatic attack. To what extent the inhibition of enzyme activity by phytate may be responsible for the overall adverse effects of phytate remains uncertain.

The phytate content of legumes is very little affected by heat treatment but can be reduced by taking advantage of the endogenous enzyme, phytase, which is located in a separate compartment of the plant tissue or by providing an exogenous source of the enzyme from microbial sources (Liener, 1987). Thus the phytate content of beans can be significantly reduced by simply allowing a slurry of the ground beans to undergo autolysis or by adding a preparation of a fungal phytase or by use of a bioreactor which incorporates an immobilized form of the enzyme of fungal origin (Ullah and Phillippy, 1988). Germination also results in an increased synthesis of phytase which causes a concomitant decrease in the level of phytate. Fermented beans likewise have reduced levels of phytate due to the action of the phytase elaborated by the microorganisms involved in the fermentation process.

The most widespread methods for phytate assay are based on the measurement of phosphorus involving preliminary separation from inorganic phosphate by precipitation with iron (Macower, 1970) or by ion-exchange (Harland and Oberleas, 1977). A colorimetric method has been recently described which permits the direct determination of phytate in crude seed extracts without preliminary purification (Vaintraub and Lapteva, 1988). Several HPLC methods for phytate determination have been proposed (Cilliers and van Niekerk, 1986) but their use requires rather sophisticated equipment.

Tannins

Oilseeds and legumes contain appreciable levels of polyphenolic compounds broadly referred to as tannins. These tannins present major problems in the feeding to non-ruminants of such plant sources as rapeseed, sunflower seed, kidney beans, and faba beans. Heat processing does little to reduce the tannin content of these materials since the tannins are relatively heat stable. The tannins are located mainly in the seed coat, and their concentration is positively correlated with the colour of the seed coat.

The negative nutritional effects of the tannins are diverse and incompletely understood (Reddy *et al.*, 1985), but the major effect is to decrease the digestibility of protein and carbohydrate. This is most likely a consequence of the interaction of tannins with either protein or starch to form enzyme-resistant substrates. Interaction with the enzymes themselves may also lead to an interference with the digestibility of these substrates. Other antinutritional effects which have been attributed to tannins include damage to the intestinal mucosa, the inherent toxicity of any tannins that might be absorbed from the intestines, and an interference with the absorption of iron.

Rats and mice adapt to dietary tannins by responding with an induced synthesis of proline-rich salivary proteins (Mehansho, Butler and Carlson, 1987). Since the latter show an increased affinity for tannins, these proteins may act as a defence mechanism against the antinutritional effects of tannins. To what extent this adaptive mechanism is operative in other non-ruminants is not known.

Since tannins are concentrated in the seed coat, preliminary dehulling constitutes the simplest technique for their removal. Soaking in water or salt solution prior to household cooking also causes a significant reduction in tannin content, provided the cooking broth is discarded. Chemical treatment with polyvinylpyrrolidone or polyethylene glycol or extraction with water-alcohol mixtures have also proved effective in reducing the tannin content of rapeseed and sunflower seed.

The most commonly used methods for the analysis of tannins are based on the colorimetric determination of total extractable polyphenols (McGrath *et al.*, 1982) or by the vanillin reaction (Price and Butler, 1977). A protein precipitation method has also been employed as a simple and rapid method for the determination of polyphenols (Hagerman and Butler, 1978).

Alkaloids

Although alkaloids are common constituents of many plants and have been implicated in many incidences of poisoning of grazing animals throughout the world (Cheeke and Shull, 1985), the alkaloids in lupins are of special concern where this legume is commonly used as a feed ingredient. The indiscriminant use of lupins should be avoided since as many as twelve different alkaloids may be present in varying concentrations and degrees of toxicity. The most common alkaloids in lupins are derivatives of quinolizidine, namely lupanine and sparteine (Figure 22.10). The growth performance of pigs and broiler hens and the egg production of laying hens are adversely affected beyond a certain threshold level of lupins in the diet (Waldroup and Smith, 1989).

Lupanine   Sparteine

Figure 22.10 Structures of the principal quinolizidine alkaloids, lupanine and sparteine, present in lupins

Boiling and steeping lupin seeds effectively reduce the alkaloid content, but it has been the genetic development of low-alkaloid or 'sweet' lupins that has increased the acceptability of lupins as a feed ingredient. Since lupins in general are quite low in trypsin inhibitors and lectins, cultivars with low alkaloid content require little or no processing other than particle size reduction by grinding or milling. However, the fact that the protein is somewhat deficient in methionine remains a problem.

Although rapid semi-quantitative colorimetric methods may be employed for the screening of lupin alkaloids (Ruiz, 1976; von Baer, Reimerdes and Feldheim, 1979), precise identification and quantitation of the various alkaloids present in lupins require the use of a technique such as capillary gas chromatography (Priddis, 1983; Meeker and Kilgore, 1987).

Conclusion

Although oilseeds and legumes contain a variety of constituents which may exert a negative nutritional effect when fed to non-ruminants, these plant materials have nevertheless proved to be valuable and practical sources of protein. This can be attributed to the fact that animal nutritionists have learned how to destroy these factors by suitable processing techniques or, in collaboration with plant geneticists, have developed strains with reduced levels of these offending substances. Nevertheless there is the ever present possibility that the consumption of improperly processed plant materials may cause severe economic losses in animal production. Strict control for the detection and quantitation of these toxic factors remains an important consideration. Aside from the analytical techniques that must be employed in this regard, an important issue that must always be addressed is what constitutes a safe level of ingestion for a particular species of animal.

References

Anderson, R.L., Rackis, J.J. and Tallent, W.H. (1979) *Soy Protein and Human Nutrition* (eds H.L. Wilcke, D.T. Hopkins and D.H. Waggle). New York, Academic Press, pp. 209–233

Brandon, D.L., Bates, A.H. and Friedman, M. (1988) Enzyme-linked immunoassay of soybean Kunitz trypsin inhibitor using monoclonal antibodies. *Journal of Food Science*, **53**, 102–106

Brandon, D.L., Bates, A.H. and Friedman, M. (1989) Monoclonal antibody-based enzyme immunoassay of the Bowman-Birk protease inhibitor of soybeans. *Journal of Agricultural and Food Chemistry*, **37**, 1192–1196

Brimer, L. and Dalgaard, L. (1984) Cyanogenic glycosides and cyanohydrins in plant tissues. Qualitative and quantitative determination by enzymatic post-column cleavage and electrochemical detection after separation by HPLC. *Journal of Chromatography*, **303**, 77–83

Cheeke, P.R. and Shull, L.R. (1985) *Natural Toxicants in Feeds and Poisonous Plants*, Westport, CT, Avi Publishing, pp. 92–172

Cilliers, J.L. and van Nickerk, P.J. (1986) LC determination of phytic acid in food by post-column colorimetric detection. *Journal of Agricultural and Food Chemistry*, **34**, 680–683

Conn, E.E. (1973) *Toxicants Occurring Natural in Plants*, Washington, DC, National Academy of Sciences, pp. 299–308

Donatucci, D.A., Liener, I.E. and Gross, C.J. (1987) Binding of navy bean (*Phaseolus vulgaris*) lectin to the intestinal cells of the rat and its effect on the absorption of glucose. *Journal of Nutrition*, **117**, 2154–2160

Fenwick, G.R., Heaney, R.K. and Mullin, W.J. (1983) Glucosinolates and their breakdown products in food and food plants. *CRC Critical Reviews in Food Science and Nutrition*, **18**, 123–201

Fillisetticozzi, T.M.C.C. and Lajola, F.M. (1988) Effect of processed soybean on the thyroid gland of rats fed low iodine diets. *Nutrition Reports International*, **37**, 983–994

Green, G.M. and Lyman, R.L. (1972) Feedback regulation of pancreatic secretion as a mechanism for trypsin-induced hypersecretion in rats. *Proceedings of the Society for Experimental Biology and Medicine*, **104**, 6–12

Hagerman, A.E. and Butler, L.G. (1978) Protein precipitation method for the quantitative determination of tannins. *Journal of Agricultural and Food Chemistry*, **26**, 809–812

Hamerstrand, G.E., Black, L.T. and Glover, J.D. (1981) Trypsin inhibitors in soy products: modification of the standard analytical procedure. *Cereal Chemistry*, **58**, 42–45

Harland, B.F. and Oberleas, D. (1977) A modified method for phytate analysis using an ion-exchange procedure: application to texturized vegetable proteins. *Cereal Chemistry*, **54**, 827–832

Hassan, F., Rothnie, N.E., Yeung, S.P. and Palmer, M.V. (1988) Enzyme-linked immunosorbent assays for alkenyl glucosinolates. *Journal of Agricultural and Food Chemistry*, **36**, 398–401

Hendricks, H.G.C.J.M., Koninkx, J.F.J.G., Draaijer, M., van Dijk, J.E., Raaijimakers, J.A.M. and Mouwen, J.M.V.M. (1987) Quantitative determination of the lectin binding capacity of small intestinal brush-border membrane. An enzyme-linked lectin sorbent assay. *Biochimica Biophysica Acta*, **905**, 371–375

Honavar, P.M., Shih, C.-V. and Liener, I.E. (1962) Inhibition of the growth of rats by purified hemagglutinin fractions isolated from *Phaseolus vulgaris*. *Journal of Nutrition*, **77**, 109–114

Honig, D.H., Hockridge, M.E., Gould, R.M. and Rackis, J.J. (1983) Determination of cyanide in soybeans and soybean products. *Journal of Agricultural and Food Chemistry*, **31**, 272–275

Howard, J. and Shannon, L. (1977) A rapid, quantitative and highly specific assay for carbohydrate-binding proteins. *Analytical Biochemistry*, **79**, 234–239

Hussein, L., Motawei, H., Nassib, A., Khalil, S. and Marquardt, R. (1986) The complete elimination of vicine and convicine from the faba beans by combinations of genetic selection and processing techniques. *Plant Foods for Human Nutrition*, **36**, 231–242

Iwai, K., Fushiki, T. and Fukuoka, S.-I. (1988) Pancreatic enzyme secretion mediated by novel peptide-monitor peptide hypothesis. *Pancreas*, **3**, 720–728

Jamalian, J. (1978) Favism-inducing toxins in broad bean (*Vicia faba*). Determination of vicine content and investigation of other non-protein nitrogenous compounds in different broad bean cultivars. *Journal of Science of Food and Agriculture*, **29**, 136–140

Kakade, M.L., Rackis, J.J., McGhee, J.E. and Puski, G. (1974) Determination of trypsin inhibitor activity of soy products: collaborative analysis of an improved procedure. *Cereal Chemistry*, **51**, 376–382

Kakade, M.L., Simons, N. and Liener, I.E. (1969) An evaluation of natural vs. synthetic substrates for measuring the antitryptic activity of soybean samples. *Cereal Chemistry*, **46**, 518–526

King, T.P., Pusztai, A. and Clarke, M.W. (1980) Immunocytochemical localization of ingested kidney bean (*Phaseolus vulgaris*) lectins in the gut. *Histochemical Journal*, **12**, 201–208

Konijn, A.M., Gershon, B. and Guggenheim, K. (1973) Further purification and mode of action of a goitrogenic material from soybean flour. *Journal of Nutrition*, **103**, 378–383

Krogdahl, A. and Holm, H. (1981) Soybean proteinase inhibitors and human proteolytic enzymes: selective inactivation of inhibitors by treatment with human gastric juice. *Journal of Nutrition*, **111**, 2045–2051

Krogdahl, A. and Holm, H. (1983) Pancreatic proteinases from man, trout, pig, cow, chicken, mink and fox. Enzyme activities and inhibition by soybean and lima bean proteinase inhibitors. *Comparative Biochemistry and Physiology*, **74B**, 403–409

Kupchella, L. and Syty, A. (1984) Determination of cyanogenic glycosides in seed by molecular absorption spectrometry. *Journal of the Association of Official Analytical Chemists*, **67**, 188–191

Liener, I.E. (1955) The photometric determination of the hemagglutinating activity of soyin and crude soybean extracts. *Archives of Biochemistry and Biophysics*, **54**, 223–231

Liener, I.E. (1958) Inactivation studies on the soybean hemagglutinin. *Journal of Biological Chemistry*, **233**, 401–405

Liener, I.E. (1983) *Chemistry and World Food Supplies: The New Frontiers* (ed. L.W. Shemitt), New York, Pergamon Press, pp. 453–463

Liener, I.E. (1986) *The Lectins, Properties, Functions, and Applications in Biology and Medicine* (eds I.E. Liener, N. Sharon and I.J. Goldstein). New York, Academic Press, pp. 527–552

Liener, I.E. (1987) *Food Biotechnology* (eds R.D. King and P.S.J. Cheetham). London, Elsevier Applied Science, pp. 249–271

Liener, I.E., Deuel, H.J. Jr. and Fevold, H.L. (1949) The effect of supplemental methionine on the nutritive value of diets containing concentrates of the soybean trypsin inhibitor. *Journal of Nutrition*, **39**, 325–339

Liener, I.E. and Hasdai, A. (1986) The effect of the long term feeding of raw soy flour on the pancreas of the mouse and the hamster. *Advances in Experimental Medicine and Biology*, **199**, 189–198

Liener, I.E. and Kakade, M.L. (1980) *Toxic Constituents of Plant Foodstuffs* (ed. I.E. Liener). New York, Academic Press, pp. 7–72

Liener, I.E., Nitsan, Z., Srisangnam, C., Rackis, J.J. and Gumbmann, M.R. (1985) The USDA trypsin inhibitor study. II. Time related biochemical changes in the pancreas of the rat. Qualitas Plantarum. *Plant Foods for Human Nutrition*, **35**, 259–274

Liener, I.E., Sharon, N. and Goldstein, I.J. (eds) (1986) *The Lectins. Properties, Functions, and Applications in Biology and Medicine*. New York, Academic Press

Loser, C., Folsch, U.R., Mustroph, D., Cantor, P., Wunderlich, U. and Cruetzfeld, W. (1988) Pancreatic polyamine concentrations and cholecystokinin plasma levels in rats after feeding raw or heat-inactivated soybean flour. *Pancreas*, **3**, 285–291

McGrath, R.M., Kaluza, W.Z., Daiber, K.H., van der Heit, W.B. and Glennie, C.W. (1982) Polyphenols of sorghum grain, their changes during malting and their inhibitory nature. *Journal of Agricultural and Food Chemistry*, **30**, 450–456

McGregor, D.I., Mullin, W.J. and Fenwick, G.R. (1983) Analytical methodology for determining glucosinolate composition and content. *Journal of the Association of the Official Analytical Chemists*, **66**, 825–832

McGuiness, M.E., Morgan, R.G.H., Levison, D.A., Frape, D.L., Hopwood, G. and Wormsley, K.G. (1980) The effect of long term feeding of soya flour on the rat pancreas. *Scandinavian Journal of Gastroenterology*, **15**, 497–502

Macower, R.U. (1970) Extraction and determination of phytic acid in beans (*Phaseolus vulgaris*). *Cereal Chemistry*, **47**, 288–295

Mager, J., Chevion, M. and Glaser, G. (1980) *Toxic Constituents of Plant Foodstuffs* (ed. I.E. Liener). New York, Academic Press, pp. 266–294

Marquardt, R.R. and Frohlich, A.A. (1981) Rapid reversed-phase HPLC method for the quantitation of vicine, convicine, and related compounds. *Journal of Chromatography*, **208**, 373–376

Meeker, J.E. and Kilgore, W.W. (1987) Identification and quantitation of the alkaloids of Lupinus latifolius. *Journal of Agricultural and Food Chemistry*, **35**, 431–433

Mehansho, H., Butler, L.G. and Carlson, D.M. (1987) Dietary tannins and salivary proline-rich proteins: interactions, induction, and defense mechanisms. *Annual Reviews of Nutrition*, **7**, 423–456

Montgomery, R.D. (1980) *Toxic Constituents of Plant Foodstuffs* (ed. I.E. Liener). New York, Academic Press, pp. 143–160

Mossaba, M.M., Shaw, G.J., Andrezejewski, D., Sphon, J.A. and Page, S.W. (1989) Application of gas chromatography/matrix isolation/Fourier transform infrared spectrometry to the identification of glucosinolates from Brassica vegetables. *Journal of Agricultural and Food Chemistry*, **37**, 367–372

Muduuli, D.S., Marquardt, R. and Guenter, W. (1982) Effect of dietary vicine and vitamin E supplementation on the productive performance of growing and laying chickens. *British Journal of Nutrition*, **47**, 53–60

Naresingh, D., Jaipersad, D. and Cheng-Yen, I. (1988) Immobilization of linamarase and its use in the determination of bound cyanide in cassava using flow injection analysis. *Analytical Biochemistry*, **172**, 89–95

Nitsan, Z. and Liener, I.E. (1976) Enzyme activities in the pancreas, digestive tract, and feces of rats fed raw or heated soy flour. *Journal of Nutrition*, **106**, 300–305

Oke, O.L. (1980) Toxicity of cyanogen glycosides. *Food Chemistry*, **6**, 97–109

Price, M.L. and Butler, L.G. (1977) Rapid visual estimation and spectrophotometric determination of tannin content of sorghum grains. *Journal of Agricultural and Food Chemistry*, **25**, 1268–1273

Priddis, C.R. (1983) Capillary gas chromatography of lupin alkaloids. *Journal of Chromatography*, **261**, 95–101

Pusztai, A., Clarke, E.M.W. and King, T.P. (1979a) Nutritional toxicity of *Phaseolus vulgaris* lectins. *Proceedings of the Nutrition Society*, **38**, 115–120

Puztai, A., Clarke, E.M.W., King, T.P. and Stewart, J. (1979b) Nutritional evaluation of kidney beans (*Phaseolus vulgaris*): Chemical composition, lectin content, and nutritional value of selected cultivars. *Journal of Science of Food and Agriculture*, **30**, 843–848

Rackis, J.J., Wolf, W.J. and Baker, E.C. (1986) *Nutritional and Toxicological Significance of Enzyme Inhibitors in Foods* (ed. M. Friedman). New York, Plenum Press, pp. 299–347

Rascon, A., Seidl, D.S., Jaffe, W.G. and Aizman, A. (1985) Inhibition of trypsins and chymotrypsins from different animal species: A comparative study. *Comparative Biochemistry and Physiology*, **82B**, 375–378

Reddy, N.R., Pierson, M.D., Sathe, S.K. and Salunkhe, D.K. (1985) Dry bean tannins. A review of nutritional implications. *Journal of the American Oil Chemists Society*, **62**, 541–549

Reddy, N.R., Sathe, S.K. and Salunkhe, D.K. (1982) Phytates in legumes and cereals. *Advances in Food Research*, **28**, 1–92

Roozen, J.P. and de Groot, J. (1987) Analysis of low levels of trypsin inhibitor activity in food. *Lebensmittel-Wissenschaften-und-Technologie*, **20**, 305–308

Ruiz, L.P. Jr. (1976) A rapid screening test for lupine alkaloids. *New Zealand Journal of Agricultural Research*, **20**, 51–52

Smith, C., van Megen, W., Twaalfhoven, L. and Hitchcock, C. (1980) The determination of trypsin inhibitor levels in foodstuffs. *Journal of Science of Food and Agriculture*, **31**, 341–350

Srinivasan, V., Mougdal, N.R. and Sarma, P.S. (1957) Goitrogenic agents in foods. I. Goitrogenic action of groundnut. *Journal of Nutrition*, **61**, 87–96

Ullah, A.H.J. and Phillippy, B.Q. (1988) Immobilization of Aspergillus ficum phytase: product characterization of the bioreactor. *Preparative Biochemistry*, **18**, 483–489

Vaintraub, I.A. and Lapteva, N.A. (1988) Colorimetric determination of phytate in unpurified extracts of seeds and products of their processing. *Analytical Biochemistry*, **175**, 227–230

von Baer, D., Reimerdes, E.H. and Feldheim, W. (1979) Methods for the determination of quinolizidine alkaloids in Lupinus mutabilis. I. Screening tests. *Zeitschrift Lebensmittel-Untersuchung und-Forschung*, **169**, 27–31

Waldroup, P.W. and Smith, K.J. (1989) *Legumes. Chemistry, Technology and Human Nutrition* (ed. R.H. Mathews). New York, Marcel Dekker, pp. 245–337

Yanatori, Y. and Fujita, T. (1976) Hypertrophy and hyperplasia in the endocrine and exocrine pancreas of rats fed soybean trypsin inhibitor or repeatedly injected with pancreozymin. *Archivum Histologicum Japonicum*, **39**, 67–78

Yang, Z.-H. (1988) Determination of glucosinolate in rapeseed meal by infrared spectrometry. *Analyst*, **113**, 356–359

23

ESTIMATION AND RELEVANCE OF RESIDUES IN ANIMAL FEEDINGSTUFFS

H. W. EVANS
B.P. Nutrition (UK) Ltd., Wincham, Northwich, Cheshire, UK

This chapter reviews some of the analytical methods available for estimating residues in animal feedingstuffs. At the same time the relevance of residues from a biological and legislative standpoint will be considered. A wide variety of chemicals are used in animal feedingstuffs for therapeutic and growth promotion purposes whilst a number of the raw materials used to produce those feeds has been treated with pest control agents. It is obvious that misuse or careless application of these chemicals could lead to unacceptably high levels in the end product for human consumption. Even with careful application and use it is inevitable that minute quantities will remain in the end product and it is important, therefore, to determine the safe levels of these chemicals rather than insist on 'zero residues'.

In the case of medicinal additives, toxicity studies are required to be carried out prior to licensing in order to determine a safe residue tolerance level. Tissue residue studies then enable calculation of the appropriate withdrawal period which is included in the product licence. It is important that finishing or withdrawal feeds are kept as clear as possible of unacceptable residues and that withdrawal periods are observed.

It is not sufficient to concern ourselves solely with residues in the final feed which may be transferred to the end product for human consumption. A number of medicinal additives are contraindicated for non-target species and such residues in these feeds can result in adverse effects in livestock. With respect to animal feeds some of the ways in which residues can occur will be considered.

Residues from prior treatment of raw materials

It is common practice for farmers to treat harvested cereals, fruit and other agricultural products to prevent infestation and spoilage during storage. The procedures used for treatment with pesticides can vary significantly in the degree of sophistication, from accurate metering to much cruder methods resulting in considerable variation in the levels of pesticide residues. In the case of cereals they may also be subjected to repeated pesticide treatment and additionally be treated with a variety of pesticides after leaving the primary producer and before reaching the final consumer. Pesticide residues are likely to be concentrated in fats derived from treated produce and, therefore, may show a significant contribution to the overall level in finished feeds in which they are incorporated.

Residues arising from misuse of medicinal additives

Some 170 licensed preparations of medicinal additives exist for incorporation into animal feeds and approximately 80 of these can only be incorporated under the direction of a veterinary surgeon. Incorporation of the wrong additive or an incorrect level of additive into a particular animal diet can lead to a situation where medicinal residues could occur in an unmedicated feed. The Medicines (Medicated Animal Feedingstuffs) Regulations 1988 were introduced to regulate the manufacture, sale and supply of medicated feed in the UK. An important part of these regulations is the requirement for all manufacturers who incorporate medicinal products to register with the Royal Pharmaceutical Society of Great Britain and the Department of Agriculture for Northern Ireland. All registered manufacturers are legally required to comply with a code of practice whose purpose is to provide guidelines for establishing good manufacturing practices, to minimize risks of cross-contamination and to ensure operator safety. The code of practice also requires records of incorporation of medicinal additives to be kept and for the manufacturer to demonstrate by the results of analysis the nature, level and homogeneity of medicinal additives incorporated into medicated feedingstuffs.

Residues arising from cross-contamination of medicated feeds

For any manufacturer producing medicated animal feeds for all species of livestock considerable care is required to avoid cross-contamination which can occur through feed becoming held up in mixers, elevators, conveying systems and storage bins.

In 1984 the United Kingdom Agricultural Supply Trade Organisation (UKASTA) issued a revised code of practice for the avoidance of cross-contamination in the manufacture of medicated animal feedingstuffs. This code of practice was produced after consultation with The Ministry of Agriculture, offers guidelines in manufacture such as scheduling of production and flushing and cleaning of plant to avoid contamination. Special care is needed in some cases where a particular medicinal additive is contraindicated for a specific animal species and Table 23.1 illustrates some such examples. The LD50 of monensin for horses is 3 mg/kg bodyweight. Assuming a horse of 100 kg weight eats 10 kg then the feed would need to contain 30 mg/kg monensin for a lethal effect. This level would be unlikely to occur by mere cross-contamination and would require misuse of the

Table 23.1 CONTRAINDICATIONS OF MEDICINAL ADDITIVES

Active constituent	Use	Contraindication
Copper	Growth promotion	Toxic to sheep
Monensin	Coccidiostat/growth promotion	Fatal to equines
Salinomycin	Coccidiostat	Fatal to equines, turkeys
Lincomycin	Treatment of swine dysentery	Depresses ruminant appetite
Tiamulin	Treatment of swine dysentery	Growth depression or death when combined with ionophores
Narasin	Coccidiostat	Fatal to equines, turkeys
Nicrazin	Coccidiostat	Adverse effect on egg production and hatchability

addition. The death of horses through monensin poisoning is more likely to occur by consumption of feed intended for other species.

The manufacture of animal feedingstuffs inevitably generates discard material which can arise from out-of-date stock, quality rejects, customer returns, sievings, flushing and cleanings, broken bags and spillages. The majority of animal feed production is in pelleted form and a good proportion of this is in bulk. This process inevitably leads to the generation of discard material which in the interests of economy the compounder will want to reuse. The code of practice recommends that reuse of medicated discards should ideally be into identical feeds or feeds for the same target species giving due consideration to the special nature of any particular medicinal additive and should not be used in feeds for vulnerable species. Given the above restrictions the code suggests that discards containing medicinal additives under the Permitted Merchants List (PML) category should be used so that medicated residues are included at not more than 5%. For discards containing medicinal additives under the Prescription Only Medicine (POM) category (i.e. those incorporated under the direction of a veterinary surgeon) the code recommends that as well as taking into consideration the above restrictions incorporation should take place so that medicated residues are included at not more than 1%.

Residues arising from carryover of medicinal additives

Traditionally fine powder formulations of many medicinal additives have been used in order to obtain accurate dispersion of the active component throughout the feed. However, in some cases fine powder formulations tend to build up an electrostatic

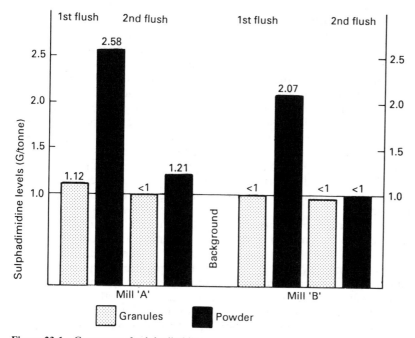

Figure 23.1 Carryover of sulphadimidine

charge on the particles during mixing which may result in the additive sticking to the sides of mixing equipment. In addition to obtaining a reduced level of medication this effect can result in carryover of the additive to the following or subsequent mixings and thus lead to residues.

Some manufacturers have formulated granular forms of medicinal additives which because of the larger particle size carry negligible electrostatic charge and so minimize sticking and carryover. The granular formulations can also be claimed to be more environmentally acceptable and safer for the operator as there is reduced dust level in the atmosphere. Particle size can, however, be critical as in some circumstances granular formulations have been found to produce less homogeneous feeds.

Figure 23.1 shows the results of two trials to demonstrate the degree of carryover found when both powder and granular forms of sulphadimidine are used at the 100 g per tonne level. In each case the level of carryover when using a granular form of sulphadimidine is at or around the background level of 1 g/tonne in the two feeds following the one which has been medicated. In the case of the powdered form of sulphadimidine the level present after the first flush is in excess of 2 g/tonne and only at the second flush is the residue level reduced to about the background level. It is recognized that a level of 2 g/tonne of sulphadimidine in feed can result in carcass residues of 0.1 mg/kg. The MAFF Veterinary Products Committee has stated that residue concentrations above 0.1 mg/kg for sulphonamides in pigmeat are unacceptable in terms of protecting human health interests.

Methodology of detection

The determination of micro-organic components in an animal feedingstuff often presents the analyst with a difficult problem due to the very complex nature of the material to be examined. In the case of compound animal feeds the presence of a mixture of macro-ingredients of animal and vegetable origin containing numerous chemical compounds in varying degrees of concentration, together with often between 20 and 30 additives in the form of vitamins, minerals and medicines, require the use of sophisticated analytical techniques. The more straightforward spectrophotometric techniques which are perfectly satisfactory for examining the component in question in isolation in a relatively pure form often cannot be applied to feedingstuffs due to interference from the numerous other components present. For accurate measurement of organic micro-ingredients it is necessary to use a technique which will enable separation of the component in question from the rest of the medium. This is particularly true when examining for residues, due to the often extremely low concentrations present which may be at or below the detection limit of the method.

Chromatography embraces a variety of extremely powerful separation techniques, some of which have been used to determine organic components in animal feedingstuffs and which have also found application in the determination of residues. The common feature of all chromatographic separation techniques is that the components of the sample mixture are distributed between two phases, one of which remains stationary while the other phase percolates through the interstices or over the surface of the fixed phase. The movement of the mobile phase results in a differential migration of the sample components. The stationary phase can be either solid or liquid and the mobile phase can be either liquid or gas thus giving several possible combinations. The types of procedure considered here are thin layer chromatography, gas liquid chromatography and high pressure liquid chromatography.

Thin layer chromatography (TLC)

Thin layer chromatography is a special field of liquid–liquid chromatography in which the stationary liquid is an adsorbed film on a thin layer coating of alumina, silica gel or other powdered material on a glass plate. The technique is very simple in that a spot of sample solution is placed near the edge of the plate which is then placed on its end in the eluting solvent. The various components are adsorbed or partitioned at different rates. The ratio of the distance travelled by the substance to the distance travelled by the solvent front is known as the retardation factor (Rf

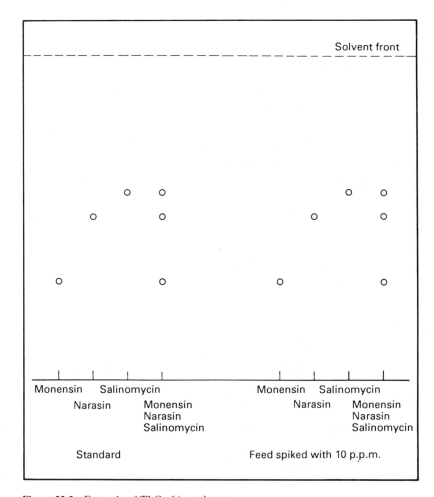

Figure 23.2 Example of TLC of ionophores

value) and is characteristic of the substance provided that the conditions are unaltered. The thin layer plate is then dried and treated with a developing reagent and the sample spot appears as a characteristic colour which can be identified and quantified against known standard concentrations. The technique is largely qualitative or semi-quantitative but can be carried out quite quickly without the need for expensive equipment or a high level of technical expertise.

One useful application of TLC has been in the identification of the ionophores used as coccidiostats in broiler feeds where carryover or cross-contamination into feeds for equines or turkeys can be dangerous. The feed sample is first extracted with methanol and the extract purified with aluminium oxide. Figure 23.2 shows an example of a thin layer chromatograph for the detection of salinomycin, monensin and narasin using vanillin as the developing agent. Monensin appears on the TLC plate as a yellowish red spot while narasin and salinomycin appear as bluish-red spots. The presence of individual ionophores can be confirmed by comparison with the standards. The technique enables detection down to 10 mg/kg but it is possible to identify down to 1 mg/kg if identification by bioautography is used after the TLC separation.

Gas liquid chromatography (GLC)

Gas liquid chromatography (or vapour phase chromatography) also finds considerable application in the detection of certain organic compounds at residue levels. The basis of the technique is that a gas or the vapour of a liquid or volatile solid passes down a column containing an inert solid impregnated with a non-volatile liquid as the stationary phase. The vapour is carried along by an inert carrier gas such as nitrogen or argon. Two types of column, packed or open tubular (capillary) are used and can vary in length from 1 to 30 m and are heated in a thermostatically controlled oven. The retention time on the column depends on how strongly the substance is attracted to the stationary phase and is characteristic

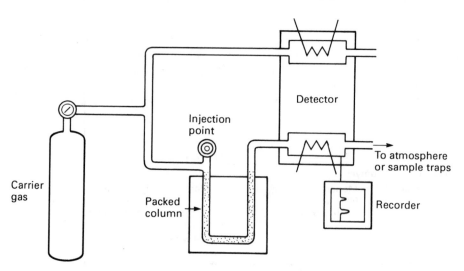

Figure 23.3 Essentials of a gas liquid chromatograph

of that substance for a given set of conditions. The sample components are separated as they pass through the column and to a detector which sends a signal to a recorder. Figure 23.3 gives an illustration of a typical apparatus.

Gas liquid chromatography has found wide application in the analysis of pesticide residues because the technique allows multi-component resolution of very small quantities to a high degree of sensitivity. Figure 23.4 illustrates one such procedure applied to animal feedingstuffs.

Initial extraction with hexane is required before a two-stage clean-up procedure. In the first clean-up stage the extracted sample is injected into a heated column containing silanized glass beads and is distilled using hexane as an eluant. The pesticide residues are collected in the distillate which is then passed through an alumina micro column again with hexane as the eluant at the second stage clean-up.

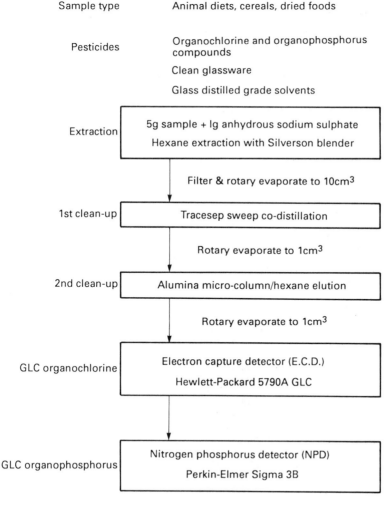

Sample type	Animal diets, cereals, dried foods
Pesticides	Organochlorine and organophosphorus compounds
	Clean glassware
	Glass distilled grade solvents
Extraction	5g sample + 1g anhydrous sodium sulphate / Hexane extraction with Silverson blender
	Filter & rotary evaporate to 10cm³
1st clean-up	Tracesep sweep co-distillation
	Rotary evaporate to 1cm³
2nd clean-up	Alumina micro-column/hexane elution
	Rotary evaporate to 1cm³
GLC organochlorine	Electron capture detector (E.C.D.) / Hewlett-Packard 5790A GLC
GLC organophosphorus	Nitrogen phosphorus detector (NPD) / Perkin-Elmer Sigma 3B

Figure 23.4 Flow chart – pesticide residue analysis

After concentration by evaporation of solvent the sample extract is ready for gas chromatography. The organo-chlorine residues are determined using an instrument with an electron capture detector whilst the organo-phosphorus pesticide malathion is determined separately using an instrument with a nitrogen phosphorus detector. Using this procedure limits of detection for the more common pesticides are 1 ppb for DDT, dieldrin, lindane (gamma HCH), heptachlor and 20 ppb for malathion. Figure 23.5 shows an example of an organochlorine pesticide residue chromatogram where the standard consists of a mixture of pesticides at the 10 ppb and 20 ppb level. The feed sample indicates those pesticides detected.

In 1988 maximum residue levels for organo-chlorine pesticides were introduced into United Kingdom legislation via schedule 5 – prescribed limits for undesirable substances of the feedingstuff regulations. These specify maximum content for a variety of pesticides in straight feedingstuffs, fats and compound feedingstuffs and are given in Table 23.2.

It is not clear whether these limits bear any relationship to potential toxicity to livestock or humans. Enforcement of the regulations could be difficult as there are as yet no statutory methods of analysis for pesticides and in many cases the maximum is close to the detection limit.

Table 23.2 PRESCRIBED MAXIMUM LEVELS OF ORGANOCHLORINE PESTICIDES

Substances	Feedingstuffs	Maximum content (mg/kg) in feedingstuffs referred to a moisture content of 12%
Aldrin - singly or combined	All feedingstuffs	0.01
Dieldrin - expressed as dieldrin	except fats	0.2
Camphechlor (Toxaphene)	All feedingstuffs	0.1
Chlordane (sum of cis and trans	All feedingstuffs	0.02
isonomers and of oxychlordane)	except fats	0.05
DDT (sum of DDT, TDE and DDE	All feedingstuffs	0.05
isomers, expressed as DDT)	except fats	0.5
Endosulphan (sum of alpha and beta	All feedingstuffs except:	0.1
isomers and of endosulphan sulphate,	Maize	0.2
expressed as endosulphan)	oilseeds	0.5
	complete feedingstuffs for fish	0.005
Endrin (sum of endrin and delta,	All feedingstuffs	0.01
keto endrin expressed as endrin)	except fats	0.05
Heptachlor (sum of heptachlor and	All feedingstuffs	0.01
of heptachlor epoxide,	except fats	0.2
expressed as heptachlor)		
Hexachlorobenzene (HCB)	All feedingstuffs	0.01
	except fats	0.2
Hexachlorocyclohexane (HCH):		
α isomer	All feedingstuffs	0.02
	except fats	0.2
β isomer	Straight feedingstuffs	0.01
	except fats	0.1
	Compound feedingstuffs	0.01
	Except for compound feeding stuffs for dairy cattle	0.005
γ isomer	All feedingstuffs	0.2
	except fats	2.0

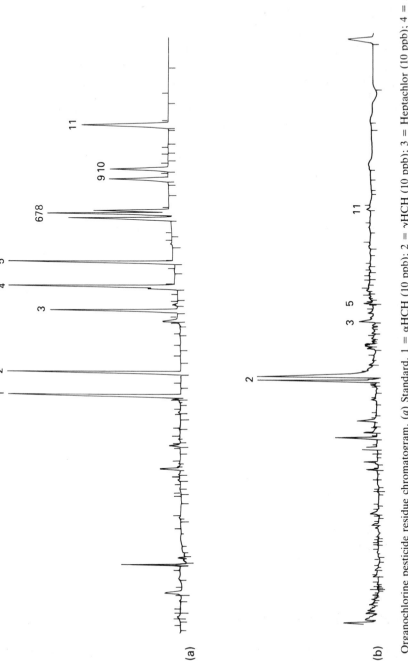

Figure 23.5 Organochlorine pesticide residue chromatogram. (*a*) Standard. 1 = αHCH (10 ppb); 2 = γHCH (10 ppb); 3 = Heptachlor (10 ppb); 4 = Aldrin (10 ppb); 5 = Heptachlor epoxide (10 ppb); 6 = Dieldrin (10 ppb); 7 = P₁P₄-DDE (10 ppb); 8 = O₁P-DDD (10 ppb); 9 = P₁P-DDD (10 ppb); 10 = O₁P-DDT (10 ppb); 11 = P₁P-DDT (20 ppb). (*b*) Sample. 2 = γHCH (20 ppb); 11 = P₁P-DDT, 1 ppb; 3 = Heptachlor, less than 1 ppb; 5 = Heptachlor epoxide, 1 ppn; 11 = P₁P-DDT, 1 ppb

In 1988 the MAFF carried out the first part of a survey into organo-chlorine pesticide residues in feedingstuffs. The results of this first part of the survey are summarized in Table 23.3. Out of 153 samples examined 25 (16%) were found to contain detectable levels of pesticides at or above the reporting limit of 0.01 mg/kg except for some instances where interferences resulted in a reporting level of 0.05 mg/kg. Four samples (3%) contained residue levels above the specified maximum. Only pesticides in the following categories were detected: dieldrin, alpha, beta and gamma isomers of hexachlorocyclohexane (HCH), hexachlorobenzene (HCB), and p,p-dichlorophenyl-dichloroethylene (DDE) (component of DDT).

Table 23.3 SURVEY OF ORGANOCHLORINE RESIDUES IN FEEDINGSTUFFS (MAFF, 1988)

Substance	Number of samples	Number positive	Number above specified maximum
Sugar beet	8	Nil	Nil
Rice bran	6	2	2
Maize gluten	11	1	1
Rapeseed	11	Nil	Nil
Wheat	11	Nil	Nil
Wheatfeed	14	4	Nil
Barley	11	Nil	Nil
Soya bean meal	12	1	1
Full fat soya	1	Nil	Nil
Sunflower	9	Nil	Nil
Maize	4	Nil	Nil
Meat and bone meal	11	1	Nil
Peas	2	Nil	Nil
Beans	2	Nil	Nil
Paddy meal	2	1	Nil
Cotton cake	2	Nil	Nil
Fats	36	15	Nil
TOTAL	153	25	4

Pesticides detected:
Dieldrin
α hexachlorocyclohexane (HCH)
β HCH
γ HCH
Hexachlorobenzene (HCB)
pp-DDE (component of DDT)

High pressure liquid chromatography (HPLC)

In high pressure liquid chromatography (variously called high performance or high speed chromatography) eluent from a solvent reservoir is filtered, pressurized and pumped through a chromatographic column. The sample solution is injected into the system and separated into components on travelling down the column. The individual components are monitored by a detector and recorded as peaks on a chart recorder or integrator. Figure 23.6 shows the main components of a high

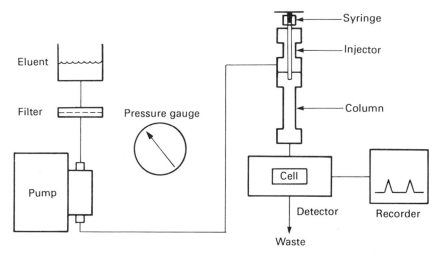

Figure 23.6 Components of a high pressure liquid chromatograph

pressure liquid chromatograph which are a high pressure pump, column, injection system and detector.

HPLC is a development of liquid chromatography and was first introduced some 20 years ago. Operating at high pressure enables the effect of higher liquid viscosities relative to gas viscosities to be overcome and give comparable analysis times to that of GLC. A reduction in the particle size of the support material of the stationary phase also leads to increased efficiency. Ultraviolet detectors have been predominantly used but fluorescence, electrochemical, refractive index and diode array detectors have been introduced to give the technique a wide range of application. Gradient elution can also be employed where the composition of the mobile phase is varied and can be useful in obtaining good resolution in complex separations.

HPLC has been widely used in pharmaceutical, biochemical, clinical and environmental analyses. In recent years it has found wide application in the analysis of animal feedingstuffs, in particular in medicinal additives where it is gradually replacing some of the more traditional spectrophotometric methods. It is also particularly well suited to the analysis of those compounds which are not readily handled by GLC. Thermally labile compounds can be analysed at ambient temperatures by HPLC and highly polar compounds can often be chromatographed without prior derivitization. Sample clean-up is usually much less of a problem with HPLC than GLC and biological fluids can often be directly injected onto an HPLC column. As aqueous solvents can be used in HPLC much sample pretreatment can often be avoided. Whilst HPLC detectors are not usually as sensitive as GLC detectors the fact that greater sample quantities can be used enables detection limits to be similar. It is usually feasible, therefore, to apply a particular HPLC method for a medicinal additive to the estimation of residue amounts of that same additive with, in some cases, modification either to sample preparation or sensitivity of detection.

One typical application of HPLC is the determination of nicarbazin in poultry feed. Nicarbazin is used to prevent coccidiosis in broiler chickens but at levels between 100 mg/kg and 200 mg/kg in finished feed adverse effects on egg production and hatchability can occur. The withdrawal period is 7 days before slaughter and it is obviously important that great care must be taken to avoid mixing nicarbazin into feed for layers and breeding stock.

The procedure requires the feed sample to be extracted with hot dimethylformamide, followed by clean-up on an alumina column. The nicarbazin is eluted from the column with ethanol and quantitated using a reverse-phase C-18 column with a methanol–water mobile phase and ultraviolet detection at 344 nm. Recoveries of 98% have been obtained at the 100 mg/kg level with a standard deviation of 3%. Recoveries of 92% were obtained with samples fortified at 0.1 mg/kg. The method is designed to operate in the 0.1–200 mg/kg range and the detection limit is 1 ng. Figure 23.7 illustrates a typical chromatogram for a feed spiked with nicarbazin at the 10 mg/kg level; a blank feed giving only the solvent peak was used. An average

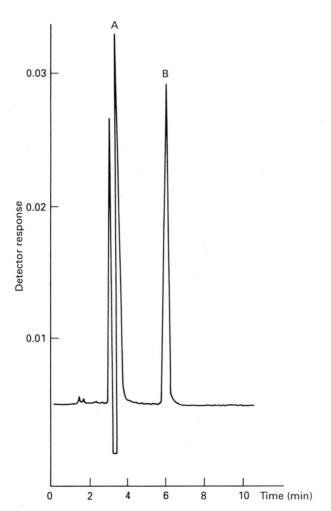

Figure 23.7 HPLC chromatogram of feed spiked with nicarbazin. A = solvent; B = nicarbazin (10 mg/kg)

recovery of 96% nicarbazin was obtained in the presence of other medicinal feed additives.

Because of the extreme sensitivity of this method it has been used to examine the degree of carryover of nicarbazin in feed to eggs which is estimated to be between 10% and 20%, i.e. 1 mg/kg nicarbazin in feed contributes 100–200 ppb in eggs. The limit of detection of the method for nicarbazin in eggs is of the order 2–5 ppb.

The demand for residue free eggs in Germany has created problems for continental feed manufacturers in achieving contamination free layers feed, so much so, that feed for broilers containing nicarbazin has to be produced on a separate plant to that producing layers feed.

The techniques of gas liquid chromatography and high pressure liquid chromatography can fairly be described as laboratory techniques requiring expensive instrumentation and relatively skilled operators. However, it can be argued that there is a real need in the study of residues for methods that are sensitive, rapid, inexpensive, easy to perform and capable of determining multi components. Owing to the very complex range of possible residues and the low concentrations it seems unlikely that multi-residue methods will ever be anything but laboratory techniques. However, development of methods based on

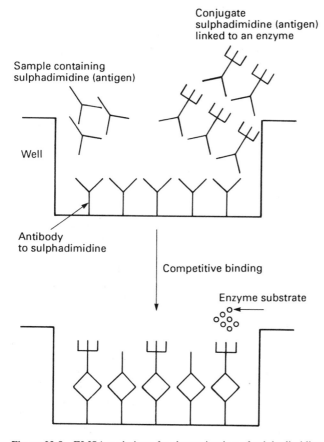

Figure 23.8   ELISA technique for determination of sulphadimidine

immunological techniques such as radio-immune assays (RIA) and enzyme linked immunosorbent assays (ELISA) represent realistic alternatives to conventional methods due to their potential versatility.

One such ELISA method has been developed for the determination of sulphadimidine in tissue, urine and feed. The principle which is illustrated in Figure 23.8 requires an antibody to sulphadimidine to be bound to a microtitre plate. Sample extract and sulphadimidine conjugated to an enzyme are added to the microtitre plate well where acting as antigens they compete for the limited number of binding sites on the antibody. The enzyme substrate is added which reacts with the bound enzyme conjugate to give a coloured product. The colour intensity is measured and is inversely proportional to the concentration of sulphadimidine in the sample. Although the method takes at least 6 hours to perform for feed samples, 28 samples can be tested at once and the capital outlay for equipment is relatively low.

This method was developed by the Veterinary Research Laboratories, Stormont, Belfast, Northern Ireland and is routinely used by them to monitor sulphadimidine levels in pig's urine. If a level above 500 ng/ml is found in the urine a tissue test is performed to confirm that the carcase contains illegal levels of sulphadimidine. Further investigations can then be carried out on the feed to establish the source of contamination. The limit above which the sample is said to be positive is 100 ng/g for tissue and 1000 ng/g for feed. Confirmation of positives is carried out using an HPLC procedure. The limit of detection for the ELISA method is 50 ng/g and intralaboratory studies give a coefficient of variation between 10–15% and interlaboratory studies of 12–15%.

Under EEC Directive 86/469 and the UK *Animals and Fresh Meat* (Examination of Residues) Regulations 1988, the MAFF have official powers to trace 'contaminated carcases' back to the farm of origin. Restrictions can be placed on suspect animals which are subject to testing. The presence of residues above the maximum limit prescribed can result in the animals being detained until the residues have fallen below this limit. However, these controls will not be implemented until maximum residue levels are adopted under EC64/433. Until that time actual enforcement will rest with member states.

In the UK official methods do not exist for all permitted medicinal additives. Some of those that do exist have been superseded by more accurate and sensitive techniques. It is desirable that a more direct effort be made to coordinate the development and validation of official methods for active components at both the usage and residue levels.

In the USA the Food and Drug Administration (FDA) and the Food Safety and Inspection Service (FSIS) play leading roles in the animal drug residue area. The FSIS monitors residues in animal tissues that are offered for human consumption while the FDA approves drugs for use in animals, establishes tolerances for drug residues in animals and leads in the approval of analytical methods for monitoring drug residues in edible animal tissues. In 1987 representatives of the FDA and FSIS formed part of a task force on animal drug methodology to establish principal needs for improved analytical procedures often working with the sponsors of the drugs.

In conclusion it seems clear that consumer demand for additive free food is likely to increase and, therefore, lead to a requirement for zero residues. The danger is that as modern analytical techniques become more sensitive and detection levels become lower, there will be an increasing demand for lower residue levels which may not be achievable with current feed manufacturing and farming practices. It is important, therefore, that residue levels are set which are based on a safe tolerance level. At present a review of product licences of medicinal additives is being carried out specifically to consider safety aspects towards man as a consumer and a handler and towards animals in terms of toxicity. It seems likely that some licences will be withdrawn if licence holders are unable to provide satisfactory data. In other cases withdrawal periods may be altered.

References

Feedingstuff Regulations (1988) S.I. 396

Hurlbut, Nightengale and Burkepile (1985) *J.O.A.C.*, **68**, 596–598

EEC Directive 86/469

The Medicines (Medicated Animal Feedingstuffs) Regulations (1988) S.I. 976

UK Animals and Fresh Meat (Examination of Residues) Regulations 1988

UKASTA Codes of Practice for Cross-contamination in Animal Feedingstuffs Manufacture

UKASTA Codes of Practice for Registered Manufactures of Medicated Animal Feedingstuffs

24

THE OCCURRENCE, DETECTION AND SIGNIFICANCE OF MYCOTOXINS IN ANIMAL FEEDING STUFFS

A.E. BUCKLE and K.A. SCUDAMORE
Agricultural Development and Advisory Service, Central Science Laboratory, Ministry of Agriculture, Fisheries and Food, Slough, UK

Introduction

Mycotoxins are toxic secondary metabolites produced by fungi under certain conditions. Several hundred mycotoxins have been described, though the majority have only been reported as being formed in laboratory cultures. Even so, about thirty mycotoxins have now been found to occur naturally in a wide range of commodities which have undergone fungal spoilage. Many mycotoxins are produced uniquely by a particular fungal species whereas other mycotoxins are produced by more than one species which can be of different genera. As mycotoxin production is dependent on environmental conditions the presence on a commodity of a fungal species able to produce a mycotoxin does not imply that the mycotoxin will necessarily be present. Conversely, the absence of any viable fungi does not imply the absence of mycotoxins.

Some crops may be invaded by fungi in the field prior to harvest and become contaminated with mycotoxins. After harvest, those fungi active in the field usually die out and tend to be replaced by other fungal species when the crop is stored. If it is processed then fungi may be killed although most mycotoxins are able to persist. This is of particular significance to animal feeding stuffs manufacturers when using commodities which have been processed or are the by-products of processing. It illustrates the need for quality control of raw materials that goes beyond an assessment of their physical appearance or moisture content. The wide range of climates and conditions under which the diverse range of raw materials used in animal feeding stuffs is produced tends to favour the growth of particular fungi in each case. As a result some commodities are particularly prone to contamination with a certain mycotoxin or combination of mycotoxins. In this chapter some of the reports describing the natural occurrence and significance of mycotoxins in the principal commodities used in animal feeding stuffs are reviewed, together with the methods used to detect them.

Occurrence

CEREALS

Cereals and cereal products make up the largest component of animal feeding stuffs. Cereals are liable to fungal spoilage before and after harvest and may therefore contain several mycotoxins produced by a combination of field and storage fungi. One of the earliest described forms of mycotoxin poisoning to affect man and livestock was attributed to the consumption of cereals, notably rye, contaminated with ergot. Ergot is a fungal body (sclerotium), produced by *Claviceps purpureum*, that is formed in place of the grain in cereals. The ergot contains various pharmacologically active substances (alkaloids), which can affect the nervous system and cause vaso-constriction of the capillaries. In livestock, ingestion of ergot can cause lameness and gangrene (Woods, A.J. *et al.*, 1966). When ingested by man, bread made of flour contaminated with ergots can also give rise to gangrene and/or convulsions. Numerous outbreaks of ergotism in man have been described for over 1000 years. In the Middle Ages it became known as St. Anthony's Fire because of the burning sensation experienced by those afflicted and the belief that a cure could be invoked with help from St. Anthony. Nowadays greater faith is put in the power of legislation to prevent harmful amounts of ergot getting into animal feeding stuffs, the maximum admissible level of ergot contamination being 0.05% (The Feeding Stuffs Regulations, 1988). Whilst ergot contamination of grain is clearly visible to the naked eye, other mycotoxins that can be present in grain are not so obvious and cannot be so readily removed by screening processes. Other indications of the possible presence of mycotoxins in grain at harvest are its physical appearance. Grain which has been invaded by fungi in the field is often shrivelled, scabby or discoloured.

Pink and scabby wheat is indicative of damage by *Fusarium* species. Wheat and barley are especially susceptible to invasion by *Fusarium* species in wet and cold field conditions. This can result in the production of various mycotoxins including zearalenone and trichothecenes such as deoxynivalenol (vomitoxin). The natural occurrence of deoxynivalenol in wheat has been reported in the winter crop grown

Table 24.1 SUMMARY OF UK PRODUCED CEREALS AND IMPORTED WHEAT FOR *FUSARIUM* MYCOTOXINS

Cereal	No. samples examined	No. samples positive	Mycotoxin	Range (mg/kg)	Reference
Feed barley (UK)	43	22	Deoxynivalenol	0.01->0.10	Gilbert *et al.*, 1983
Malt barley (UK)	42	13	Deoxynivalenol	0.01-0.10	Gilbert *et al.*, 1983
Wheat (UK)	199*	32	Deoxynivalenol	0.01-0.5	Osborne and Willis, 1984
Wheat (imported)	33*	23	Deoxynivalenol	0.02->0.5	Osborne and Willis, 1984
Wheat (UK)	31	20	Deoxynivalenol	0.004-0.312	Tanaka *et al.*, 1986
Wheat (UK)	31	17	Nivalenol	0.004-0.670	Tanaka *et al.*, 1986
Wheat (UK)	31	6	Zearalenone	0.001-0.0003	Tanaka *et al.*, 1986
Barley (Scotland)	8	4	Deoxynivalenol	0.001-0.081	Tanaka *et al.*, 1986
Barley (Scotland)	8	2	Nivalenol	0.027-1.140	Tanaka *et al.*, 1986
Wheat (England)	10	0	Zearalenone	–	MAFF, 1987
Wheat (England and Wales)	55†	6	Deoxynivalenol	0.080-0.750	MAFF, 1987
	55†	3	Fusarenol	0.140-0.570	MAFF, 1987

*Analyses for nivalenol, fusarenon-X, neosolaniol, diacetoxyscirpenol, HT-2 toxin and T-2 toxin were also done with negative results within the limits of detection.
†Analyses for nivalenol, diacetoxyscirpenol and T-2 toxin were also done with negative results within the limits of detection.

in Ontario in 1980 (Trenholm *et al.*, 1981) and in wheat from the Mid Western USA in 1982 (Hagler *et al.*, 1984). In the UK several surveys of home-grown cereals and imported wheat for *Fusarium* mycotoxins have been undertaken and their findings are summarized in Table 24.1.

Deoxynivalenol was frequently detected but the levels present were very much lower than reported in North American wheat and there is no evidence that such low levels of deoxynivalenol in animal feeding stuffs can exert any demonstrable harmful effects. Judged on the available information there is little indication, apart from one exception to be discussed later, that contamination of home-grown cereals with *Fusarium* mycotoxins might pose any real hazard to livestock production in the UK.

A different picture exists for those mycotoxins produced by species of *Aspergillus* and *Penicillium* which colonize stored grain unless it has been dried to below 15% moisture content. Fungal species of these genera can produce many mycotoxins which include aflatoxin, ochratoxin A, citrinin, sterigmatocystin and the naphthoquinones. On stored grain one of the most commonly reported mycotoxins is ochratoxin A. Its production by *Aspergillus ochraceus* in a cereal based laboratory culture was first reported by Van der Merwe (1965), and its production by *Penicillium viridicatum* Westling was demonstrated by Van Walbeek *et al.* (1969).

The first reported natural occurrence of ochratoxin A on wheat was discovered by Scott (1970) in Canadian wheat that had heated up in farm storage bins. It was subsequently found in 11 out of 848 samples of USA wheat randomly selected by the Grain Division of the Agricultural Marketing Service, USDA (Shotwell *et al.*, 1976). In Europe it was first reported by Krogh *et al.* (1973) in cereals grown in Denmark. It was found together with another mycotoxin, citrinin, in a survey of cereals used in pig feeds. It has also been reported to occur in barley and oats sampled in Sweden in 1972 (Krogh *et al.*, 1974) and in Polish cereals (Szebiotko *et al.*, 1981). In the UK, home-grown cereals used in animal feeds have been examined for mycotoxins by the Agricultural Development and Advisory Service (ADAS) since 1976 using a thin layer chromatography technique (Patterson and Roberts, 1979). The majority of samples examined were received in the course of advisory work; others were collected from farm stores in June and July 1980 in a survey of barley for mycotoxins. Many of the advisory samples were examined when investigating disease problems in livestock or for quality control when the grain was being used in poultry feed. In a few cases, analysis of mouldy grain was requested by farmers to assess if it might be safely fed to livestock. Between 1976 and 1987 over 1400 samples of home-grown barley, wheat and oats were analysed. The incidence of mycotoxins is shown in Table 24.2. Ochratoxin A and citrinin were the most frequently detected and often occurred together since they can be produced by the same mould *Penicillium viridicatum* which was frequently isolated from mouldy cereals. Sterigmatocystin was detected occasionally, usually co-occurring with other mycotoxins. It is produced by *Aspergillus versicolor* which was often isolated. The discovery of aflatoxin B_1 in barley on a farm in Wales (Hacking and Biggs, 1979) was particularly significant because it was hitherto not known to occur in home-grown cereals. In this instance the affected barley had been inadequately treated with propionic acid used as a moist grain preservative and had subsequently been colonized by *Aspergillus flavus*.

The available data suggest that ochratoxin A and citrinin are the most commonly occurring mycotoxins in home-grown cereals, but it should be recognized that the

Table 24.2   THE OCCURRENCE OF MYCOTOXINS IN HOME-GROWN CEREALS IN
ENGLAND AND WALES, 1976–1987 (EXCLUDINJG 1983) EXAMINED BY ADAS

Cereal	No. of samples examined	Percentage positive samples				
		AfB₁	Cit	Och A	Ster	Zen
Barley	890	2	5	12	3	2
Wheat	491	2	15	14	5	2
Oats	89	1	9	9	4	1

AfB$_1$ = Aflatoxin B$_1$; Cit = Citrinin; Och A = Ochratoxin A; Ster = Sterigmatocystin; Zen =
Zearalenone.

methods applied to screening cereals are limited to known mycotoxins for which
standards are available. Examination of the TLC plates under UV light often
reveals many other fluorescing substances of unknown identity. A more detailed
study of some of these compounds from mouldy home-grown cereals has revealed
another class of compounds known as the naphthoquinones, which are produced by
a number of *Penicillium* species, notably *Penicillium aurantiogriseum*. These toxins
have been detected in several samples of UK produced grain which has undergone
mould spoilage (Scudamore *et al.*, 1986).

In summary, home-grown cereals are a potential source of mycotoxins in animal
feeding stuffs. The most commonly occurring mycotoxins are those produced by
Aspergillus and *Penicillium* species in storage because of inadequate drying. Trace
amounts of the *Fusarium* mycotoxins, mostly deoxynivalenol, have been found in
several surveys but this mycotoxin is not known to have any harmful effects on
livestock at the levels reported to occur in home-grown cereals.

MAIZE AND MAIZE BY-PRODUCTS

Maize and maize by-products can be contaminated with a variety of mycotoxins.
There are numerous reports of the worldwide occurrence of aflatoxin and *Fusarium*
toxins in maize. Surveillance of maize produced in the USA for aflatoxin has been
undertaken by Shotwell *et al.* (1969, 1970, 1971, 1973, 1980) and has revealed
aflatoxin present at levels of up to 0.308 mg/kg. The incidence of aflatoxin
contamination in these surveys ranged from less than 1% in a survey of 1977 maize
to 35% in maize produced in 1969 and 1970. Contamination of maize with aflatoxin
occurs following invasion by *Aspergillus flavus*. This process can be initiated in the
growing crop by insect damage often caused by the European corn borer (*Ostrinia
nubilalis*) in North America (Lillehoj *et al.*, 1976). In years of drought insect
damage is often greater and increases the possibility of aflatoxin contamination. *A.
flavus* can continue to grow on maize after harvest and produce additional
aflatoxin, if it has not been dried to below 13% moisture content. Aflatoxin
contamination of maize is a worldwide problem. Post-harvest spoilage of maize in
Thailand is considered to be the most important cause of aflatoxin contamination.
A survey of maize on sale for human consumption from market towns and villages
in Thailand during 1967–1969 revealed 35% samples to contain aflatoxin B$_1$ at an
average level of 0.4 mg/kg (Shank *et al.*, 1972). In Uganda 40% of samples were
contaminated at an average level of 0.133 mg/kg (Stoloff, 1976).

Maize by-products are often used in animal feeding stuffs. When maize is wet-milled any aflatoxin present can become concentrated in some of the by-products. About 4% of the aflatoxin is recovered in the steep water and 10% is retained in the gluten. The 'gluten feed', which comprises a blend of the steep water, fibre and spent germ meal can contain up to 80% of the aflatoxin initially present. Less than 1% of the aflatoxin is recovered in the starch. Both the gluten and 'gluten feed', which are used in animal feeding stuffs, could therefore contain over 90% of the aflatoxin originally present in the maize (Table 24.3) (Bennett and Anderson, 1978).

Table 24.3 THE DISTRIBUTION OF AFLATOXIN IN WET-MILLED MAIZE FRACTIONS FROM NATURALLY CONTAMINATED MAIZE (Bennet and Anderson, 1978)

Fraction	Yield % of grain	Aflatoxin B_1 (μg/kg)	Percentage of total
Whole maize	100	120	
Steepwater and solubles	7.2	610	39.5
Germ	5.2	140	6
Fibre	12.5	340	38
Gluten	10.7	140	13
Middling	9.2	2.5	2.5
Starch	53.7	2.2	1

Maize is also susceptible to pre-harvest invasion by *Fusarium* species and can become contaminated with *Fusarium* mycotoxins such as zearalenone, deoxynivalenol and T-2 toxin. A survey of maize produced in the USA detected zearalenone in 6 out of 576 samples at levels ranging from 0.45–0.80 mg/kg (Shotwell *et al.*, 1971). In another survey of USA maize in 1972, following climatic conditions favouring the growth of *Fusarium*, zearalenone was found in 17% of 223 samples at levels ranging from 0.1 to 5.0 mg/kg (Eppley, 1974). It has also been reported on maize grown in France (Collet and Regnier, 1976) and Yugoslavia (Balzar *et al.*, 1976).

Deoxynivalenol can also occur in maize. In a survey of maize produced in the USA in 1977 it was detected in 24 out of 52 samples at levels ranging from 0.5 to 10.0 mg/kg (Vesonder *et al.*, 1978). Deoxynivalenol can also co-occur with zearalenone as demonstrated by the re-examination of nine zearalenone positive samples from the survey carried out by Eppley *et al.* in 1972 (Vesonder *et al.*, 1979). The occurrence of mycotoxins on maize and maize by-products, mostly from North America, imported into Sweden during 1982–1984 for use in animal feeding stuffs has been investigated. A total of 104 samples of whole maize, maize gluten, maize meal and maize grits were analyzed. Zearalenone, deoxynivalenol and aflatoxin B_1 were detected in respectively 14%, 28% and 14% of samples. Somewhat surprisingly only in a few cases were two mycotoxins found to contaminate the same sample (Olsen *et al.*, 1986). This finding is in contrast to experience with distillers' maize imported into England in 1983–1984. This product is the dried residue remaining after the alcoholic fermentation of maize. Often poor quality mould damaged grain is used for this industrial process. Some mycotoxins are able to withstand alcoholic fermentation and become concentrated in the insoluble residue known as distillers' maize which is used as an animal feed. This product has been examined by ADAS for mycotoxins and out of 16 samples analysed, 8, 4 and 3 respectively contained deoxynivalenol, zearalenone and aflatoxin, sometimes in combination.

Lesser known mycotoxins have also been associated with maize; these include secalonic acid D and citreoviridin. Secalonic acid D has been reported as a natural contaminant of maize dust (Ehrlich *et al.*, 1982). It is produced by *Penicillium oxalicum*, which is one of the principal fungal contaminants occurring in maize in the Mid-Western United States. Secalonic acid D was detected in 9 out of 12 samples of dust collected from grain elevators at levels ranging from 0.3 to 4.5 mg/kg. Citreoviridin is a neurotoxic mycotoxin produced by *Eupenicillium ochrosalmoneum* which invades maize before harvest. Analysis of eight samples of maize harvested in Southern Georgia, USA in 1983 revealed six to contain citreoviridin at levels ranging from 0.012–2.750 mg/kg (Wicklow *et al.*, 1988).

OIL SEEDS

It was the occurrence of aflatoxin in groundnut (Allcroft and Carnaghan, 1962) that led to the discovery of mycotoxins. Since then several other oil seeds used in animal feeding stuffs have been found to contain aflatoxin.

Groundnuts have been examined on numerous occasions and have frequently been found to contain high levels of aflatoxin, often exceeding 100 µg/kg. When the plants are affected by drought they are especially susceptible to insect damage and fungal invasion by *Aspergillus flavus* leading to aflatoxin contamination before harvest. Fungal spoilage can continue during drying if it is prolonged beyond five days and can also occur during storage if the groundnuts have not been dried to 9% moisture content or less. Further spoilage is also possible if bagged groundnuts awaiting shipment are exposed to rain. Aflatoxin is not the only mycotoxin found in groundnuts; they can also be contaminated with cyclopiazonic acid, a metabolite produced by *Penicillium cyclopium* and *Aspergillus flavus*. These mycotoxins were both detected by Lansden and Davidson (1983) in peanuts (groundnuts) produced in the South Eastern USA.

Cottonseed is widely used in animal feeding stuffs after removal of the oil when it is known as cottonseed cake. Aflatoxin is the only mycotoxin reported to occur in cottonseed and it has been found at levels of up to 300 mg/kg in USA cottonseed (Marsh *et al.*, 1973). The contamination of cottonseed with aflatoxin B_1 in the USA has necessitated controls over its use in feeding stuffs for dairy cattle in order to limit its carry over into the milk as aflatoxin M_1, which must not exceed 0.5 µg/litre (Price *et al.*, 1985).

Palm kernels and the related oilseed babassu are also liable to aflatoxin contamination. Surveys of palm kernels produced in West African countries have identified a high incidence of positive samples containing up to 1600 µg/kg (Cornelius, 1984; Cornelius and Maduagwu, 1986). A survey of palm kernel expeller cake in the UK identified 98/98 samples positive for aflatoxin B_1 at levels ranging from 10 to 400 µg/kg (MAFF, 1987).

Soya is a major constituent of animal feeding stuffs in many countries but fortunately it has rarely been found contaminated with mycotoxins. A survey of 866 samples of soya beans produced in the USA in 1964 and 1965 identified only two positive samples containing 7 and 10 µg/kg aflatoxin B_1 (Shotwell *et al.*, 1969). In another survey of USA produced soya beans in 1975 aflatoxin was not detected in 180 samples examined (Shotwell *et al.*, 1977).

Other oil seeds found to be contaminated with aflatoxin are copra and sunflower seeds. Imports of copra into the USA were discontinued after surveillance by the FDA detected aflatoxin in 63 of 72 samples of copra (Stoloff, 1976). A survey of copra imported into Finland also found it to be contaminated with 10–100 µg/kg aflatoxin (Krogh *et al.*, 1970). Traces (less than 10 µg/kg) of aflatoxin have been found in 3 out of 10 samples of sunflower seeds examined in the UK (MAFF, 1987).

PEAS

Dried peas are occasionally used in animal feeding stuffs and a few samples have been examined by ADAS. One sample of mouldy peas taken from a farm store in 1979 was found to contain aflatoxin B_1 (25 µg/kg), sterigmatocystin (1600 µg/kg) and ochratoxin A (<250 µg/kg) (unpublished observations, A.E. Buckle). More recently another sample of peas examined by ADAS was also found to contain aflatoxin B_1, sterigmatocystin and ochratoxin A (A. Lyne, unpublished observations).

ANIMAL FEEDING STUFFS

Compound feeding stuffs have often been found to contain mycotoxins due to the inclusion of contaminated ingredients. Aflatoxin, ochratoxin A and zearalenone are the most commonly reported. Maize has been identified as the source of zearalenone in mixed feeds in Canada (Funnel, 1979) and aflatoxin in poultry feeds in Saudi Arabia (Ewaidah, 1988). Groundnut was identified as the source of aflatoxin in compound feeding stuffs examined in Queensland, Australia (Connole *et al.*, 1981). In the UK groundnut was identified as the major source of aflatoxin contamination in dairy feeding stuffs. Concern about the levels of aflatoxin M_1 detected in milk and dairy products in the UK (MAFF, 1980) were attributed to the presence of aflatoxin B_1 in dairy feeding stuffs. This led to a temporary ban on the import of groundnut into the UK in 1981. This ban was later relaxed in 1982 to allow the import of groundnut provided that the aflatoxin B_1 content did not exceed 50 µg/kg. Since December 1988 this level has been raised to 200 µg/kg in line with all EEC countries. (The Feedingstuffs Regulations, 1988.)

Detection

Analytical methods for the determination of mycotoxins in human food or animal feedingstuffs may be required for a number of reasons. For instance if a statutory or advisory limit has been set it follows that a reliable and tested method must be available to enforce this limit effectively. However, even in the absence of any such limit, the grain trade or feed compounder handling raw materials may still require assurance that these commodities are of the desired quality and this may include a requirement for mould and mycotoxin assessment. Veterinary investigation of problems occurring in the production or health of farm livestock may on occasion strongly suggest implication of a toxic agent in the feed supplied and one of a number of possibilities to consider is the presence of mycotoxins. Thus, the ability to screen for the presence of a number of mycotoxins is essential in attempting to identify the causative agent. Lastly the availability of suitable methodology is essential for carrying out studies on the natural occurrence of mycotoxins and the levels at which they may contaminate human food or animal feeding stuffs.

In the UK no statutory limits have been set for mycotoxins in human food although a suggested limit of 10 µg/kg total aflatoxin (the sum of aflatoxins B_1, B_2, G_1 and G_2) has been proposed for nuts and nut products (MAFF, 1987). Limits have however been set in the UK for aflatoxin B_1 in animal feeding stuffs and these are summarized in Table 24.4. No agreed approach worldwide has been adopted for setting limits for mycotoxins and Van Egmond (1989) has reported the results of a survey carried out in 1986/7 of the limits existing worldwide. He established that approximately 60 nations had set or proposed limits for aflatoxins in human food. Fewer countries had regulations for aflatoxin in animal feedingstuffs but where they exist most are in the range 10 to 50 µg/kg which may be either for aflatoxin B_1 alone in some instances or for total aflatoxins in others. When food containing aflatoxin B_1 is ingested by cattle the mycotoxin is metabolized to aflatoxin M_1, and can be excreted in the milk. Limits for this metabolite have also been set in some countries. However, very few countries have set limits for mycotoxins other than for aflatoxin. Those that were identified by Van Egmond are given in Table 24.5. Of these mycotoxins, limits have been set by eight countries for patulin, chiefly in fruit juice, while ochratoxin A is controlled in four countries though not in the UK despite being the most commonly found mycotoxin in home-grown stored grain.

Aflatoxins are in general considered to be the most important of the mycotoxins in view of their high toxicity, particularly that of aflatoxin B_1, and their widespread occurrence in nature. In the UK, however, climatic conditions do not favour the development of aflatoxins and they rarely occur in cereal and other crops grown in this country (Buckle, 1983) as discussed earlier. However, they are still of major concern to the feed compounders and the farming industry as many imported raw

Table 24.4   PRESCRIBED LIMITS FOR AFLATOXIN B_1 GIVEN IN THE UNITED KINGDOM FEEDING STUFFS REGULATIONS, 1988

Substances	Feeding stuffs	Maximum content (mg/kg) of feeding stuffs referred to a moisture content of 12%
Aflatoxin B_1	Straight feeding stuffs	0.05
	Complete feeding stuffs for cattle, sheep and goats (except dairy animals, calves, lambs and kids)	
	Complete feeding stuffs for pigs	0.05
	and poultry (except piglets and chicks)	0.02
	Other complete feeding stuffs	0.01
	Complementary feeding stuffs for cattle, sheep and goats (except complementary feeding stuffs for dairy animals, calves and lambs)	0.05
	Complementary feeding stuffs for pigs and poultry (except young animals)	0.03
	Other complementary feeding stuffs	0.01
	Groundnut, copra, palm-kernel, cotton seed, babassu, maize and products derived from the processing thereof	0.2

Table 24.5 LIMITS FOR MYCOTOXINS OTHER THAN AFLATOXINS

Mycotoxin	Range (μg/kg)	Country	Commodities
Deoxynivalenol	5–4000	Canada, Roumania, USA, USSR	Wheat, wheat products
Ochratoxin A	1–50	Brazil, Denmark, Czechoslovakia, Roumania	Beans, cereals, pork kidneys, infant foods
Patulin	20–50	Eight countries	Fruit juices, all foods, feedstuffs
Phomopsin	5	Australia	Lupin products
T-2 toxin	100	USSR	Grains
Zearalenone	30–1000	Brazil, Roumania, USSR	Cereals, fats, beans

materials, such as maize, palm kernels and groundnut, which are used as components of animal feedingstuffs are prone to contamination with aflatoxins.

Mycotoxins are natural products which may be toxicologically active at very low levels and their determination in food commodities presents the analyst with a particularly demanding task. The precision of reliable quantitative analytical results is therefore time consuming and expensive so it is important that the objectives for carrying out such analyses are clearly defined before they are undertaken. For example, it might be decided that no detectable amount of a mycotoxin would be permitted in a particular animal feeding stuff. In this instance the level of sensitivity for detecting the mycotoxin would need to be agreed but apart from this there would be no requirement for quantitative results and a simple screening method would be the most appropriate. Conversely, enforcement of a statutory limit would necessitate the availability and use of a reliable, fully quantitative analytical procedure. The difference in cost of the two analytical procedures could be considerable.

To appreciate the difficulties involved in developing suitable analytical methods each step in the procedure can be considered in turn. This is, in fact, how the analyst would approach the problem. He would then test and evaluate each stage before making the candidate method available for scrutiny by other analysts. Publication of the method would follow and finally the method should be subjected to trials undertaken simultaneously by a number of accredited laboratories.

To ensure that any analysis is meaningful it is essential that the sample tendered for analysis is representative of the bulk from which it has been taken. This is sometimes very difficult to achieve because of the way moulds colonize crops or commodities. Fungi often grow in damp patches and this may result in very uneven mould growth which is reflected in the distribution of any subsequent mycotoxin production. On receipt of a sample, mycotoxins must be extracted and this is usually done by grinding or macerating with a solvent or mixture of solvents. The choice of solvent will depend both on the properties of the individual mycotoxins and the commodity in which they are present. The effectiveness of a solvent is usually checked by adding a known amount of pure mycotoxin to an uncontaminated sample of the appropriate commodity (spiking) and determining how much of this can be recovered and accounted for by the method under evaluation. However, a mycotoxin which has developed naturally may be bound by physical or chemical forces much more firmly than that of a spiked standard so the use of such a test procedure may give unjustified confidence in the effectiveness of the method. This is illustrated by the data presented in Table 24.6. Each solvent appears to give acceptable recoveries of aflatoxins spiked onto peanuts, maize or mixed feeds. However, extraction is less effective for all the solvents tested when the mycotoxin arises naturally. In this study, chloroform or acetonitrile appear to be more effective than methanol.

Because mycotoxins are natural products and are often present in very low amounts, many other food constituents or other fungal metabolites may be extracted simultaneously and clean-up of the extract solution metabolites and separation of the compounds of interest is necessary prior to detection and quantitation. This clean-up procedure may be complex and time consuming. Sensitive and selective detection systems are then required especially where clean-up has not been particularly effective. Finally it is important to ensure that the identity of the suspected mycotoxin is beyond question and confirmatory procedures are important although not invariably used.

Table 24.6   EFFICIENCY OF THREE SOLVENTS FOR EXTRACTION AND RECOVERY OF AFLATOXIN B_1 FROM NATURALLY CONTAMINATED OR SPIKED COMMODITIES

Sample	Aflatoxin B_1 (μg/kg)				
	Natural	Added	After extraction by:		
			Methanol	Chloroform	Acetonitrile
Animal feed	<1	10	7.2	9.0	9.8
Dairy feed	16	0	5.8	13.2	14.8
Maize 1	<1	10	7.2	9.0	9.8
Maize 2	52	0	8	38	47
Peanut butter 1	<1	200	188	175	206
Peanut butter 2	200	0	67	212	204

(Data from M. Howell, RHM Research and Engineering Ltd, High Wycombe, Bucks.)

A large number of analytical methods have been developed for determination of aflatoxins and a selection of these is given in Table 24.7 (Egan, 1982). The mini-column method is relatively simple but non-quantitative and has been used widely as a rapid screening method for detection of total aflatoxins in a variety of samples. Because random screening usually results in most samples proving negative, such a method may provide the most cost effective approach in the circumstances. Any samples found positive may be subsequently re-examined if required, using a quantitative method such as the BF or the CB/ISO/BSI method. This latter group of methods are all very similar, being based on extraction using chloroform and water. They have been applied widely for determination of aflatoxins in animal feeding stuffs and either TLC or HPLC can be used for detection and quantitation. The Patterson and Roberts (1979) screening method which is based on TLC can be used for detection of a number of different mycotoxins including aflatoxins. It has been widely used in the UK by a number of institutes including the regional laboratories of ADAS.

Table 24.7 METHODS OF ANALYSIS FOR AFLATOXINS

Method	Scope	Status	Limits (μg/kg)
Mini column	Nuts, maize, mixed feeds	S	10
CB/ISO/BSI	Nuts, oilseeds, grain, mixed feeds	Q	2–10
Patterson/Roberts	Many products	S (TLC)	1–10
		SQ (HPLC)	0.5–5
BF	Nut products, peanuts	Q (TLC)	5
		Q (HPLC)	1
Immunoassays	Nuts/cereals	S/SQ	1–5
Immunoaffinity columns	Many products?	S→Q	1–10

S = Screening method; SQ = Semi-quantitative method; Q = Quantitative method.

Recently, specific antisera to aflatoxins have been raised which have opened up the possibility of highly specific and rapid screening methods. These antibodies are now being exploited commercially in rapid immunoassay test kits with which it is theoretically possible to screen a large number of samples at minimal cost. Expertise is still required in obtaining satisfactory results and a full evaluation of these methods is required internationally before they are likely to be widely accepted, though this technology represents a major breakthrough for mycotoxin analysis. These antibodies have been bound to inert base materials and

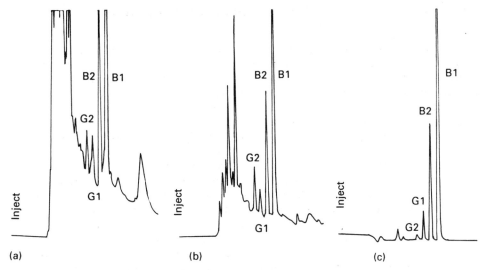

Figure 24.1 HPLC chromatogram comparing different clean-up procedures for determination of aflatoxin in a compound animal feeding stuff. (*a*) Column chromatography. (*b*) Gel permeation chromatography. (*c*) Immunoaffinity column

incorporated into immunoaffinity columns. These can be used to clean-up crude extracts prepared by any of the established extraction methods. The dirty extracts are added directly to the column without any prior treatment. Only aflatoxins are held in the column by the antibodies while other contaminants pass through. Any aflatoxin present can then be eluted and determined by conventional methods, e.g. TLC or HPLC. Figure 24.1 shows how effective these columns can be when compared with more traditional methods.

Studies were carried out at the ADAS Central Science Laboratory at Slough and form part of a commissioned research and development programme aimed at developing analytical methods and investigating the mycotoxins which may occur naturally in home-grown cereals. Part of this study is concerned with developing and improving methods for detection and determination of a large number of mycotoxins. In the course of this work a technique known as Gel Permeation Chromatography is being studied as a clean-up method for a range of mycotoxins. Its use for the determination of aflatoxins is shown in Figure 24.1, and while not quite as effective as the immunoaffinity columns it still represents a marked improvement on previous methods. Its use has recently been extended to the simultaneous determination of ochratoxin A, citrinin, zearalenone, zearalenol and sterigmatocystin, and further mycotoxins are being investigated.

Discussion of methods in this chapter has been confined mostly to those for determination of aflatoxins. However when extracts of naturally contaminated cereals or animal feeding stuffs are examined using TLC or HPLC many other fungal metabolites are indicated by the large number of TLC spots seen on the plates or by the number of peaks on chromatograms. Most of these metabolites are of unknown identity and their toxicological significance is not understood. While methods exist for some of these, many cannot be determined at present.

The significance of mycotoxins in animal feeding stuffs

The presence of mycotoxins indicates that the feed and/or its constituents have at some stage been subjected to fungal spoilage. Whether those mycotoxins present exert an adverse effect when consumed by livestock will obviously depend on the amount of mycotoxin present, the quantity fed and the age and class of stock to which it is fed. Some mycotoxins such as deoxynivalenol (vomitoxin) cause feed rejection or induce vomiting in pigs (Vesonder *et al.*, 1976, 1981), whereas poultry appear unaffected by consuming feed containing up to 5000 μg/kg of deoxynivalenol (Hamilton *et al.*, 1985).

The toxicity of mycotoxins to all classes of livestock has been demonstrated experimentally with much higher concentrations of mycotoxins than would occur in naturally contaminated products. However, these experiments generally use single and pure mycotoxins, whereas naturally contaminated products may often contain a mixture of mycotoxins which can act synergistically to enhance their toxicity at low concentrations (Huff *et al.*, 1988). Therefore the toxicity threshold is likely to be lower in practice than indicated by some experimental data. It is not intended to review the vast quantity of experimental data on the subject in this chapter. Instead, a few cases where mycotoxins in the feed have been implicated or were strongly suspected to have produced disease or adversely affected production are given as examples.

Cattle

The now classic and much quoted outbreak of Turkey X Disease (Blount, 1961) led to the discovery of aflatoxin in groundnut. About the same time groundnut from Brazil was implicated with a mysterious illness affecting a herd of Hereford x Friesian steers of 1½–2 years (Clegg and Bryson, 1962). Their livers were abnormal showing extensive bile duct proliferation. The groundnut was toxic when fed to ducklings and contained *Aspergillus flavus*. Although, at the time, the groundnut was not examined for aflatoxin it appears, with the benefit of hindsight, that it was most probably present and the authors were unknowingly the first to describe the symptoms of aflatoxicosis in cattle. This case bears a close resemblance to that reported in a herd of steers in Georgia, USA (Colvin *et al.*, 1984) which had been fed maize found to contain 1500 μg/kg aflatoxin (aflatoxin B_1, 1380 μg/kg and aflatoxin B_2, 120 μg/kg). Some of the steers lost condition and died and their livers showed bile duct proliferation which is characteristic of aflatoxicosis.

Aflatoxin in dairy feeding stuffs can reduce milk yields. An investigation by ADAS into the cause of lower than predicted yields from a Friesian herd fed a complete diet identified the groundnut component which contained aflatoxin B_1 (average 680 μg/kg) as the most likely cause of the problem (Stuart-Jones and Ewart, 1979). The groundnut comprised 7% of the dry matter of the diet. The dry matter intake was lower than expected, many cows scoured and milk production was below predicted levels. When the groundnut was omitted from the complete diet the scouring was reported to have stopped, appetites improved and milk production increased sharply though failing to reach the predicted level for the herd.

A case of aflatoxicosis in calves on a farm in Sweden has been attributed to the consumption of barley found to contain up to 8200 μg/kg of aflatoxin B_1 (Holmberg *et al.*, 1983). The barley had been treated with formic acid as a grain preservative which had failed to prevent spoilage by *Aspergillus flavus*. Ochratoxin A (1000 μg/kg) was also detected in the barley.

Sterigmatocystin is another mycotoxin sometimes present in mouldy cereals. It is produced by *Aspergillus vericolor* and its chemical structure is similar to aflatoxin. It has been found in home-grown grain in the UK on several occasions by ADAS (Buckle, 1983) and in a mixed feed for dairy cows containing maize, cottonseed and a protein mix in the USA (Vesonder and Horn, 1985). The cattle eating this feed exhibited bloody diarrhoea, with a subsequent loss of milk production, some of the animals died. The feed contained a mixed fungal flora of *Aspergillus* species with *Aspergillus versicolor* and *Aspergillus candidus* dominant. A sample of the feed contained 7.75 mg/kg of sterigmatocystin.

Pigs

There are many reported cases of mycotoxicosis in pigs which have been attributed to the consumption of feed naturally contaminated with mycotoxins including aflatoxin (Ketterer *et al.*, 1982), ochratoxin A (Krogh *et al.*, 1973) and zearalenone (Blaney *et al.*, 1984). Maize is an important constituent of pig feeds and reference to its contamination with zearalenone, which is produced by some species of *Fusarium*, has been made earlier. The cases reported by Blaney (1984) occurred in Northern Queensland, Australia and are good examples of the harmful oestrogenic effects of zearalenone on gilts; in these cases the sources of the zearalenone were maize and sorghum.

Consumption of these cereals produced enlarged teats and the characteristic signs of oestrus such as red, swollen vulvas. In several cases both rectal and vaginal prolapses occurred resulting in the death of 25 pigs on one of the farms. There have been many other similar cases of hyperoestrogenic activity in gilts attributed to zearalenone contamination of cereals reported in Europe and North America. An interesting example of the legal consequences of selling feeding stuffs contaminated with mycotoxins is provided by an alleged case of mycotoxicosis in pigs that took place on a farm in Canada (Schiefer and O'Ferrall, 1981). Prolapse of the rectum and uterus occurred in an unspecified number of animals, some of them aborted in late pregnancy and others had red swollen vulvas. Several months after these symptoms were first noticed the feed company supplying the farm had mycotoxin analyses done on some of the feeds with negative results for several mycotoxins including zearalenone. When the farmer refused to pay for the feed, the feed company took the case to court which, on the basis of the Sale of Goods Legislation, ruled that the company supplying the feed was responsible for the losses that occurred even though the cause of the disease was never clearly established.

Poultry

Numerous cases of mycotoxicosis in poultry caused by eating naturally contaminated feeding stuffs have been reported in many countries. They have been attributed to various mycotoxins including aflatoxin (Blount, 1961) and ochratoxin A of which nine separate cases occurring in the USA are described by Hamilton *et al.* (1982). Ochratoxin A is nephrotoxic and causes the kidneys to become swollen and pale. When present in layers feed it has been associated with reduced egg production and poor egg shell quality. In broiler feed it has been associated with poor growth rate and a reduction in feed conversion efficiency.

Reference has previously been made to the occurrence of *Fusarium* mycotoxins in cereals produced in the UK. Although they have been found at low levels on occasions there is little evidence that they occur in sufficient amounts in home-grown cereals to cause problems in livestock production in the UK. Notwithstanding the findings of these surveys there is some evidence implicating *Fusarium* toxins on wheat grown in Scotland and imported maize that was fed to broilers. Several cases of sub-optimal growth, poor feathering and behavioural abnormalities occurred during the winter of 1980–81 on broiler farms in Scotland. Both the wheat and maize contained diacetoxyscirpenol and deoxynivalenol (which were confirmed by gas chromatography). Extracts of the feed from farms with affected birds were cytotoxic as were several pure *Fusarium* mycotoxins examined, whereas feed samples from unaffected birds showed no cytotoxicity (Robb *et al.*, 1982).

An interesting case of mycotoxicosis causing the estimated loss of 9075 sandhill cranes (*Grus canadensis*), that occurred in the vicinity of a major roost site in Texas, USA has also been attributed to *Fusarium* mycotoxins. The cranes had fed on waste peanuts from which *Fusarium* was the major fungal genus isolated. An isolate of *F. compactum* from the waste peanuts produced the trichothecene isoneosolaniol when cultured on peanuts; culture extracts were highly toxic to HEp2 cells. Extracts of the waste peanuts were also cytotoxic and contained neosolaniol, alothough isoneosolaniol was not detected. Isoneosolaniol showed greater cytotoxicity than neosolaniol. The authors concluded that the most logical cause of the sandhill crane intoxication appeared to be due to contamination of the peanuts with trichothecenes produced by the *Fusarium* species present (Cole *et al.*, 1988).

This case is particularly significant because it is the first time that *Fusarium* mycotoxins have been reported on peanuts, though it should be emphasized that they had been discarded and exposed to rehydration by snow and rain in conjunction with freezing and thawing. It illustrates the possibilities for mycotoxin production under extreme conditions that may lead to the formation of mycotoxins hitherto unreported as naturally occurring. It also adds to our knowledge about the toxic effects of trichothecene mycotoxins and serves to caution against the use of commodities which have been subjected to fungal damage even when no known mycotoxins can be identified. The significance of finding a mycotoxin in a feeding stuff does not necessarily reside in the harm that could arise from ingesting the mycotoxin which may only be present in 'harmless' amounts. It should be borne in mind that there is always a risk of other more potent, yet unidentified, mycotoxins being present. It is therefore prudent to screen routinely all new materials used in animal feeding stuffs and ensure they remain mycotoxin-free whenever possible.

References

Allcroft, R. and Carnaghan, R.B.A. (1962) Groundnut toxicity – *Aspergillus flavus* toxin (aflatoxin) in animal products: Preliminary communication. *The Veterinary Record*, **74**, 863–864

Balzar, I., Muzic, S. and Bogdanic, C. (1976) Natural contamination of corn with mycotoxins in Yugoslavia. Presented at *The Third IUPAC Symposium 'Mycotoxins in Foodstuffs'*, Paris, France (Abstract No. 5)

Bennett, G.A. and Anderson, R.A. (1978) Distribution of aflatoxin and/or zearalenone in wet-milled corn products: A review. *Journal of Agricultural and Food Chemistry*, **26**, 1055–1060

Blaney, B.J., Bloomfield, R.C. and Moore, C.J. (1984) Zearalenone intoxication of pigs. *Australian Veterinary Journal*, **61**, 24–27

Blount, W.P. (1961) Turkey 'X' disease. *Journal of the British Turkey Federation*, **9**, 55–58, 61–77

Buckle, A.E. (1983) The occurrence of mycotoxins in cereals and animal feedstuffs. *Veterinary Research Communications*, **7**, 171–186

Clegg, F.G. and Bryson, H. (1962) An outbreak of poisoning in store cattle attributed to Brazilian groundnut meal. *The Veterinary Record*, **74**, 992–994

Cole, R.J., Dorner, J.W., Gilbert, J., *et al.* (1988) Isolation and identification of trichothecenes from *Fusarium compactum* suspected in the aetiology of a major intoxication of sandhill cranes. *Journal of Agricultural and Food Chemistry*, **36**, 1163–1167

Collet, J.C. and Regnier, J.M. (1976) Contamination par mycotoxines de mais conserve en cribs et visiblement alteres. Presented at *The Third IUPAC Symposium 'Mycotoxins in Foodstuffs'*, Paris, France (Abstract No. 5)

Colvin, B.M., Harrison, L.R., Gosser, H.S. and Hall, R.F. (1984) Alfatoxicosis in feeder cattle. *Journal of the American Veterinary Medical Association*, **184**, 956–958

Connole, M.D., Blaney, B.J. and McEwan, T. (1981) Mycotoxins in animal feeds and toxic fungi in Queensland 1971–80. *Australian Veterinary Journal*, **57**, 314–318

Cornelius, J.A. (1984) News item: aflatoxins in palm kernels. *Oil Palm News*, **28**, 3–5

Cornelius, J.A. and Maduagwu, E.N. (1986) The possible effect of aflatoxin contamination on the marketing of palm kernels from West Africa. *International Biodeterioration Supplement*, **22**, 95–101

Egan, H. (ed.) (1982) In *Environmental Carcinogen Selected Methods of Analysis. Volume 5 – Some Mycotoxins*. IARC Scientific Publications No. 44. Lyon, International Agency for Research on Cancer

Eppley, R.M., Stoloff, L., Trucksess, M.W. and Chung, C.W. (1974) Survey of corn for *Fusarium* toxins. *Journal of the Association of Official Analytical Chemists*, **57**, 632–635

Ehrlich, K.C., Lee, L.S., Ciegler, A. and Palmgren, M.S. (1982) Secalonic acid D: natural contaminant of corn dust. *Applied and Environmental Microbiology*, **44**, 1007–1008

Ewaidah, E.H. (1988) Survey of poultry feeds for aflatoxins from the Riyadh region. *Arab Gulf Journal of Scientific Research Agricultural and Biological Sciences*, **B6**, 1–7

Funnell, H.S. (1979) Mycotoxins in animal feedstuffs in Ontario. *Canadian Journal of Comparative Medicine*, **43**, 243–246

Gilbert, J., Shepherd, M.J. and Startin, J.R. (1983) A survey of the occurrence of the trichothecene mycotoxin deoxynivalenol (Vomitoxin) in UK grown barley and in imported maize by combined gas chromatography – mass spectrometry. *Journal of the Science of Food and Agriculture*, **34**, 86–92

Hacking, A. and Biggs, N.R. (1979) Aflatoxin B_1 in barley in the UK. *Nature (London)*, **282**, 128

Hagler, W.M., Jr., Tyczkowska, K. and Hamilton, P.B. (1984) Simultaneous occurrence of deoxynivalenol, zearalenone and aflatoxin in 1982 scabby wheat from the Mid-Western United States. *Applied and Environmental Microbiology*, **47**, 151–154

Hamilton, P.B., Huff, W.E., Harris, J.R. and Wyatt, R.D. (1982) Natural occurrences of ochratoxicosis in poultry. *Poultry Science*, **61**, 1832–1841

Hamilton, R.M.G., Trenholm, H.L., Thompson, B.K. and Greenhalgh, R. (1985) The tolerance of white leghorn and broiler chicks, and turkey poults to diets that contained deoxynivalenol (vomitoxin) – contaminated wheat. *Poultry Science*, **64**, 273–286

Holmberg, T., Pettersson, H., Nilsson, N.G., Goransson, B. and Grossman, R. (1983) A case of aflatoxicosis in fattening calves caused by aflatoxin formation in inadequate formic acid treated grain. *Zentralblatt fur Veterinarmedizin Reihe A*, **30**, 656–663

Huff, W.E., Kubena, L.F., Harvey, R.B. and Doerr, J.A. (1988) Mycotoxin interactions in poultry and swine. *Journal of Animal Science*, **66**, 2351–2355

Ketterer, P.J., Blaney, B.J., Moore, C.J., McInnes, I.S. and Cook, P.W. (1982) Field cases of aflatoxicosis in pigs. *Australian Veterinary Journal*, **59**, 113–117

Krogh, P., Hald, B. and Korpinen, E.L. (1970) Occurrence of aflatoxin in groundnut – and copra – products imported to Finland. *Nordiskveterinarmedicin*, **22**, 584–589

Krogh, P., Hald, B. and Pedersen, E.J. (1973) Occurrence of ochratoxin A and citrinin in cereals associated with mycotoxic porcine nephropathy. *Acta Pathologica et Microbiologica Scandinavica, Section B*, **81**, 689–695

Krogh, P., Hald, B., Englund, P., Rutqvist, L. and Swahn, O. (1974) Contamination of Swedish cereals with ochratoxin A. *Acta Pathologica et Microbiologica Scandinavica, Section B*, **82**, 301–302

Landsden, J.A. and Davidson, J.I. (1983) Occurrence of cyclopiazonic acid in peanuts. *Applied and Environmental Microbiology*, **45**, 766–769

Lillehoj, E.B., Fennell, D.I. and Kwolek, W.F. (1976) *Aspergillus flavus* and aflatoxin in Iowa corn before harvest. *Science*, **193**, 495–496

MAFF (1980) *Survey of Mycotoxins in the United Kingdom: The Fourth Report of the Steering Group on Food Surveillance*. The Working Party on Mycotoxins. Food Surveillance Paper No. 4. HMSO, London

MAFF (1987) *Mycotoxins: The Eighteenth Report of the Steering Group on Food Surveillance*. The Working Party on Naturally Occurring Toxicants in Food: Sub-Group on Mycotoxins. Food Surveillance Paper No. 18, HMSO, London

Marsh, P.B., Simpson, M.E., Craig, G.O., Donoso, J. and Ramey, H.H., Jr. (1973) Occurrence of aflatoxins in cotton seeds at harvest in relation to location of growth and field temperatures. *Journal of Environmental Quality*, **2**, 276–281

Olsen, M., Pettersson, H., Sandholm, K., Holmberg, T., Rutqvist, L. and Kiessling, K.H. (1986) The occurrence of aflatoxin, zearalenone and deoxynivalenol in maize imported into Sweden. *Swedish Journal of Agriculture*, **16**, 77–80

Osborne, B.G. and Willis, K.H. (1984) Studies into the occurrence of some trichothecene mycotoxins in UK home-grown wheat and in imported wheat. *Journal of the Science of Food and Agriculture*, **35**, 579–583

Price, R.L., Paulson, J.H., Lough, O.G., Gingg, C. and Kurtz, A.G. (1985) Aflatoxin conversion by dairy cattle consuming naturally-contaminated whole cottonseed. *Journal of Food Protection*, **48**, 11–15

Patterson, D.S.P. and Roberts, B.A. (1979) Mycotoxins in animal feedstuffs: sensitive thin layer chromatographic detection of aflatoxin, ochratoxin A, sterigmatocystin, zearalenone and T-2 toxin. *Journal of the Association of Official Analytical Chemists*, **62**, 1265–1267

Robb, J., Kirkpatrick, K.S. and Norval, M. (1982) Association of toxin-producing fungi with disease in broilers. *The Veterinary Record*, **111**, 389–390

Scott, P.M., Van Walbeek, W., Harwig, J. and Fennell, D.I. (1970) Occurrence of a mycotoxin, Ochratoxin A, in wheat and isolation of ochratoxin A and citrinin producing strains of *Penicillium viridicatum*. *Canadian Journal of Plant Science*, **50**, 583–585

Scudamore, K.A., Atkin, P.M. and Buckle, A.E. (1986) Natural occurrence of the naphthoquinone mycotoxins, xanthomegnin, viomellein and vioxanthin in cereals and animal feedstuffs. *Journal of Stored Products Research*, **22**, 81–84

Shank, R.C., Wogan, G.N., Gibson, J.B. and Nondasuta, A. (1972) Dietary aflatoxins and human liver cancer. II. Aflatoxins in market foods and foodstuffs of Thailand and Hong Kong. *Food and Cosmetics Toxicology*, **10**, 61–69

Schiefer, H.B. and O'Ferrall, B.K. (1981) Alleged mycotoxicosis in swine: Review of a court case. *Canadian Veterinary Journal*, **22**, 134–139

Shotwell, O.L., Hesseltine, C.W., Burmeister, H.R., Kwolek, W.F., Shannon, G.M. and Hall, H.H. (1969) Survey of cereal grains and soybeans for the presence of aflatoxin: II. Corn and soybeans. *Cereal Chemistry*, **46**, 454–463

Shotwell, O.L., Hesseltine, C.W., Goulden, M.L. and Vandegraft, E.E. (1970) Survey of corn for aflatoxin, zearalenone and ochratoxin. *Cereal Chemistry*, **47**, 700–707

Shotwell, O.L., Hesseltine, C.W., Vandegraft, E.E. and Goulden, M.L. (1971)

Survey of corn from different regions for aflatoxin, ochratoxin, and zearalenone. *Cereal Science Today*, **16**, 266–273

Shotwell, O.L., Hesseltine, C.W. and Goulden, M.L. (1973) Incidence of aflatoxin in Southern corn, 1969–1970. *Cereal Science Today*, **18**, 192–195

Shotwell, O.L., Goulden, M.L. and Hesseltine, C.W. (1976) Survey of US wheat for ochratoxin and aflatoxin. *Journal of the Association of Official Analytical Chemists*, **59**, 122–124

Shotwell, O.L., Goulden, M.L., Bennett, G.A., Plattner, R.D. and Hesseltine, C.W. (1977) Survey of 1975 wheat and soybeans for Aflatoxin, Zearalenone and Ochratoxin. *Journal of the Association of Official Analytical Chemists*, **60**, 778–783

Shotwell, O.L., Bennett, G.A., Goulden, M.L., Shannon, G.M., Stubblefield, R.D. and Hesseltine, C.W. (1980) Survey of 1977 midwest corn at harvest for Aflatoxin. *Cereal Foods World*, **25**, 12–14

Stoloff, L. (1976) Occurrence of mycotoxins in food and feeds. In *Mycotoxins and Other Fungal Related Food Problems* (ed. J.V. Rodricks). Washington, DC, American Chemical Society, pp. 23–50 (Advances in Chemistry Series 149)

Stuart-Jones, M.G. and Ewart, J.M. (1979) Effects on milk production associated with consumption of decorticated extracted groundnut meal contaminated with aflatoxin. *The Veterinary Record*, **105**, 492–493

Szebiotko, K., Chelkowski, J., Dopierala, G., Godlewska, B. and Radomyska, W. (1981) Mycotoxins in cereal grain. Part 1. Ochratoxin, citrinin, sterigmatocystin, penicillic acid and toxigenic fungi in cereal grain. *Die Nahrung*, **25**, 415–421

Tanaka, T., Hasegawa, A., Matsuki, Y., Lee, U.S. and Ueno, Y. (1986) A limited survey of *Fusarium* mycotoxins nivalenol, deoxynivalenol and zearalenone in 1984 UK harvested wheat and barley. *Food Additives and Contaminants*, **3**, 247–252

Trenholm, H.L., Cochrane, W.P., Cohen, H., *et al.* (1981) Survey of vomitoxin contamination of the 1980 white winter wheat crop in Ontario, Canada. *Journal of American Oil Chemists Society*, **58**, 992A–994A

Van der Merwe, K.J., Steyn, P.S. and Fourie, L. (1965) Mycotoxins II. The constitution of ochratoxins A, B, and C, metabolites of *Aspergillus ochraceus*. *Journal of the Chemical Society*, **1985**, 7083–7088

Van Egmond, H.D. (1989) Current situation on regulations for mycotoxin. Overview of tolerances and status of standard methods of sampling and analysis. *Food Additives and Contaminants*, **6**, 139–188

Vesonder, R.F., Ciegler, A., Rogers, R.F., Burbridge, K.A., Bothast, R.J. and Jensen, A.H. (1978) Survey of 1977 crop year preharvest corn for vomitoxin. *Applied and Environmental Microbiology*, **36**, 885–888

Vesonder, R.F., Ciegler, A., Rohwedder, W.K. and Eppley, R. (1979) Re-examination of 1972 midwest corn for vomitoxin. *Toxicon*, **17**, 658–660

Vesonder, R.F., Ellis, J.J. and Rohwedder, W.K. (1981) Swine refusal factors elaborated by *Fusarium* strains and identified as trichothecenes. *Applied and Environmental Microbiology*, **41**, 323–324

Vesonder, R.F. and Horn, B.W. (1985) Sterigmatocystin in dairy cattle feed contaminated with *Aspergillus vesicolor*. *Applied and Environmental Microbiology*, **49**, 234–235

Wicklow, D.T., Stubblefield, R.D., Horn, B.W. and Shotwell, O.L. (1988) Citreoviridin levels in *Eupenicillium ochrosalmoneum* – infested maize kernels at harvest. *Applied and Environmental Microbiology*, **54**, 1096–1098

Woods, A.J., Bradley Jones, J. and Mantle, P.G. (1966) An outbreak of gangrenous ergotism in cattle. *The Veterinary Record*, **78**, 742–749

25

ANIMAL PATHOGENS IN FEED

M. HINTON
Department of Veterinary Medicine, University of Bristol, Langford House, Langford, Avon, UK
and
M.J. BALE
Department of Microbiology, University of Bristol, Medical School, Bristol, UK

Introduction

Several important virus diseases of farm animals may be transmitted *via* the feed if it should contain contaminated animal products which have not been rendered safe by either heat treatment or some other means. This risk is well recognized for the notifiable diseases such as foot and mouth disease, swine fever and swine vesicular disease. As a consequence the handling and treatment of animal products destined for animal feed has been controlled in the United Kingdom by several pieces of legislation including the Waste Food Order 1973, the Processed Animal Protein Order 1989, which superseded the Diseases of Animals (Protein Processing) Order 1981, and the Meat (Sterilising and Staining) Regulations 1984 although the way in which they have been enforced by both central and local government departments has not been without its critics.

Serious bacterial infections in farm animals derived from the feed are relatively uncommon and this is reflected by the fact that the indices of the principal veterinary textbooks concerned with infectious diseases of farm animals (e.g. Buxton and Fraser, 1971; Gillespie and Timoney, 1981; Blood and Radostits, 1989), and the relevant headings in *Index Veterinarius* contain few specific references to feed borne bacterial disease.

The recent outbreaks in the United Kingdom of (1) bovine spongiform encephalopathy (BSE) and (2) of *Salmonella enteritidis* infections in poultry and in humans, who have either handled or consumed contaminated poultry meat or eggs, has emphasized the potential of contaminated animal feed for the dissemination of pathogens among farm livestock. BSE is now a notifiable disease and the Bovine Spongiform Encephalopathy (No. 2) Order 1988 requires, among other things, that protein derived from ruminant carcasses is not included in the feed of ruminant species. As far as salmonella infections are concerned the Ministry of Agriculture has issued codes of practice for the manufacturers of animal feed which aim to minimize the risks of spread of infections in poultry, and a new order, the Processed Animal Protein Order 1989, has been implemented. All businesses which process animal protein must now be registered and samples of processed material must be tested for salmonella on a daily basis.

Viral pathogens in feed

Diseases such as foot and mouth disease, swine fever and swine vesicular disease have been eradicated from the United Kingdom and it is not intended to consider them further in this review.

A variant of Newcastle Disease (ND) virus which primarily affects pigeons was associated with outbreaks of ND in poultry in the United Kingdom in 1984. The virus spread to the birds *via* feed which had been contaminated with virus excreted by feral pigeons. The problem was overcome by bird-proofing feed stores, heat treating the feed given to poultry and the vaccination of susceptible stock (Wilson, 1986).

BSE was first recognized in 1985 and has since been recognized in most counties in the United Kingdom. The epidemiology has been studied in detail by Wilesmith *et al.* (1988) and they concluded that the disease is probably caused by the same agent responsible for scrapie in sheep and that it is transmitted *via* the feed with cattle being first exposed to the pathogen when calves in 1981/82.

Bacterial pathogens in feed

The principal bacterial causes of gastro-enteritis in farm livestock include enteropathogenic *Escherichia coli*, *Clostridium perfringens*, the salmonellas and *Mycobacterium paratuberculosis* in cattle and the first two together with *Tryponema hyodysenteriae* in pigs and *Cl. perfringens* in poultry. However, with the exception of salmonella infections, there is little epidemiological evidence to suggest that the feed is a major source of these pathogens for animals.

In theory feed may become contaminated with any one of a number of pathogens. For example *Erysipelothrix insidiosa* has been recovered from Irish fish meal (Buxton and Fraser, 1971) and *Cl. perfringens* from pelleted feed (Greenham *et al.*, 1987) but it is really only two, apart from the salmonellas, which are important, although one of these (*Bacillus anthracis*) may frequently be contracted from the environment and in the case of the other (*Clostridium botulinum*) it is pre-formed toxin in the feed, rather than the organism itself, which causes illness.

BACILLUS ANTHRACIS

B. anthracis, the cause of anthrax, is a spore forming organism and it is these which are particularly resistant to adverse environmental conditions and act as a source of infection for animals.

Anthrax is principally a disease of cattle, in which it causes sudden death. It has a world-wide distribution although it is not considered to be endemic in temperate countries such as the United Kingdom. The spores may be imported in animal products containing bone meal and cattle become infected following consumption of contaminated feed. Anthrax occurs sporadically in the United Kingdom and there are usually only one or, at the most, a few cases in an outbreak, although if due care is not taken with the disposal of the carcase, the farm environment may become contaminated with spores and these can be responsible for recrudescences of infection in subsequent years.

The incidence of infection in the United Kingdom has declined to very low levels during the last decade and it is suggested that this is due in part to the inclusion of less potentially dangerous material, for example imported bone meal in animal feed.

The methods available for the isolation of *B. anthracis* have been reviewed by Carman, Hambleton and Melling (1985).

CLOSTRIDIUM BOTULINUM

Botulism is an uncommon disease of farm livestock and develops following the consumption of pre-formed toxin elaborated by *Cl. botulinum*, an organism which, like *B. anthracis*, can form spores which can survive for long periods in the environment. The organism may multiply and the toxin develop in decaying vegetation or in animal carcases and botulism may be seen in poultry reared on deep litter in which carcases have become buried as well as in cattle grazing pastures treated with poultry litter containing carcases.

The inclusion of dried poultry waste, ensiled poultry litter and brewers grains in the diets of cattle have all led to deaths due to botulism (Blood and Radostits, 1989; Neill, McLoughlin and McIlroy, 1989).

Botulism is unlikely to develop following the consumption of manufactured feed, however, since the toxin is heat labile and will be destroyed during the manufacturing process whilst the toxin is unlikely to be produced in feed after manufacture since the water activity of feed is below the minimum required for the growth of *Cl. botulinum* (Smart and Roberts, 1977).

Methods for the enumeration of *Cl. botulinum* and the detection of the toxin in feed and poultry litter have been described by Smart and Roberts (1977).

SALMONELLA SPECIES

The subject of salmonella infections in farm animals has been reviewed recently (Linton and Hinton, 1988). In all there are over 2000 serovars and infections in animals may be associated with clinical illness (salmonellosis) or they may be subclinical and remain undetectable unless laboratory investigations are undertaken (e.g. Wray, Todd and Hinton, 1987).

Cattle

Salmonellosis occurs more frequently in cattle than in any of the other farm animal species. In the United Kingdom it is associated principally with two endemic serovars namely *S. dublin*, a host adapted serovar, and *S. typhimurium*. The feed is not considered to be the principal source of either serovar, however, since they either spread from animal to animal directly or they persist in the environment in which the animal is reared.

On the other hand, the other serovars, which are responsible for a relatively small number of disease incidents each year (Linton and Hinton, 1988), are probably introduced onto farms in the feed (Richardson, 1975; Jones *et al.*, 1982).

Poultry

The two host-adapted serovars, *S. pullorum* and *S. gallinarum*, the cause of pullorum disease and fowl typhoid respectively, are not spread by the feed. Both serovars have been eradicated from the majority of poultry flocks in the United Kingdom by the use of a blood test coupled with the slaughter of infected animals and the institution of appropriate hygienic precautions.

The remaining serovars can be divided into two unequally sized groups; those which are invasive (e.g. *S. enteritidis*, *S. thompson* and *S. typhimurium*), and which cause illness in young birds, and those, which comprise the majority, which are non-invasive and rarely cause illness in the birds although they may be associated with food poisoning in humans.

The invasive serovars may be isolated occasionally from animal feed and, although this is probably not the primary source of these infections, birds may become colonized with them following the consumption of contaminated feed (Xu, Pearson and Hinton, 1988; Hinton *et al.*, 1989). The non-invasive serovars may however be isolated regularly from feed (see Williams, 1981 for an extensive review), and birds, particularly young chicks, may become infected following the consumption of feed containing small numbers of these salmonellas (Schleifer *et al.*, 1984; Hinton, 1988).

Resistance to infection appears to develop with age (Milner and Shaffer, 1952; Sadler, Brownell and Fanelli, 1969; Impey, Mead and Hinton, 1987) although the precise conditions required for infection to establish in older birds with a 'mature' gut flora have yet to be established. Certainly food poisoning can develop in fit and healthy humans following the eating of foods containing very small numbers of salmonellas and presumably similar, hitherto undefined, conditions must obtain which allow them to establish in the gastrointestinal tract of mature chickens.

Antibacterial drug resistant bacteria in feed

Drug resistance is common among the coliforms and enterococci of farm animals (Hinton, Kaukas and Linton, 1986) and, clearly, resistant organisms in these and other genera may be transmitted to other animals *via* the feed if this is contaminated with faecal bacteria (Howells and Joynson, 1975). The significance of this risk either to farm animal or human health is however not known.

Bacterial pathogens in silage

Silage is a fermented grass product of low pH. It is used in winter as a roughage feed for cattle and sheep and also for horses. Its consumption may be associated with two bacterial diseases, namely botulism and listeriosis.

CLOSTRIDIUM BOTULINUM

Botulism has been diagnosed in cattle and horses fed 'big bale' silage, the centres of which provide a satisfactory environment for the multiplication of *Cl. botulinum* organisms.

LISTERIA MONOCYTOGENES

Listeria monocytogenes is widely distributed in the soil and on vegetation and can survive in silages with low pH (<4.0) and multiply when the pH is high (>5.5), as is found in poor quality silage. Listeriosis in ruminants may present as either abortion, encephalitis or mastitis in individual adults and the development of the disease has been regularly associated with the consumption of silage. *L. monocytogenes* has also been isolated from animal feed (Skovgaard and Morgen, 1988). Methods for isolating this organism have been reviewed by Gitter (1985) and Watkins (1985). If antibiotics are included as selective agents in the media then it may be necessary to incubate agar media at 30°C rather than 37°C (Curtis, Nichols and Falla, 1989). Enzyme immunoassays for the detection of *Listeria* in foods have also been described (Mattingly *et al.*, 1988).

Indicator organisms

The Processed Animal Protein Order 1989 requires that feed is screened specifically for the presence of salmonellas. As only a small amount of feed is cultured there is a very real risk that consignments of feed contaminated with salmonellas will be deemed to be free of infection since the pathogen may be present in low numbers or irregularly dispersed in the material. In order to reduce this risk feed can be cultured for index or indicator organisms rather than for the pathogens themselves, although this is not required by law.

The culture of marker organisms does not replace direct tests for the pathogens and their use is not without its problems. The subject has been reviewed by Mossel (1982), Elliot and Collswell (1985), Pike and Ridgeway (1985) and Cox, Keller and van Schothorst (1988) with Mossel (1982) proposing that these bacteria should be classified as either 'index organisms', which give a measure of the risk of the occurrence of a pathogen, or 'indicator organisms', which provide an indication of the bacteriological quality of the product.

The most common organisms sought in this respect are the coliforms, which serve as indicator organisms, and *E. coli* which may be used as an index organism. The problem of detecting *E. coli* in foods, rather than animal feeds, has been reviewed by Abbis and Blood (1982) and they concluded that its value may be limited by the shortcomings of the cultural techniques available, particularly if the bacterial cells are severely stressed. In addition *E. coli* cannot be used satisfactorily as an index organism when (1) there is no knowledge of the microbiological status of the environment, (2) when the species occurs as a component of the natural factory environment and (3) when its numbers have no direct relationship with the numbers of the pathogen that may be present.

The enterococci (faecal streptococci) may also be used as indicators of faecal contamination. They are generally more resilient than the coliforms and therefore are likely to persist for longer than the salmonellas in feed. This may limit their use as index organisms for this pathogen. Similarly the *Bacillus* spp. have been employed as potential index organisms for *B. anthracis*.

Van Schothorst and Oosterom (1984) and Cox *et al.* (1988) summarized the results of studies carried out in rendering plants in Holland and they concluded that in this circumstance the enumeration of the Enterobacteriacae was of value for assessing the benefits of good manufacturing practices and also for determining the sites for critical control points for preventing bacterial multiplication.

Methods of isolation and detection of salmonellas

Animal feed can be expected to be contaminated with bacteria of many genera including possibly the salmonellas. These may be present in feed ingredients of animal origin including fish meal and also vegetable materials. The subject of their isolation from animal feed has been summarized by Williams (1981) and Pietzsch (1984).

The salmonellas in feed may be present in very low numbers and may be sublethally injured as a consequence of insults such as osmotic shock and heat stress (see Hurst, 1977; Ray, 1979 and Andrew and Russell, 1984, for reviews on sublethal injury). This is in distinction to those present in samples collected from patients with salmonellosis in which the organisms are likely to be numerous and actively metabolizing and this has to be taken into account when formulating a protocol for culturing samples of feed (Hartman, 1979). The principal methods in current use involve three stages, namely, pre-enrichment, selective enrichment and selective cultivation on solid medium (e.g. Edel and Kampelmacher, 1969; Hoben, Ashton and Peterson, 1973; Smith, 1977) although, despite the use of standardized methods based on this approach, variation in the success rate of different laboratories can be expected (Edel and Kampelmacher, 1969).

SAMPLE SIZE

When bacteria are present in low numbers and are also irregularly dispersed in the material, the size and number of samples of each batch that are cultured will have a crucial influence on the results obtained. When large numbers of samples have to be screened a balance has to be struck so that the maximum amount of benefit is derived from the smallest amount of work.

Silliker and Gabis (1973) compared several different sampling protocols and concluded that for a total weight of 1500 g of dried food there was little to choose between culturing 60 × 25 g samples or 3 × 500 g samples.

The Processed Animal Protein Processing Order 1981 requires that 2 × 25 g of an aggregate sample of each feed is cultured. The ways by which the aggregate sample for both loose and bagged animal protein are collected differ and are dependent on the total size of the consignment.

INHIBITORY SUBSTANCES IN FEED

Antibacterial agents in feed, for example furazolidone, may interfere with the isolation of salmonellas (Bailey *et al.*, 1985) although this problem has not been evaluated in any detail for other agents. The neutralization of antibiotics by the use of monoclonal antibodies has been developed for use with body fluid specimens (Sierra-Madero *et al.*, 1988) and these may possibly have a role when feed containing therapeutic concentrations of antibiotics have to be cultured.

PRE-ENRICHMENT AND SELECTIVE ENRICHMENT

In order to allow sub-lethally injured bacteria a chance to resuscitate they are incubated first in a non-selective broth although, when feed is added to pre-enrichment broth, the bacteria may suffer osmotic shock during rehydration. Van Schothorst *et al.* (1979) recommended that this could be reduced by adopting a two-stage process in which the dry feed, in this case dried egg, is added to a small volume of broth initially (1 : 2) before being made up to the final volume (1 : 9). Subsequently D'Aoust and Sewell (1986) failed to confirm the value of this approach when culturing animal feed. They concluded that variations in the isolation rate were probably a reflection of the uneven distribution of small numbers of salmonellas in the feed.

Several broths have been recommended for pre-enrichment (e.g. D'Aoust, 1981; Juven *et al.*, 1984) and it is essential that they contain no inhibitory substances since these may inhibit sublethally injured salmonellas (Wernery, Leach and Pangumen, 1982). It is also important to allow a sufficient time for the cells to repair (D'Aoust, 1981) because the lag phase for injured bacteria is necessarily prolonged (Mackey and Derrick, 1982). This observation is clearly important in the context of formulating protocols for regulatory monitoring (Read, 1979) and the development of rapid methods for detecting salmonellas.

The dynamics of bacterial growth during pre-enrichment has been studied by Beckers *et al.* (1987) who found that although salmonellas multiplied during incubation, they did so at a lower rate than other members of the Enterobacteriaceae. As a consequence large numbers of non-salmonella organisms may be transferred to the selective enrichment broth and if these are not inhibited the salmonellas may not grow sufficiently ($>10^3$/ml) to be recovered on the selective agar despite their presence. This problem can be minimized by transferring only small volumes of the pre-enrichment broth to the selective broth and final dilutions of 1 : 100 (Fricker and Girdwood, 1985) and 1 : 1000 (van Schothorst and Renaud, 1983; Rhodes and Quesnel, 1986) have been recommended.

There are many potential selective enrichment broths available which contain inhibitory substances such as sodium selenite, tetrathionate and the dye brilliant green or compounds which are preferentially metabolized by the salmonellas such as mannitol and trimethylamine-N-oxide. The relative efficiencies of various formulations have been compared by many investigators (e.g. D'Aoust, 1981, Vassiliades, 1983; Juven *et al.*, 1984; Easter and Gibson, 1985; Fricker and Girdwood, 1985; Rhodes and Quesnel, 1986; Gibson, 1987). The selective process during enrichment can be further enhanced by incubating the broths at 43°C (Harvey and Price, 1968).

SELECTIVE CULTIVATION ON AGAR

Following incubation the enrichment broth is subcultured onto selective agar. There are many formulations available for this purpose with those incorporating bile salts, the dye brilliant green and indicators such as phenol red being used commonly (Moats, 1981).

The ultimate diagnosis is confirmed using biochemical and serological tests. The latter frequently involves a slide or a card agglutination test in which the organisms are either agglutinated directly by specific antisera or they agglutinate with inert particles, for example latex, on which the antibodies are adsorbed. The advantage of the latex particle system is that both the pre-enrichment and enrichment broth can be screened for the presence of salmonellas which means a presumptive diagnosis can be made in a correspondingly shorter time.

DETECTION OF SALMONELLAS BY ELECTRICAL MEASUREMENTS

Impedance microbiology is being used increasingly by the food industry to assist quality control. The detection of salmonellas in naturally and artificially contaminated food materials has been evaluated using a number of different media for pre-enrichment and selective enrichment (Easter and Gibson, 1985; Gibson, 1987; Arnott *et al.*, 1988; Pugh *et al.*, 1988; Bullock and Frodsham, 1989).

This method would appear to be efficient and rapid although detailed evaluation of its use for animal feed has yet to be published. The capital cost of the equipment is high but the running costs are likely to be relatively modest.

OTHER METHODS

The use of automated techniques for the rapid identification of salmonellas has been reviewed by Hartman and Minnich (1981) while a two-day procedure involving selective motility enrichment techniques has been recently described (Holbrook *et al.*, 1989).

There are a number of other methods which have been developed for either the isolation or detection of salmonellas and the literature on a number of these including fluorescent antibody staining, enrichment serology, radiometric and immunoassay methods and the use of bacterio-phage has been reviewed by Ibrahim and Fleet (1985).

In the context of animal feeds enzyme immunoassays (EAI) are probably the most promising. These have been used for detecting salmonellas in both artificially and naturally contaminated food and feeds (e.g. Minnich, Hartman and Heimsch, 1982; Aleixo, Swaminathan and Minnich, 1984; Todd *et al.*, 1987; Harford, 1987; Prusak-Sochaczewski and Luong, 1989). Test kits are available as is specialized equipment which allows large numbers of samples to be processed routinely by diagnostic laboratories.

The samples are cultured initially using pre-enrichment and selective enrichment broths since it is essential to have sufficient salmonellas to provide a strong positive reaction. It may also be necessary to adjust the base-line optical density reading for different feeds or feed components in order to minimize the problem of false-positive results (Todd *et al.*, 1987).

The use of DNA probes for the diagnosis of salmonella infections has also been developed and encouraging results have been reported (Flowers *et al.*, 1987; Dovey and Towner, 1989). As with the EAI it is necessary to have sufficiently large numbers of bacteria present in the final broth culture although this problem may ultimately be overcome by the use of the DNA polymerase chain reaction (e.g. Olive, 1989).

Control

The subject of reducing or eliminating bacteria from feed was reviewed by Williams (1981). In general the methods referred to in this section will be effective, to a greater or lesser extent, against vegetative bacterial cells and ineffective against bacterial spores, for example those of *B. anthracis*.

HEAT TREATMENT

The thorough cooking of feed in an autoclave is an effective way of killing bacterial and viral pathogens and is required by law in the United Kingdom (Waste Food Order, 1973) before certain animal by-products can be incorporated into animal feeds.

The heat used during pelleting is only sufficient to pasteurize the feed and, although this will reduce the numbers of vegetative bacteria by up to 1000-fold (Stott, Hodgson and Chaney, 1975), it is unwise to assume that this process, even when it involves 'steam conditioning', will guarantee the production of feed free from pathogens since coliforms may be recovered from pelleted feed (Stott *et al.*, 1975; Cox, Bailey and Thomson, 1983) and pelleted feed has been associated with the development of salmonella infections in broiler chickens (Hinton and Linton, 1988).

A new process called the 'Anaerobic Pasteurizing Conditioning System' has been described recently which is claimed to be more efficient than 'traditional' heat-treatment processes because it permits the attainment of high mash temperatures without the mash becoming too moist (McCapes *et al.*, 1989).

IRRADIATION

Irradiation has been used to eliminate salmonellas from feed (see Williams, 1981) and its use after pelleting has been proposed as a way of producing salmonella-free feed (Mossel, van Schothorst and Kampelmacher, 1967).

The effectiveness of irradiation is dose-dependent such that the use of an inadequate dose will only delay the colonization of broiler chickens with salmonellas and not prevent it (Hinton, Al-Chalaby and Linton, 1987).

Irradiation is used in the production of specialized diets, for example those for laboratory animals. Its use on a large scale for farm animal feeds is impracticable and would add greatly to the cost of the final product.

CHEMICAL DISINFECTION

Neither heat treatment nor irradiation will protect the feed against possible recontamination during subsequent distribution and storage. An alternative technique which provides protection against recontamination is to add chemical disinfectants to the feed (Hinton and Linton, 1988). The agents used must be non-toxic at the concentrations used and remain undegraded in the feed until it is consumed, when they should preferably be metabolized to obviate any problems with residues. The short chain fatty acids, for example formic and propionic acids, fulfil these criteria and their use has been shown to be effective in reducing the incidence of salmonella infections in chicks consuming artificially contaminated acid-treated feed (Hinton and Linton, 1988).

The acids only cause a small reduction in the numbers of salmonellas in the dry feed (Duncan and Adams, 1972; Vanderwal, 1979; Banton, Parker and Dunn, 1984; Hinton and Linton, 1988; Humphrey and Lanning, 1988) and this suggests that they are either inactive or only slightly active against bacteria in these circumstances. The feed becomes hydrated after it has been consumed by the bird and it is probably at this stage that the acids exert their antibacterial effect since they are only effective when the water activity is high enough to permit bacterial multiplication (Hinton, Cherrington and Chopra, unpublished observations).

It is essential to give acid treated feed throughout the rearing period since there is no beneficial effect in using it once the birds have become infected (Hinton and Linton, 1988). Similarly the acids are ineffective in preventing infection if the feed contains very large numbers of organisms although their efficiency can be improved by increasing the concentration of acid added to the feed (Vanderwal, 1979; van Staden, van der Made and Jordaan, 1980; Hinton and Linton, 1988), although this may ultimately reduce diet palatability for the birds (Cave, 1984).

PRODUCT MANAGEMENT

The way that the feed mill is operated is crucial if animal pathogens in the feed are to be controlled. Each plant must be evaluated by appropriate specialists including veterinarians and microbiologists who have a thorough understanding of infectious animal disease. It is not intended to go into this matter in detail in this review. As has already been stated some aspects of the subject are covered by legislation and others are dealt with by Codes of Practice. Suffice it to say the successful control of the pathogens requires feed mills with a workable layout, common sense, an understanding of the ways of microbes, attention to detail and 'good housekeeping' through the application of appropriate Hazard Analysis Critical Control Point (HACCP) procedures which ensure the dry storage of feed, the physical separation of raw materials and finished products and the correct operation of equipment.

If this approach does not produce the desired effect then the whole production process will require a thorough investigation.

Conclusions

Many tens of thousands of tonnes of manufactured feed are consumed by farm animals each year and yet the incidence of specific diseases from this source is extremely low.

Salmonellas in feed are the principal problem, not because they cause disease in the animals consuming the feed, but because they may ultimately cause food-poisoning in people who either handle or consume the products derived from the animals concerned.

The screening of feed destined for poultry for salmonellas is now required by law in the United Kingdom. The traditional methods of culturing these organisms are cumbersome and labour intensive and are not amenable to automation. However, alternative techniques have been developed for the food industry, for example impedance microbiology and enzyme immunoassays, and these methods will have a role in the screening of feed for bacterial pathogens in the future, as may novel techniques such as the use of DNA probes and the polymerase chain reaction.

End-product testing is not an efficient way of evaluating production processes and so the control of pathogens in feed requires carefully planned, and properly supervised, intervention at several points in the chain of production and distribution. The raw materials must be of good quality and chosen with care while the adoption of HACCP procedures will improve the efficiency of any control methods that are adopted.

The traditional techniques of microbial control, such as the use of heat and disinfectant agents, have a part to play in this process but their effect may be negated if suitable controls are not included on the rearing farms and in the abattoirs.

References

Abbis, J.S. and Blood, R.M. (1982) The detection of *Escherichia coli* in foods. In *Isolation and Identification Methods for Food Poisoning Organisms* (eds J.E.L. Correy, D. Roberts and F.A. Skinner). London, Academic Press, pp. 217–226

Aleixo, J.A.G., Swaminathan, B. and Minnich, S.A. (1984) *Salmonella* detection in foods and feeds in 27 hours by an enzyme immunoassay. *Journal of Microbiological Methods*, **2**, 135–145

Andrew, M.H.E. and Russell, A.D. (eds) (1984) *The Revival of Injured Microbes*. London, Academic Press

Arnott, M.L., Gutteridge, C.S., Pugh, S.J. and Griffiths, J.L. (1988) Detection of salmonellas in confectionery products by conductance. *Journal of Applied Bacteriology*, **64**, 409–420

Bailey, J.S., Juven, B.J., Cox, N.A. and Thomson, J.E. (1985) Recovery of inoculated *Salmonella* from poultry feed containing furazolidone. *Poultry Science*, **64**, 1670–1672

Banton, C.L., Parker, D. and Dunn, M. (1984) Chemical treatment of feed ingredients. *Journal of the Science of Food and Agriculture*, **35**, 637

Beckers, H.J., Van de Heide, J., Fenigsen-Narucka, U. and Peters, R. (1987) Fate of salmonellas and competing flora in meat sample enrichments in buffered peptone water and in Muller–Kauffmann's tetrathionate medium. *Journal of Applied Bacteriology*, **62**, 97–104

Blood, D.C. and Radostits, O.M. (1989) *Veterinary Medicine*, 7th edn, Baillière Tindall, London

Bullock, R.D. and Frodsham, D. (1989) Rapid impedance detection of salmonellas and confectionery using modified LICNR broth. *Journal of Applied Bacteriology*, **66**, 385–391

Buxton, A. and Fraser, G. (1977) *Animal Microbiology*, Vol. 1, Blackwell Scientific Publications, Oxford

Carman, J.A., Hambleton, P. and Melling, J. (1985) *Bacillus anthracis*. In *Isolation and Identification of Micro-organisms of Medical and Veterinary Importance* (eds C.H. Collins and J.M. Grange). London, Academic Press, pp. 207–214

Cave, A.G. (1984) Effect of dietary propionic and lactic acids on feed intake by chicks. *Poultry Science*, **63**, 131–134

Cox, L.J., Keller, N. and Van Schothorst, M. (1988) The use and misuse of quantitative determinations of Enterobacteriacae in food microbiology. *Journal of Applied Bacteriology*, **65**, Symposium Suppl. 237S–249S

Cox, N.A., Bailey, J.S. and Thomson, J.E. (1983) *Salmonella* and other *Enterobacteriacae* found in commercial poultry feed. *Poultry Science*, **62**, 2169–2175

Curtis, G.D.W., Nichols, W.W. and Falla, T.J. (1989) Selective agents for listeria can inhibit their growth. *Letters in Applied Microbiology*, **8**, 169–172

D'Aoust, J.Y. (1981) Update on pre-enrichment and selective enrichment conditions for detection of *Salmonellae* in food. *Journal of Food Protection*, **44**, 369–374

D'Aoust, J.Y. and Sewell, A.M. (1986) Slow rehydration for detection of *Salmonella* spp. in feeds and feed ingredients. *Applied and Environmental Microbiology*, **51**, 1220–1223

Dovey, S. and Towner, K.J. (1989) A biotinylated DNA probe to detect bacterial cells in artificially contaminated foodstuffs. *Journal of Applied Bacteriology*, **66**, 43–47

Duncan, M.S. and Adams, A.W. (1972) Effects of a chemical additive and of formaldehyde-gas fumigation on *Salmonella* in poultry feeds. *Poultry Science*, **51**, 797–802

Easter, M.C. and Gibson, D.M. (1985) Rapid and automated detection of salmonella by electrical methods. *Journal of Hygiene*, **94**, 245–262

Edel, W. and Kampelmacher, E.H. (1969) Salmonella isolation in nine European laboratories using a standardised technique. *Bulletin of the World Health Organization*, **41**, 297–306

Elliot, E.L. and Colwell, R.R. (1985) Indicator organisms for estuarine and marine waters. *FEMS Microbiology Reviews*, **32**, 61–79

Flowers, R.S., Mozola, M.A., Curiale, M.S., Gabis, D.A. and Silliker, J.H. (1987) Comparative study of a DNA hybridization method and the conventional culture procedure for detection of *Salmonella* in foods. *Journal of Food Science*, **52**, 781–785

Fricker, C.R. and Girdwood, R.W.A. (1985) A note on the isolation of salmonellas from environmental samples using three formulations of Rappaport's broth. *Journal of Applied Bacteriology*, **58**, 343–346

Gibson, D.M. (1987) Some modifications to the media for rapid automated detection of salmonellas by conductance measurement. *Journal of Applied Bacteriology*, **63**, 299–304

Gillespie, J.H. and Timoney, J.F. (1981) *Hagen and Bruner's Diseases of Domestic Animals*, 7th edn. Cornell University Press, London

Gitter, M. (1985) Listeriosis in farm animals in Great Britain. In *Isolation and Identification of Micro-organisms of Medical and Veterinary Importance* (eds C.H. Collins and J.M. Grange). London, Academic Press, pp. 191–200

Greenham, L.W., Harber, C., Lewis, E. and Scullion, F.T. (1987) *Clostridium perfringens* in pelleted feed. *Veterinary Record*, **120**, 557

Harford, J.P. (1987) An evaluation of a commercially available enzyme immunoassay test for the rapid detection of salmonellae in food and environmental samples. *Epidemiology and Infection*, **99**, 127–136

Hartman, P.A. (1979) Modification of conventional methods of recovery of injured coliforms and Salmonellae. *Journal of Food Protection*, **42**, 356–361

Hartman, P.A. and Minnich, S.A. (1981) Automation for rapid identification of Salmonellae in food. *Journal of Food Protection*, **44**, 385–393

Harvey, R.W.S. and Price, T.H. (1968) Elevated temperature incubation of enrichment media for the isolation of salmonellas from heavily contaminated materials. *Journal of Hygiene*, **66**, 377–381

Hinton, M. (1988) Salmonella infection in chicks following the consumption of artificially contaminated feed. *Epidemiology and Infection*, **100**, 247–256

Hinton, M., Al-Chalaby, Z.A.M. and Linton, A.H. (1987) Field and experimental investigations into the epidemiology of *Salmonella* infections in broiler chickens. In *Elimination of Pathogenic Organisms from Meat and Poultry* (ed. F.J.M. Smulders), Amsterdam, Elsevier, pp. 27–36

Hinton, M., Kaukas, A. and Linton, A.H. (1986) The ecology of drug resistance in enteric bacteria. *Journal of Applied Bacteriology*, **61**, Symposium Supplement 77S–92S

Hinton, M. and Linton, A.H. (1988) Control of salmonella infections in broiler chickens by the acid treatment of their feed. *Veterinary Record*, **123**, 416–421

Hinton, M., Pearson, G.R., Threlfall, E.J., Rowe, B., Woodward, M. and Wray, C. (1989) Experimental *Salmonella enteritidis* infection in chicks. *Veterinary Record*, **124**, 223

Hoben, D.A., Ashton, D.H. and Peterson, A.C. (1973) A rapid presumptive procedure for the detection of *Salmonella* in foods and food ingredients. *Applied Microbiology*, **25**, 123–129

Holbrook, R., Anderson, J.M., Baird-Parker, A.C., Dodds, L.M., Sawhney, D., Stuchbury, S.H. and Swaine, D. (1989) Rapid detection of salmonella in foods – a convenient two-day procedure. *Letters in Applied Microbiology*, **81**, 139–142

Howells, C.H.L. and Joynson, D.H.M. (1975) Possible role of animal feeding-stuffs in spread of antibiotic-resistant intestinal coliforms. *Lancet*, **i**, 156–157

Humphrey, T.J. and Lanning, D.G. (1988) The vertical transmission of salmonellas and formic acid treatment of chicken feed. *Epidemiology and Infection*, **100**, 43–49

Hurst, A. (1977) Bacterial injury: a review. *Canadian Journal of Microbiology*, **23**, 936–944

Ibrahim, G.F. and Fleet, G.H. (1985) Detection of salmonellae using accelerated methods. *International Journal of Food Microbiology*, **2**, 259–272

Impey, C.S., Mead, G.C. and Hinton, M. (1987) Influence of continuous challenge *via* the feed on competitive exclusion of salmonellas from broiler chicks. *Journal of Applied Bacteriology*, **63**, 139–146

Jones, P.W., Collins, P., Brown, G.T.H. and Aitkin, M. (1982) Transmission of *Salmonella mbandaka* to cattle from contaminated feed. *Journal of Hygiene*, **88**, 255–263

Juven, B.J., Cox, N.A., Bailey, J.S., Thomson, J.E., Charles, O.W. and Shutze, J.V. (1984) Recovery of *Salmonella* from artificially contaminated poultry feeds in non-selective and selective broth media. *Journal of Food Protection*, **47**, 299–302

Linton, A.H. and Hinton, M. (1988) Enterobacteriacae associated with animals in health and disease. *Journal of Applied Bacteriology*, **65**, *Symposium Series*, 71S–85S

McCapes, R.H., Ekperigin, H.E., Cameron, W.J., Ritchie, W.L., Slagter, J., Strangeland, V. and Nagaraja, K.V. (1989) Effect of a new pelleting process on the level of contamination of poultry mash by *Escherichia coli* and salmonella. *Avian Diseases*, **33**, 103–111

Mackey, B.M. and Derrick, C.M. (1982) The effect of sublethal injury by heating, freezing, drying and gamma-radiation on the duration of the lag phase of *Salmonella typhimurium*. *Journal of Applied Bacteriology*, **53**, 243–251

Mattingly, J.A., Butman, B.T., Plank, M.C., Durham, R.J. and Robison, B.J. (1988) Rapid monoclonal antibody-based enzyme-linked immunosorbent assay for detection of *Listeria* in food products. *Journal of the Association of Official Analytical Chemists*, **71**, 679–681

Milner, K.C. and Shaffer, M.F. (1952) Bacteriologic studies of experimental salmonella infections in chicks. *Journal of Infectious Diseases*, **90**, 81–96

Minnich, S.A., Hartman, P.A. and Heimsch, R.C. (1982) Enzyme immunoassay for detection of Salmonellae in foods. *Applied and Environmental Microbiology*, **43**, 877–883

Moats, W.A. (1981) Update on *Salmonella* in foods: selective plating media and other diagnostic media. *Journal of Food Protection*, **44**, 375–380

Mossel, D.A.A. (1982) Marker (index and indicator) organisms in food and drinking water. Semantics, ecology, taxonomy and enumeration. *Antonie van Leeuwenhoek*, **48**, 609–611

Mossel, D.A.A., Van Schothorst, M. and Kampelmacher, E.H. (1967) Comparative study on decontamination of mixed feeds by radicidation and by pelletisation. *Journal of the Science of Food and Agriculture*, **18**, 362–367

Neill, S.C., McLoughlin, M.F. and McIlroy, S.G. (1989) Type C botulism in cattle being fed ensiled poultry litter. *Veterinary Record*, **124**, 558–560

Olive, D.M. (1989) Detection of enterotoxigenic *Escherichia coli* after polymerase chain reaction amplification of thermostable DNA polymerase. *Journal of Clinical Microbiology*, **27**, 261–265

Pietzsch, O. (1984) Methods of the detection of salmonella – standardization and harmonization of the procedure. In *Proceedings of the International Symposium on Salmonella*, July 19–20, 1984. Louisiana, USA

Pike, E.B. and Ridgway, J.W. (1985) EC directives for water quality. *Journal of Applied Bacteriology*, **159**, *Symposium Supplement*, 105S–126S

Prusak-Sochaczewski, E. and Luong, J.H.T. (1989) An improved ELISA method for the detection of *Salmonella typhimurium*. *Journal of Applied Bacteriology*, **66**, 127–135

Pugh, S.J., Griffiths, J.L., Arnott, M.L. and Gutteridge, C.S. (1988) A complete protocol using conductance for rapid detection of salmonellas in confectionery materials. *Letters in Applied Microbiology*, **7**, 23–27

Ray, B. (1979) Methods to detect stressed microorganisms. *Journal of Food Protection*, **42**, 346–355

Read, R.B. (1979) Detection of stressed microorganisms – implications for regulatory monitoring. *Journal of Food Protection*, **42**, 368–369

Rhodes, P. and Quesnel, L.B. (1986) Comparison of Muller–Kauffmann tetrathionate broth with Rappaport-Vassiliadis (RV) medium for the isolation of salmonellas from sewage sludge. *Journal of Applied Bacteriology*, **60**, 161–167

Richardson, A. (1975) Outbreaks of bovine salmonellosis caused by serotypes other than *Salmonella dublin* and *Salmonella typhimurium*. *Journal of Hygiene*, **74**, 195–203

Sadler, W.W., Brownell, J.R. and Fanelli, M.J. (1969) Influence of age and inoculum level on the shed pattern of *Salmonella typhimurium* in chickens. *Avian Diseases*, **13**, 793–803

Schleifer, J.H., Juven, B.J., Beard, C.W. and Cox, N.A. (1984) The susceptibility of chicks to *Salmonella montevideo* in artificially contaminated poultry feed. *Avian Diseases*, **28**, 497–503

Sierra-Madero, J.G., Caulfield, M.J., Hall, G.S. and Washington, J.A. (1988) Detection of bacteria in the presence of antibiotics by using specific monoclonal antibodies to neutralize the antibiotics. *Journal of Clinical Microbiology*, **26**, 1904–1906

Silliker, J.H. and Gabis, D.A. (1973) ICMSF methods studies. I. Comparison of analytical schemes for detection of *Salmonella* in dried foods. *Canadian Journal of Microbiology*, **19**, 475–479

Skovgaard, N. and Morgen, C.A. (1988) Detection of *Listeria* spp. in faeces from animals, in feeds and in raw foods of animal origin. *International Journal of Food Microbiology*, **6**, 229–242

Smart, J.L. and Roberts, T.A. (1977) An outbreak of type C botulism in broiler chickens. *Veterinary Record*, **100**, 378–380

Smith, P.J. (1977) A standard technique for the isolation of *Salmonella* from animal feeds. *Journal of Hygiene*, **79**, 449–461

Stott, J.A., Hodgson, J.E. and Chaney, J.C. (1975) Incidences of Salmonellae in animal feed and the effect of pelleting on content of Enterobacteriacae. *Journal of Applied Bacteriology*, **39**, 41–46

Todd, L.S., Roberts, D., Bartholemew, B.A. and Gilbert, R.J. (1987) Assessment of an enzyme immunoassay for the detection of salmonellas in foods and animal feedstuffs. *Epidemiology and Infection*, **93**, 301–310

Vanderwal, P. (1979) Salmonella control of feedstuffs by pelleting or acid treatment. *World's Poultry Science Journal*, **35**, 70–78

Van Schothorst, M. and Oosterom, J. (1984) Enterobacteriacae as indicators of good manufacturing practices in rendering plants. *Antonie van Leeuwenhoek*, **50**, 1–6

Van Schothorst, M. and Renaud, A.M. (1983) Dynamics of salmonella isolation with modified Rappaport's medium (R10). *Journal of Applied Bacteriology*, **54**, 209–215

Van Schothorst, M., Van Leusden, F.M., De Gier, E., Rijnierse, V.F.M. and Veen, A.J.D. (1979) Influence on reconstitution on isolation of *Salmonella* from dried milk. *Journal of Food Protection*, **42**, 936–937

Van Staden, J.J., Van der Made, H.N. and Jordaan, E. (1980) The control of bacterial contamination in carcass meal with propionic acid. *Onderstepoort Journal of Veterinary Research*, **47**, 77–82

Vassiliades, P. (1983) The Rappaport-Vassiliades (RV) enrichment medium for the isolation of salmonellas: an overview. *Journal of Applied Bacteriology*, **54**, 69–76

Watkins, J. (1985) Isolation of *Listeria monocytogenes*. In *Isolation and Identification of Micro-organisms of Medical and Veterinary Importance* (eds C.H. Collins and J.M. Grange), London, Academic Press, pp. 201–206

Wernery, U., Leach, A. and Pangumen, M. (1982) Comparative quantitative study of sublethally damaged salmonella in heated and unheated chicken feed. *Deutsche Tierarztliche Wochenschrift*, **89**, 499–450

Wilesmith, J.W., Wells, G.A.H., Cranwell, M.P. and Ryan, J.B.M. (1988) Bovine spongiform encephalopathy: Epidemiological studies. *Veterinary Record*, **124**, 638–644

Williams, J.E. (1981) Salmonellas in poultry feeds – a world wide review. *World's Poultry Science Journal*, **37**, 6–25, 97–105

Wilson, G.W.C. (1986) Newcastle disease and paramyxovirus 1 of pigeons in the European Community. *World's Poultry Science Journal*, **42**, 143–153

Wray, C., Todd, J.N. and Hinton, M. (1987) Epidemiology of *Salmonella typhimurium* infection in calves: excretion of *S. typhimurium* in the faeces of calves in different management systems. *Veterinary Record*, **121**, 293–296

Xu, Y.M., Pearson, G.R. and Hinton, M. (1988) The colonization of the alimentary tract and visceral organs of chicks with salmonellas following challenge via the feed: bacteriological findings. *British Veterinary Journal*, **144**, 403–410

LIST OF POSTER PRESENTATIONS

THE EFFECT OF SEPARATED HULL AND COTYLEDONS FROM WHITE-
AND COLOURED-FLOWERED FIELD BEANS ON CHICK METABOLISM
C. Wareham, J. Wiseman and D.J.A. Cole
University of Nottingham, Nottingham

THE EFFECT OF SUPPLEMENTARY ENZYMES ON SOME NUTRITION-
ALLY IMPORTANT CHARACTERISTICS OF WHEAT
Eija Helander
Finnish Sugar Company, Helsinki, Finland
Johan Inborr
Finnfeeds International Ltd., Redhill, Surrey

THE USE OF FEED EVALUATION TECHNIQUES IN THE DEVELOP-
MENT OF A NEW FEED RAW MATERIAL
S.L. Woodgate
Prosper De Mulder Ltd., Doncaster
J.M. Everington
Agricultural Development and Advisory Service, Feed Evaluation Unit, Stratford-
on-Avon
C. Morgan
East of Scotland College of Agriculture, Aberdeen
J. McNab
IGAP, Edinburgh

EVALUATION OF THE EFFECT OF FREE FATTY ACID CONTENT ON
THE APPARENT METABOLIZABLE ENERGY VALUE OF PALM OIL
FED TO BROILERS
F. Salvador and J. Wiseman
University of Nottingham, Nottingham

THE METABOLIZABLE ENERGY CONTENTS OF DISTILLERY BY-PRODUCTS
R.T. Pass
Pentlands Scotch Whisky Research Ltd., Edinburgh
P.J.S. Dewey
Rowett Research Institute, Aberdeen
R.M. Livingstone
Rowett Research Services, Ltd., Aberdeen

A COMPARISON OF NEUTRAL DETERGENT CELLULASE (NCD) AND OTHER METHODS FOR THE PREDICTION OF GRASS SILAGE ORGANIC MATTER DIGESTIBILITY
M.S. Kridis
West of Scotland College
D.I. Givens
Agricultural Development and Advisory Service, Feed Evaluation Unit, Stratford-on-Avon
N.W. Offer and G.D. Barber
West of Scotland College

RELATIONSHIP BETWEEN THE APPARENT DIGESTIBILITY OF THE DRY MATTER OF HAYS AND THEIR RUMEN DEGRADATION CHARACTERISTICS IN SHEEP
Dolores Carro, S. Lopez, J.S. Gonzalez and F.J. Ovejero
University of Léon, Animal Production Department, Léon, Spain

EFFECT OF FEEDING ENSILED POULTRY OFFAL ON THE GROWTH PERFORMANCE OF GROWING PIGS AND ON MEAT QUALITY AND FATTY ACID COMPOSITION OF THE PORK
T.A. Van Lunen and R.L. Wilson
Agriculture Canada, Nappan, Nova Scotia, Canada
V.A. Hindle and L. Sebek
IVVO, The Netherlands

THE USE OF NON-ADDITIVE PARAMETERS IN LEAST COST DIET FORMULATION
A.G. Munford
University of Exeter, Exeter
P.C. Garnsworthy
University of Nottingham, Nottingham

A STUDY OF THE RELATIONSHIP BETWEEN LACTIC ACID AND pH VALUE IN GRASS SILAGE MADE ON FARMS
T. O'Meara and D.G. Gilheany
Chemistry Department, Maynooth College, County Kildare, Ireland
R.K. Wilson
Teagasc, Grange Research Centre, Dunsany, County Meath, Ireland

LIST OF PARTICIPANTS

Adamson, A.H.	Agricultural Development and Advisory Service, Government Buildings, Westbury On Trym, Bristol BS10 6NJ
Allen, J.D.	Frank Wright Ltd., Blenheim House, Blenheim Road, Ashbourne, Derbyshire DE6 1EE
Aman, P.	Swedish University of Agricultural Science, Department of Animal Nutrition and Management, Box 7024, S-75007, Uppsala, Sweden
Anderson, D.M.	Nova Scotia Agricultural College, Department of Animal Science, PO Box 550, Truro, Nova Scotia, Canada
Anderson, K.R.	Peter Hand Animal Health Ltd., 15/19 Church Road, Stanmore, Middlesex HA7 4AR
Bailey, S.	ADAS, Nobel House, 17 Smith Square, London SW1P 3JR
Baker, C.W.	Ministry of Agriculture, Fisheries and Food, Staplake Mount, Starcross, Exeter EX6 8PE
Bale, M.J.	University of Bristol, Department of Microbiology, School of Medical Sciences, University Walk, Bristol BS8 1TD
Barnes, R.	IGAP, Hurley, Maidenhead, Berks
Batterham, E.S.	New South Wales Agriculture and Fisheries, North Coast Agricultural Institute, Wollongbar, NSW 2480, Australia
Beames, R.	University of British Columbia, Department of Animal Science, Vancouver, BC Canada V6T 2A2
Bonnici, J.	School of Agriculture, 581 King Street, Aberdeen AB9 1UD
Booles, D.	Waltham Centre for Pet Nutrition, Freeby Lane, Waltham on the Wolds, Melton Mowbray, Leics LE14 4RT

Boorman, K.N.	University of Nottingham, School of Agriculture, Sutton Bonington, Loughborough, Leics LE12 5RD
Brett, P.A.	Paul Brett Associates, PO Box 434, Chester CH3 5YX
Brown, A.C.G.	Lilly Research Centre Ltd., Erlwood Manor, Windlesham, Surrey GU20 6PH
Buckle, A.	Agricultural Development and Advisory Service, Central Science Laboratory, London Road, Slough SL3 7HJ
Buckle, A.E.	Agricultural Development and Advisory Service, Central Science Laboratory, London Road, Slough SL3 7HJ
Buttery, P.J.	University of Nottingham, School of Agriculture, Sutton Bonington, Loughborough, Leics LE12 5RD
Carre, B.	INRA, Centre de Tours, Station de Recherches Avicoles, Nouzilly 37380, France
Carro, M.D.	University of Leon, Department of Animal Production, 24007 Leon, Spain
Chamberlain, D.G.	Hannah Research Institute, Ayr KA6 4HL
Chesson, A.	Rowett Research Institute, Greenburn Road, Bucksburn, Aberdeen AB2 9SB
Churchman, D.L.	Dalgety Agriculture Ltd, The Promenade, Clifton, Bristol BS8 3NJ
Clark, N.	Waltham Centre for Pet Nutrition, Freeby Lane, Waltham-on-the Wolds, Melton Mowbray, Leics LE14 4RT
Close, W.H.	IGAP, Church Lane, Shinfield, Reading RG2 9AQ
Cogan, D.	Butterworth Scientific Ltd., Westbury House, PO Box 63, Bury Street, Guildford, Surrey GU2 5BH
Cole, D.J.A.	University of Nottingham, Sutton Bonington, Loughborough, Leics LE12 5RD
Cole, M.	IGAP, Hurley, Maidenhead, Berks
Crehan, M.	Nutec Ltd., Eastern Avenue, Lichfield, Staffs WS13 7SE
De Paz Sanchez, F.	Saprogal S.A., Auda. Pel Ejercito 2-2, 15006 La Corina, Spain
Edmunds, B.K.	Intermol, King George Dock, Hull HU9 5PR
Ekern, A.	Agricultural University of Norway, Department of Animal Science, PO Box 25, 1432 AAS-NLH, Norway
Emmans, G.	East of Scotland College of Agriculture, APAD, Bush Estate, Penicuik, Midlothian EH26 0QE
Evans, H.W.	B.P. Nutrition Ltd., Wincham, Northwich, Cheshire CW9 6DF
Everington, J.	ADAS Feed Evaluation Unit, Alcester Road, Stratford-on-Avon, Warks CV37 9RQ
Fernandez, J.A.	National Institute of Animal Science, Department for Research in Pigs & Horses, Foulum, PO Box 39, DK-8830 Tjele, Denmark

Fisher, C.	IGAP, Roslin, Midlothian EH25 9PS
Fitt, T.	Colborn-Dawes Nutrition Ltd, Heanorgate, Heanor, Derbys DE7 7SG
Garnsworthy, P.C.	University of Nottingham, School of Agriculture, Sutton Bonington, Loughborough, Leics LE12 5RD
Garrido, A.	Dpto Produccion Animal, ETSIA Apdo 3048, 14080 Corboda, Spain
Gill, B.P.	North of Scotland School of Agriculture, 581 King Street, Aberdeen AB9 1UD
Givens, D.I.	ADAS Feed Evaluation Unit, Alcester Road, Stratford-on-Avon, Warks CV37 9RQ
Gjefson, T.	Norske Felleskjop, Lille Grensen 7, 0159 Oslo 1, Norway
Gonzalez, J.S.	University of Léon, Animal Production Department, 24007 Léon, Spain
Gorst, P.J.	W. & J. Pye Ltd, Fleet Square, Lancaster, Lancs LA1 1HA
Green, M.	*Farmers Weekly*, 1, Throwley Way, Sutton, Surrey SM1 4QQ
Greenaway, K.	Waltham Centre for Pet Nutrition, Freeby Lane, Waltham on the Wolds, Melton Mowbray, Leics LE14 4RT
Hall, A.C.	Agricultural Development and Advisory Service, Room 5, Wing 15, Block 2, Government Buildings, Lawnswood, Leeds LS16 5PY
Hindle, S.	Peter Hand Animal Health Ltd., Unit 32, Marathon Place, Moss Side Industrial Estate, Leyland, Lancs
Horwood, P.	Butterworth Scientific Ltd., Westbury House, PO Box 63, Bury Street, Guildford, Surrey GU2 5BH
Huang, K.Y.	Agri Nutrition Asia Pte Ltd., Taiwan
Huhtanen, P.	University of Helsinki, Department of Animal Husbandry, SF-00710 Helsinki, Finland
Hyam, J.M.	Nutral S.A., PO Box 58, Colmenar Viejo, Madrid, Spain
Inborr, J.C.	Finnfeeds International Ltd, Forum House, 41–51 Brighton Road, Redhill, Surrey RH1 6YS
Jesus, L.A.	Nanta S.A., Vista Alegre, 4y6, 28019 Madrid, Spain
Johnson, J.V.	Waltham Centre for Pet Nutrition, Freeby Lane, Waltham on the Wolds, Melton Mowbray, Leics LE14 4RT
Jones, A.	MAFF, Room 303, Ergon House, 17 Smith Square, London SW1P 3HX
Jones, E.	BOCM Silcock Ltd., Basing View, Basingstoke, Hampshire RG21 2EQ
Jones, T.	J. Bibby Agriculture Ltd, Adderbury, Banbury, Oxon OX17 3HL
Kennedy, D.	ADAS, MAFF, Woodthorne, Wolverhampton, WV6 8TQ

Kennelly, R.P.	Department of Agriculture and Food, Kildare Street, Dublin 2, Ireland
Kjeldsen, J.N.	Danske Slagterier, Axelborg, Axletorv 3, 1609 Copenhagen V, Denmark
Lawrence, T.L.J.	University of Liverpool, Department of Animal Husbandry, Veterinary Field Station, Neston, South Wirral L64 7TE
Lee, L.T.K.	Gold Coin Singapore Pte Ltd., 14 Jalan Tepong, Singapore 2261
Liener, D.L.	University of Minnesota, St. Paul, MN 55108, USA
Liener, I.E.	University of Minnesota, Department of Biochemistry, St. Paul, MN 55108, USA
Longland, A.	IGAP, Church Lane, Shinfield, Reading, Berks
Low, A.G.	IGAP, Shinfield, Reading, Berks
Lowe, J.A.	Gilbertson and Page, PO Box 321, Welwyn Garden City, Herts AL7 1LF
Lucey, P.	Ballyclough Co-op Creamery Ltd., Mallow, County Cork, Ireland
McNab, J.M.	IGAP, Poultry Department, Roslin, Midlothian EH25 9PS
Madsen, A.	National Institute of Animal Science, Department for Research in Pigs and Horses, Foulum, PO Box 39, DK 8830 Tjele, Denmark
Malandra, F.	Sildamin SPA, 27010 Sostegno Di Spessa, Parid, Italy
Marchetti, S.	University of Bologna, Via San Giacomo, 11, 10126 Bologna, Italy
Martin, P.A.	Hannah Research Institute, Ayr, Scotland KA6 5HL
Mills, C.	University of Nottingham, School of Agriculture, Loughborough, Leics LE12 5RD
Munday, H.S.	Waltham Centre for Pet Nutrition, Freeby Lane, Waltham on the Wolds, Melton Mowbray, Leics LE14 4RT
Munford, A.G.	University of Exeter, Department MSOR, Streatham Court, Rennes Drive, Exeter EX4 4PU
Newcombe, J.O.	University of Nottingham, Sutton Bonington, Loughborough, Leics LE12 5RD
Norton, G.	University of Nottingham, School of Agriculture, Sutton Bonington, Loughborough, Leics LE12 5RDA
Oldham, J.D.	ESCA, APAD Bush Estate, Penicuik, Midlothian, EH26 0QE
Panciroli, A.	University of Bologna, Via San Giacomo, 11, 40126 Bologna, Italy
Pass, R.	Pentland Scotch Whisky Research Ltd., 84, Slateford Road, Edinburgh

Pastuszewska, B.	Institute of Animal Physiology and Nutrition, 05-110 Jablonna, Poland
Pauwelyn, P.L.L.	RADAR, Dorpstraat 4, B-9800 Deinze, Belgium
Pietri, A.	Ist. Scienze Nutrizione, Facolta Di Agraria UCSC Via Emilia Parmense, 84, 29100 Piacenza, Italy
Rainbird, A.L.	Waltham Centre for Pet Nutrition, Freeby Lane, Waltham on the Wolds, Melton Mowbray, Leics LE14 4RT
Rantanen, A.	Alko Ltd, SF-05200 Rajamaki, Finland
Robertson, S.	Hannah Research Institute, Ayr KA6 4HL
Rowan, T.G.	University of Liverpool, Department of Animal Husbandry, Vet Field Station, Neston, Wirral, Cheshire L64 7TE
Salvador Torres, F.	University of Nottingham, School of Agriculture, Sutton Bonington, Loughborough, Leics LE12 5RD
Santoma, G.	Cyanamid Iberica, Madrid, Spain
Sayer, M.P.	Cargill Malacca SDN BMD, PO Box 292, 75750 Melaka, Malaysia
Schaper, Ing S.	Centraal Veevoederbureau, PO Box 1076, 8200 BB Lelystad, The Netherlands
Scudamore, K.	ADAS, Central Science Laboratory, London Road, Slough SL3 7HJ
Shannon, D.W.F.	MAFF, Room G17, Nobel House, 17 Smith Square, London SW1P 3JR
Smith, F.H.	University College of Dublin, Faculty of Veterinary Medicine, Ballsbridge, Dublin 4, Ireland
Steg, Ir A.	IVVO, Institute for Livestock Feeding and Nutrition Research, PO Box 168, 8200 AD Lelystad, Netherlands
Subba, M.D.B.	Pakhribas Agric. Centre, C/o BTCO, PO Box 106, Kathmandu, Nepal
Sundstoel, F.	Agricultural University of Norway, PO Box 25, 1432 AAS-NLH, Norway
Suokanto, M.	Suomen Rehu Oy, PO Box 105, 00241 Helsinki, Finland
Tamminga, S.	Agricultural University, Department of Animal Nutrition, Wageningen, Netherlands
Taylor, A.J.	BOCM Silcock Ltd., Basing View, Basingstoke, Hampshire RG21 2EQ
Tenneson, M.E.	Peter Hand Animal Health, Unit 25, Marathon Place, Moss Side Industrial Estate, Leyland, Lancs PR5 3QN
Thomas, C.	West of Scotland College, Auchincruive, Ayr KA6 5HW
Thomas, P.C.	West of Scotland College, Auchincruive, Ayr KA6 5HW

Thompson, F.	Rumenco, Stretton House, Burton-on-Trent, Staffs DE13 0DW
Thompson, J.K.	University of Aberdeen, School of Agriculture, 581 King Street, Aberdeen AB9 1UD
Tilton, J.	North Dakota State University, Fargo, N.D. 58105, USA
Tuori, M.	University of Helsinki, Department of Animal Husbandry, SF-00716, Finland
Unsworth, E.F.	DANI, Food and Agricultural Chemistry Division, Newforge Lane, Belfast BT9 5PX
Van den Broecke, J.R.A.	Eurolysine, 16 Rue Ballu, 75009 Paris, France
Van der Aar, P.	CLO-Inst. for Animal Nutrition, De Schothorst, Meerkoetenweg 26, 8218 NA Lelystad, The Netherlands
Van der Honing, Y.	IVVO, PO Box 160, 8200 AD Lelystad, The Netherlands
Van Lunen, T.	Agriculture Canada, Research Branch, Experimental Farm, Nappan, Nova Scotia, Canada BOL 1CO
Van Straalen, W.	IVVO, 8200 AD Lelystad, The Netherlands
Vernon, B.	Pauls Agriculture Ltd., PO Box 39, 47 Key Street, Ipswich, Suffolk IP4 1BX
Williams, A.P.	IGAP, Hurley, Maidenhead SL6 5LR, Berks
Wilson, J.	Unilever Research, Colworth Laboratory, Sharnbrook, Bedford MK44 1LQ
Wilson, R.	Agriculture Canada, Nappan Experimental Farm, Nappan, Nova Scotia, BOl 1CO, Canada
Wilson, R.	Grange Research Centre, Teagasc, Dunsany, County Meath, Ireland
Wilson, S.	Prosper de Mulder, Ings Road, Doncaster
Wilson, S.	Pauls Agriculture Ltd., PO Box 39, 47 Key Street, Ipswich, Suffolk IP4 1BX
Wiseman, J.	University of Nottingham, School of Agriculture, Sutton Bonington, Loughborough, Leics LE12 5RD
Woodgate, S.	Prosper de Mulder, Ings Road, Doncaster
Zebrowska, T.	Institute of Animal Physiology and Nutrition, 05-110 Jablonna, Instytucka 4, Poland

INDEX